Lecture Notes in Electrical Engineering

Volume 294

For further volumes:
http://www.springer.com/series/7818

About this Series

"Lecture Notes in Electrical Engineering (LNEE)" is a book series which reports the latest research and developments in Electrical Engineering, namely:

- Communication, Networks, and Information Theory
- Computer Engineering
- Signal, Image, Speech and Information Processing
- Circuits and Systems
- Bioengineering

LNEE publishes authored monographs and contributed volumes which present cutting edge research information as well as new perspectives on classical fields, while maintaining Springer's high standards of academic excellence. Also considered for publication are lecture materials, proceedings, and other related materials of exceptionally high quality and interest. The subject matter should be original and timely, reporting the latest research and developments in all areas of electrical engineering.

The audience for the books in LNEE consists of advanced level students, researchers, and industry professionals working at the forefront of their fields. Much like Springer's other Lecture Notes series, LNEE will be distributed through Springer's print and electronic publishing channels.

Valery Sklyarov · Iouliia Skliarova
Alexander Barkalov · Larysa Titarenko

Synthesis and Optimization of FPGA-Based Systems

Valery Sklyarov
Iouliia Skliarova
Department of Electronics,
Telecommunications and Informatics
University of Aveiro
Aveiro
Portugal

Alexander Barkalov
Larysa Titarenko
Institute of Informatics and Electronics
University of Zielona Góra
Zielona Góra
Poland

ISSN 1876-1100 ISSN 1876-1119 (electronic)
ISBN 978-3-319-04707-2 ISBN 978-3-319-04708-9 (eBook)
DOI 10.1007/978-3-319-04708-9
Springer Cham Heidelberg New York Dordrecht London

Library of Congress Control Number: 2013958443

Printed on acid-free paper

Springer is part of Springer Science+Business Media (www.springer.com)

Preface

Field-Programmable Gate Arrays (FPGAs) were invented by Xilinx in 1985, i.e., less than 30 years ago. The influence of FPGAs on many directions in engineering is growing continuously and rapidly. There are many reasons for such progress and the most important are the inherent configurability of FPGAs and their relatively cheap development cost. Forecasts suggest that the impact of FPGAs will continue to grow and the range of applications will increase considerably in the future. Recent field-configurable microchips incorporate multicore processors and reconfigurable logic appended with a number of frequently used devices such as digital signal processing slices and block memories. FPGA-based systems can be synthesized and implemented in general-purpose computers using integrated design environments. Experiments and explorations of such systems are commonly based on prototyping boards linked to the same environment.

It is widely known and proven that FPGAs can be applied efficiently in a vast variety of engineering applications. One reason for this is that growing system complexity makes it very difficult to ship designs without errors. Hence, it is essential to be able to fix errors after fabrication, which can be done significantly easier with customizable devices.

The complexity of contemporary chips is increasing exponentially with time and the number of available transistors grows faster than the ability to design meaningfully with them. This situation is a well-known design productivity gap, which is increasing continuously. Therefore, design productivity will be the real challenge for future systems. Although in unit production volumes and revenue, Application-Specific Integrated Circuits (ASICs) and Application-Specific Standard Products (ASSPs) surpass FPGAs, forecasts of FPGA design start numbers are currently ahead of ASIC/ASSP design starts. Thus, the high involvement of FPGAs in new designs of circuits and systems and the need for better design productivity undoubtedly require huge engineering resources, which are the major output of technical universities, and this book is intended to provide assistance for the relevant courses.

FPGAs still operate at lower clock frequencies than general-purpose computers and ASICs. The cost of the most advanced devices is high. Cheaper microchips operate at clock frequencies that are lower than in inexpensive computers that are widely used. One of the most important applications of FPGAs is improving the performance of implemented systems. To achieve acceleration with devices that are generally slower, parallelism needs to be applied extensively.

The book pursues two main objectives and is composed of two parts. The **first part** with appendices A and B (written by Valery Sklyarov and Iouliia Skliarova) introduces the concepts of the design of digital systems using contemporary Field-Programmable Gate Arrays and presents the recent results of the authors in FPGA-based high-performance accelerators. This part is composed of five chapters with the main objective of extending topics that are traditionally included within digital systems in a way that enables FPGA-based design to be discussed, illustrated by examples, and supported by experiments with relatively cheap prototyping boards that are widely available. The **second part** of the book (written by Alexander Barkalov and Larisa Titarenko) includes four chapters and covers more theoretical aspects of finite state machines (FSMs) with the main objective of reducing FPGA basic resources (slices or look-up tables), minimizing delays in the circuits, and achieving greater optimization of fundamental components in FPGAs.

The following features set the book apart from others in the field:

1. It provides easily understandable introductory sections (appropriate even for the first-year students in the area) that are gradually extended to more advanced topics covering the novel techniques that are proposed and disseminated by the authors and demonstrated in numerous examples from practical applications.
2. Fully synthesizable hardware-description language specifications (VHDL, in particular) for the majority of the circuits and systems described are presented ready to be tested and incorporated in practical engineering designs, which is indispensable for both undergraduate and postgraduate university students.
3. A number of practical designs based on the proposed models and methods for complete applications are discussed from areas such as data processing, combinatorial search, and computations relying on the model of a hierarchical finite state machine.
4. Exploring models and methods that involve not only core reconfigurable logical elements but also a number of embedded blocks (e.g., memories and digital signal processing slices) and template-based circuits.

The book provides the following additional features:

1. The design examples have been tested in three prototyping boards with Xilinx and Altera FPGAs. The latest Nexys-4 board from Digilent with the recent Artix-7 FPGA from the Xilinx 7 series and the well-known Digilent Atlys board with the Xilinx Spartan-6 FPGA were used for the majority of the examples. Many projects were also tested in the DE2-115 board with the Altera Cyclon-IVe FPGA that was developed especially for education and is popular in university courses.
2. All the VHDL examples from the book are available online at http://sweet.ua.pt/skl/Springer2014.html. The website also provides the latest updated projects. These projects can be downloaded and tested and evaluated immediately. Each example includes a brief description, the VHDL code, a user constraints file, and a bitstream for the selected FPGA.

The chapters in the book contain the following material:

Chapter 1 introduces FPGA architectures by presenting the general structure of modern devices and explaining the core elements and the most important embedded blocks, such as memory and digital signal processing slices. A few typical FPGA-based design scenarios are discussed that cover the phases of specification, supplying physical constraints, implementation, configuration, and finally, testing. In this introductory chapter, design specifications are presented at the schematic level, where a circuit is constructed either from components available in vendor-specific libraries, user-defined blocks, or from properly customized intellectual property cores. A number of simple examples are given that are ready to be tested in FPGA-based prototyping boards. The three prototyping boards used in the book are characterized briefly and the general ideas for interaction with circuits and systems implemented in FPGAs are introduced. All the processing steps are explained through numerous examples.

Chapter 2 presents a concise introduction to synthesizable VHDL that is sufficient for the design methods and examples given in subsequent chapters to be understood without much background knowledge. The main objective of this chapter is to explain the basis of VHDL modules and their specification capabilities without going into detail. There are many excellent books dedicated to VHDL that may be used to complement this book. Our primary target is the synthesis and optimization of FPGA-based circuits and systems and VHDL is just an instrument that is used in the book to describe the desired functionalities and structures. Thus this chapter only provides the minimum necessary to allow subsequent chapters to be read without additional material, and to enable all the proposed examples to be understood and tested with the FPGA-based prototyping boards.

Chapter 3 begins with a brief description of widely used simple combinational and sequential circuits. Many examples are given with implementations of the circuits in FPGAs. Next, various optimization techniques are discussed with special emphasis on broad parallelism, which is very important for FPGA-based applications. More complicated digital circuits and systems are introduced, such as parallel networks for sorting and searching, hamming weight counters/comparators, concurrent vector processing units, and advanced finite state machines. The circuits are designed so that operations over multiple data items can be executed concurrently. Network-based solutions, such as sorting and counting networks in particular, and the efficient mapping of circuits to FPGA primitives (look-up tables) are examples. A number of alternative competing methods are discussed and evaluated. All the circuits and systems are described in VHDL, implemented and tested in FPGAs, and finally evaluated by applying various criteria. Many of the novel solutions proposed are parameterized, which permits very complex projects to be developed in FPGAs for solving advanced problems in several areas, such as data processing and combinatorial search.

Chapter 4 begins with examples that demonstrate how commercially available intellectual property cores can be embedded in different designs. In particular, arithmetical circuits constructed from digital signal processing slices, and parameterized memory blocks that provide support for data buffering (such as

FIFO—first input first output), are described. More details on digital signal processing slices are then given and it is shown how these may be used efficiently in practical circuits such as hamming weight counters/comparators. The major part of this chapter is dedicated to interactions between a host computer and FPGA-based prototyping boards through the Digilent enhanced parallel port and the UART (Universal Asynchronous Receiver and Transmitter) interfaces. Complete details of the communication modules are described, including both software for general-purpose computers that was developed in the C++ language, and hardware for FPGAs. The next section makes use of the designed modules for projects that involve such interactions for different purposes. A more complicated design for a network-based iterative data sorter from Chap. 3 is implemented and tested in this way as a complete fully functioning example. The chapter concludes with a brief description of programmable systems-on-chip (PSoC) that combine an embedded processing system with a reconfigurable logic, which can lead to more efficient implementations of the applications. Proposals for mapping the designs from Chap. 3 to the PSoC are given and discussed.

Chapter 5 gives an overview of the design techniques based on hierarchical and parallel specifications. First, hierarchical graph-schemes (HGSs) are introduced that enable complex digital control algorithms to be decomposed and described efficiently. A module, described by an HGS, is the fundamental entity that provides the basis for the technique, and is an autonomous, complete, and potentially reusable component. A module has to be designed such that: (1) it can be verified independently of other modules; (2) it possesses a well-defined external interface so it can be reused in different specifications. It is shown that a set of HGSs (modules) can be implemented in a hierarchical finite state machine (HFSM) with a stack memory. Many VHDL examples are given that demonstrate that HFSMs permit the execution of hierarchical algorithms and provide support for recursion if required. Various types of HFSMs are described and synthesizable VHDL templates for these are given that can be customized for particular problems. Parallel specifications and parallel HFSMs are also discussed. Many fully functioning VHDL examples for all the types of HFSMs above are presented and evaluated. It is also shown how software programs can be mapped to hardware with the aid of HFSM models. Finally, a variety of HFSM optimization techniques are proposed.

Chapter 6 is devoted to the problems of optimization of Moore FSM's logic circuits implemented with FPGAs. The general characteristic is given for methods of functional and structural decomposition. Distinctive features of FPGA are analyzed allowing the number of look-up table (LUT) elements in logic circuits of Moore FSMs to be decreased. The classification of optimization methods are given for Moore FSM including: (1) the transformation of state codes into codes of the classes of pseudo-equivalent states (PES); (2) presentation of state codes as concatenations of codes of PES and collections of microoperations; (3) replacement of logical conditions (input variables of FSM) with additional variables. All discussed methods are illustrated by examples.

Chapter 7 deals with design of Moore FSMs based on using embedded memory blocks (EMB). The methods of trivial EMB-based implementation of logic circuits of both Moore and Mealy FSMs are discussed. In this case, only one EMB is enough for implementing the circuit. Next, the optimization methods are discussed based on replacement of logical conditions as well as encoding of the collections of microoperations. The considered methods are based on encoding the rows of FSM's structure table. All these methods lead to two-level models of Mealy FSMs and to three-level models of Moore FSMs. Next, these methods are combined together for further optimization of hardware in FSM logic circuits. The last section considers applying PES-based methods in EMB-based Moore FSMs. All discussed methods are illustrated by examples.

Chapter 8 is devoted to optimization of logic circuits of EMB-based FSMs. First of all, the design methods based on the replacement of logical conditions are discussed for both Moore and Mealy FSMs. Next, the proposed optimization methods are presented. These methods are based on splitting the set of logical conditions. This approach allows decreasing the number of LUTs in the circuit of the block of replacement of logical conditions. In the case of Moore FSM, the optimization methods are based on optimal state assignment, as well as the transformation of state codes into codes of the classes of PES. All discussed methods are illustrated by examples.

Chapter 9 is devoted to using the datapath for decreasing the number of LUTs in logic circuits of FPGA-based Moore FSMs. Firstly, the principle of operational implementation of interstate transitions is proposed. It is based on the usage of operational elements (adders, counters, shifters, and so on) for calculating codes the states of transitions. Next, the organization of FSM with operational implementation of interstate transitions is discussed. An example is given for application of the proposed method. Next, the base structure of synthesis process is proposed for Moore FSM with operational implementation of interstate transitions. The structure of the synthesis process depends on initial conditions such as set of operations or codes of FSM states. The typical structures are discussed for the operational automaton executing the transitions. Next, the method is shown based on mixture of traditional and proposed approaches for calculation of the codes of states of transitions. The last part of the chapter discusses the efficiency of the proposed solutions.

Appendix A contains a short description used in the book synthesizable VHDL constructions and reserved words in alphabetical order with examples.

Appendix B offers a number of synthesizable VHDL specifications that provide support for many projects in the part I of the book. All the examples are presented so that they can be tried out and examined directly.

The book can be used as supporting material for university courses that involve FPGA-based design, such as "Digital design," "Computer architecture," "Electronics," "Embedded systems," "Reconfigurable computing," "Communications," and "FPGA-based systems." It will also be helpful in engineering practice and research activity in areas where FPGA-based circuits and systems are planned to be designed and investigated. It is important to note that the presented fully functioning VHDL projects (that are also available online at http://sweet.ua.pt/skl/Springer2014.html) can be used directly in many research and engineering applications.

Contents

Abbreviations

ACP	Accelerator Coherency Port
ALM	Adaptive Logic Modules
API	Application Programming Interface
APSoC	All Programmable System-on-Chip
ARM	Advanced RISC Machine
ASCII	American Standard Code for Information Interchange
ASIC	Application-Specific Integrated Circuit
ASMBL	Advanced Silicon Modular Block
ASSP	Application-Specific Standard Product
AXI	Advanced eXtensible Interface
BCD	Binary-Coded Decimal
BCT	Block of Code Transformer
BIMF	Block of Input Memory Functions
BMO	Block of MicroOperations
BOT	Block of Operations of Transitions
BRLC	Block of Replacement of Logical Conditions
BST	Block of State Transformer
BV	Binary Vector
CAD	Computer-Aided Design
CC	Combinational Circuit
CLB	Configurable Logic Block
CMO	Collection of Microoperations
CMT	Clock Management Tiles
CN	Carry Network
CNB	Carry Network Block
CPLD	Complex Programmable Logic Device
CT	Counter
DCM	Digital Clock Manager
DDR	Double Data Rate
DSP	Digital Signal Processing
EG	Equivalent Gate
EMB	Embedded Memory Block
EMBer	Logic Circuit Consisting of EMBs

EPP	Enhanced Parallel Port
FA	Full Adder
FIFO	First Input First Output
FPGA	Field-Programmable Gate Array
FPLD	Field-Programmable Logic Device
FSM	Finite State Machine
FSMD	Finite State Machine with Data path
GFT	General Formula of Transitions
GPI	General-Purpose Interface
GSA	Graph-Scheme of Algorithm
HA	Half Adder
HDL	Hardware Description Language
HDMI	High-Definition Multimedia Interface
HFSM	Hierarchical Finite State Machine
HGS	Hierarchical Graph-Scheme
HID	Human Interface Device
HW	Hamming Weight
HWC	Hamming Weight Comparator
IGCD	Iterative Greatest Common Divisor
IP	Intellectual Property
ISE	Integrated Software Environment
JTAG	Joint Test Action Group
LAB	Logic Array Block
LC	Logical Condition
LCC	Linear Chain of Classes of PES
LCS	Linear Chain of States
LE	Logic Element
LED	Light Emitting Diode
LSB	Least Significant Bit
LUT	Look-Up Table
LUTer	Logic Circuit Consisting of LUTs
MI	Microinstruction
MLAB	Memory Logic Array Block
MMCM	Mixed-Mode Clock Manager
MO	Microoperation
MSB	Most Significant Bit
OA	Operational Automaton
OAT	Operational Automaton of Transitions
OLC	Operational Linear Chain
OP	Operational Part
PAL	Programmable Array Logic
PB	Parallel Branch
PC	Personal Computer
PEO	Pseudoequivalent Outputs

PES	Pseudoequivalent States
PHFSM	Parallel Hierarchical Finite State Machine
PL	Programmable Logic
PLA	Programmable Logic Arrays
PLD	Programmable Logic Device
PLL	Phase-Locked Loop
PLR	PipeLine Register
Pmod	Peripheral Module
PROM	Programmable Read-Only Memory
PS	Processing System
PSoC	Programmable Systems-on-Chip
RAM	Random-Access Memory
RG	Register
RGCD	Recursive Greatest Common Divisor
RISC	Reduced Instruction Set Computer
ROM	Read-Only Memory
RTL	Register-Transfer Level
SBF	System of Boolean Functions
SDC	Sequential Digital Circuit
SDK	Software Development Kit
SIMD	Single Instruction Multiple Data
SHWC	Simplest Hamming Weight Counter
SOP	Sum-Of-Products
SPI	Serial Peripheral Interface
ST	Structure Table
STT	Synthesizable Table of Transitions
UART	Universal Asynchronous Receiver/Transmitter
UCF	User Constraints File
USB	Universal Serial Bus
VHDL	VHSIC Hardware Description Language
VHSIC	Very High Speed Integrated Circuits
XDC	Xilinx Design Constraints
XST	Xilinx Synthesis Technology

Conventions

1. VHDL keywords are shown in **bold font**.
2. VHDL comments are shown in the following font: -- this is a comment.
3. The most important concepts are shown in *italic font*.
4. VHDL is not case sensitive language and thus UPPERCASE and lowercase letters may be used interchangeably.
5. Many examples in the book need libraries which are not explicitly shown and have to be included (see details in Sect. 2.6 and ***library*** in appendix A).

Xilinx®, Artix®, ISE®, LogiCore®, Spartan®, Virtex®, Vivado®, Zynq® are registered trademarks of Xilinx Inc. Chipscope, CORE Generator are trademarks of Xilinx Inc. Adept, Atlys, and Nexys-4 are trademarks of Digilent, Inc. Altera®, Stratix® and Cyclone® are registered trademarks of Altera Corp. Other product and company names mentioned may be trademarks of their respective owners.

The research results reported in this book were supported by the European Union through the European Regional Development Fund, FEDER through the Operational Program Competitiveness Factors—COMPETE, and National Funds through FCT—Foundation for Science and Technology in the context of the projects FCOMP-01-0124-FEDER-022682 (FCT reference PEst-C/EEI/UI0127/2011) and Incentivo/EEI/UI0127/2013.

Part I
Design of Digital Circuits and Systems on the Basis of FPGA

Chapter 1
FPGA Architectures, Reconfigurable Fabric, Embedded Blocks and Design Tools

Abstract This chapter introduces FPGA architectures by presenting the general structure of modern devices and explaining the core elements and the most important embedded blocks, such as memory and digital signal processing slices. A few typical FPGA-based design scenarios are discussed that cover the phases of specification, supplying physical constraints, implementation, configuration and finally, testing. In this introductory chapter, design specifications are presented at the schematic level, where a circuit is constructed either from components available in vendor-specific libraries, user-defined blocks, or from properly customized intellectual property cores. A number of simple examples are given that are ready to be tested in FPGA-based prototyping boards. The three prototyping boards used in the book are characterized briefly and the general ideas for interaction with circuits and systems implemented in FPGAs are introduced. All the processing steps are explained through numerous examples.

1.1 Introduction to FPGA

Field-Programmable Gate Arrays (FPGAs) were invented less than 30 years ago and their influence on different directions in engineering is growing continuously and extremely fast. There are many reasons for such progress and the most important of them is an inherent configurability and relatively cheap development cost.

In accordance with forecasts, the impact of FPGAs on different development directions will continue to grow and the number of such directions will be extended in future. When FPGAs were first proposed, they were predominantly used for implementing simple random and glue logic [1]. Nowadays, even undergraduate students are capable of constructing complex digital devices on the basis of FPGAs.

The world's first FPGA XC2064™ was introduced and shipped in 1985 by Xilinx. It offered 800 gates (85,000 transistors, 128 logic cells, 64 Configurable

V. Sklyarov et al., *Synthesis and Optimization of FPGA-Based Systems*,
Lecture Notes in Electrical Engineering 294, DOI: 10.1007/978-3-319-04708-9_1,

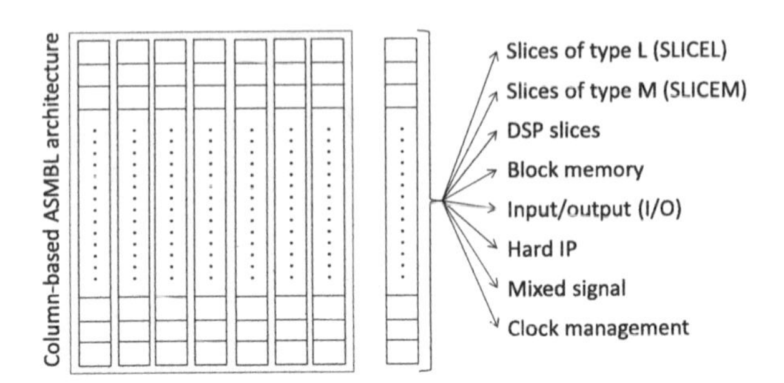

Fig. 1.1 Basic architecture of the 7 series Xilinx FPGA

Logic Blocks—CLBs with two 3-input look-up tables, max clock frequency 50 MHz). The chip was sold for $55 and produced on a 2.0 μ process [2]. Recent field-configurable micro-chips can be seen as a mixture of traditional gate arrays and ASIC (Application-Specific Integrated Circuit) components (such as ARM dual-core Cortex-A9) where the development of software and hardware can be done relatively independent of each other (e.g. *Zynq* all programmable system-on-chip [3]). FPGA complexity has reached 6.8 billion transistors [4], clock frequency exceeds gigahertz, and the most advanced technology is 20 nm [5] (14 and 10 nm processes are expected to be announced in 2014 [5]).

FPGA-targeted Computer-Aided Design (CAD) systems permit different specifications, tools, and components (such as hardware and system-level description languages, design templates, intellectual property (IP) cores, soft/hard build-in blocks) to be linked and combined within a single project. The relevant circuits can be synthesized, implemented, and tested in the same environment installed on a general-purpose computer with connected through standard interfaces (e.g. USB, PCI express, wireless) FPGA-based prototyping boards/systems.

Nowadays, the way to evolve higher performance systems from a general-purpose computer, proposed more than 50 years ago [6], has been finally implemented in reality. Advances in FPGA technologies and architectures are clearly shown in [7]. From 1985 to 2013 FPGAs grew 100,000 times in capacity and became significantly faster. Two largest companies Altera and Xilinx continue to dominate on the market [8].

Figure 1.1 depicts the recent 7 series Xilinx FPGA column-based ASMBL (Advanced Silicon Modular Block) architecture [9]. The core configurable elements are slices that contain look-up tables (LUT), flip-flops and supplementary logic. A CLB consists of two slices and will be described in more detail in Sect. 1.2. DSP blocks are efficient for digital signal processing. They execute multiplication, addition, subtraction, logical and other operations over up to 48 bit operands. Some of FPGA columns contain block memories, hard intellectual property (IP) cores, input/output blocks, clock distributers and mixed signal managers. We will describe these blocks later on in this chapter.

Different FPGA elements can be:

1. Configured to implement the desired functionality.
2. Flexibly interconnected with each other.

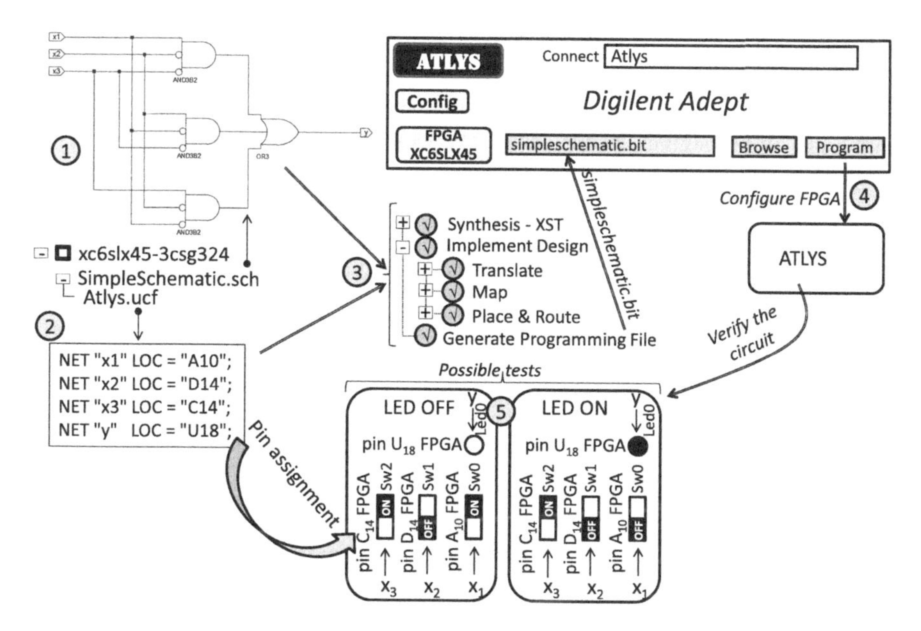

Fig. 1.2 One possible scenario of FPGA-based design

For example, an available 6-input/1-output slice LUT enables any Boolean function of 6 variables to be implemented. Configuration permits a particular function to be realized. A DSP block can be configured to implement a variety of arithmetic and logic operations. Besides it provides many additional useful features for digital signal and other types of processing that will be discussed later. Interconnections set up links between internal pins of different elements. Customization (configuring elements and interconnections) is done through reloading a bitstream to FPGA. The details will be given on examples below. Since the development of circuits and systems does not involve complex technological processes, FPGA are very appropriate for prototyping and verifying different design ideas.

Figure 1.2 illustrates one possible scenario of FPGA-based design using Xilinx Integrated Software Environment (ISE release version 14.7), Digilent Adept software [10] and Atlys prototyping board [11] containing xc6slx45 FPGA of Xilinx Spartan-6 family.

Different sources for the project are available in the ISE and we will use schematic editor (see point 1 in Fig. 1.2) and describe a circuit with 3 inputs (x_1, x_2, x_3) and 1 output (y) detecting exactly one value '1' in a 3-bit input vector ("$x_1x_2x_3$"). Thus, y = '1' for any vector from the set {"001","010',"100"}, otherwise y = '0'. The circuit is saved in the file *SimpleSchematic.sch*. Let us connect three inputs (x_1, x_2, x_3) and one output (y) of the circuit with external FPGA pins that are in turn connected with switches Sw0, Sw1, Sw2, and LED (Light-Emitting Diode) Led0 on the Atlys board. Such connections are indicated in a user constraints (implementation constraints) file (UCF) which is entitled *Atlys.ucf* and shown in point 2 of Fig. 1.2, where A10, D14, C14, and U18 are names of FPGA external pins connected with the switches and the LED (see point 5 in Fig. 1.2). The NET keyword is used to apply constraints to specific signals (to the input and output signals in our case). The LOC keyword defines where a design element can be placed within the device. The detailed information about constraints can be found in [12]. Our ISE project (shown in between points 1 and 2 in Fig. 1.2) indicates the chosen FPGA (*xc6slx45-3csg324*), the top-level module (*SimpleSchematic.sch*) and the UCF (*Atlys.ucf*) which specifies pin assignment for the top-level module.

At the next phase synthesis, implementation and generate programming file steps are applied to our project (see point 3 in Fig. 1.2). The generated file *simple-schematic.bit* may now be used to configure the FPGA, which can be done either directly from the ISE or in board-targeted software, such as Digilent Adept [10] (see point 4 in Fig. 1.2). At the last step (see point 5 in Fig. 1.2) we verify the circuit functionality in FPGA using the onboard switches Sw0, Sw1, Sw2 to supply values of inputs x_1, x_2, x_3 and the onboard Led0 to examine the result (i.e. the value of y). The circuit is so simple that it is implemented in just one slice LUT and there are totally 27,288 such LUTs available in the chosen FPGA.

The designed circuit can be taken as a component for new projects and thus, a hierarchy will be involved. Suppose, we would like to analyze three groups of signals (x_1,x_2,x_3), (x_4,x_5,x_6), and (x_7,x_8,x_9) and detect that exactly one group contains exactly one value '1'. The circuit that implements such functionality from the schematic editor of ISE is shown in Fig. 1.3.

At the beginning we create a component that contains the circuit shown in Fig. 1.2 (see point 1). This can be done in the ISE using the option *Create Schematic Symbol*. Since the name of the entity in point 1 of Fig. 1.2 is *SimpleSchematic*, the name of the component is the same (see Fig. 1.3). The component has to be connected within the new designed circuit much like it is done for library primitives (gates) in point 1 of Fig. 1.2. Thus, the desired functionality is described. The *Atlys.ucf* needs to be modified as follows:

```
NET "x1" LOC = "A10";    # Sw0
NET "x2" LOC = "D14";    # Sw1
NET "x3" LOC = "C14";    # Sw2
NET "x4" LOC = "P15";    # Sw3
NET "x5" LOC = "P12";    # Sw4
NET "x6" LOC = "R5";     # Sw5
NET "x7" LOC = "T5";     # Sw6
NET "x8" LOC = "E4";     # Sw7
NET "x9" LOC = "P3";     # BTND – onboard button available on the Atlys
NET "y" LOC = "U18";     # Led0
```

Here the symbol # permits comments to be provided and the comments shown above characterize links of the circuit pins with the FPGA external pins connected to the onboard switches and the button. The available onboard components and FPGA pins are connected in the printed circuit of the board and information about such connections is available from the Atlys board documentation [11].

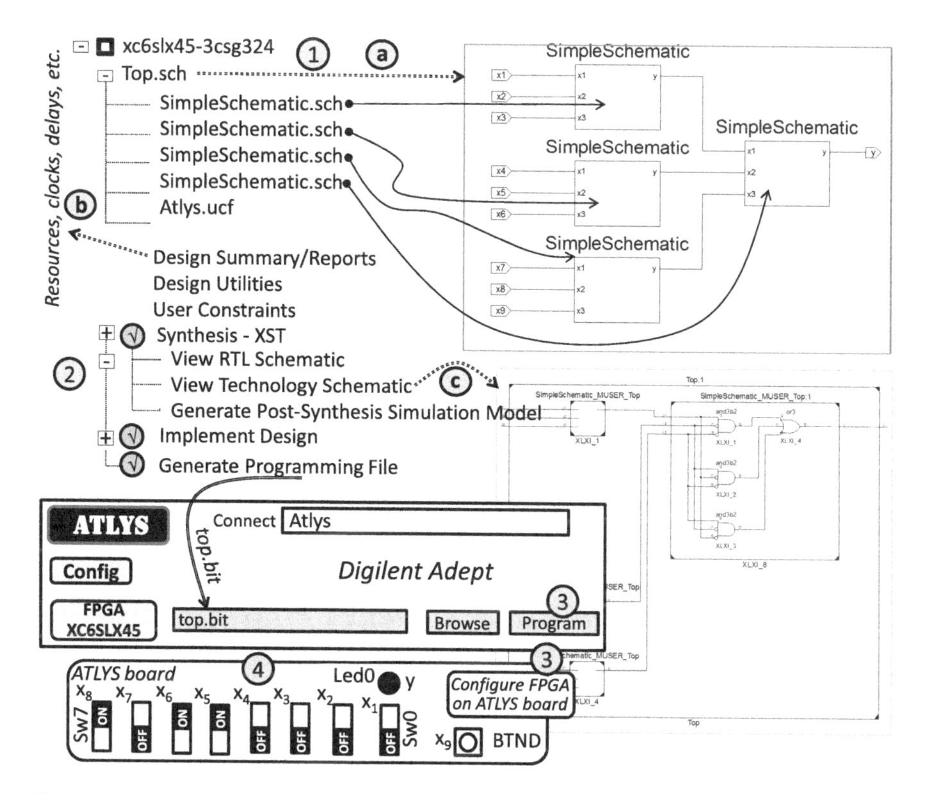

Fig. 1.3 Hierarchical design and analysis of the results

Point 1 in Fig. 1.3 illustrates the project structure with the top-level module *Top.sch*. It is clearly seen that the structure is hierarchical in which the top-level module (*Top.sch*) is composed of four lower level modules (*SimpleSchematic.sch*). A dashed arrow line *a* points to the circuit *Top.sch* copied from the ISE schematic editor. Point 2 refers to the ISE design steps that have already been briefly discussed. In the *Design Summary/Reports* (see point *b* in Fig. 1.3) different characteristics of the circuit are summarized, particularly the used resources (now three LUTs are required in two FPGA slices), and delays (the maximum combinational path delay is 9.1 ns). We can use many other options, for example, *View RTL Schematic* (RTL is a Register Transfer Level). For our project schematic is taken as a design entry and it is the same as in the schematic editor (see the circuit indicated by dashed arrow line *c*). However, schematic can also be built by ISE tools from specifications in hardware description languages (HDL), which are generally considered to be more productive and efficient than schematic descriptions. Configuring (programming) the FPGA is provided in point 3. Verification of the designed circuit on the Atlys board is done in point 4.

Although the two projects in Figs. 1.2 and 1.3 are indeed trivial, they demonstrate the essential steps for FPGA-based design, which are also common for complex systems. A similar technique can be used for Altera Quartus environment (later on we will demonstrate some examples in Quartus 13 Web edition software for Altera FPGAs). For example, the *block editor* of Quartus enables schematic of the design to be created. Note, that although diagrams, such as in point 1 of Fig. 1.2, are illustrative for simple circuits, they become confused, difficult for verification and error prone for complex designs, for which HDL becomes more preferable. We will make an introduction to VHDL—Very-high speed integrated circuit HDL in Chap. 2. Now let us characterize basic FPGA components with more details.

1.2 The Basis of FPGA Devices

Configurable logic blocks (CLBs) are the main logic resources for implementing digital circuits. We will discuss such blocks that are used in recent FPGAs of the major two companies: Altera and Xilinx.

1.2.1 Configurable Logic Blocks of Xilinx FPGAs

We consider CLBs for the recent 7 series FPGAs that are also very similar to the popular Spartan-6 family of FPGAs. A CLB is composed of *2 slices* that are connected to a switch matrix for access to the general routing matrix [9]. Every slice contains: (1) *four LUTs*; (2) *eight* edge-triggered D-type *flip-flops*, four of which can also be configured as level-sensitive latches; (3) *multiplexers*; and (4) *carry logic* for arithmetic circuits. Up to 16:1 multiplexer can be implemented in one slice using built in multiplexers and LUTs.

There are two types of slices: SLICEL and SLICEM. Each CLB has either two SLICELs or a SLICEL and a SLICEM. SLICEM provides support for two additional operations: storing data in the slice that in this case may be used to compose a distributed RAM; and shifting up to 32-bit data.

Each slice LUT has 6 independent inputs ($x_0,\ldots,x_5$), 2 independent outputs O_5 and O_6 and can be configured to implement: (1) any Boolean function of up to 6 variables $x_0,\ldots,x_5$; (2) any two Boolean functions of up to 5 shared variables $x_0,\ldots,x_4$ and x_5 has to be set to high level; (3) any two Boolean functions of up to 3 and 2 separate variables. The propagation delay is independent of the function implemented in the LUT.

Let us consider an example of LUT(6,1) with 6 inputs x_5,x_4,x_3,x_2,x_1,x_0 and 1 output y. The LUT will be used to implement a parity function for 6-bit binary vector $x_5x_4x_3x_2x_1x_0$ in such a way that the Hamming weight of the vector $x_5x_4x_3x_2x_1x_0y$ is odd (the Hamming weight of a binary vector is the number of values '1' in the vector). The truth table for the function y is given in Table 1.1.

The column *Hex* contains hexadecimal vector that is used for INIT attribute in the ISE environment (it is accessed through the *Object Properties* described below). The vector begins with the value marked with an asterisk in Table 1.1:

Table 1.1 Truth tables for configuring LUT(5, 2) and LUT(6, 1)

$x_4x_3x_2x_1x_0$	y_1y_0	Hex	$x_5x_4x_3x_2x_1x_0$	y	Hex	$x_5x_4x_3x_2x_1x_0$	y	Hex
00000	00	ca	000000	1	9	100000	0	6
00001	01		000001	0		100001	1	
00010	10		000010	0		100010	1	
00011	11		000011	1		100011	0	
00100	11	35	000100	0	6	100100	1	9
00101	10		000101	1		100101	0	
00110	01		000110	1		100110	0	
00111	00		000111	0		100111	1	
01000	01	a5	001000	0	6	101000	1	9
01001	10		001001	1		101001	0	
01010	01		001010	1		101010	0	
01011	10		001011	0		101011	1	
01100	11	59	001100	1	9	101100	0	6
01101	00		001101	0		101101	1	
01110	10		001110	0		101110	1	
01111	01		001111	1		101111	0	
10000	01	a5	010000	0	6	110000	1	9
10001	10		010001	1		110001	0	
10010	01		010010	1		110010	0	
10011	10		010011	0		110011	1	
10100	01	ab	010100	1	9	110100	0	6
10101	11		010101	0		110101	1	
10110	00		010110	0		110110	1	
10111	11		010111	1		110111	0	
11000	01	65	011000	1	9	111000	0	6
11001	10		011001	0		111001	1	
11010	11		011010	0		111010	1	
11011	00		011011	1		111011	0	
11100	01	ab* **	011100	0	6	111100	1	9 *
11101	11		011101	1		111101	0	
11110	00		011110	1		111110	0	
11111	11		011111	0		111111	1	

(i.e. the first digit is 9_{16}): **9669699669969669**$_{16}$ and it represents the binary vector for the output y: **1001** 0110 0110 **1001** 0110 **1001** **1001** 0110 0110 **1001** **1001** 0110 **1001** 0110 0110 **1001**$_2$ (for easier comparison of hexadecimal and binary vectors, digits **9** are shown in **bold** and digits 6 are shown in a normal font).

Figure 1.4 demonstrates configuring the LUT(6,1) in the schematic editor of the ISE. The vector **9669699669969669**$_{16}$ is assigned to the INIT attribute accessed through the *Object Properties* (point mouse to the LUT in the schematic editor and right mouse button click to change *Object Properties*). Figure 1.5 demonstrates configuring LUT(5,2) to implement the functions y_0 and y_1 shown in Table 1.1.

Now hexadecimal vector composed of 16 hexadecimal digits is split in two 8-digit sub-vectors in such a way that the first sub-vector configures the first

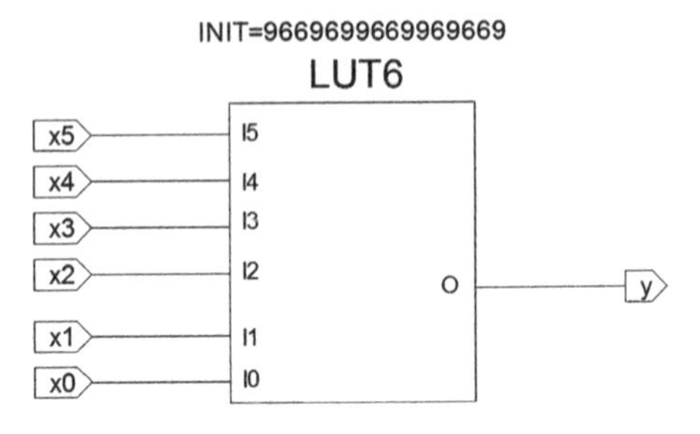

Fig. 1.4 Configuring LUT(6,1) using Xilinx primitive LUT6

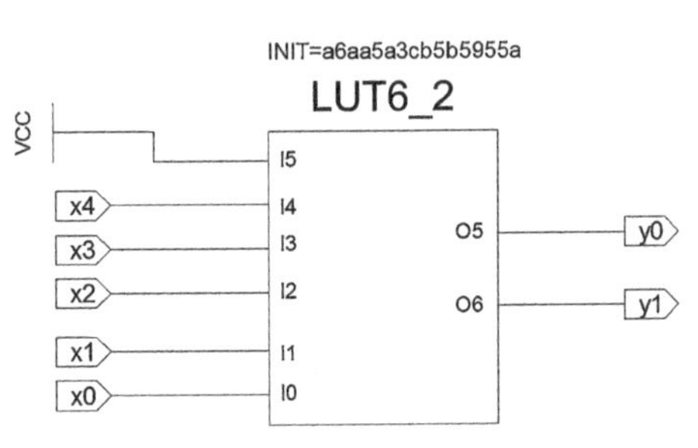

Fig. 1.5 Configuring LUT(5,2) using Xilinx primitive LUT6_2

function and the second sub-vector configures the second function. For our example in Table 1.1 the following vector is built: $a6aa5a3cb5b5955a_{16}$. The first sub-vector $a6aa5a3c_{16}$ enables the function y_1 to be configured. The second sub-vector $b5b5955a_{16}$ enables the function y_0 to be configured.

An asterisk (*) in Table 1.1 indicates the beginning of the first sub-vector and two asterisks (**) in Table 1.1 indicate the beginning of the second sub-vector. The circuits in Figs. 1.4 and 1.5 can easily be verified supplying inputs from switches and displaying values of the functions on LEDs. In Fig. 1.5 a 6-input LUT is taken but the most significant input is set to high ('1') by supplying VCC signal and configuring the LUT(5,2) to implement two different Boolean functions of 5 shared variables.

LUTs in SLICEMs can implement a synchronous (distributed) RAM/ROM with single, dual or quad ports. We can also configure SLICEM as up to 32-bit shift register not requiring slice flip-flops. The register enables serial data on its input to be delayed on its output from 1 to 32 clock cycles. The number of clocks for delay is controlled by a dedicated 5-bit input vector. Figure 1.6 gives an example of a circuit that contains a LUT-based 256×1 ROM and a shift register with variable size (from 1 to 32).

The Xilinx primitive ROM256x1 is programmed with the following INIT attribute: 0f0f0f0f0f0f0f0f0f0f0f0f0f0f0f0f0f0f0f0f0f0f0f0f0f0f070301013731. Now the value '1' is written at the address 0 (see the right part of the vector above), the value '0' at the address 1, etc. The block *clock_divider* outputs a clock signal with frequency approximately equal to 1 Hz and the VHDL code for such divider is presented in Appendix B. The Xilinx primitive CB8CE is an 8-bit binary counter, which generates addresses for the ROM, incrementing them

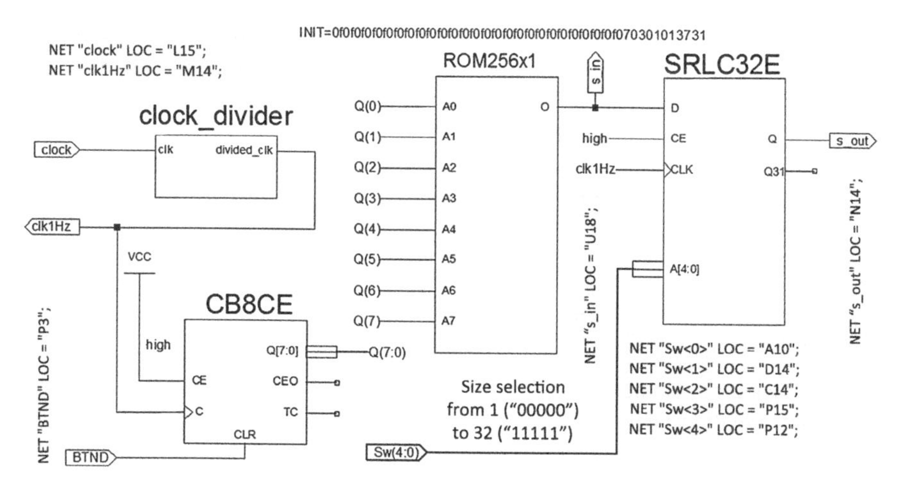

Fig. 1.6 Using LUT-based (distributed) memory (Xilinx primitive ROM256x1) and a shift register (Xilinx primitive SRLC32E)

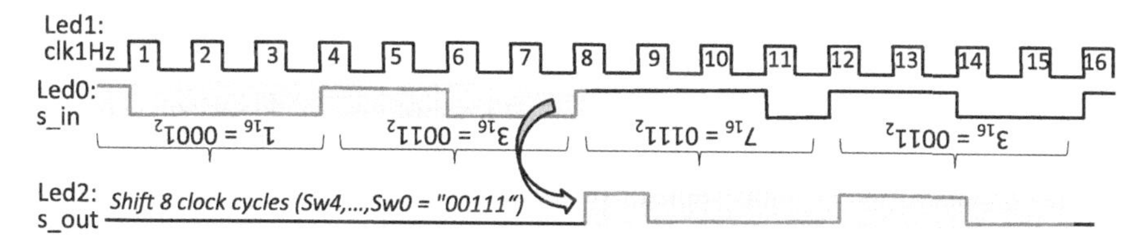

Fig. 1.7 Waveforms of signals displayed on the LEDs Led1 (clock with frequency 1 Hz), Led0 (data from the ROM), Led2 (data from the shift register)

approximately every second. Connections between the bus *Q(7:0)* and the lines *Q(7),…,Q(0)* are provided by names much like it is done for the lines *clk1 Hz* and *high* (i.e. non-shown in Fig. 1.6 wires with equal names are, in fact, connected). The project has been tested in the Atlys board. Onboard switches Sw4,…,Sw0 are used to set the shift register size which can be defined from 1 ("00000") to 32 ("11111"). All necessary connections with FPGA pins (see grey rectangular areas) are shown in Fig. 1.6 in form of lines of the file *Atlys.ucf*. The onboard button *BTND* supplies the *reset* signal (just for the counter, because the shift register does not need to be cleared). Three onboard LEDs (Led2, Led1, and Led0) are used for verification of the project. Led1 gets the clk1Hz signal (clock with a frequency approximately equal to 1 Hz). Thus, all other signals can be verified sequentially relatively to the low-frequency clock (see waveforms in Fig. 1.7). Switches Sw4,…,Sw0 permit delay of the Led2 (shift register output s_out) relatively to the Led0 (shift register input s_in) to be set. For example, if Sw4,…,Sw0 are assigned to the value "00111" then the delay is 8 clock cycles and waveforms for such a case are shown in Fig. 1.7. The project is implemented on 12 slices from which one slice is used for the ROM and one slice is used for the shift register. A number of other useful LUT configurations are considered in [9]. It is important to note that LUT-based memories can be used as configurable combinational circuits enabling functionality to be changed during execution time.

The number of slices in Spartan-6 family FPGAs varies from 600 to 23,038. The number of slices in 7 series FPGAs varies from 10,250 to 305,400.

1.2.2 Logic Elements of Altera FPGAs

We consider logic elements for recent Stratix-V series FPGAs with core reconfigurable fabric called logic array block (LAB) [13] composed of adaptive logic modules (ALM), which can be configured to implement logic, arithmetic, and memory functions. Half of the available LABs may be used as memory LAB (MLAB).

Each ALM contains different LUT-based resources and can implement any Boolean function of up to six variables. Besides, a number of other types of Boolean functions F(n,m) describing circuits with n inputs and m outputs can be implemented such as F(4,3) and F(5,2).

ALMs operate in four possible modes [13]:

1. Normal mode enables two Boolean functions of up to 5 variables or one Boolean function of up to 6 variables to be implemented. Besides, 8 available data inputs allow certain Boolean functions with more than 6 variables to be realized.
2. Extended mode permits the result of the implemented Boolean function to be registered.
3. Arithmetic mode uses four 4-input LUTs for pre-adder logic connected with two dedicated full adders.
4. Shared arithmetic mode permits 3 input additions to be implemented. The details are given in [13].

Each ALM in an MLAB can be programmed as either a 64×1 or a 32×2 block. Since each MLAB supports a maximum of 640 bits it can be configured as either a 64×10 or a 32×20 simple dual-port static RAM.

For some examples of this book we will use the DE2-115 prototyping board with Altera Cyclone-IV FPGA [14]. LABs of this FPGA contain groups of logic elements (LE) and one LAB consists of 16 LEs. Each LE contains: 4-input LUT which can implement any Boolean function of 4 variables; a flip-flop, which in [14] is called a programmable register; a carry and a register (a flip-flop) chains connections.

LEs operate in normal and arithmetic modes. The first mode is efficient for general logic applications and combinational functions. The second mode is more appropriate for adders, counters, accumulators, and comparators.

Thus, the primary reconfigurable resources of Xilinx and Altera FPGAs are based on LUTs. The simplest element of Altera FPGAs is LE/ALM and it contains fewer resources than a Xilinx FPGAs slice, which is the simplest element in Xilinx FPGAs. The most advanced recent devices of both companies include 6-input LUTs, which can be configured for implementing logic, memory, and arithmetic functions.

The number of LEs in Cyclone-IV FPGAs varies from 6,725 to 149,760. The number of LEs in Stratix-V FPGAs is from 236K to 952K. In this book we will mainly use Xilinx FPGA of Spartan-6 and Artix-7 families. The majority of examples can easily be converted to Altera FPGAs and we will consider some examples for Altera Cyclone-IVe devices.

1.3 Embedded Blocks

In addition to basic reconfigurable logic described in the previous section contemporary FPGAs contain numerous embedded blocks which can be observed in the basic architecture of the Xilinx 7 series FPGA in Fig. 1.1 (similar embedded blocks are available for Altera FPGAs [13]). We will discuss such blocks and their use in different projects on examples of embedded memories and DSP slices for Spartan-6 family and 7 series of Xilinx FPGAs.

1.3.1 Embedded Memories

Embedded memory blocks, or Block RAMs, are widely available in modern FPGAs and are used for efficient data storage and buffering. FPGAs of Spartan-6 family contain from 12 to 268 Block RAMs each of which stores up to 18 Kb of data and can be configured as either two independent 9 Kb RAMs or one 18 Kb RAM. Each RAM is addressable through two ports, but can also be configured as a single-port RAM. The width of the two ports of a 18 Kb RAM is configurable independently of each other as 16K $\times$ 1, 8K $\times$ 2, 4K $\times$ 4, 2K $\times$ 8, 1K $\times$ 16, 512 $\times$ 32 (when no parity bits are used) or 16K $\times$ 1, 8K $\times$ 2, 4K $\times$ 4, 2K $\times$ 9, 1K $\times$ 18, 512 $\times$ 36 (when parity bits are used). Data can be written to either or both ports and can be read from either or both ports [15]. Each port has its own address, data in (input data), data out (output data), clock, clock enable, and write enable. The read and write operations are synchronous and require an active clock edge. Block RAMs are organized in columns within an FPGA device (see Fig. 1.1) and can be interconnected to create wider and deeper memory structures. It is possible to specify Block RAMs characteristics and to initialize memory content in VHDL code, which will be shown in the next chapter. Intellectual Property (IP) core generator and schematic library primitives can also be used (Chap. 4 gives some examples).

The 7 series FPGAs contain from 135 to 1,880 Block RAMs each of which stores up to 36 Kb of data. FPGAs provide support for 36 and 18 Kb block RAMs

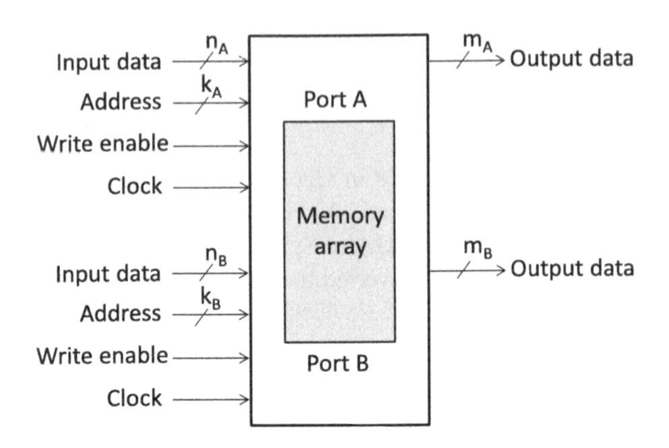

Fig. 1.8 A simplified structure of block RAM

[16] with built-in FIFO (first input first output) logic. Each 36 Kb block RAM can be configured as 32K × 1, 16K × 2, 8K × 4, 4K × 8, 2K × 16, 1K × 32, or 512× 64 (when no parity bits are used) or 32K × 1, 16K × 2, 8K × 4, 4K × 9, 2K × 18, 1K × 36, or 512 × 72 (when parity bits are used). An additional Block RAM features in the 7 series devices is an opportunity to use output registers. A simplified structure of block RAM is presented in Fig. 1.8, where n_A/n_B is the size of input data for the port A/B, m_A/m_B is the size of output data for the port A/B, k_A/k_B is the size of address for the port A/B.

Each block RAM contains two completely independent ports that share the same memory array for write and read operations (i.e. true dual-port memory can be built). Potential conflicts during write operations need to be avoided and this issue is addressed in [16].

We have already mentioned that Block RAM (36 Kb for the 7 series devices or 18 Kb for the Spartan-6 family devices) can be decomposed into two independent block RAMs (18 Kb for the 7 series devices or 9 Kb for the Spartan-6 family devices), each of which behaves similarly to the initial block. Several block RAMs can compose larger memory if required.

Each memory access (a read or a write) in the devices [15, 16] is controlled by a clock. All inputs, data, address, clock enables, and write enables are registered. Clocking the address means that data remain unchanged until the next clock cycle.

Let us consider now two simple examples of using single and dual-port embedded memories for Spartan-6 FPGA available on the Atlys prototyping board. Block RAMs will be created by the Xilinx *LogiCore* block memory generator [17]. First we add a new source in the Xilinx ISE (option *Project → New Source*) and then select IP and a name *SinglePort*. The core generator will be launched. Let us leave all the options (listed in 6 steps) unchanged except the following: *Memory type* (step 2) has to be defined as *Single Port RAM*, *Write Width* (step 3) is assigned to *8*, *Write Depth* (step 3) is assigned to *65536* (i.e. we would like to create a memory from several block RAMs with the total size of 64 KB) and *Load Init File* option (step 4) is *checked* (i.e. we would like to upload an initialization file of type COE). A COE is a text file (created, for instance, in Notepad) indicating *memory_initialization_radix* (valid values are 2, 10, or 16) and

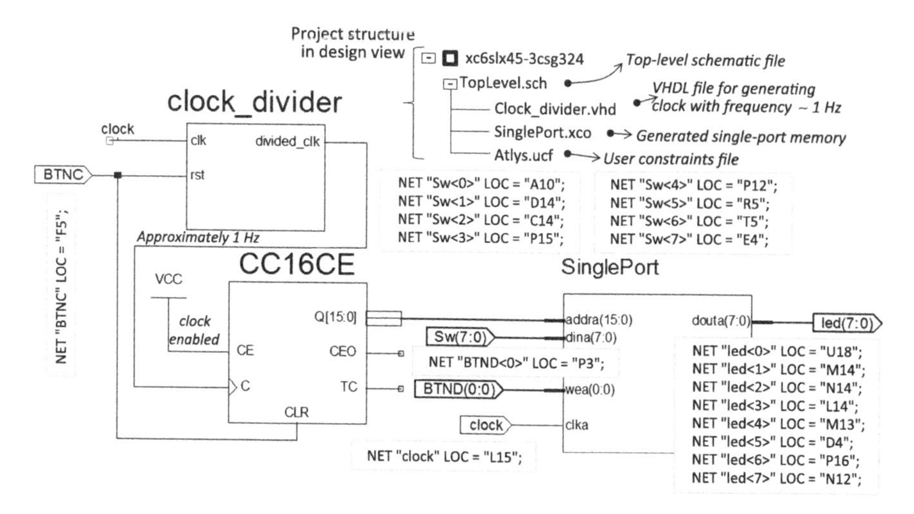

Fig. 1.9 Simple example with memory built from a single port block RAMs (CC16CE is the Xilinx library primitive for a 16-bit binary counter)

memory_initialization_vector which contains values for each memory element (that in our case is an 8-bit word). Any value has to be written in radix defined by the *memory_initialization_radix*. The following example presents a valid COE file which will be used for our project (additional details can be found in [17]):

```
memory_initialization_radix = 16;
memory_initialization_vector =
00, 18, 3c, 7e, ff, 7e, 3c, 18, 00,
80, 40, 20, 10, 08, 04, 02, 01, 00,
01, 02, 04, 08, 10, 20, 40, 80, 00,
80, c0, 60, 30, 18, 0c, 06, 03, 01, 00,
80, c0, e0, 70, 38, 1c, 0e, 07, 03, 01, 00;
```

Coefficients are separated by a space, a comma, or by placing one value in each line with a carriage return. A semicolon indicates the end of specification line such as memory_initialization_radix = 16;. In our example the first 48 bytes of memory will be filled in from the COE file above and the remaining bytes are assigned to FF_{16} (the option *Fill Remaining Memory Locations* available at step 4 is checked and the value FF is chosen). The button *Show* permits the contents of the COE file to be displayed and examined.

After generation a primitive for the single port memory can be used in the ISE schematic editor much like any other Xilinx library primitive. Figure 1.9 gives an example of a trivial circuit which permits to read from RAM and to display on the Atlys onboard LEDs the sequence partially shown in Fig. 1.10 (much like Fig. 1.6 the relevant constraints of the *Atlys.ucf* are given in Fig. 1.9).

The sequence in Fig. 1.10 was specified in the first (00, 18, 3c, 7e, ff, 7e, 3c, 18, 00) and in the second (80, 40, 20, 10, 08, 04, 02, 01, 00) lines of the COE file above. If the button *BTND* is pressed data from the onboard switches are written to the RAM

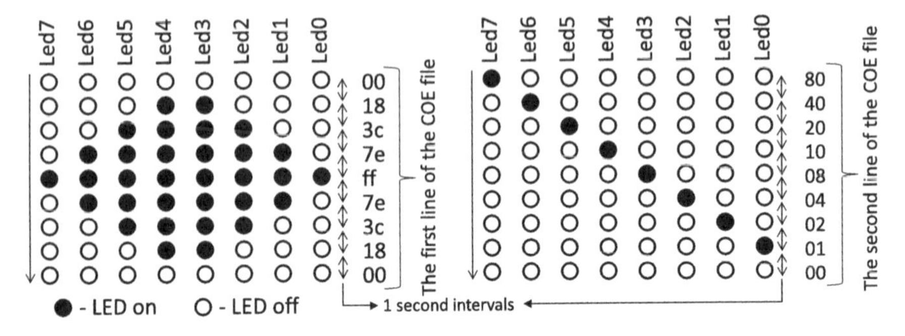

Fig. 1.10 Initialization (COE) file specifies visual sequence on the LEDs

and displayed on the LEDs. Thus, we can examine both write and read operations. From the ISE *Design Summary* we can see that 32 RAM blocks have been allocated. A similar memory can be defined as an HDL component.

Let us build now a simple dual-port memory. The steps are similar to the steps for the single port memory. The majority of options of the *LogiCore* generator are unchanged except the following: *Memory type* (step 2) has to be defined as *Simple Dual Port RAM*, *Write Width* for the port A (step 3) is assigned to *8*, *Write Depth* for the port A (step 3) is assigned to *8192* (i.e. we would like to create memory from several block RAMs with the total size 8 KB), *Write Width* for the port B (step 3) is assigned to *1*, and *Load Init File* option (step 4) is *checked*. Thus, the first port A is configured as 8,192 × 8 and the second port B—as 65,536 × 1. The following COE file is used (note that now the radix is chosen to be 2):

```
memory_initialization_radix = 2;
memory_initialization_vector =
11001010, 00001100, 00001111, 00001111,
11111111, 00000000, 11111111, 00000000;
```

Figure 1.11 gives an example of a trivial circuit which permits to initialize 8,192 × 8 RAM by byte values from the COE file above through the first port A and then to read from RAM and to display on the rightmost Atlys onboard LED Led0 signals from the single output of the second port B of 65,536 × 1 RAM.

In our example the first 8 bytes of memory will be filled in from the COE file above and the remaining bytes are assigned to FF_{16} (the option *Fill Remaining Memory Locations* available at step 4 is checked and the value FF is chosen). The clock signal with the reduced frequency (approximately 1 Hz) is displayed on Led1. Thus we can see changes of Led0 (i.e. the second port B output) relatively to Led1 (i.e. relatively to clock with the reduced frequency) and these changes are shown in Fig. 1.12. The desired sequence has been specified in the COE file above by binary 8-bit vectors. From the ISE *Design Summary* we can see that 4 RAM blocks have been allocated. The second example clearly demonstrates that two ports of the same block RAMs can have different aspect ratio (8,192 × 8 for the port A and 65,536 × 1 for the port B). Data are written to the memory by

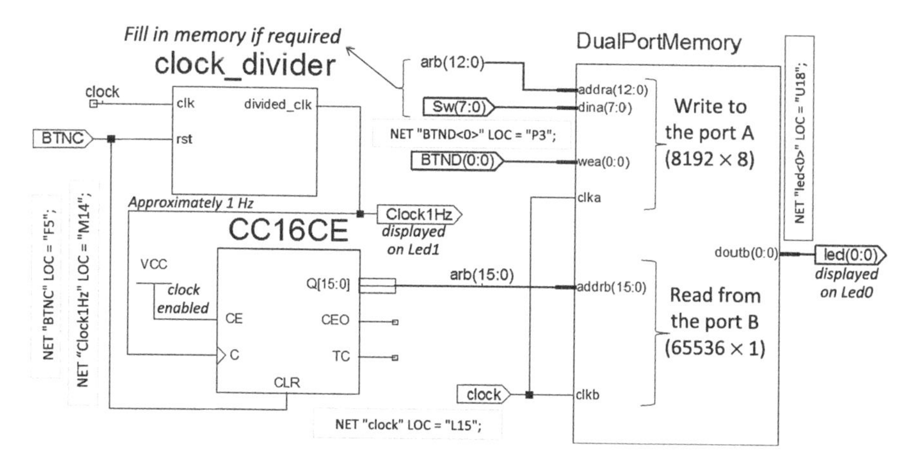

Fig. 1.11 Example with memory to be built from simple dual-port block RAMs (CC16CE is the Xilinx library primitive for a 16-bit binary counter)

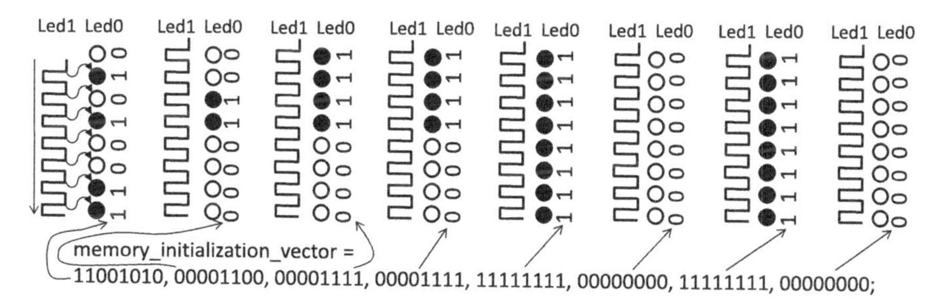

Fig. 1.12 Initialization (COE) file specifies visual sequence on the LED Led0

bytes and read by bits. Thus, many useful converters can be created directly in memory without the need for additional logic. All necessary details can be found in [15–17].

1.3.2 Embedded DSP Slices

Devices of Spartan-6 family include from 8 to 180 digital signal processing slices DSP48A1 which support several functions, including multiplier, multiplier–accumulator, pre-adder/subtracter followed by a multiply accumulator, multiplier followed by an adder, wide bus multiplexers, magnitude comparator, and wide counter [18]. These types of functions are frequently required in DSP applications. It is also possible to connect multiple DSP48A1 slices to form wide math functions, DSP filters, and complex arithmetic without the use of general FPGA logic

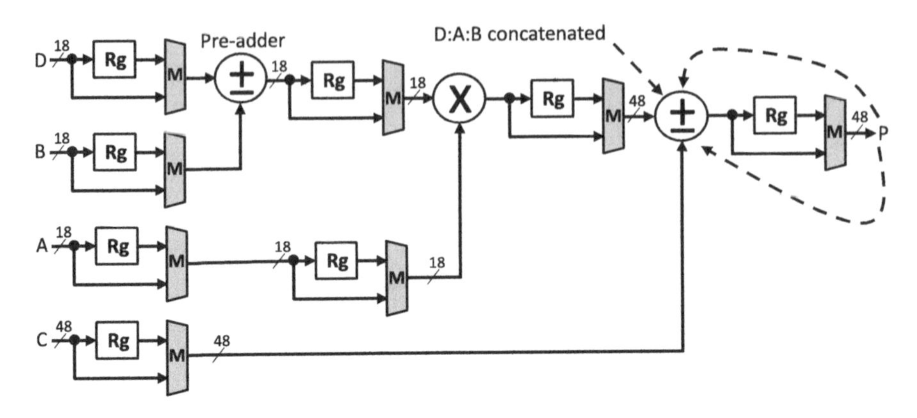

Fig. 1.13 A simplified architecture of the DSP48A1 slice [18]

which leads to lower power consumption and higher performance. Basically, the DSP48A1 slice contains an 18-bit input pre-adder followed by an 18 × 18 two's complement multiplier and a 48-bit sign-extended adder/subtracter/accumulator. A simplified architecture of the slice [18] is presented in Fig. 1.13, where A, B, D are 18-bit operands, C is a 48-bit operand and P is a 48-bit result. The gray-colored multiplexers M are configurable and they block or unblock registers *Rg* that can be used in a pipeline. Concatenated lines D[11:0], A[17:0], B[17:0] can be taken directly as an operand of the right-hand adder/subtracter controlled by a multiplexor that is not shown in Fig. 1.13. The result P can also be used as an operand of the adder/subtracter. The slice DSP48A1 has a mode input which permits to specify the desired function of individual components, such as whether the adders realize an addition operation, a subtraction operation, or are disabled, how to connect carry signals, how to build a pipeline and others. DSP slices are organized in vertical DSP columns (see Fig. 1.1) and can be easily interconnected without the use of general routing resources. These components can be instantiated and configured with the aid of the ISE tools as will be illustrated in Sects. 4.1, 4.2 of the book. IP *LogiCore* generator can also be used.

The DSP48E1 slice [19] (see Fig. 1.14) extends functionality of the DSP48A1 slice and improves characteristics. There are from 240 to 3,600 such slices available in the 7 series FPGAs. The multiplier is organized as 25 × 18. The register A is extended up to 30 bits. The adder/subtracter was replaced with an arithmetic logic unit and, thus, many bitwise logical functions can be executed over up to 48-bit operands. Besides a pattern detector and 17-bit shifter are added.

Let us consider a simple example demonstrating potential use of DSP slices for the Spartan-6 FPGA available on the Atlys prototyping board. DSP will be created and configured by the Xilinx *LogiCore* DSP48 macro [17]. Much like it was done in the previous section, first we add a new source in the Xilinx ISE (option *Project → New Source*) and then select IP and indicate the name *DSP_slice*. The core generator will be launched. At the first step let us specify arithmetic

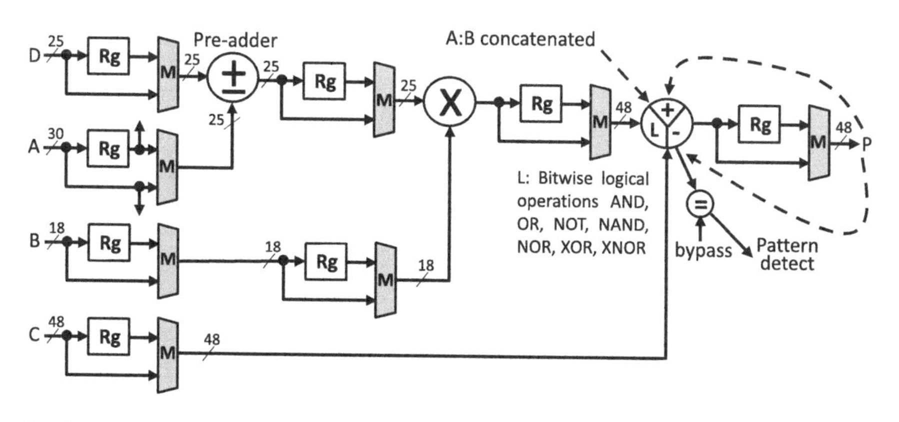

Fig. 1.14 A simplified architecture of the DSP48E1 slice [19]

instructions shown in Fig. 1.15 inside the block DSP_slice. They can be selected by codes on the inputs *sel(1:0)* in which the left and the right bits are provided by the buttons *BTNL* and *BTNR*, accordingly.

Let us define the size 3 for all four operands A, B, C, and D. The most significant bit (2) represents signal and if it is equal to '0' the number is positive otherwise negative. Since D(2) is always equal to 0 in Fig. 1.15, the value of D is assigned to be always positive. Other operands A, B, and C can be either positive or negative and the sign may be chosen by the onboard buttons *BTNU*, *BTNC*, and *BTND*, respectively (see Fig. 1.15). For any button shown in Fig. 1.15 the value '1' is produced when the button is pressed. Two-bit operands without signs are taken from the onboard switches as it is shown in Fig. 1.15. The result is displayed on the LEDs Led6,…,Led0. Thus, if all the switches are ON then the results are equal to the following values:

- "0010101" when sel = "00" because $(3 + 3)*3 + 3 = 21_{10} = 0010101_2$;
- "0000110" when sel = "01" because $3 + 3 = 6_{10} = 0000110_2$;
- "0010010" when sel = "10" because $(3 + 3)*3 = 18_{10} = 0010010_2$;
- "0001001" when sel = "11" because $(3 + 3) + 3 = 9_{10} = 0001001_2$.

If the button *BTND* is pressed then the operand C becomes negative with the value -1 (two's complement code is used). Thus, when sel = "00", the result is "00010001": $(3 + 3)*3-1 = 17_{10} = 00010001_2$. From the ISE *Design Summary* we can see that one DSP slice is used. The Sects. 4.1, 4.2 and Appendix B will give more complicated examples which will explore many additional DSP slice capabilities. It will be shown that DSP slices can be more effectively used as components in HDL code. Since for the 7 series FPGAs DSP slices can execute bitwise logical operations, they are effective for solving combinatorial problems over binary vectors and matrices. Besides, arithmetic operations over 48-bit operands can be presented in one DSP slice in form of four independent operations over 12-bit operands.

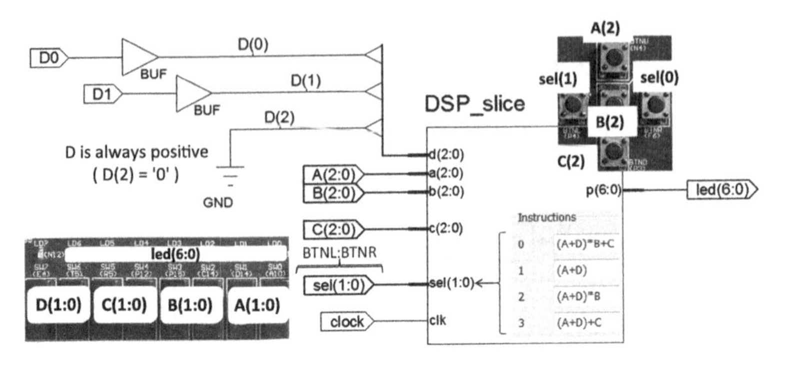

Fig. 1.15 Trivial use of a DSP slice for Spartan-6 FPGA

1.4 Clock Distributions and Resets

To guarantee efficient clock distribution FPGAs include dedicated clock inputs, buffers, and routing. These resources are used automatically by CAD tools.

To provide for high-performance clocking, devices of Spartan-6 family include from 2 to 6 Clock Management Tiles (CMT). Every CMT is composed of 2 Digital Clock Managers (DCMs) and one Phase-Locked Loop (PLL). CMTs are used to phase shift a clock signal, to eliminate the *clock skew* (difference between arrival times of a clock edge to various devices that compose a given circuit), to multiply or divide clock frequency, to synthesize a new clock frequency, and to convert an incoming clock signal to a different I/O standard [20].

Some features of clocking for Spartan-6 FPGAs are unique to the relevant architecture and they have been replaced with the new 7 series FPGAs clocking structures [21]. The PLL is a subset of the mixed-mode clock manager (MMCM). Some clocking primitives for Spartan-6 FPGAs were removed and some were replaced. The details are given in [21].

Reset is a synchronous or asynchronous signal that sets necessary storage elements to a desired state. It is noted in [22] that regardless of the type (synchronous or asynchronous) the reset signal needs to be synchronized with the clock. This permits potential metastable state of flip-flops to be avoided. Besides, in some circuits (such as state machines and counters) reset of all flip-flops has to be deasserted on the same clock edge to eliminate eventual transitions to illegal states. According to [22] active-high resets enable better device utilization and improve performance.

The reset bridge circuit [22] shown in Fig. 1.16 provides a mechanism to assert reset asynchronously (and it properly functions even without a valid clock) and to deassert reset synchronously.

When a Xilinx FPGA is configured or reconfigured, every cell (including flip-flops and block RAMs) is initialized much like it is done on a global reset, i.e. all storage will be set to their specified initial states. Thus, global reset is not always required. From [22] we can see that the design tools synthesize initialization of

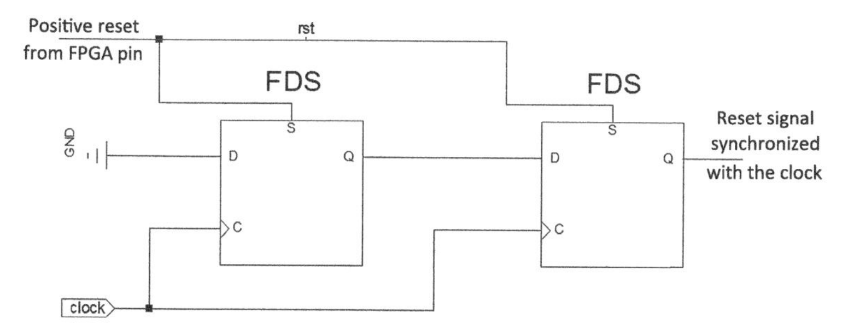

Fig. 1.16 Generation of the reset signal synchronized with the clock

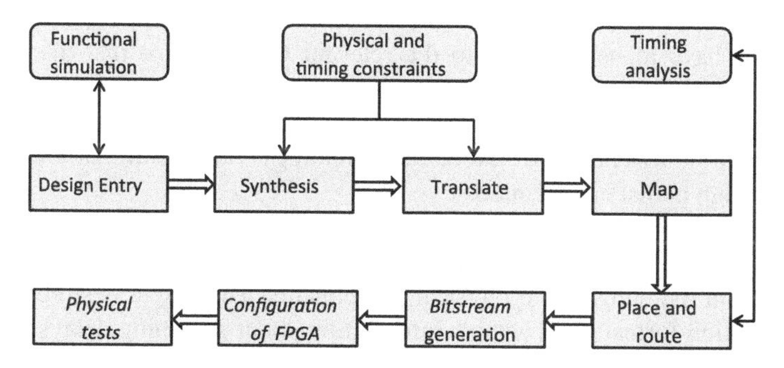

Fig. 1.17 Typical FPGA-based design flow

signals, such as in the following VHDL line, which assigns the values 0 to all eight elements of the signal rg (the details will be given in the next chapter):

```
signal rg: std_logic_vector (7 downto 0) := (others <= '0');
```

The initialization value of the *rg* signal (that is all zeros above) becomes the INIT value loaded into the relevant flip-flops during configuration. Similarly block RAMs are initialized and it was shown with the COE files in Sect. 1.3.1 above. Since many embedded components use just synchronous reset capabilities, synchronous resets enhance FPGA utilization. Thus, almost always synchronous resets should be used rather than asynchronous resets [22].

1.5 Design Tools

A typical FPGA-based design flow is illustrated in Fig. 1.17.

We start with the design entry which can be accomplished using a variety of methods such as the considered above schematic and HDL which will be discussed in Chap. 2. Once the desired functionality of a given circuit or system is specified, we can model its behavior through *functional simulation*. This means that the circuits can be checked for correct *functionality* assuming that all their components react instantly to changes on their inputs, i.e. functional simulation does not take into account the timing characteristics of electronic elements that will be used to implement the circuits. If a problem is detected the designer has to return back to the circuit specification and provide the required changes.

During the simulation process of simple circuits it is possible to generate and apply inputs and to observe outputs manually. For larger designs, *test benches* are usually created. A *test bench* is a program/specification, normally developed in the same language as the circuit under test that automatically applies inputs to the circuit and eventually compares the circuit's outputs with the expected values.

Once the design has simulated correctly, we can advance to the circuit synthesis. At this level, all the characteristics of a specific device (such as package, speed grade, etc.) have to be provided to the relevant CAD tool so that the synthesis could be done. The synthesis is performed by CAD tools and converts the design entry into a set of components (such as LUTs, flip-flops, memories, DSP slices, etc.) that can be assembled in the target technology. As a result, an architecture-specific design netlist is generated.

The translate phase merges all the synthesized netlists and the physical and timing constraints to produce a generic database file. The map phase groups logical symbols from the netlists into physical components. The output is stored in a circuit description format and contains information about switching delays. The map phase reports an error if the design exceeds the available resources or the user-specified timing constraints are violated. The place and route phase performs the placement and routing of the mapped symbols on the physical FPGA device and verifies the timing constraints once again. Translation, mapping, placement and routing compose the design implementation phase which converts the synthesized logical design into available device resources. It may mean selecting and programming individual LUTs and finding ways to connect them within the physical constraints of the target FPGA. The output of this stage is a *bitstream* file that can be uploaded to the selected device.

After implementation, the circuit can be analyzed for timing performance, device resource utilization, and power consumption. The results of this analysis may force the designer to go back and make changes either in the circuit specification, design constraints, or synthesis strategies and optimization goals in the relevant CAD tools. Afterwards synthesis and/or implementation need to be rerun.

Once you are satisfied with the results of implementation, the generated *bitstream* file can be uploaded to the FPGA to configure the device properly. Then, the resulting physical circuit undergoes final in-circuit tests.

In this book, we will mainly use the release version 14.7 of Xilinx ISE for fulfilling all the design steps, from circuit specification to simulation, synthesis, and implementation. Other CAD tools can also be involved in a similar manner since the underlying ideas are exactly the same, just the relevant software environments are different. For some examples we will also use the Quartus II 13 Web edition software for Altera FPGAs.

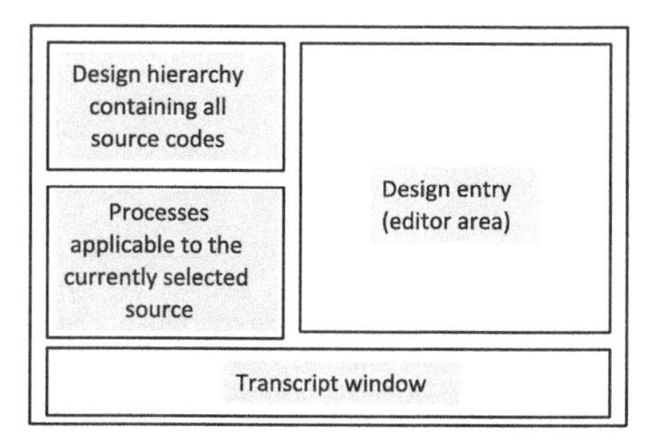

Fig. 1.18 Default interface of the ISE

To summarize the design flow that has already been discussed in the previous sections let us consider once again all the steps that need to be accomplished in the ISE in order to implement a simple full adder (FA) and test it in the Atlys prototyping board. Note that there is a free version of ISE called WebPACK available for download from the Xilinx website [23].

We start with creating a new project (option *File → New Project*) by specifying its name, location, and then all the characteristics of the target FPGA. Please note that we have selected "Spartan6" as the FPGA family and also indicated the device XC6SLX45 and package CSG324. These data can be found in the documentation that accompanies your device [11] and are also written on the proper microchip. Schematic is again chosen as a top-level source type (design entry).

Once you create a new project, a default ISE interface appears which is divided in 4 areas (see Fig. 1.18):

- *Design hierarchy* is used to display files associated with the current project and their hierarchical organization. We have not created any file yet, therefore this area is initially empty.
- *Processes* area is context-sensitive and always shows processes that are available for the currently selected source. A process can be started by double click on its name. The processes permit us to synthesize a particular design entry file, to specify user constraints, to generate programming file and so on.
- *Editor* area on the right-hand side provides support for design entry in a variety of forms, such as schematic editor, text editor for VHDL, etc.
- *Transcript* window shows progress of processes and error/warning messages that might appear during synthesis/implementation phases.

To specify a schematic file for a full adder let us add a new design entry to our project. It can be done choosing options: *Project → New Source… → Schematic* and specifying a name, for example, FA. Suppose we would like to create a simple hierarchical design and to compose the *FA* of two half adders (*HA*) and an *OR* gate. Thus, let us create another schematic source with the name *HA*. Figure 1.19a depicts a circuit for the *HA* described with the aid of the ISE library primitives *XOR2* and *AND2* (see [24] for details). At the next step we create a schematic symbol for the half adder which can be used as a user-library primitive much like

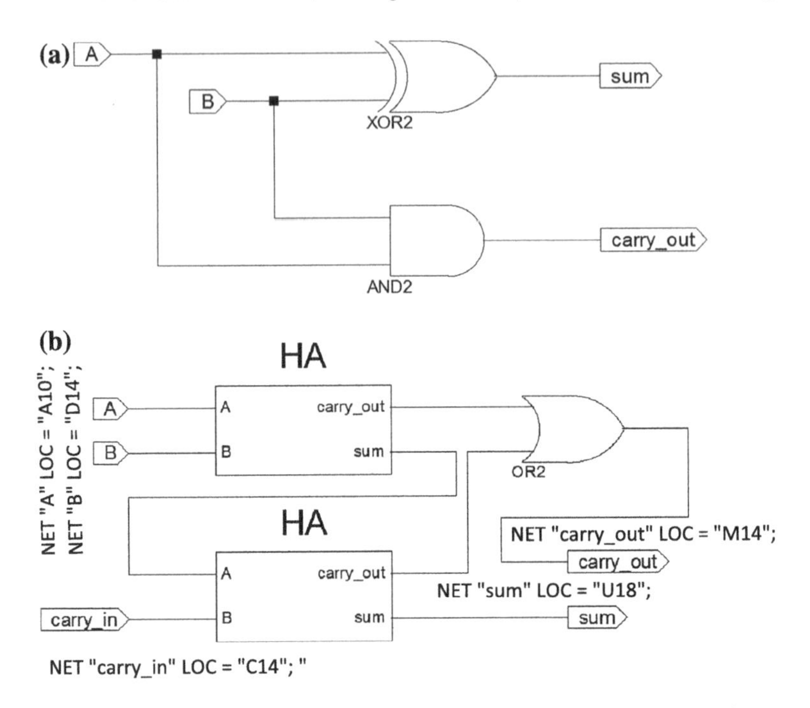

Fig. 1.19 Using schematic editor to describe a full adder: half-adder (**a**); full adder (**b**)

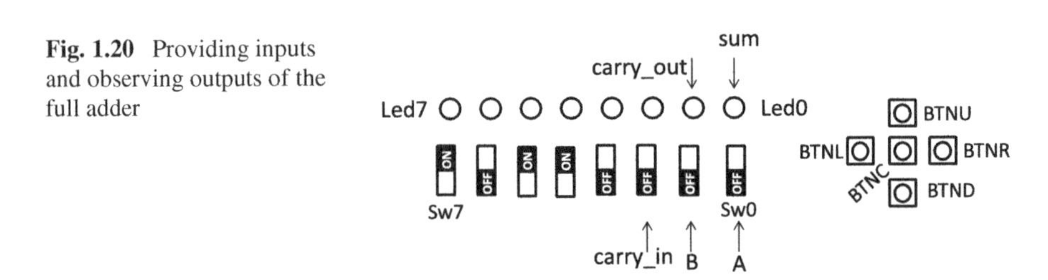

Fig. 1.20 Providing inputs and observing outputs of the full adder

the ISE library primitives. Figure 1.19b describes the full adder composed of two *half adders* (i.e. our previously created primitives) and an *OR2 gate* (the existing ISE library primitive).

To test our adder in the Atlys prototyping board we need to route the adder's inputs and outputs to some board components that would allow us to interact with the circuit. For instance, we could specify inputs with three switches and observe the output values on two LEDs as exemplified in Fig. 1.20. Lines of the *Atlys.ucf* file with the necessary constraints are given directly in Fig. 1.19b. The exact pin locations can be found by consulting the Atlys prototyping board's documentation [11]. Please note that the complete UCF file for the Atlys board is available online [25]. When you would like to implement your design on a different board or with a different FPGA, you will need to modify not only the target FPGA but also the UCF file accordingly. In the UCF file all lines that start with the symbol

"#" are comments and the syntax of the remaining lines for the Atlys board is the following:

```
NET "Input or output name in the top module" LOC = "name of the PIN to connect to";
```

Now the steps shown in Fig. 1.17 are executed and a bitstream for the project is generated. Uploading the bitstream to the Atlys board can be done in the Adept software [10] (see Fig. 1.2 as an example) or directly from the ISE selecting the option Configure Target Device in the *processes* area. In the last case iMPACT is launched which enables the generated bitstream to be uploaded to the Atlys board. Finally the project can be tested in hardware. From the ISE *Design Summary* we can see that the circuit in Fig. 1.19b occupies just 2 FPGA slices (from 6,822 available slices).

Simulation can be accomplished with the aid of a test bench. Since a test bench can efficiently be described in HDL we will consider this feature in the next chapter (see Sect. 2.7).

It is not strictly necessary to run *Synthesize*, *Implement Design*, and *Generate Programming File* processes one by one sequentially. Instead you can proceed in the ISE directly to the *Generate Programming File* (or to the *Configure Target Device*) option which will execute all the required previous processes automatically.

For the Atlys board [11] (and further for the Nexys-4 board [26]) we will mainly use the design steps (1–8) briefly illustrated in Fig. 1.21 and described in the corresponding points below:

1. The name, location and working directory of the project are introduced.
2. FPGA family (such as Spartan-6), FPGA device (e.g. XC6SLX45), package (e.g. CSG324), speed (e.g. −3) and preferred language (e.g. VHDL) are selected
3. New or existing project sources (design entries) are specified. In this book we will use the main project sources indicated in Fig. 1.21.
4. Basically a project specifies design hierarchy in which a top-level module invokes lower-level modules created from different sources. We can apply at this step top-down, bottom-up or mixed design strategy.
5. Optional functional simulation is executed at different hierarchical levels. We will discuss this feature in the next chapter.
6. Activating *Generate Programming File* (or *Configure Target Device*) process executes sequentially all the major design steps and finally (if the project is correct) generates a *bitstream* (see Fig. 1.21).
7. The generated *bitstream* is uploaded to the FPGA. Different opportunities can be used for such purposes and we will discuss them in the next section. An important feature of reconfigurable systems is a possibility of prototyping, evaluation of the implemented designs, experiments and comparisons with the aid of numerous available FPGA-based boards.
8. The project is tested in hardware much like it has been done in the previous sections. In the subsequent chapters we will use prototyping boards briefly

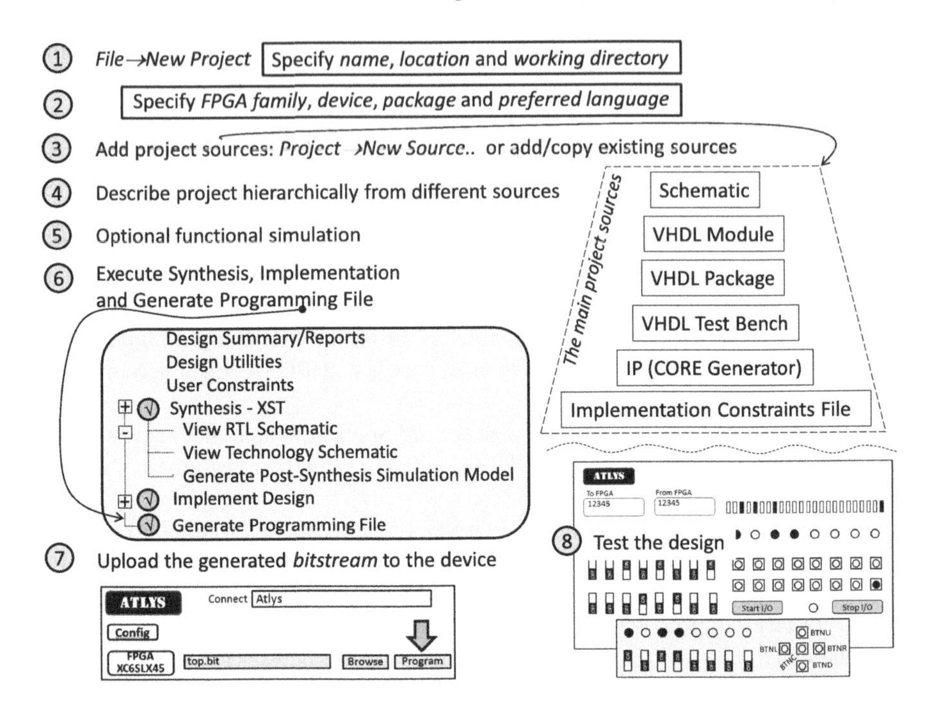

Fig. 1.21 Design steps used in the book

discussed in the next section. Verification of the implemented circuits can be done with the aid of different methods and tools, such as run-time signal analyzers (e.g. the Xilinx ChipScope), onboard and externally attached peripheral devices, existing interfaces that enable communications with higher-level computational systems, and others. Some of such methods and tools used in the book will be briefly described in Sect. 1.7.

The majority of projects described in the book were also implemented and tested in Xilinx Vivado design suite (version 2013.4). Each VHDL project available online at http://sweet.ua.pt/skl/Springer2014.html contains a compressed file that may include: (1) only Atlys files for ISE (if only the UCF file containing the word *Atlys* is given); (2) only Nexys-4 files for ISE (if only the UCF file containing the word *Nexys* is given); (3) Atlys and Nexys-4 files for ISE (if there is a directory *ISE* and *Atlys/Nexys4* sub-directories in the directory *ISE*); (4) Vivado Nexys-4 files (if there is a directory *Vivado*). At the beginning, the compressed file needs to be decompressed into a set of components that may include: (1) schematic file(s) **.sch*; (2) generated by IP core **.xco* file(s) in sub-directory *ipcore_dir*; (3) **.vhf* file(s) that are VHDL specifications generated by ISE from schematic(s); (4) user constraints file **.ucf* for ISE; (5) Xilinx design constraints file **.xdc* for Vivado; (6) VHDL file(s) **.vhd*; (7) bitstreams **.bit* for programming FPGA.

Any project in ISE may be built as follows: (1) create a new project; (2) add copy of sources such as **.sch*, **.vhd*, **.ucf*, and **.xco* files and the latter is taken from the directory *ipcore_dir*; (3) copy available initialization files (such as *coe* and *txt*) in the project directory; (4) if there is a schematic file then open this file and update if requested; (5) run synthesis, implementation and generate programming file; (6) configure target device (upload bitstream to the target device) and test the design in FPGA. Please note, that peripheral components (switches, buttons, LEDs) for Atlys and Nexys-4 boards are frequently not the same (check correspondence between the design ports and pre-connected FPGA pins).

Any project in Vivado may be built as follows: (1) create a new project; (2) add copy of sources such as *.*vhf*, *.*vhd*, *.*xdc*, and *.*xco* files; (3) right mouse click on *.*xco* file (if this file is available) and upgrade IP; (4) run synthesis, implementation and generate programming file; (5) configure target device (upload bitstream to the target device using hardware manager) and test the design in FPGA. Please note that a few small changes are done in VHDL files for Vivado projects, which can be seen at http://sweet.ua.pt/skl/Springer2014.html.

An example demonstrating migration of ISE projects to Vivado projects can be found at the end of Appendix B. The most important points for such migration are:

- All the described in the book source files may be added to Vivado projects with the exception of schematic (*.*sch*) files. However the files *.*vhf* generated by ISE from schematic files can be used instead of *.*sch* files. Thus, all the projects from Chap. 1 may also be tested in Vivado;
- ISE UCF files have to be converted to XDC (Xilinx Design Constraints) format which may be done for the examples in the book as follows: (1) open PlanAhead for ISE Design Suite; (2) in PlanAhead open ISE project for Nexys-4 (if the design is in schematic you need to add *.*vhf* file(s) manually); (3) run synthesis and open synthesized design; (4) in Tcl Console of the PlanAhead run command *write_xdc <directory>/<name>.xdc* (for example, *write_xdc c:/tmp/Nexys4.xdc* where the sub-directory *tmp* has to be previously created in the directory *c:*); (5) use the generated *.*xdc* file (from *c:/tmp*) in Vivado. Additional details are given in Appendix B. Note that the complete XDC file for the Nexys-4 board can be downloaded from Digilent Web site: http://www.digilentinc.com/Data/Products/NEXYS4/Nexys4_Master_xdc.zip.

1.6 Implementation and Prototyping

There are a number of FPGA-based prototyping boards available in the market that simplify the process of FPGA configuration and provide support for testing user circuits and systems in hardware. All the examples in the first chapter of the book were prepared for the Atlys prototyping board [11] manufactured by Digilent, which includes one Xilinx FPGA xc6slx45 of Spartan-6 family [27]. Examples in subsequent chapters will use 3 prototyping boards: the Nexys-4 [26], the Atlys [11], and the DE2-115 [28] that are briefly characterized below.

The Nexys-4 board manufactured by Digilent [26] contains one FPGA Artix-7 xc7a100t from the 7 series of Xilinx [29]. Almost all examples in the Chaps. 2–5 and appendices A, B have been implemented and tested in the Nexys-4. VHDL codes, user constraints files, and bitstreams for all projects of the book are available online at http://sweet.ua.pt/skl/Springer2014.html. The following onboard devices will be involved (all necessary details about these devices can be found in [26]):

1. Xilinx Artix-7™ FPGA xc7a100t-csg324 [29];
2. USB-JTAG and USB-UART;
3. 100 MHz clock oscillator;
4. Eight 7-segment displays;
5. 16 slide switches;
6. 16 user LEDs;
7. 5 user buttons;
8. Pmod expansion connectors;
9. USB host connector.

The FPGA on the Nexys-4 board can be configured using methods [26]. In this book we will configure the board with the aid of the following two modes:

- From the ISE environment (the option *Configure Target Device*) and iMPACT tools through the USB JTAG/UART;
- From a USB memory stick attached to the USB host connector.

Please note that configuration of the board from Adept software is not supported and the Digilent component *IOExpansion* [30] (considered below for some examples with the Atlys board) cannot be used.

Many examples in the Chaps. 2–5 will also be tested in the Atlys prototyping board [11]. The following onboard devices will be involved (all necessary details about these devices can be found in [11]):

1. Xilinx Spartan-6 xc6slx45-csg324 FPGA [27];
2. USB-UART and USB port for programming and data transfer;
3. 100 MHz clock oscillator;
4. 8 slide switches;
5. 8 user LEDs;
6. 5 user buttons;
7. Reset button.

In this book we will configure the Atlys board with the aid of the following two modes:

- From the ISE environment (the option *Configure Target Device*) and iMPACT tools through the USB JTAG/UART;
- From the Digilent Adept software [10] through the USB JTAG/UART.

The Atlys board has a limited number of onboard user switches and LEDs but it is supported by the Adept software and may interact with a host PC through

Table 1.2 Characteristics of the Xilinx FPGAs on the Atlys, Nexys-4, and ZedBoard

Board	FPGA/APSoC	N_s	N_{LUT}	N_{ff}	N_{DSP}	N_{BR}	M_{Kb}
Atlys	xc6slx45	6,822	27,288	54,576	58	116	2,088
ZedBoard	xc7z020	13,300	53,200	106,400	220	140	5,040
Nexys-4	xc7a100t	15,850	63,400	126,800	240	135	4,860

Table 1.3 Characteristics of the Altera FPGA on the DE2-115 board

Board	FPGA	N_{LE}	M_{Kb}	N_{EB}
DE2-115	4CE115	114,480	3,888	266

a virtual window enabling many virtual peripheral elements to be attached [11], which is convenient for simple tests of the developed circuits. Availability of the virtual devices is the main point in favor of the Atlys board for some subsequent examples of the book.

Almost all examples from Chap. 2 and some examples from Chaps. 3–5 have also been tested in the DE2-115 board [28] containing one Altera Cyclone-IVe EP4CE115 FPGA. The main objective is to demonstrate that the majority of the projects are technology independent and may be implemented in FPGAs of different companies. The following onboard devices will be involved (all necessary details about these devices can be found in [28]):

1. Altera Cyclone-IV EP4CE115F29C7 FPGA [14];
2. USB Blaster port for FPGA programming;
3. 50 MHz clock oscillator;
4. Eight 7-segment displays;
5. 18 slide switches;
6. 26 user LEDs (18 red and 8 green);
7. 4 user buttons.

The only method of FPGA programming from a host computer that has been used in the book is through the USB Blaster port.

In Sect. 4.5 we will also briefly describe the Xilinx all programmable systems-on-chip (APSoC) and, in particular, Zynq family. One of such microchips (xc7z020) is available on the ZedBoard [31]. Since the device xc7z020 incorporates the Xilinx Artix-7 FPGA, all the examples of the book can directly be used and, besides, the majority of them have been tested in the ZedBoard. However, we will not consider in the subsequent chapters the relevant implementations.

Tables 1.2 and 1.3 give some details about Xilinx (see Table 1.2) and Altera (see Table 1.3) FPGAs available on the referenced above boards. Here: N_s is the number of FPGA slices; N_{LUT} is the number of FPGA LUTs; N_{ff} is the number of FPGA flip-flops; N_{DSP} is the number of DSP slices DSP48A1/DSP48E1 for xc6slx45/(xc7a100t/xc7z020) microchip; N_{BR} is the number of 18/36 Kb block RAMs for xc6slx45/(xc7a100t/xc7z020) microchip; M_{Kb} is the size of embedded block RAM memory in Kb; N_{LE} is the number of logic elements; N_{EB} is the number of embedded multipliers with 18-bit operands (i.e. 18×18).

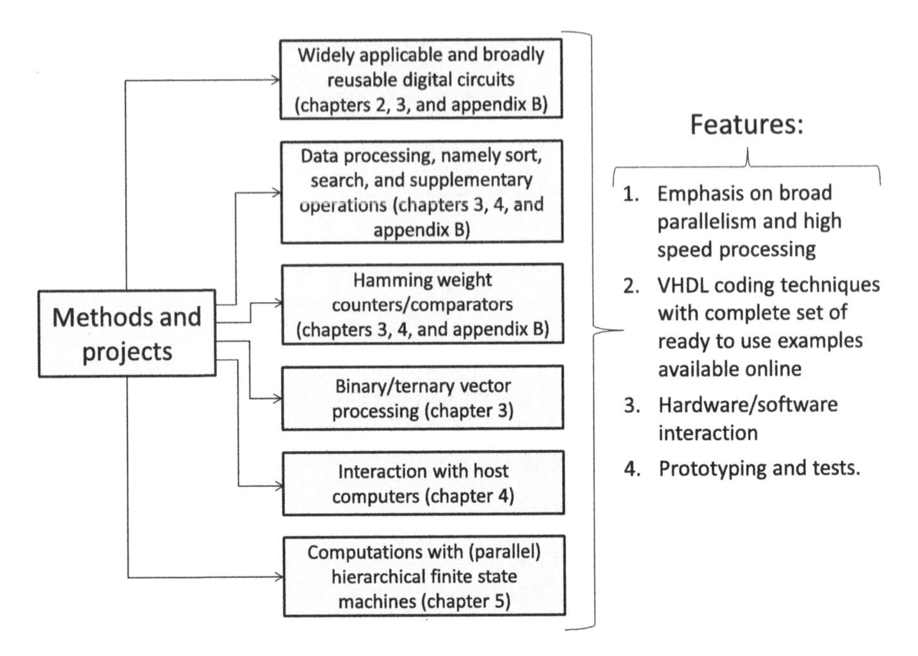

Fig. 1.22 Methods and projects described in the book

The projects for Xilinx FPGAs will be created in the Xilinx ISE 14.7 software. The total number of available components and the number of components used in a particular project can be found from the ISE *Design Summary/Reports*. If you check projects that have already been developed in the previous sections you can see that the number of occupied FPGA resources was negligible comparing with the total number of available resources. Thus, the selected boards (although being low-cost) enable complex circuits and systems to be developed. Design projects for Altera FPGA will be created in the Quartus II version 13 Web Edition software.

The methods and projects described in the subsequent chapters cover a few areas that are outlined in Fig. 1.22 with indication of the chapters where the relevant topics have been discussed. Special attention will be paid to features shown on the right-hand side of Fig. 1.22.

Each project will be verified in hardware with the aid of different methods and tools that will be described in the next section.

Since two prototyping boards, Atlys [11] and Nexys-4 [26] from Digilent, will be used for all the examples in the book, let us overview these boards with more details. Besides of the described above onboard devices, there are many additional components that permit the developed projects to be expanded and new more advanced circuits and systems to be designed. This feature is especially valuable for education. We briefly describe below the basic capabilities and layouts of the Atlys and Nexys-4 boards (with permission of Digilent Inc.).

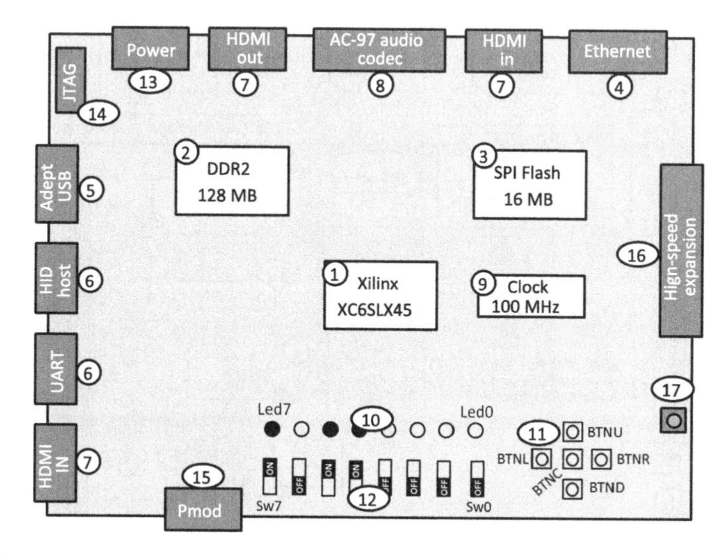

Fig. 1.23 The simplified layout of the Atlys prototyping board from Digilent

The Atlys [11] has the following main components and connectors (see Fig. 1.23):

1. Xilinx Spartan-6 xc6slx45 FPGA;
2. 128 MB DDR2 (Double Data Rate) with 16-bit wide data;
3. 16 MB ($\times$ 4) SPI (Serial Peripheral Interface) Flash for configuration and data storage;
4. 10/100/1000 Ethernet;
5. USB2 (Universal Serial Bus) port for programming and data transfer;
6. USB-UART (Universal Asynchronous Receiver/Transmitter) and USB-HID (Human Interface Device) port (for mouse/keyboard);
7. HDMI (High-Definition Multimedia Interface) video input and output ports;
8. AC-97 Codec (coder-decoder) with line-in, line-out, mic, and headphone;
9. 100 MHz oscillator clock source;
10. 8 user LEDs;
11. 5 push buttons;
12. 8 slide switches;
13. Power connector and power-on LED indicator;
14. 2 $\times$ 7 programming JTAG (Joint Test Action Group) connector;
15. Pmod (Peripheral module) expansion connector (2 $\times$ 6);
16. High-speed expansion connector;
17. Reset button.

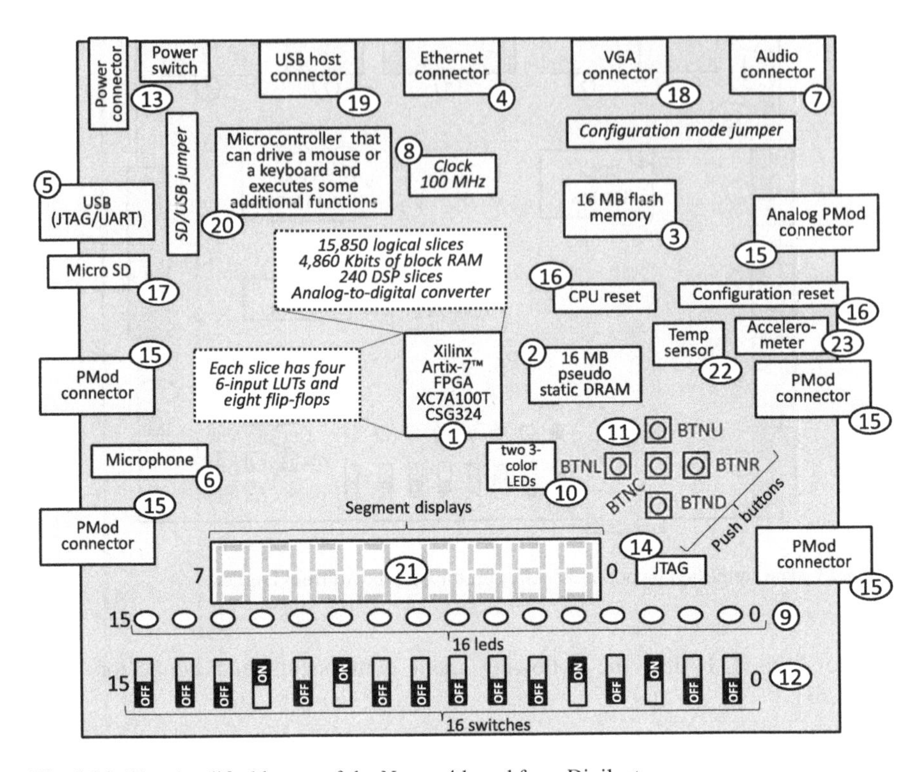

Fig. 1.24 The simplified layout of the Nexys-4 board from Digilent

The FPGA on the board can be configured using one of the following three methods [11]:

- from a USB-connected PC via Adept USB-JTAG port (see 5 in Fig. 1.23) or directly from the JTAG connector (see 14 in Fig. 1.23);
- from the SPI Flash (see 3 in Fig. 1.23), provided the configuration file has been previously stored in the flash memory;
- from a USB memory stick connected to the USB HID port (see 6 in Fig. 1.23).

The onboard jumpers (not shown in Fig. 1.23) permit the required configuration mode to be selected (the details can be found in [11]).

The Nexys-4 [26] includes one Artix-7 xc7a100t FPGA from the 7 series of Xilinx. The main components and connectors are the following (see Fig. 1.24):

1. Xilinx Artix-7™ FPGA xc7a100t-csg324;
2. 128 Mb = 16 MB cellular RAM;
3. 128 Mb = 16 MB SPI (quad-SPI) Flash;
4. 10/100 Ethernet;
5. USB-JTAG programming and USB-UART;
6. Microphone;
7. Audio connector;

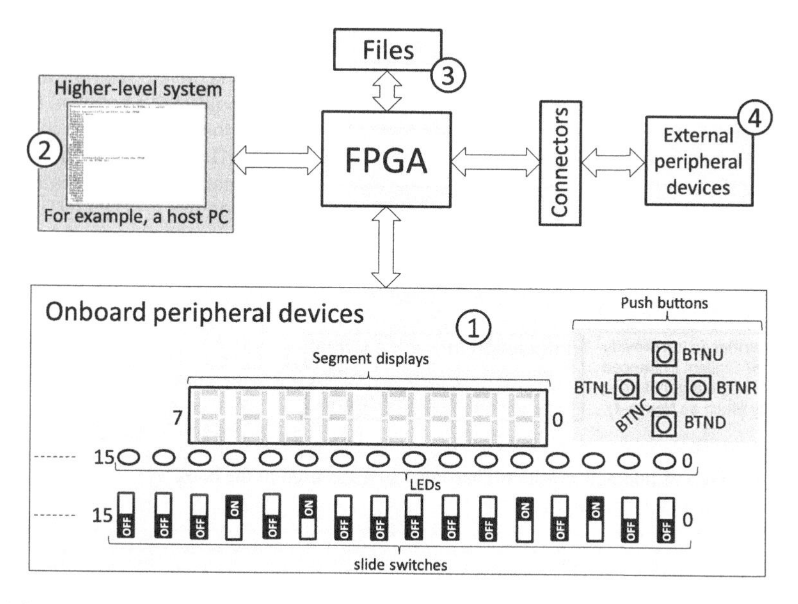

Fig. 1.25 Different types of interactions with external devices/systems used in the book

8. 100 MHz clock oscillator;
9. 16 user LEDs;
10. Two 3-color user LEDs;
11. 5 user buttons;
12. 16 slide switches;
13. Power connector and power-on LED indicator;
14. JTAG port;
15. 5 Pmod expansion connectors (2 × 6);
16. Two reset buttons;
17. Micro SD card slot;
18. VGA connector;
19. USB host connector;
20. Microcontroller;
21. Eight 7-segment displays;
22. Temperature sensor;
23. Accelerometer.

The FPGA on the Nexys-4 board can be configured using one of the following four methods: Quad-SPI, SD Card, USB JTAG, or USB memory stick. The onboard jumpers shown in Fig. 1.24 permit the required configuration mode to be selected. The details are given in [26].

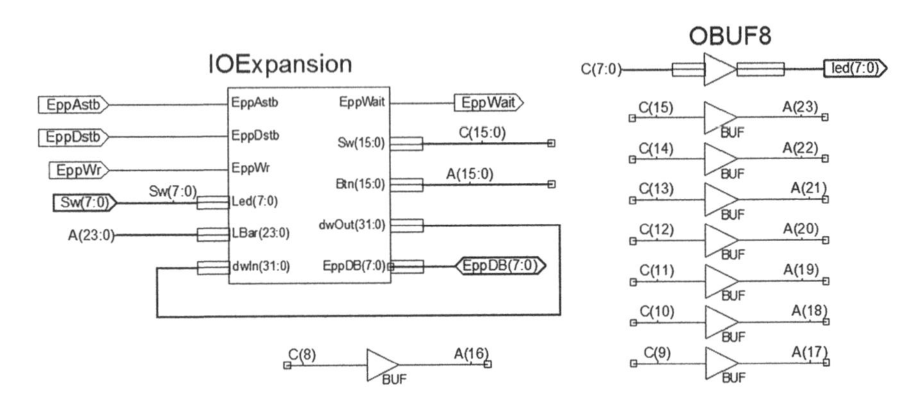

Fig. 1.28 Project for interaction with a host PC

problems with different dimensions. Although such technique is very requested it is not always possible especially when devices with different incompatible architectures are used. Indeed, many generic characteristics established for FPGAs of one company that involve dedicated libraries, primitives and embedded blocks are not equally applicable (or not applicable at all) to FPGAs of another company that involve dedicated libraries, primitives and embedded blocks of different types. In the subsequent chapters all projects that involve Xilinx primitives and libraries can only be implemented in FPGAs from Xilinx with compatible architectures. They can also be used for other FPGAs but necessary modifications have to be provided. Besides, effectiveness of resources and performance for the implemented circuits may become different.

It is clear from Sect. 1.6 that the number of onboard peripheral devices is limited. In the subsequent chapters we would like to implement and evaluate such circuits and systems that require significantly larger number of inputs and outputs that often exceed the number of available FPGA pins. The following two techniques will be applied: (1) the designs are evaluated with the aid of auxiliary circuits that supply the input signals and analyze the output signals (e.g. random number generators, comparators and counters); (2) through an interaction with a higher-level system (e.g. a host PC).

Let us discuss the point 2 with a bit more details. Three types of interactions will be explored and they are shown in Fig. 1.26 for Digilent prototyping boards.

The first type is targeted to such prototyping boards that do not provide support for the Digilent Enhanced parallel port (EPP) data transferring capability. The developed software and hardware modules are described in Sects. 4.3 and 4.4.

The second type may be used for prototyping boards that do provide support for the Digilent EPP data transferring capability (see Sects. 4.3 and 4.4).

The third type enables interactions through a virtual window managed by the Digilent Adept software, involving the relevant Digilent *IOExpansion* component described in VHDL. Since this type is entirely based on the Digilent products [30] and will be needed for some examples in the subsequent chapters with the Atlys board, we will present here a bit more details.

The primary objective is to expand the number of available input/output devices. This feature is offered by any Digilent board which provides support for EPP [10]. For example, Atlys, Nexys-2 and Nexys-3 boards do support this feature but the Nexys-4 does not.

The expanded input/output controls include:

- 24 light bars;
- 8 LEDs;
- 16 buttons;
- 16 slide switches;
- 32-bit data to be sent from a host PC to the FPGA (can be specified in either binary, hexadecimal, or decimal both signed and unsigned formats);
- 32-bit data to be received in the host PC from the FPGA (can be displayed in either binary, hexadecimal, or decimal both signed and unsigned formats).

In this section we will demonstrate two simple projects that use the Digilent *IOExpansion* VHDL component and the latter can be downloaded from [30]. The first project executes the following operations (see Fig. 1.27 for additional details):

1. Receives a 32-bit data item from the host PC virtual window and sends it back to the host PC.
2. Receives a 16-bit vector from the virtual window buttons and displays the vector on the first 16 light bars (they are yellow and red) located on the right-hand side of 24 light bars in the virtual window.
3. Receives an 8-bit vector from the PC window lower switches and displays the vector on the LEDs available on the Atlys board.
4. Receives an 8-bit vector from the PC window upper switches and displays the vector on the last 8 light bars (they are green) located on the left-hand side of 24 light bars in the PC window.
5. Receives an 8-bit vector from the switches available on the Atlys board and displays the vector on the PC window LEDs.

The top-level design entry of the project can be prepared in the schematic editor of the ISE as it is shown in Fig. 1.28. At the beginning, VHDL file of the *IOExpansion* component has to be downloaded from [30] and a schematic symbol *IOExpansion* has to be created in the ISE. Other components (OBUF8 and BUF) are the Xilinx library primitives.

The project requires the following user constraints file:

```
NET "EppAstb"      LOC = "B9";     # UCF for the Atlys board
NET "EppDstb"      LOC = "A9";
NET "EppWr"        LOC = "C15";
NET "EppWait"      LOC = "F13";
NET "EppDB<0>"     LOC = "A2";
NET "EppDB<1>"     LOC = "D6";
```

```
NET "EppDB<2>"    LOC = "C6";
NET "EppDB<3>"    LOC = "B3";
NET "EppDB<4>"    LOC = "A3";
NET "EppDB<5>"    LOC = "B4";
NET "EppDB<6>"    LOC = "A4";
NET "EppDB<7>"    LOC = "C5";
NET "led<0>"      LOC = "U18";   # remove for the second project
NET "led<1>"      LOC = "M14";   # remove for the second project
NET "led<2>"      LOC = "N14";   # remove for the second project
NET "led<3>"      LOC = "L14";   # remove for the second project
NET "led<4>"      LOC = "M13";   # remove for the second project
NET "led<5>"      LOC = "D4";    # remove for the second project
NET "led<6>"      LOC = "P16";   # remove for the second project
NET "led<7>"      LOC = "N12";   # remove for the second project
NET "Sw<0>"       LOC = "A10";   # remove for the second project
NET "Sw<1>"       LOC = "D14";   # remove for the second project
NET "Sw<2>"       LOC = "C14";   # remove for the second project
NET "Sw<3>"       LOC = "P15";   # remove for the second project
NET "Sw<4>"       LOC = "P12";   # remove for the second project
NET "Sw<5>"       LOC = "R5";    # remove for the second project
NET "Sw<6>"       LOC = "T5";    # remove for the second project
NET "Sw<7>"       LOC = "E4";    # remove for the second project
```

To test the circuit the following steps have to be carried out:

1. Generate the bitstream (see the steps indicated in Fig. 1.21), connect the Atlys board to the host PC through the appropriate USB socket and upload the generated bitstream to the board using the Adept software.
2. Select *I/O Ex* tab available in the virtual window.
3. Press *Start I/O* button (see Fig. 1.27a).
4. Interact with the board. Some examples are shown in Fig. 1.27a and 1.27b. It is assumed that the black button in the virtual window is pressed with the mouse.

The second project (see Fig. 1.29) enables the full adder (described in Sect. 1.5) to be tested in the virtual window.

At the beginning, a schematic symbol *FA* has been created in the ISE for the circuit of the full adder in Fig. 1.19b. Since now we do not need the onboard LEDs and switches, 16 lines have to be removed from the UCF file above (they are marked with comments # *remove for the second project*). The same steps 1–4 as before have to be done. The functionality of the full adder is tested using the virtual window switches and LEDs as it is shown in Fig. 1.30. Note that some connections between unused in the project ports are actually done in Fig. 1.29 (for example virtual LED2 is always OFF because it is connected to the ground).

Adept software provides also support for interaction of prototyping boards with a host PC using some other options [10, 11].

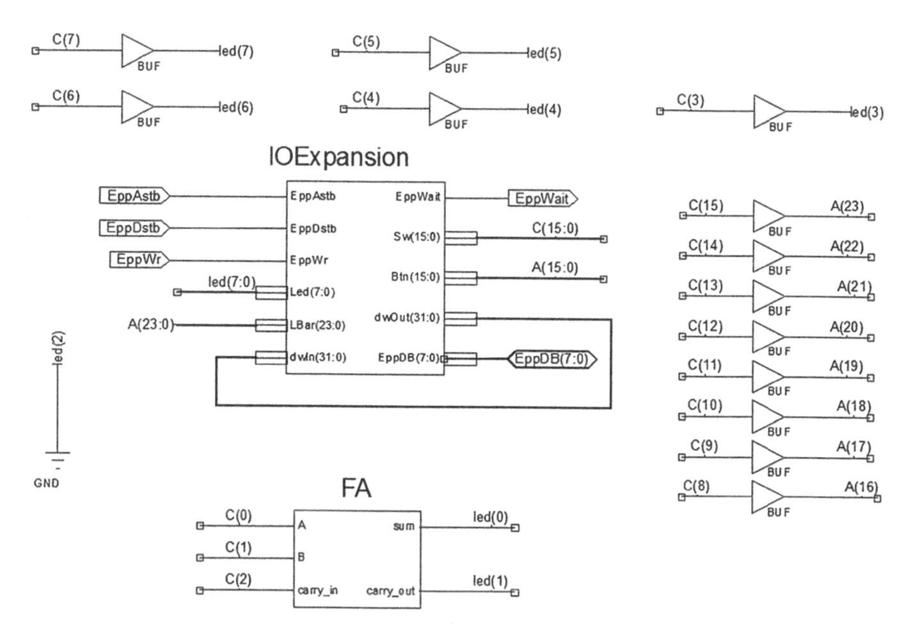

Fig. 1.29 Project for testing the full adder from Sect. 1.5 in the virtual window

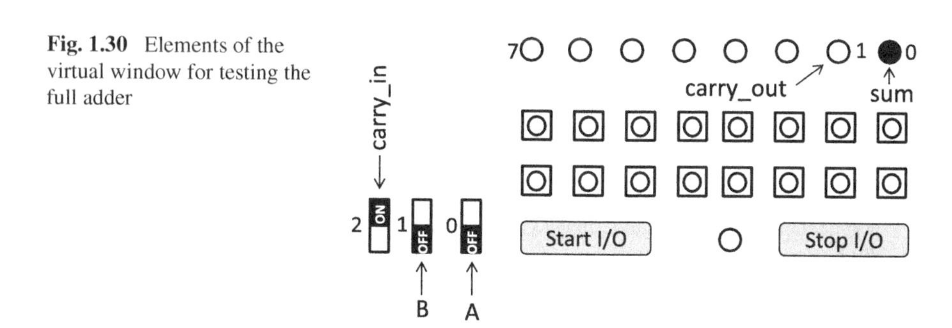

Fig. 1.30 Elements of the virtual window for testing the full adder

Schematic design entries were chosen for the first chapter just to provide an introduction to FPGA-based circuit without describing many supplementary topics that are normally required. Capabilities of schematic diagrams are limited. All the used components including those from the *LogiCore* can be described in HDL in a more compact and understandable form. Hierarchical design may similarly be applied. Editing HDL files is easier and the languages permit to work not only with structural (this form is used by schematic editors) but also with behavioral and mixed (behavioral plus structural) specifications. Besides, HDL and schematic diagrams can be combined within the same project if required. The next chapter gives a brief introduction to VHDL and all circuits and systems in the subsequent chapters will be described in this language.

References

1. Hauck S (1998) The roles of FPGAs in reprogrammable systems. Proc IEEE 86(4):615–638
2. Roelandts W (1999) 15 years of innovation. Xcell J 32(2):4–8
3. Santarini M (2011) Zynq-7000 EPP sets stage for new era of innovations. Xcell J 75(2):8–13
4. Xilinx Press Releases (2011) Xilinx ships world's highest capacity FPGA and shatters industry record for number of transistors by 2X. http://press.xilinx.com/2011-10-25-Xilinx-Ships-Worlds-Highest-Capacity-FPGA-and-Shatters-Industry-Record-for-Number-of-Transistors-by-2X. Accessed 10 Oct 2013
5. Altera Corp. (2013) Expect a breakthrough advantage in next-generation FPGAs. http://www.altera.com/literature/wp/wp-01199-next-generation-FPGAs.pdf. Accessed 10 Oct 2013
6. Estrin G (1960) Organization of computer systems—the fixed plus variable structure computer. In: Proceedings of the western joint computer conference, New York, 1960
7. Skliarova I, Sklyarov V, Sudnitson A (2012) Design of FPGA-based circuits using hierarchical finite state machines. TUT Press, Tallinn
8. SourceTech411 (2013) Top FPGA companies for 2013. http://sourcetech411.com/2013/04/top-fpga-companies-for-2013/. Accessed 10 Oct 2013
9. Xilinx Inc. (2012) 7 series FPGAs configurable logic block. http://www.xilinx.com/support/documentation/user_guides/ug474_7Series_CLB.pdf. Accessed 10 Oct 2013
10. Digilent Inc. (2010) Digilent Adept software. http://www.digilentinc.com/Products/Detail.cfm?NavPath=2,66,828&Prod=ADEPT2. Accessed 10 Oct 2013
11. Digilent Inc. (2013) Atlys™ board reference manual. http://www.digilentinc.com/Data/Products/ATLYS/Atlys_rm.pdf. Accessed 19 Nov 2013
12. Xilinx Inc. (2013) Constraints guide (UG625). http://www.xilinx.com/support/documentation/sw_manuals/xilinx14_5/cgd.pdf. Accessed 10 Oct 2013
13. Altera Corp. (2013) Stratix V device handbook. http://www.altera.com/literature/hb/stratix-v/stratix5_handbook.pdf. Accessed 10 Oct 2013
14. Altera Corp. (2013) Cyclone-IV devices handbook. http://www.altera.com/literature/hb/cyclone-iv/cyclone4-handbook.pdf. Accessed 10 Oct 2013
15. Xilinx Inc. (2011) Spartan-6 FPGA block RAM resources user guide. http://www.xilinx.com/support/documentation/user_guides/ug383.pdf. Accessed 10 Oct 2013
16. Xilinx Inc. (2012) 7 series FPGAs memory resources user guide. http://www.xilinx.com/support/documentation/user_guides/ug473_7Series_Memory_Resources.pdf. Accessed 10 Oct 2013
17. Xilinx Inc. (2012) LogiCORE IP block memory generator v7.3 product guide. http://www.xilinx.com/support/documentation/ip_documentation/blk_mem_gen/v7_3/pg058-blk-mem-gen.pdf. Accessed 10 Oct 2013
18. Xilinx Inc. (2011) Spartan-6 FPGA DSP48A1 slice user guide. http://www.xilinx.com/support/documentation/user_guides/ug369.pdf. Accessed 10 Oct 2013
19. Xilinx Inc. (2013) 7 series DSP48E1 slice user guide. http://www.xilinx.com/support/documentation/user_guides/ug479_7Series_DSP48E1.pdf. Accessed 10 Oct 2013
20. Xilinx Inc. (2013) Spartan-6 FPGA clocking resources user guide. http://www.xilinx.com/support/documentation/user_guides/ug382.pdf. Accessed 10 Oct 2013
21. Xilinx Inc. (2013) 7 series FPGAs clocking resources user guide. http://www.xilinx.com/support/documentation/user_guides/ug472_7Series_Clocking.pdf. Accessed 10 Oct 2013
22. Srikanth E (2011) How do i reset my FPGA? Xcell J 76(3):44–49
23. Xilinx Inc. (2013) ISE WebPACK design software. http://www.xilinx.com/products/design-tools/ise-design-suite/ise-webpack.htm. Accessed 10 Oct 2013
24. Xilinx Inc. (2012) ISE in-depth tutorial. http://www.xilinx.com/support/documentation/sw_manuals/xilinx14_3/ise_tutorial_ug695.pdf. Accessed 10 Oct 2013
25. Digilent Inc. (2010) Master UCF file for Atlys. http://www.digilentinc.com/Data/Products/ATLYS/AtlysGeneralUCF.zip. Accessed 10 Oct 2013

26. Digilent Inc. (2013) Nexys-4™ reference manual. http://www.digilentinc.com/Data/Products/NEXYS4/Nexys4_RM_VB1_Final_3.pdf. Accessed 9 Nov 2013
27. Xilinx Inc. (2011) Spartan-6 family overview. http://www.xilinx.com/support/documentation/data_sheets/ds160.pdf. Accessed 10 Oct 2013
28. Terasic technologies Inc. (2010) DE2-115 user manual. http://www.terasic.com.tw/cgi-bin/page/archive.pl?Language=English&CategoryNo=139&No=502&PartNo=4. Accessed 10 Oct 2013
29. Xilinx Inc. (2013) 7 series FPGAs overview. http://www.xilinx.com/support/documentation/data_sheets/ds180_7Series_Overview.pdf. Accessed 10 Oct 2013
30. Digilent Inc. (2009) Adept I/O expansion reference design. http://www.digilentinc.com/Products/Detail.cfm?NavPath=2,66,828&Prod=ADEPT2. Accessed 9 Nov 2013
31. Avnet Inc. (2013) ZedBoard (Zynq™ evaluation and development) hardware user's guide. http://www.zedboard.org/sites/default/files/documentations/ZedBoard_HW_UG_v1_9.pdf. Accessed 10 Oct 2013

Chapter 2
Synthesizable VHDL for FPGA-Based Devices

Abstract This chapter presents a concise introduction to synthesizable VHDL that is sufficient for the design methods and examples given in subsequent chapters to be understood without much background knowledge. The main objective of this chapter is to explain the basis of VHDL modules and their specification capabilities without going into detail. There are many excellent books dedicated to VHDL that may be used to complement this book. Our primary target is the synthesis and optimization of FPGA-based circuits and systems and VHDL is just an instrument that is used in the book to describe the desired functionalities and structures. Thus this chapter only provides the minimum necessary to allow subsequent chapters to be read without additional material, and to enable all the proposed examples to be understood and tested with the FPGA-based prototyping boards.

2.1 Introduction to VHDL

VHSIC (Very High Speed Integrated Circuits) Hardware Description Language (VHDL) was created as a result of USA government sponsored program in 1980s [1]. The language has been standardized in 1987 (with revisions done in 1993, 2002, and 2008) and is widely adopted by designers.

The target of this section is to provide a brief introduction to VHDL through simple examples. The main objective is to describe such constructions that will be used for FPGA projects in the book. VHDL is a complex language with wide-ranging specifications not all of which are synthesizable. Subsequent sections of this chapter will present just a basis for using VHDL in FPGA design. For deeper study of the language the books [1, 2] are recommended.

A specification of a digital circuit in VHDL includes two major parts: an *entity declaration* which is a definition of the circuit interface (where all the external circuit connections are declared), and an *architecture body* where a description of internal functionality is given. There are three types of architecture: *structural*, *behavioral*, and *mixed*.

V. Sklyarov et al., *Synthesis and Optimization of FPGA-Based Systems*, Lecture Notes in Electrical Engineering 294, DOI: 10.1007/978-3-319-04708-9_2,

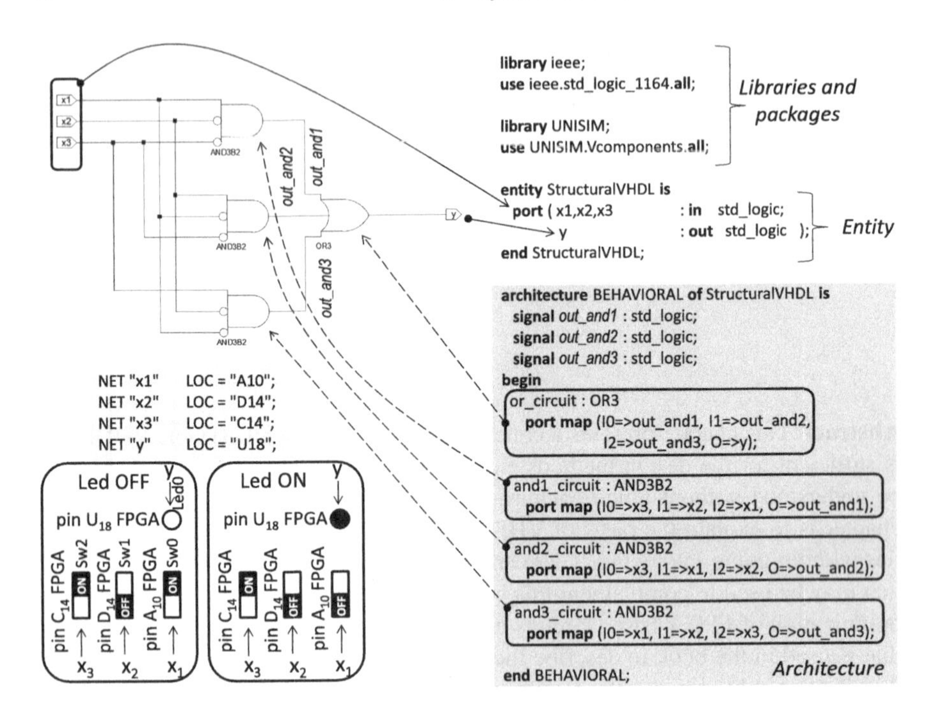

Fig. 2.1 Structural VHDL for the circuit shown in Fig. 1.2

Structural architecture provides all necessary internal connections between the circuit components that are either library primitives or previously developed circuits. Figure 2.1 demonstrates a structural VHDL description of the circuit firstly shown as a schematic entry in Fig. 1.2.

The first two lines of VHDL code identify a standard library, IEEE, and a package, std_logic_1164, which contains important definitions needed for our specification. In particular, we would like to use the type std_logic and the associated operations defined in that package. The type std_logic includes 9 values ('U'—uninitialized, 'X'—unknown, '0'—0, '1'—1, 'Z'—high impedance, 'W'—weak unknown, 'L'—weak 0, 'H'—weak 1, '–'—don't care) that allow signals to be modeled with strong, weak and high-impedance strengths. For now, from these 9 values we need just two: '0' and '1' (logic values are enclosed in quotation marks to distinguish them from the numbers 0 and 1). VHDL is not case sensitive language. That is why we can use the name STD_LOGIC instead of std_logic.

The second two lines of VHDL code identify a library, UNISIM.vcomponents, (with the package vcomponents) which contains the component declarations for the Xilinx primitives and defines models needed for simulation.

As you can see from Fig. 2.1 there are three sections in VHDL code:

1. Specification of libraries and packages that are intended to be used.
2. Specification of interface (entity).
3. Specification of architecture.

The components OR3 and AND3B2 are Xilinx library primitives and they correspond to the relevant schematic symbols in Fig. 1.2. The declared internal signals out_and1, out_and2, and out_and3 are needed to describe internal connections between the library primitives (there are totally 3 instances and1_circuit, and2_circuit and and3_circuit of the primitive AND3B2 and one instance or_circuit of the primitive OR3). Connections are shown by comma delimited lines in parenthesis after the **port map** keywords, for example, **port map** (I0=>x3, I1=>x2, I2=>x1, O=>out_and1). The component AND3B2 is defined in the UNISIM library (the file *unisim_VCOMP.vhd*) as follows:

```
component AND3B3
 port (O          : out   std_ulogic;  -- std_ulogic is unresolved type [1] similar to std_logic
       I0, I1, I2 : in    std_ulogic);
end component;
```

The VHDL keyword **signal** permits signals to be declared in the declarative part of an **architecture** (between the head of the **architecture** and the keyword **begin**). Signals in VHDL are similar to wires in hardware circuits.

Keywords (reserved words) here and later in the book are shown in **bold** font. In VHDL two successive hyphens (--) denote a single-line comment and they are shown in such font. Each port is given a name (e.g. O, I0, I1, I2) and is either an input (**in**) or an output (**out**). Other types (namely **inout** and **buffer**) are also allowed and they are described in Appendix A. For every port we specify the associated type which states the range of values that can be used on that port. In the example above each port is of type std_ulogic. Please note that the specification of each port is followed by a semicolon except for the last port. A signal of the type std_ulogic is similar to std_logic but it does not contain predefined resolution functions (the details can be found in [1, 3]). The names O, I0, I1, I2 of the interface signals in the component declaration above appear in the mapping line: **port map** (I0=>x3, I1=>x2, I2=>x1, O=>out_and1). The latter involves a named association where each component port I0, I1, I2, O (see the component AND3B3 above) is associated with x3, x2, x1 and out_and1 signals from the entity where the component is used (see the StructuralVHDL entity in Fig. 2.1). Internal signals (used for connections just within the entity StructuralVHDL) are explicitly declared as (Fig. 2.1):

```
signal out_and1 : std_logic;   -- signal and component declarations appear in the declarative
signal out_and2 : std_logic;   -- part of architecture which is between the keywords
signal out_and3 : std_logic;   -- architecture...of...is and begin (see example in Fig. 2.1)
```

Besides of the named association a positional association can be used, which will be considered in another example of structural specification below and is also described in Appendix A (see *Aggregate*).

Behavioral architecture represents the desired *functionality* of a circuit in an abstract way similar to general-purpose programming languages. However, VHDL statements differ in many aspects mainly because of inherent to hardware

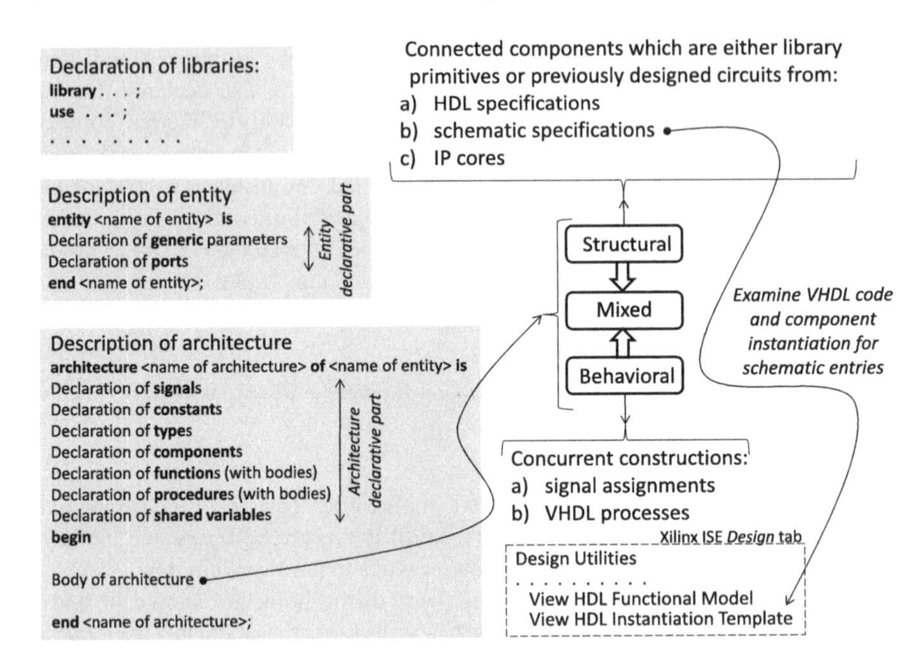

Fig. 2.2 A simplified structure of elements for a VHDL module

description languages concurrency and advanced operations manipulating individual bits and sets of bits.

For the considered above structural architecture an equivalent behavioral specification can be done as follows:

```
library ieee;                    -- note that the UNISIM library is not needed now
use ieee.std_logic_1164.all;

entity BehavioralVHDL is    -- the entity name (such as BehavioralVHDL) is chosen by the designer
  port (x1, x2, x3 : in   std_logic;
        y           : out  std_logic);
end BehavioralVHDL;

architecture behavioral of BehavioralVHDL is
begin -- and/not/or are VHDL logical operators for AND/NOT/OR logical operations
y <=   (x1 and not x2 and not x3) or (not x1 and x2 and not x3) or
       (not x1 and not x2 and x3);              -- <= is VHDL signal assignment operator
end behavioral;
```

Functionality of the synthesized circuit is exactly the same. *Structural* and *behavioral* specifications complement each other and may have different effectiveness for different projects. Thus, it is reasonable to combine them within a *mixed architecture*, which is composed of both *behavioral* and *structural* specifications. For complex projects such *mixed architecture* can often be seen as the most frequently used.

Figure 2.2 gives a simplified structure of elements for a VHDL module (design entry in VHDL) which nevertheless is sufficient for an introductory level.

Up to now we have not described yet many keywords shown in Fig. 2.2:

- **generic** enables compact scalable and parameterizable designs to be described (see Sect. 2.5 for details and Appendix A);
- **constant** permits objects with unchangeable values to be declared (see Sect. 2.2 for details and Appendix A);
- **type** is used to declare new types including arrays and enumerations (see Sect. 2.2 for details and Appendix A);
- **function** and **procedure** (subprograms) allow pieces of code to be used multiple times in a design (see Sect. 2.4 for details and Appendix A);
- shared variable is an extension of **variable**, allowing inter-process communication. Note that variable cannot be declared directly in architecture and it is declared in a process or in a subprogram (function or procedure). Variable is assigned using the := operator.
- **process** is a concurrent statement with such behavior that is described by sequential statements (see also Sect. 2.3 and Appendix A).

Subsequent sections of this chapter will present details about indicated above and other VHDL keywords (reserved words). A summary about the use of different reserved words is given in Appendix A.

A code below demonstrates a *behavioral* VHDL specification for a half-adder discussed in Sect. 1.5. The external interface and the truth table of the half-adder are shown in Fig. 2.3.

```
library IEEE;
use IEEE.std_logic_1164.all;

entity half_adder is
  port (A         : in std_logic;
        B         : in std_logic;
        carry_out : out std_logic;
        sum       : out std_logic);  -- there is no semicolon following the specification
end half_adder;                       -- of the last port

architecture half_adder_behavior of half_adder is
begin
      sum       <= A xor B;          -- xor is a VHDL keyword for XOR logical operation
      carry_out <= A and B;          -- and is a VHDL keyword for AND logical operation
end half_adder_behavior;
```

Each port of the half-adder is given a name (A, B, carry_out, sum). The architecture is entitled half_adder_behavior and is associated with the half_adder entity. These names can be chosen arbitrary but have to respect VHDL syntax rules, i.e. a user identifier can only include alphanumerical symbols and the underline character _ must start with a letter, may not include two consecutive underline characters, and may not have an underline character at the end.

The next example presents the complete *mixed* VHDL specification of a full adder composed of two structural components (half-adders) and a behavioral description of a two-input OR gate: carry_out <= s2 **or** s3; (see also Fig. 1.19b).

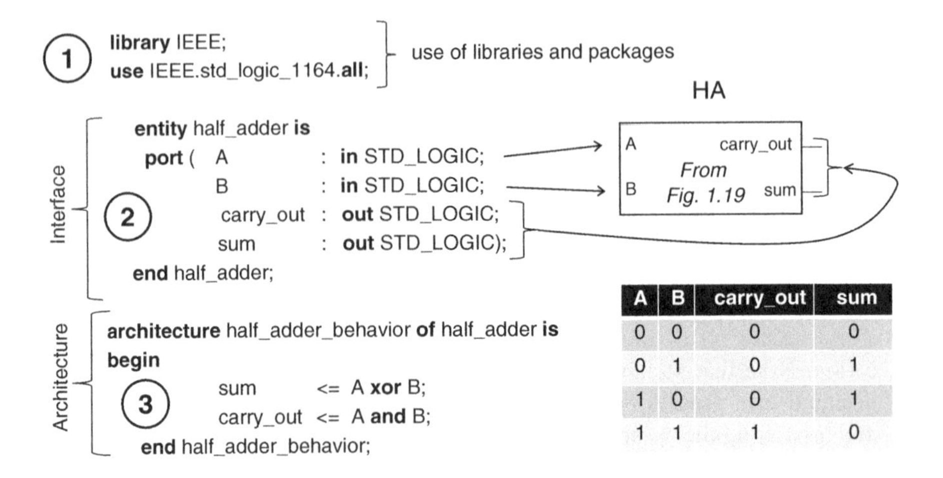

A	B	carry_out	sum
0	0	0	0
0	1	0	1
1	0	0	1
1	1	1	0

Fig. 2.3 Specification in VHDL and the truth table of a half-adder

```
library IEEE;
use IEEE.std_logic_1164.all;

entity FULLADD is
port ( A, B, carry_in         : in  std_logic;
       sum, carry_out         : out std_logic );
end FULLADD;

architecture STRUCT of FULLADD is
signal s1, s2, s3 : std_logic;

component half_adder
 port(A,B                     : in  std_logic;
       carry_out, sum         : out std_logic);
end component;

begin
       u1:  half_adder   port map(A, B, s2, s1);
       u2:  half_adder   port map(s1, carry_in, s3, sum);
       carry_out <= s2 or s3;
end STRUCT;
```

The component half_adder is described explicitly using the VHDL keyword **`component`**. If we comment the lines:

```
component half_adder
 port( A,B                    : in  std_logic;
       carry_out, sum         : out std_logic);
 end component;
```

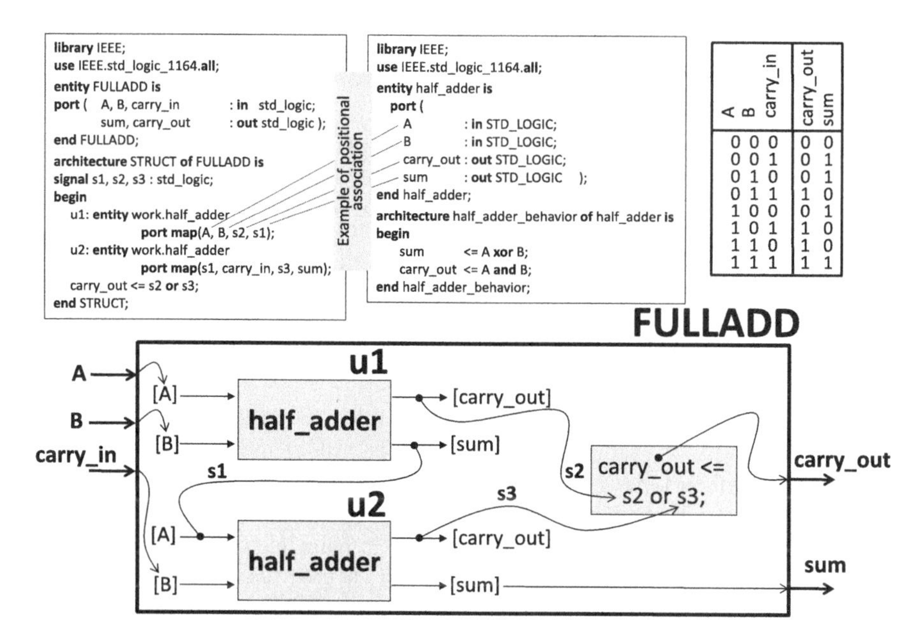

Fig. 2.4 Structural VHDL description of a full adder

the following error appears: *<half_adder> is not declared*. However, since all VHDL modules are compiled (by default) to a library with the name *work* we can use the half-adder component directly from the library as follows:

```
architecture STRUCT of FULLADD is
  signal s1, s2, s3 : std_logic;
begin   -- getting the half_adder from the library work in the construction: entity work.half_adder
      u1:  entity work.half_adder          port map(A, B, s2, s1);
      u2:  entity work.half_adder          port map(s1, carry_in, s3, sum);
      carry_out <= s2 or s3;
end STRUCT;
```

Now the code does not have errors and the resulting circuit works exactly the same as the circuit in Fig. 1.19b. The connections of the components are done through the respective external (A, B, carry_in, sum, carry_out) and internal (s1, s2, s3) signals that are associated with components' ports by positions (i.e. a positional association has been used). For example, the half adder has 4 ports A, B, carry_out, sum. In the component *u1* they are connected with external signals A, B and internal signals s2, s1, accordingly. In the component *u2* they are connected with s1 (internal), carry_in (external), s3 (internal), and sum (external) signals. All other details should be understandable from Fig. 2.4 (see also Appendix A).

The examples above illustrate the general organization of *structural*, *behavioral*, and *mixed* VHDL specifications. In the next sections of this chapter we will present more details about different VHDL constructions paying the main attention to comprehensive examples that can be directly synthesized, implemented and tested in FPGA-based circuits.

There are two appendices A and B in this book. Appendix A explains informally a variety of synthesizable constructions and VHDL keywords listed alphabetically. Appendix B includes some coding examples for frequently needed modules.

To conclude this section we would like to explicitly indicate that the book is not about VHDL and only a subset of this language is used to describe functionality of the considered FPGA-targeted circuits and systems. There are some limitations assumed in the book and they are listed below:

1. Only two values '0' and '1' from the allowed values of std_logic type are used.
2. For the majority of examples unsigned vectors with element values '0' and '1' are used and their type is declared as std_logic_vector. There are just a few examples with the types signed and unsigned (see the next section and Appendices).
3. Taking into account the assumptions 1 and 2, in many examples below the type std_logic_vector is used in the same way as an unsigned type although the latter might be more correct, for instance, for such operations as comparison, arithmetical, and some others. This way does not give rise to any problem for the resulting (synthesized and implemented) circuits that are presented in the book and it permits the number of conversion functions to be minimized. This is done because we would like to pay the primary attention to the design methods and the described circuits but not to supplementary constructions, which often make the code more difficult to analyze and understand.
4. Many design methods described in the book are equally applicable to signed vectors and if required the necessary (minimal) changes can easily be done assuming that the given examples are firstly well understood and tested.

2.2 Data Types, Objects and Operators

We consider the following VHDL basic data types: (1) *enumerated* (including pre-defined and user-defined); (2) *bit vector*; (3) *integer*; and (4) *record*.

Pre-defined enumerated types are: (1) bit (with possible values '0' and '1'); boolean (with possible values false and true); and (3) std_logic defined in the *IEEE std_logic_1164* package (with 9 possible values 'U', 'X', '0', '1', 'Z', 'W', 'L', 'H', '-' described in the previous section).

User-defined enumerated types are frequently introduced for naming states of finite state machines, for example:

```
type FSM_states is (begin, run, end); -- begin, run, end are user-defined names of FSM states
```

Bit vector is (1) a standard bit_vector type with elements of the type bit, and (2) defined in the *IEEE std_logic_1164* package std_logic_vector with elements of type std_logic. Std_logic and std_logic_vector are the most frequently used types in the book. Two examples are given below:

```
signal sw      :    std_logic_vector(3 downto 0);
signal my_bit  :    bit_vector(2 to 3);
```

The first example declares a vector sw with 4 elements: sw(3), sw(2), sw(1), sw(0). If, for example, sw <= "1100" then sw(3) is '1', sw(2) = '1', sw(1) = '0', sw(0) = '0'. If for the second example my_bit <= "01" then my_bit(2) is '0', and my_bit(3) is '1'. Single-bit values are written in between single quotes while multi-bit values are specified with double quotes.

Integer type enables an integer to be declared. The range of the `integer` values can explicitly be defined, for example:

```
signal my_int : integer range 3 to 8; -- allowed values now are only 3, 4, 5, 6, 7, and 8
```

Record type permits a set of data with different types to be combined in a named structure, for example:

```
type user_defined_record is record  -- the name of the structure is user_defined_record
  data1 : std_logic_vector(7 downto 0);      -- record fields
  data2 : integer range 0 to 7;              -- a field can also be of type record
end record;
```

Data types can form an array. Although any number of dimensions can be chosen it is frequently recommended to limit them, for example, by 3 in [3]. The following type declares an array named my_array of 16 integers with possible values 0, 1, 2, 3, 4:

```
type my_array is array (0 to 15) of integer range 0 to 4;
```

The following line declares a two dimensional array containing 4 sets of integers:

```
type my_table is array (3 downto 0) of my_array; -- the type my_array is declared above
```

We consider here three VHDL objects that are *signals, variables* and *constants*.

Signals are declared in the declarative part of architecture (shown in Fig. 2.2 between the lines **`architecture`**... and **`begin`**) with the keyword **`signal`** and used within that architecture.

Variables are declared in the declarative part of a process or a sub-program (function or procedure) with the keyword **`variable`** and used within that process or sub-program. We will discuss processes and sub-programs a bit later in this section.

Constants are declared in the declarative part of architecture, process, or sub-program (function or procedure) with the keyword **`constant`**. Declarative part of a process, a function or a procedure is placed between the lines **`process`**... /**`function`**.../**`procedure`**... and **`begin`**).

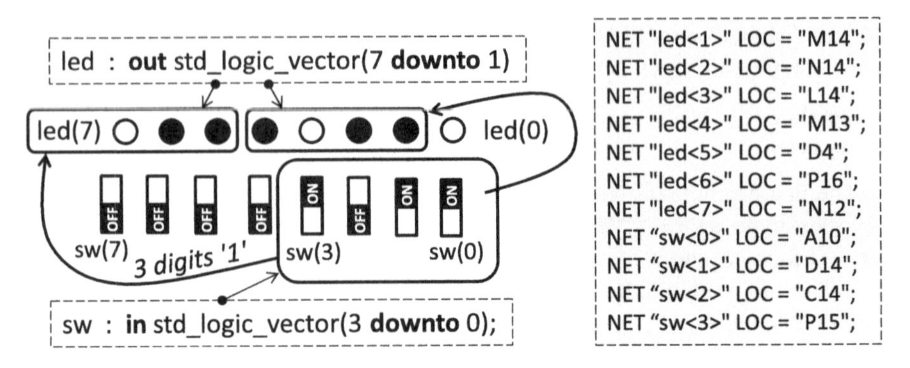

Fig. 2.5 UCF and functionality of the project with the entity types_and_objects for the Atlys board

Let us consider a complete example:

```
library IEEE;  -- in future VHDL modules we will assume including these libraries
use IEEE.STD_LOGIC_1164.all;
use IEEE.STD_LOGIC_ARITH.all;          -- see also appendix A and section 2.6
use IEEE.STD_LOGIC_UNSIGNED.all;       -- for conversion functions

entity types_and_objects is  -- sw and led are signals from switches and to LEDs
  port (sw   : in std_logic_vector(3 downto 0);
        led  : out std_logic_vector(7 downto 1));
end types_and_objects;

architecture Behavioral of types_and_objects is
  type my_array is array (0 to 15) of integer range 0 to 4;
  constant Hamming_weight : my_array := (0,1,1,2,1,2,2,3,1,2,2,3,2,3,3,4);
  signal index : integer range 0 to 15;
begin
  led(4 downto 1)      <= sw;
  index                <= conv_integer(sw(3 downto 0));
  led(7 downto 5)      <= conv_std_logic_vector(Hamming_weight(index), 3);
end Behavioral;
```

Here, conv_integer (casting std_logic_vector type to integer type) and conv_std_logic_vector (casting integer type to std_logic_vector type of size n where n is the second argument) are conversion functions for which we need to include additional packages indicated in the code above. The line:

```
constant Hamming_weight : my_array := (0,1,1,2,1,2,2,3,1,2,2,3,2,3,3,4);
```

declares and initializes a constant Hamming_weight which is a one-dimensional array of integers. Each integer with index i_{10} is a Hamming weight of i_2, i.e. the number of values '1' in the binary vector i_2. Indeed, if $i_{10} = 5$, then my_array(5) = 2, and i_2 = "0101" contains 2 digits '1'. The one-dimensional array is a new type my_array declared in the line: `type my_array is array (0 to 15) of integer range 0 to 4;`. Figure 2.5 demonstrates the user constraints file (UCF) for our project and the project functionality.

In subsequent VHDL modules we will also use the following derived data types (they are also described in Appendix A):

- natural that declares integers with nonnegative values (0,1,2,...);
- positive which is the same as natural without the value 0 (1,2,...);
- unsigned declares unsigned vectors based on std_logic type and is defined, for example, in the VHDL package std_logic_arith (see also Sect. 2.6);
- signed declares signed vectors based on std_logic type and is defined, for example, in the VHDL package std_logic_arith (see also Sect. 2.6);
- character is a 7-bit ASCII code;
- string(positive) is an array of characters.

The following two lines give declaration examples for a character and a string:

```
signal my_string : string(1 to 3); -- declaration of signal my_string of type string(1 to 3)
signal my_char  : character;    -- declaration of signal my_char of type character
```

The following lines give examples of assignments which can be done in an architecture body:

```
my_char <= '3';                -- my_char receives the ASCII code of digit 3
my_string(1) <= '5';           -- my_string(1) receives the ASCII code of digit 5
my_string(2) <= my_char;       -- my_string(2) receives the value of my_char
my_string(3) <= '9';           -- my_string(3) receives the ASCII code of digit 9
led <= std_logic_vector(conv_unsigned(character'pos(my_char), 8));
```

The last line finds position of my_char in ASCII table (character'pos(my_char)), then converts the position to an 8-element unsigned vector of std_logic (conv_unsigned(<position>,8)) and finally converts the unsigned vector to std_logic_vector (std_logic_vector(<unsigned vector>)) which is assumed to be displayed on eight onboard LEDs.

The following operators will be used in examples of this book:

1. Arithmetical: + (addition), − (subtraction), * (multiplication), / (division). Often, division is supported only if the right operand is a power of 2 [3].
2. Assignment: <= (for signals) and := (for variables).
3. Concatenation: &.
4. Logical: **and**, **nand**, **nor**, **not**, **or**, **xor**, **xnor** (see appendix A for details).
5. Relation: = (equal to), /= (not equal to), < (less than), <= (less than or equal to), > (greater than), > = (greater than or equal to).
6. Shift: **sll** (logic shift left), **srl** (logic shift right), **sla** (arithmetic shift left), **sra** (arithmetic shift right), **rol** (rotate left), **ror** (rotate right). Examples and additional explanations are given in Appendix A. We would prefer to use logically equivalent operators (see *Shift operators* in Appendix A).
7. Others: **abs** (absolute value), **rem** (remainder), **mod** (modulo), ** (power if the left operand is 2). Frequently, the operations **rem** and **mod** are supported only if the right operand is a constant power of 2 [3].

Using the majority of the operators is clear. So, we will consider below just a part of them. The first VHDL module is given below:

```
entity abs_rem_mod is   -- the project was tested in the ISE 14.7 and Atlys board
port ( sw                                : in std_logic_vector(7 downto 0);
       led                               : out std_logic_vector(7 downto 0);
       BTNU, BTNC, BTND, BTNL, BTNR       : in std_logic); -- onboard buttons in the Atlys
end abs_rem_mod;

architecture Behavioral of abs_rem_mod is
  signal result  : integer range 0 to 16;
  signal but     : std_logic_vector(4 downto 0);
begin
  but <= BTNU & BTNC & BTND & BTNL & BTNR;       -- concatenation of five signals
  result <= 16 when conv_integer(sw(3 downto 0)) = 0 else   -- 16 indicates "divide by 0"
          conv_integer(sw(7 downto 4)) mod conv_integer(sw(3 downto 0))
          when but = "00001" else                 -- only BTNR is pressed
          conv_integer(sw(7 downto 4)) rem conv_integer(sw(3 downto 0))
          when but = "00010" else                 -- only BTNL is pressed
          conv_integer(sw(7 downto 4)) / conv_integer(sw(3 downto 0))
          when but = "00100" else                 -- only BTND is pressed
          abs(-10) when but = "01000" else        -- abs(-10) = 10 (only BTNC is pressed)
          abs(5) when but = "10000" else 0;       -- abs(5) = 5 (only BTNU is pressed
  led <= conv_std_logic_vector(result, 8);
end Behavioral;
```

We introduced here **when** ... **else** conditional signal assignment which allows more operators to be described in a compact code. The conditional assignment has the following general form:

```
<name> <= <expression> when <condition> else <expression>;
```

which can be repeated any number of times. For example **mod** operator will be applied if and only if but = "00001", i.e. only one BTNR button is pressed. Indeed, the signal but is a concatenation (&) of 5 signals from the onboard buttons (BTNU & BTNC & BTND & BTNL & BTNR). Some operations are explained in comments above and some others are shown in Table 2.1. For example, using a modulo (A **mod** B) operator permits the result to be changed from A = 0 up to the value B–1 and then again from 0 to the value B–1 until the final allowed value A is reached (exact definition of the described above operators is given in [1]). As you can see from Table 2.1 division (/) and remainder (**rem**) give correct results in the Xilinx ISE 14.7 for any integer operands (the document [3] indicates that the respective operations are only supported if the second operand is a power of 2 or both operands are constants). The operator with an asterisk in Table 2.1 (**mod***, **rem***, and /*) are applied to the first positive and to the second negative arguments:

```
conv_integer(sw(7 downto 4)) mod (-conv_integer(sw(3 downto 0)))
conv_integer(sw(7 downto 4)) rem (-conv_integer(sw(3 downto 0)))
conv_integer(sw(7 downto 4)) / (-conv_integer(sw(3 downto 0)))
```

Table 2.1 The results of **`mod`**, **`rem`** and division (/) operations

A = sw(7:4)	B = sw(3:0)	Mod	mod*	rem (rem*)	/ (/*)
0000_2 (0_{10})	**mod, rem, / :**	(0000_2) 0_{10}	(00000_2) 0_{10}	(0000_2) 0_{10}	0 (0)
0001_2 (1_{10})	0101_2 (5_{10})	(0001_2) 1_{10}	(11100_2) -4_{10}	(0001_2) 1_{10}	0 (0)
0010_2 (2_{10})		(0010_2) 2_{10}	(11101_2) -3_{10}	(0010_2) 2_{10}	0 (0)
0011_2 (3_{10})	**mod*, rem*, /***	(0011_2) 3_{10}	(11110_2) -2_{10}	(0011_2) 3_{10}	0 (0)
0100_2 (4_{10})	(-5_{10}),	(0100_2) 4_{10}	(11111_2) -1_{10}	(0100_2) 4_{10}	0 (0)
0101_2 (5_{10})	i.e. the sign is forced to be changed	(0000_2) 0_{10}	(00000_2) 0_{10}	(0000_2) 0_{10}	1 (−1)
0110_2 (6_{10})		(0001_2) 1_{10}	(11100_2) -4_{10}	(0001_2) 1_{10}	1 (−1)
0111_2 (7_{10})		(0010_2) 2_{10}	(11101_2) -3_{10}	(0010_2) 2_{10}	1 (−1)
1000_2 (8_{10})		(0011_2) 3_{10}	(11110_2) -2_{10}	(0011_2) 3_{10}	1 (−1)
1001_2 (9_{10})		(0100_2) 4_{10}	(11111_2) -1_{10}	(0100_2) 4_{10}	1 (−1)
1010_2 (10_{10})		(0000_2) 0_{10}	(00000_2) 0_{10}	(0000_2) 0_{10}	2 (−2)
1011_2 (11_{10})		(0001_2) 1_{10}	(11100_2) -4_{10}	(0001_2) 1_{10}	2 (−2)
1100_2 (12_{10})		(0010_2) 2_{10}	(11101_2) -3_{10}	(0010_2) 2_{10}	2 (−2)
1101_2 (13_{10})		(0011_2) 3_{10}	(11110_2) -2_{10}	(0011_2) 3_{10}	2 (−2)
1110_2 (14_{10})		(0100_2) 4_{10}	(11111_2) -1_{10}	(0100_2) 4_{10}	2 (−2)
1111_2 (15_{10})		(0000_2) 0_{10}	(00000_2) 0_{10}	(0000_2) 0_{10}	3 (−3)

Since the result of (A **`mod`** B) has the same sign as B and **`abs`**(*result*)<**`abs`**(B), the result of (A **`mod`** B) is different from (A **`mod`** (−B)). The result of (A **`rem`** B) has the same sign as A and, thus, (A **`rem`** B) = (A **`rem`** (−B)). Clearly (A/B) ≠ (A/(−B)). Table 2.1 (where two's complement codes are used for negative numbers and for positive numbers just the absolute values are given) presents various examples for different values of the first operand (A) and B = 5_{10} with different signs for the latter one (positive: 5_{10} and negative: -5_{10}). The column / (/*) contains only decimal values. Additional details are given in Appendix A.

2.3 Combinational and Sequential Processes

VHDL process is a concurrent statement which is described by sequential statements. Almost always in this book we consider processes with a sensitivity list that appears within parentheses after the **`process`** keyword (it is recommended for greater flexibility, in particular, by the document [3]). A few examples of processes without a sensitivity list are given in Appendix A (see *on* and *until*). A process is activated if any of sensitivity list signals is changed (i.e. in case of event on these signals). For simulation purposes (see Sect. 2.7) processes with **`wait`** statement without a sensitivity list will also be used (it is not allowed to include both a sensitivity list and a **`wait`** statement). Additional details can be found in Appendix A.

2.3.1 Combinational Processes

A process is a *combinational* when all signals/variables assigned in the process explicitly receive new values every time the process is executed [3]. Thus, the sensitivity list must contain: (1) all signals in conditional statements, and (2) all signals on the right-hand side of assignment operators (<= or :=). If any value needs to be stored from the previous execution of the process the latter cannot be *combinational*.

There are a number of VHDL constructions that can be used in a process. Some of them (primarily needed for this book) will be described on examples below. The following combinational process tests if the value of an input vector sw is between the given low and high bounds (**if** (sw > low) **and** (sw < high) **then** led <= sw;) or less than the low bound (**elsif** sw < low **then** led <= **not** sw;):

```
entity TestCombProc is   -- simplified syntax rules for processes are given in appendix A
port ( sw     : in  std_logic_vector(7 downto 0);          -- onboard switches
       led    : out std_logic_vector(7 downto 0));         -- onboard LEDs
end TestCombProc;

architecture Behavioral of TestCombProc is
  constant low  : integer := 5;
  constant high : integer := 10;
begin

cp1: process(sw)        -- cp1 (combinational process 1) is an optional label
begin   -- A simplified syntax rule for if...elsif...else...end if statement is given in appendix A
  if (sw > low) and (sw < high) then led <= sw;
  elsif sw < low then led <= not sw;
  else led <= (others => '0');
  end if;
end process cp1;        -- cp1 (combinational process 1) is an optional label

end Behavioral;
```

If the value of sw is greater than low and less than high then this value is displayed on the onboard LEDs. If sw<low then the values of all sw elements are inverted (applying the **not** operator) and displayed on the LEDs. Otherwise all LEDs are OFF. The statement led <= (**others** => '0'); assigns to zero all elements of the signal led (corresponding to all LEDs OFF). The following conditional assignments (either the first or the second) directly used in the architecture body instead of the cp1 process execute exactly the same operations:

```
led <= sw when (sw > low) and (sw < high) else          -- the first conditional assignment
       not sw when sw < low else (others => '0');        -- see also Appendix A
with conv_integer(sw) select        -- the second (alternative) conditional assignment
     led <= sw          when low+1 to high-1,
     not sw             when low-1 downto 0,
     (others => '0')    when others;       -- see also Appendix A
```

If statement can be replaced with **case** statement in the following process cp2 below which implements similar to the process cp1 functionality:

```
cp2: process(sw) -- A simplified syntax rule for case statement is given in Appendix A
begin
  case conv_integer(sw) is
      when low+1 to high-1   => led  <= sw;
      when low-1 downto 0   => led  <= not sw;
      when others            => led  <= (others => '0');
  end case;
end process cp2;
```

The next combinational process cp3 can be used to find out the Hamming weight—HW (i.e. the number of ones) in the sw.

```
cp3: process(sw) -- numerous examples with for statement are given in appendix A
variable HammingWeightCount : integer range 0 to 8;
begin
  HammingWeightCount := 0;
  for i in sw'range loop  -- HW for sw(7), sw(6), ... , sw(0)
      if sw(i) = '1' then HammingWeightCount := HammingWeightCount+1;
      end if;
  end loop;
  led <= conv_std_logic_vector(HammingWeightCount,8);
end process cp3;
```

The line **for** i **in** sw'**range loop** begins a loop that is implemented combinationally and causes replication of the logic described in the loop body. Index i does not need to be declared and it is incremented in a range of the vector sw (i.e. 7 **downto** 0 in the order: 7,6,5,4,3,2,1,0). Besides of range we will use some other VHDL attributes shown in Appendix A (see *Attribute*). Let us consider some examples:

```
for i in sw'left downto sw'right+4 loop      -- HW for sw(7 downto 4): i.e. for i values 7,6,5,4
for i in sw'reverse_range loop               -- the order of i values is: 0,1,2,3,4,5,6,7
for i in sw'length-4 downto 0 loop           -- HW for sw(4 downto 0), because the length is 8
for i in 5 downto 3 loop                     -- the order of i values is: 5,4,3
```

The following combinational process cp4 demonstrates using the **exit** statement that allows the subsequent index values in the loop to be skipped:

```
cp4: process(sw)
  variable left_1, right_1 : integer range 0 to 8;
begin
  left_1 := 8; right_1 := 8; -- the value 8 is chosen to indicate all zeros in the sw
  for i in sw'range loop  -- exit as soon as the first '1' from the left is encountered
      if sw(i) = '1' then left_1 := i; exit;
      end if;
```

```
  end loop;
  for i in sw'reverse_range loop -- exit as soon as the first '1' from the right is found
      if sw(i) = '1' then right_1 := i; exit;     -- see also exit in Appendix A
      end if;
  end loop;
  led(7 downto 4) <= conv_std_logic_vector(left_1, 4);
  led(3 downto 0) <= conv_std_logic_vector(right_1, 4);
end process cp4;
```

The keyword **next** permits to terminate the loop with the current index value and to continue the loop with the next index value. Note that any iteration with a particular index value is not a cycle in a sequential circuit. Each iteration replicates the logic in the loop body described between the **loop** and **end loop** lines. The loop **while** (also available in VHDL) can be used similarly to the loop **for**. The details are given in Appendix A.

A process may use signals and variables. There is an important difference between them. Assignments (:=) of variables are done immediately (without delays) unlike signal assignments (<=) that are done when the process suspends. The statements in the process are executed sequentially (from the top to the bottom). If there are some mutually reassigned signals in a process they are not updated immediately. For example if A, B are integer signals initialized with the values A = 10 and B = 20:

```
A <= 5;          -- initialized before with the value 10
B <= A;          -- initialized before with the value 20
```

then at the end of the process (with single invocation) B = 10 (but not 5) because the above assignments of A and B are done at the same time at the end of the process (i.e. when the process suspends). Thus, B = 10 (the initial value of A) and A = 5 (the assigned value in the statement A <= 5 above).

In some practical applications iterative invocations of the same statement are required, for example, the statement A <= A + 1 can be executed in a combinational process with a loop such as **for** or **while**. The results are obviously wrong with the signal A because of the following: (1) the signal A has to be included in the process sensitivity list (because it appears on the right-hand side in the expression above); (2) any change of A (any event on A) forces reinvocation of the same process; (3) a combinational loop is created and this is a wrong for our example. Since variables are assigned immediately, a similar process with variables does not give rise to any problem. Let us consider the following example:

```
entity TestLoops is
   port ( led_signal       : out std_logic_vector (3 downto 0);
          led_variable     : out std_logic_vector (3 downto 0);
          sw               : in std_logic_vector(7 downto 0) );
end TestLoops;
```

```
architecture Behavioral of TestLoops is
  signal count_sig : integer range 0 to 15;
begin

use_of_signals: process(sw, count_sig)      -- this process gives definitely wrong results
begin           -- warnings in ISE about a combinational loop are displayed
  count_sig <= 0;
  optional_label: for i in sw'range loop      -- DO NOT USE SIGNALS IN SUCH LOOPS
        if(sw(i) = '1') then count_sig <= count_sig+1;  -- this is definitely wrong
        end if;
  end loop optional_label;
  led_signal <= conv_std_logic_vector(count_sig, 4);
end process use_of_signals;

use_of_variables: process(sw)     -- this process gives correct results
variable count_var        : integer range 0 to 15;
begin
  count_var := 0;
  optional_label: for i in sw'range loop                  -- this loop is correct
        if(sw(i) = '1') then count_var := count_var+1;  -- now this line is correct
        end if;
  end loop optional_label;
  led_variable <= conv_std_logic_vector(count_var, 4);
end process use_of_variables;

end Behavioral;
```

It is easy to examine that the first process use_of_signals gives wrong results and the second process use_of_variables gives correct results.

2.3.2 Sequential Processes

A process is *sequential* if some previously assigned signals keep their previous values and, thus, are not explicitly assigned in a new process execution [3]. We mainly consider *clock-edge-triggered sequential processes* with a sensitivity list and with an eventual *synchronous* reset that can be described as follows:

```
<optional label:> process(clock)    -- clock is the name of the clock signal
< optional declarative part>
begin
  if rising_edge(clock) then          -- the same as: if clock'event and clock = '1' then
     <sequential (possibly conditional) statements>
  end if;
end process <optional label>;
```

The rising_edge statement can be replaced with a falling_edge statement:

```
if falling_edge(clock) then          -- the same as: if clock'event and clock = '0' then
```

The following example demonstrates communication between several sequential processes. The first process sp1 together with a conditional assignment (marked with --**) describe a circuit that reduces the frequency of the clock (clk):

```
sp1: process(clk)
begin
  if rising_edge(clk) then internal_clock <= internal_clock+1; end if;
end process sp1;                                   -- sw is a 3-bit vector (2 downto 0)
divided_clk <= internal_clock(internal_clock'left - conv_integer(sw))        --**
                when falling_edge(clk);                                       --**
```

The following declarations have to be done in the architecture declarative part:

```
signal internal_clock      : unsigned(how_fast downto 0); -- how_fast = 30
signal positive_reset      : std_logic; -- this signal will be needed in examples below
signal divided_clk         : std_logic;
```

Since internal_clock is a 31-bit unsigned vector (std_logic_vector can also be used) and the signal divided_clk takes (internal_clock'left - conv_integer(sw)) bit in the vector internal_clock, the frequency of the clock clk is divided by $2^{\text{how\_fast+1-conv\_integer(sw)}}$. If conv_integer(sw) = 0 then the base frequency for the Atlys board (which is 100 MHz) is divided by 2^{31} = 2,147,483,648. Thus, the clock period of the divided_clk becomes ~21.5 s. If conv_integer(sw) = 7 then the base frequency is divided by 2^{31-7} = 16,777,216. Thus, the clock period becomes ~0.16 s. The greater the value of sw the higher frequency (the shorter period) of the divided_clk is provided.

Conditional signal assignment (marked with –** in the code above) can be replaced by the following lines in the sp1 process body:

```
if falling_edge(clk) then
      divided_clk <= internal_clock(internal_clock'left - conv_integer(sw));
end if;
```

The next sequential process sp2 describes functionality of a binary counter:

```
sp2: process (divided_clk)  -- signal count keeps the result of the counter
begin
  if rising_edge(divided_clk) then -- using divided_clk enables the results to be observed visually
      if positive_reset = '1' then count <= (others=>'0'); -- synchronous reset of the counter
      else
        if count_enable = '1' then  -- increment/decrement of the counter
                if increment='1' then  count <= count + 1;
                else                   count <= count - 1;
                end if;
        end if;
      end if;
  end if;
end process sp2;
```

Here, count_enable is the enable signal for the counter and increment permits either the counter increment (increment = ‘1’) or decrement (increment = ‘0’) to be selected.

The last sequential process sp3 describes functionality of a shift register:

```
sp3: process (divided_clk)           -- signal shift keeps the result of the register
begin                                -- the size of shift is chosen to be (6 downto 0)
  if rising_edge(divided_clk) then -- using divided_clk enables the results to be observed visually
       if positive_reset = '1' then shift <= (others=>'0'); -- reset of the register
       else
         if load_enable = '1' then  shift <= count;  -- loading the register
         elsif right = '1' then  -- shift right/left of the register
                 shift <= shift(0) & shift(5 downto 1);
         else
                 shift <= shift(4 downto 0) & shift(5);
         end if;
       end if;
  end if;
end process sp3;
```

Here, load_enable is the enable signal for the register (allowing the value of the count from the counter to be loaded) and the signal right permits either the shift right (right = '1') or the shift left (right = '0') to be selected.

The code below includes all the processes described above:

```
entity sequential_processes is           -- pins are given below for the Atlys board
  generic (how_fast: integer := 30 ); -- generic how_fast constant with the default value 30
port ( clk                : in std_logic;                   -- clock 100 MHz    – pin L15
       load_enable        : in std_logic;                   -- signal from sw(6) – pin T5
       count_enable       : in std_logic;                   -- signal from sw(7) – pin E4
       increment          : in std_logic;                   -- signal from sw(3) – pin P15
       right              : in std_logic;                   -- signal from sw(4) – pin P12
       count_shift        : in std_logic;                   -- signal from sw(5) – pin R5
       sw                 : in std_logic_vector(2 downto 0); -- pins C14, D14, A10
       rst                : in std_logic;                   -- RESET button     – pin T15
       led                : out std_logic_vector(7 downto 0)); -- see pins in Fig. 2.5 above
end sequential_processes;
architecture Behavioral of sequential_processes is
  signal internal_clock    : unsigned(how_fast downto 0);
  signal positive_reset    : std_logic;
  signal divided_clk       : std_logic;
  signal shift, count      : std_logic_vector(5 downto 0);
begin
  positive_reset <= not rst;  -- the onboard RESET button for the Atlys produces 0 when pressed
   -- the described above sp1 process
   -- the described above sp2 process
   -- the described above sp3 process
  led(7 downto 2) <= count when count_shift = '1' else shift;  -- the results of count or shift
  led(1) <= '0';  -- LED1 is set to OFF
  led(0) <= divided_clk;   -- divided_clk with the selected by sw frequency
  divided_clk<=internal_clock(internal_clock'left-conv_integer(sw))
                  when falling_edge(clk);
end Behavioral;
```

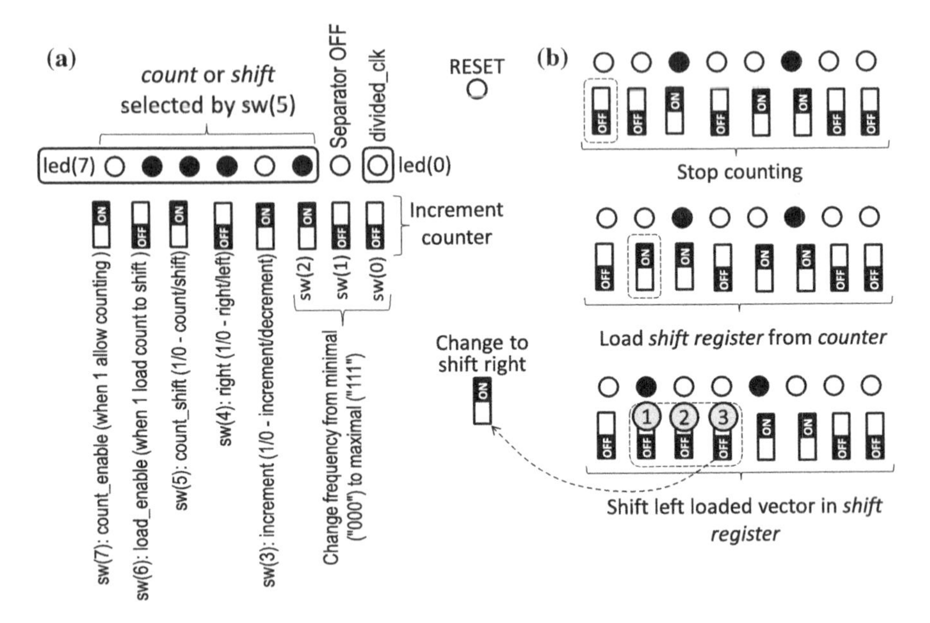

Fig. 2.6 Test of the project with sequential processes: Links with the board components **a** and the results of the test (**b**)

Figure 2.6 demonstrates how the results of the project above can be tested.

We already mentioned in the previous section that a process may use signals and variables and that there is an important difference between them. Figure 2.7 gives an additional example of a sequential process in which the block marked with 1 is executed just once. There are two signals A and B in the process test_assign. These signals are updated only when the process suspends. Thus, in the **if** statement within the process test_assign the signals led(1) and led(2) are assigned the previous values of A and B, which is perhaps not the result that you might expect.

```
if B = '1' then     A <= B;           B <= A;
                    led(1) <= A;      led(0) <= B;
end if;
```

If variables would be used instead of signals they would be assigned immediately and, thus, led(1) would receive the updated value of A and led(2) would receive the updated value of B.

In conclusion let us consider a complete example with two processes: test_variable with a variable vA; and test_signal (looking similarly) with a signal sA.

```
entity TestProc is
port ( clk      : in std_logic;
       sw       : in std_logic_vector(3 downto 0);
       led      : out std_logic_vector(7 downto 0));
end TestProc;
```

```
architecture Behavioral of TestProc is
  signal sA              : std_logic_vector(3 downto 0) := (others =>'0');
  signal divided_clk     : std_logic;
begin  -- the lines of the test_variable process are similar to the lines of the test_signal process
test_variable: process(divided_clk)
  variable vA : std_logic_vector(3 downto 0) := (others =>'0');
begin  -- the functionality of the test_variable and the test_signal processes is not the same
  if rising_edge(divided_clk) then
       vA := sw(3 downto 0);          -- a new value is assigned without delay
       led(7 downto 4)   <= vA;       -- the new value is displayed
  end if;
end process test_variable;

test_signal: process(divided_clk)
begin
  if rising_edge(divided_clk) then
       sA <= sw(3 downto 0);          -- a new value is assigned
       led(3 downto 0) <= sA;         -- the new value is delayed until the next activation
  end if;                             -- of the test_signal process
end process test_signal;

low_freq: entity work.clock_divider
          port map (clk, divided_clk);

end Behavioral;
```

If values of the switches sw3, sw2, sw1, sw0 are changed then these changes first appear on LEDs 7,6,5,4 and only after one period of the clock signal divided_clk — on LEDs 3,2,1,0. Such functionality can easily be examined because the clock frequency is divided (by the clock_divider) up to a visual scale (1 Hz or so).

As follows from the previous examples and explanations, using signals in loops might give problems. For example, if the variable HammingWeightCount is replaced with a signal in the combinational process sp3 in Sect. 2.3 then the functionality will be different from what we might expect (and eventually wrong). Many potential problems of such kind in combinational processes are recognized by synthesis tools which produce warnings about combinatorial (combinational) loops. Thus, the designers are informed. For sequential processes (like shown above and in Fig. 2.7) there is no reason for warnings but in many cases the functionality is different from what we might expect.

2.4 Functions, Procedures, and Blocks

Functions and procedures are used for blocks of codes that need to be invoked multiple times in the design. They permit such functionality to be described that is similar to combinatorial processes. A *function* is always terminated with a `return` statement and enables a single value to be computed and returned. Simplified syntax rules for functions and procedures are given in Appendix A. Note, that input

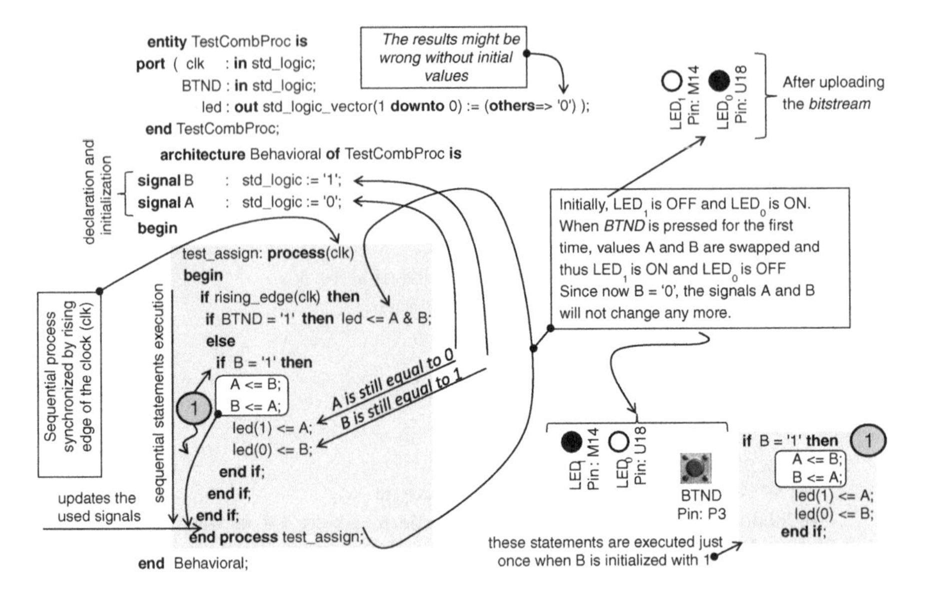

Fig. 2.7 An example demonstrating how a process test_assign is executed

parameters can be unconstrained, i.e. they do not have bounds. Let us describe a function HammingWeight that implements operations of the process sp3 in Sect. 2.3:

```
function HammingWeight (input: std_logic_vector) return integer is
  variable HammingWeightCount : integer range 0 to input'length;
begin          -- the "input" parameter is unconstrained above because bounds are not declared
  HammingWeightCount := 0;
  for i in input'range loop
      if input(i) = '1' then HammingWeightCount := HammingWeightCount+1;
      end if;
  end loop;
  return HammingWeightCount;
end HammingWeight;
```

The code of the function (such as that is shown above) needs to be defined in the declarative part of architecture.

A function can have more than one argument and may activate another function. For example, the following function HammingWeightComparator has three arguments and calls the first function HammingWeight:

```
function HammingWeightComparator (input: std_logic_vector;
          thresholdLow: integer; thresholdHigh: integer) return Boolean is
begin
  if HammingWeight(input) < thresholdLow          then return false;
  elsif HammingWeight(input) > thresholdHigh      then return false;
  else                                            return true;
  end if;
end HammingWeightComparator;
```

The code below presents a complete description of a module that invokes the functions HammingWeight and HammingWeightComparator.

```
entity TestFunctions is
port ( BTND   : in std_logic;                        -- signals from the onboard BTND
       sw     : in std_logic_vector(7 downto 0);     -- signals from the onboard switches
       led    : out std_logic_vector(7 downto 0));   -- signals to the onboard LEDs
end TestFunctions;

architecture Behavioral of TestFunctions is
  -- the code of the function HammingWeight given above
  -- the code of the function HammingWeightComparator given above
begin -- invocations of the functions are shown below on simple examples
  led(6 downto 0)<=conv_std_logic_vector(HammingWeight(sw),7) when BTND='0'
       else conv_std_logic_vector(HammingWeight(not sw(7 downto 4)), 7);
  led(7) <= '1' when HammingWeightComparator(sw, 3, 6) = true else '0';
end Behavioral;
```

It is allowed for a function to use signals that do not appear in the list of the function arguments. However, in such case the function has to be declared as *impure* (all functions are *pure* by default). Let us remove the first argument from the function HammingWeightComparator and examine the following code:

```
impure function HammingWeightComparator -- error without the use of the impure keyword
(thresholdLow: integer; thresholdHigh: integer) return Boolean is
begin
  -- the lines from the function HammingWeightComparator given above
end HammingWeightComparator;
```

The line for led(7) in the TestFunctions entity above has to be also changed (because now there are just 2 arguments) as follows: led(7) <='1' **when** HammingWeightComparator(3,6) = true **else**'0';. Now the functionality is exactly the same as before.

The keyword **impure** is an option for a function that extends the scope of variables and signals declared outside of the function that become available in the function. Thus, an *impure function* (in contrast to a *pure function*) may return different values for the same arguments (much like it is shown in the example above).

A function can receive and return values with user-defined types. Let us consider the following example:

```
entity FunctionSort is    -- this function was tested for the Nexys-4 board
port ( sw     : in std_logic_vector(15 downto 0);         -- the onboard switches
       led    : out std_logic_vector(15 downto 0));       -- the onboard LEDs
end FunctionSort;

architecture Behavioral of FunctionSort is
  type array4vect is array (0 to 3) of std_logic_vector(3 downto 0); -- user-defined type
  signal my_array        : array4vect;
```

```
function sort (Data_in : in array4vect) return array4vect is
  variable data_l1      : array4vect;
  variable data_l2      : array4vect;
  variable Data_out     : array4vect;

begin
  for i in 0 to 1 loop
     if data_in(i*2) <= data_in(i*2+1) then
                 Data_l1(i*2) := data_in(i*2+1);       Data_l1(i*2+1) := data_in(i*2);
     else        Data_l1(i*2) := data_in(i*2);         Data_l1(i*2+1) := data_in(i*2+1);
     end if;
  end loop;
  for i in 0 to 1 loop
     if data_l1(i) <= data_l1(i+2) then
                 Data_l2(i) := data_l1(i+2);           Data_l2(i+2) := data_l1(i);
     else        Data_l2(i) := data_l1(i);             Data_l2(i+2) := data_l1(i+2);
     end if;
  Data_out(i*3) := data_l2(i*3);
  end loop;
  if data_l2(1) > data_l2(2) then
                 Data_out(1) := data_l2(1);            Data_out(2) := data_l2(2);
  else           Data_out(1) := data_l2(2);            Data_out(2) := data_l2(1);
  end if;
  return Data_out;
 end sort;

begin

my_array <= (sw(15 downto 12), sw(11 downto 8), sw(7 downto 4), sw(3 downto 0));

(led(15 downto 12), led(11 downto 8), led(7 downto 4), led(3 downto 0)) <=
      sort(my_array);

end Behavioral;
```

The function implements a combinational even–odd merge sorting network for four 4-bit data items. It is not important now how the even–odd merge sorting network is coded in the function. Such networks will be described in Sect. 3.4.1. We would only like to demonstrate how to use input and return parameters of user-defined type (*e.g.* array4vect type in the code above). The presented example is ready to be tested in the Nexys-4 board with 16 onboard switches and 16 onboard LEDs. Data items are taken from groups of 4 switches as it is shown above in the assignment to my_array. The results are displayed on LEDs divided in similar groups (4 LEDs in each group shown in the statement above where the function sort is called. Data items are displayed in descending order (the maximum value on led(15 **downto** 12) and the minimum value on led(3 **downto** 0)).

Procedures differ from functions because they permit more than one object to be produced. The following example demonstrates the use of a procedure left1_right1 which finds the first and the last position '1' in the supplied vector (sw). The number of each position is indicated relatively to the right-hand switch starting with 1 (i.e. the right-hand switch is assumed to be 1 and not 0 to avoid all zeros on the LEDs when this switch is ON) (see Fig. 2.8).

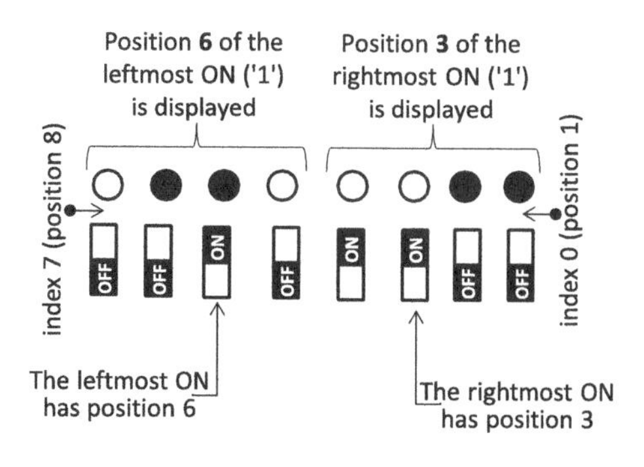

Fig. 2.8 An example demonstrating how to test the procedure

```
entity TestProcedure is              -- see Fig. 2.8 for additional explanations
port ( sw       : in  std_logic_vector(7 downto 0);    -- the onboard switches
       led      : out std_logic_vector(7 downto 0));   -- the onboard LEDs
end TestProcedure;

architecture Behavioral of TestProcedure is

procedure left1_right1
  (signal sw     : in std_logic_vector;
  -- sw is an input vector (all parameters are unconstrained; see appendix A)
  signal f_left  : out std_logic_vector; -- f_left is the first result (the leftmost value 1 in the sw)
  signal f_right : out std_logic_vector) is
  -- f_right is the second result (the rightmost value 1 in the sw)
  variable first_left, first_right       : integer range 0 to 8;
begin              -- initially the leftmost and the rightmost positions of '1' are assigned to be 0

  first_right := 0;   first_left := 0;

  for i in sw'range loop  -- the first loop finds the leftmost position of '1' (from N-1 downto 0)
      if sw(i) = '1' then first_left := i+1; exit;  -- the range of first_left is from N downto 1
      end if;
  end loop;    -- f_left below receives the value of the leftmost '1' in the given vector

  f_left <= conv_std_logic_vector(first_left, 4);

  for i in sw'reverse_range loop   -- the second loop finds the rightmost '1' (from 0 to N-1)
      if sw(i) = '1' then first_right := i+1; exit;    -- the range of first_right is from 1 to N
      end if;
  end loop;    -- f_right below receives the value of the rightmost '1' in the given vector

  f_right <= conv_std_logic_vector(first_right,4);

end left1_right1;                                             -- end of the procedure

  signal first_left, first_right          : std_logic_vector(3 downto 0);

begin
  left1_right1(sw, first_left, first_right);  -- use of the procedure left1_right1
  led(7 downto 4) <= first_left;   -- in this example the vector is taken from 8 switches and the
  led(3 downto 0) <= first_right; -- results are displayed on groups of LEDs (7,6,5,4) and (3,2,1,0)
end Behavioral;
```

If we declare the procedure like the following:

```
procedure left1_right1 (  sw : in std_logic_vector;
     -- sw is an input vector (all parameters are unconstrained; see appendix A)
                          f_left: out std_logic_vector;
     -- f_left is the first result (the leftmost value 1 in the sw))
                          f_right: out std_logic_vector) is
     -- f_right is the second result (the rightmost value 1 in the sw)
```

then the synthesis tools will report an error saying that the output arguments must be variables whereas the parameters supplied to the procedure sw, first_left and first_right were declared as signals in the entity TestProcedure above. However, the procedure may be called in a process for the parameters first_left and first_right declared as variables like the following:

```
process (sw)   -- note that the signal sw does not appear on the left-hand side of assignments in the
               -- procedure left1_right1 and the signal declaration does not give rise to any problem
  variable first_left, first_right : std_logic_vector(3 downto 0);
begin -- pay attention to the correct use of operators <= and := in the procedure left1_right1
  left1_right1(sw, first_left, first_right);
  led(7 downto 4) <= first_left;
  led(3 downto 0) <= first_right;
end process;
```

Let us consider another example in which a procedure finds the minimum and the maximum values in a set of data items used for the function FunctionSort above:

```
entity ProcMaxMin is     -- this function was tested for the Nexys-4 board
port ( sw        : in std_logic_vector(15 downto 0);    -- the onboard switches
       led       : out std_logic_vector(7 downto 0));   -- the onboard LEDs
end ProcMaxMin;

architecture Behavioral of ProcMaxMin is
  type array4vect is array (0 to 3) of std_logic_vector(3 downto 0);
  signal my_array          : array4vect;
  procedure max_min (  signal Data_in      : in array4vect;
                       signal max_v        : out std_logic_vector;
                       signal min_v        : out std_logic_vector)   is
  variable data_out        : array4vect;

  begin
    for i in 0 to 1 loop
       if data_in(i*2) <= data_in(i*2+1) then
                 Data_out(i*2) := data_in(i*2+1);  Data_out(i*2+1) := data_in(i*2);
       else      Data_out(i*2) := data_in(i*2);    Data_out(i*2+1) := data_in(i*2+1);
       end if;
    end loop;
    if Data_out(0) > Data_out(2) then          max_v <= Data_out(0);
    else                                       max_v <= Data_out(2);
    end if;
```

```
      if Data_out(3) < Data_out(1) then          min_v <= Data_out(3);
      else                                        min_v <= Data_out(1);
      end if;
   end max_min;
begin

my_array <= (sw(15 downto 12), sw(11 downto 8), sw(7 downto 4), sw(3 downto 0));

max_min(my_array, led(7 downto 4), led(3 downto 0));

end Behavioral;
```

The method used to find the maximum and the minimum values in a combinational circuit is described in Sect. 3.6 (see Fig. 3.16). We would only like to demonstrate here how to use different types of procedures. The presented example is ready to be tested in prototyping boards with 16 onboard switches and 8 onboard LEDs. Data items are taken similarly to the function FunctionSort above. The results are displayed on LEDs divided in groups: led(7 **downto** 4) for the maximum value and led(3 **downto** 0) for the minimum value.

Blocks are concurrent statements that enable designs to be partitioned. They are intended to clarify hierarchical structure of VHDL modules and (although are not widely used) may be helpful for some projects. A simplified syntax rule for block statements is given in appendix A. We will not use blocks in the subsequent chapters and only minimum details about them are given below. Let us partition the described above module with two functions HammingWeight and HammingWeightComparator in two blocks labeled block_with_one_function and block_with_another_function.

```
entity TestBlock is
port ( sw        : in std_logic_vector(7 downto 0);     -- onboard switches
       led       : out std_logic_vector(7 downto 0));   -- onboard LEDs
end TestBlock;

architecture Behavioral of TestBlock is
  signal HW : integer range 0 to 8;
begin
  block_with_one_function: block is            -- the first line of the first block
   -- code of the function HammingWeight given above
  begin
      led(6 downto 0) <= conv_std_logic_vector(HammingWeight(sw), 7);
      HW <= HammingWeight(sw);
  end block block_with_one_function;           -- the last line of the first block

  block_with_another_function: block is        -- the first line of the second block
   -- code of the impure function HammingWeightComparator given above
  begin  -- see example available at the Internet (http://sweet.ua.pt/skl/Springer2014.html)
      led(7) <= '1' when HammingWeightComparator(3,6) = true else '0';
  end block block_with_another_function;       -- the last line of the second block
end Behavioral;
```

Functionality of the partitioned design is the same as before. New signal HW in the architecture declarative part is used to supply the result of the first block to the second block.

A block statement may include a *guarded signal* assignment that allows the assignment only when the guard condition in the block is true. Let us consider an example:

```
entity TestBlockGuarded is
port ( clk              : in std_logic;
       enableBTND       : in std_logic;      -- the onboard BTND button
       BTNU             : in std_logic;      -- the onboard BTNU button
       sw               : in std_logic_vector(7 downto 0);    -- onboard switches
       led              : out std_logic_vector(7 downto 0));  -- onboard LEDs
end TestBlockGuarded;

architecture Behavioral of TestBlockGuarded is
  signal shift_rg       : std_logic_vector(7 downto 0);
  signal divided_clk    : std_logic;
begin
  -- the block below copies sw to LEDs when BTND=1 and shifts the copied values left
  -- when BTND=BTNU=1
  my_block:    block (enableBTND='1' and rising_edge(divided_clk)) is
  begin        -- the guarded assignment below is done only if the condition above is true
      shift_rg <= guarded sw when BTNU = '0' else shift_rg(6 downto 0) & shift_rg(7);
  end block my_block;    -- the end of the block

  led <= shift_rg;       -- the value of shift_rg is displayed on the onboard LEDs

  -- the clock divider below reduces the clock frequency to observe the changes of the LEDs visually
  low_freq: entity work.clock_divider port map(clk, divided_clk);   -- see appendix B

end Behavioral;
```

If the onboard button *BTND* is pressed the states of the onboard switches are copied to the shift_rg; if, in addition, the onboard button BTNU is pressed the copied values are shifted left on each rising edge of the divided_clk.

2.5 Generics and Generates

Generic statements provide support for scalable designs through supplying such parameters as sizes of vectors, ranges of values, and numbers of repetitive elements. Generics are declared with default values in the entity declarative part. The first example shows the use of different types of generics.

```
entity TestGenerics is     -- it is assumed to be used for the Atlys board
  generic( name            : string := "7954321";-- generic parameters with default values
           position        : integer := 2;         -- indicated after the characters ":="
           max_length      : integer := 7;
           my_char0        : character := '0';
```

```
            my_char9        : character := '9';
            MSL             : integer  := 4;
            bool_value      : Boolean := true);
    port (led               : out std_logic_vector(2*MSL-1 downto 0));
  end TestGenerics;

  architecture Behavioral of TestGenerics is
    signal tmp : Boolean := false;
    begin
    assert (MSL <= 4)               -- if MSL > 4 the message "wrong size for LEDs" is displayed
    report "wrong size for LEDs"    -- the message indicated here is displayed if MSL > 4
    severity FAILURE;               -- severity can be NOTE, FAILURE, WARNING and ERROR
    assert position <= name'length  -- check the position
    report "position is wrong"
    severity FAILURE;                        -- severity FAILURE terminates the synthesis
    assert name'length <= max_length         -- check the maximal length
    report "max length is wrong"
    severity WARNING;   -- for severity WARNING the warning message "max length is wrong"
                        -- (if activated) appears in the Design Summary/Reports
    led(2*MSL-1 downto MSL) <=std_logic_vector(conv_unsigned
            ((character'pos(name(position))-character'pos(my_char0)), MSL));
    tmp <= bool_value when character'pos(name(position)) >
                      character'pos(my_char9) else not bool_value;
    led(MSL-1) <= '1' when tmp else '0';
    led(MSL-2 downto 0) <= conv_std_logic_vector(name'length,MSL-1); -- name'length =7
  end Behavioral;
```

The result on the LEDs is the value 10010111. The first 4 digits (1001) is the difference in the positions in the ASCII table of the characters '9' and '0'. The next bit is 0 because the position of '9' is not greater than the position of '9' (the second character in the string "7954321" is '9' and my_char9 is '9'). The last 3 bits (111) represent the length of the string "7954321".

The generic line name: string : = "7954321"; defines a generic parameter name which is a string with the default value "7954321" (see ***literal*** in appendix A). The leftmost character '7' in "7954321" has the position 1 and the rightmost character '1' has the position 7. The part character'pos(name(position)) in the expression above uses the pos attribute (see ***attribute*** in appendix A). For our example with the default value of the position (i.e. 2) the result of character'pos(name(2)) = character'pos('9') returns the position of the character '9' in the ASCII table, which is $57_{10} = 39_{16}$. It can be verified in the following statement:

```
led(2*MSL-1 downto 0) <=
            std_logic_vector(conv_unsigned( (character'pos(name(2)) ), 8 ));
```

displaying on the LEDs the value "00111001" which is a binary equivalent of $57_{10} = 39_{16}$. The conv_unsigned and std_logic_vector provide the necessary conversion and casting. The similar result can also be obtained in the following statement:

```
led(2*MSL-1 downto 0) <= conv_std_logic_vector(character'pos(name(2)), 8);
```

which produces the LEDs value “00111001”.

It is clearly seen from the code above that the design is scalable. Indeed, it is sufficient to change generic parameters to customize the module for the proper needs. For example, the tmp signal indicates if a character in the name is below the position of the character ‘9’ in the ASCII table. If we change the default value of my_char9 from 9 to, for instance, 5 then a character is checked relatively to the position of the character ‘5’.

The assert statement ensures that some constraints are satisfied. For example in the following fragment:

```
assert position <= name'length      -- check position
report "position is wrong"
severity FAILURE;
```

it is checked if the position is less than or equal to the name'length. If the condition (less or equal: <=) is not satisfied then synthesis is terminated (because of the option **severity** FAILURE;) and the message “position is wrong” is displayed. Similarly other errors and warnings may be discovered and they are shown in the comments above.

We can now use the entity TestGenerics as a component of a higher level entity, for instance:

```
entity NowForNexys4Board is       -- it is assumed to be used for the Nexys-4 board
generic (name  : string := "FBCD"; -- the default value "7954321" was changed to "ABCD"
         new_position    : integer := 3; -- the default value 2 was changed to 3
         max_length      : integer := "FBCD"'length; -- the default value 7 was changed to 4
         my_char_F       : character := 'F'; -- the default value '0' was changed to 'F'
         -- the default value '9' for the my_char9 was unchanged
         MSL             : integer  := 8);   -- the default value '4' was changed to '8'
         -- the default value true for the bool_value was unchanged
port (led                : out std_logic_vector(2 * MSL-1 downto 0));
end NowForNexys4Board;

architecture Behavioral of NowForNexys4Board is -- the code is adjusted for the Nexys-4
begin
  assert (MSL <= 8)       -- now the MSL is tested for the value 8
  report "wrong size for LEDs"
  severity FAILURE;
  assert new_position <= name'length -- the name new_position is used instead of the position
  report "position is wrong"
  severity FAILURE;
  assert name'length <= max_length
  report "max length is wrong"
  severity WARNING;
  To_test: entity work.TestGenerics -- unchanged generics my_char9 and bool_value are
                                    -- not used in the generic map statement below
    generic map (name => name, position=> new_position,
                 max_length => max_length,  my_char0 => my_char_F, MSL => MSL)
    port map (led => led);

end Behavioral;
```

As you can see the code above is now used for the Nexys-4 board and the onboard LEDs show the following values: 1111110110000100 (the construction **generic map** permits the default generic values to be replaced with new generic values). The first eight bits 11111101 represent two's complement representation of -3_{10} that is the difference in the positions of 'C' (i.e. 67_{10}) and 'F' (i.e. 70_{10}) in the ASCII table (i.e. position of 'C' minus position of 'F'). Please note, that all the generic names that were not used in the generic map statement were left unchanged.

The second example uses generic parameters for the HammingWeight function described in the Sect. 2.4. Let us create a schematic symbol for the project shown in Fig. 1.6 in Chap. 1. At the beginning we need to add a copy of schematic source from Sect. 1.2.1 (see Fig. 1.6) to a new project, i.e. create a new project and select options *Project → Add Copy of Source...* in the ISE and add the file *DistTop.sch* from the previous project. At the next step let us add a new source that is a top level module. Then under the *Design Utilities* option double click on *View HDL Instantiation Template* and copy the following code to the top module:

```
UUT: DistTop port map (-- UUT is a label and we remind that VHDL is not case sensitiv
        s_in => ,
        clk1Hz => ,
        Sw => ,
        s_out => ,
        clock => ,
        BTND => );
```

Finally the top-level module TestGenerics1Top needs the following code:

```
entity TestGenerics1Top is
generic( number_of_bits : integer  := 48;      -- generic parameters with default values
         max_bits        : integer := 52;
         bits_sr  : std_logic_vector(4 downto 0) := (4 downto 2 => '0', others=>'1');
         rst             : std_logic := '0');
port (   clk             : in std_logic;
         led             : out std_logic_vector(7 downto 0));
end TestGenerics1Top;

architecture Behavioral of TestGenerics1Top is
  signal Rg : std_logic_vector(number_of_bits-1 downto 0):=(others=> '0');
  signal s_in, clk1Hz, s_out : std_logic;
  signal limit : integer range 0 to max_bits + conv_integer(bits_sr) := 0;
  -- code of the function HammingWeight given above in section 2.4
  begin

  process(clk1Hz) -- the process takes bits from the output s_out of the project from Fig. 1.6
  begin           -- and pushes them to the shift register RG
    if rising_edge(clk1Hz) then
      if limit <= (max_bits + conv_integer(bits_sr)) then -- less than or equal operator <=
         limit <= limit+1;                                -- assignment operator <=
         Rg <= Rg(number_of_bits-2 downto 0) & s_out;
      else Rg <= Rg;
```

```
        end if;  -- after (max_bits+conv_integer(bits_sr)) clock periods the Rg will contain max_bits
     end if;     -- shifted values. Note that bits_sr bits are skipped because the LUT-based shift register
  end process;  -- involves the bits_sr delay (see details in Fig. 1.7: sw(4 downto 0) = bits_sr)

  led(7 downto 3) <= conv_std_logic_vector(HammingWeight(Rg), 5);

  led(2) <= s_out;     led(1) <= clk1Hz;

  UUT: entity work.DistTop
     port map( s_in => led(0),            -- see also map in Appendix A
              clk1Hz => clk1Hz, Sw => bits_sr, s_out => s_out, clock => clk, BTND => rst);
     end Behavioral;
```

As can be seen from Fig. 1.6 the LUT-based 256 × 1 ROM is initialized with the INIT value: 0f0f0f0f0f0f0f0f0f0f0f0f0f0f0f0f0f0f0f0f0f0f0f0f0f**f070301013731**. These 64 hexadecimal digits represent $64 \times 4 = 256$ binary digits (bits). We want to consider max_bits = 52 least significant bits (they are shown above in **bold** font) and extract the last number_of_bits = 48 bits (i.e. the most recently copied bits to the register Rg underlined in the INIT value above). The module counts the Hamming weight in the underlined digits and copies the result to the led(7 **downto** 3). All the remaining LEDs are used exactly the same as in Fig. 1.6. Thus, for our default **generic** values the result is: led(7 **downto** 3) = 10010, i.e. 18 values 1 in **f070301 01373**$_{16}$ = 111100000111000000110000000100000001001101110011$_2$. Changing generic parameters number_of_bits and max_bits permits the Hamming weights to be computed for different sub-vectors within the indicated above INIT value.

The generic parameter

```
bits_sr  : std_logic_vector(4 downto 0) := (4 downto 2 => '0', others=>'1');
```

involves a named association in which the elements 4, 3, 2 receive the value '0' and the remaining elements receive the value '1' (the details can be found in appendix A).

The *generate* construction is employed to instantiate an array of components. The following code presents an example in which a ripple adder with a generic size N is created from the full adders described in Sect. 2.1.

```
entity Top is    -- it is assumed to be used for the Atlys board
generic( N       : integer := 4);       -- the default value of N is 4
port(      Op1   : in std_logic_vector(N-1 downto 0);
           Op2   : in std_logic_vector(N-1 downto 0);
           led   : out std_logic_vector(N downto 0));
end Top;

architecture Behavioral of Top is
  assert N <= 4
  report "cannot be used for the Atlys board because there are just 8 switches"
  severity FAILURE;
  signal carry_in          : std_logic_vector(N downto 0);
```

```
  signal carry_out        : std_logic_vector(N-1 downto 0);
  signal sum              : std_logic_vector(N-1 downto 0);
begin

carry_in(0) <= '0';       -- carry in signal for the least significant full adder is zero

generate_adder:           -- an initial line with the label generate_adder at the beginning
for i in 0 to N-1 generate  -- "for" is used to generate a network from connected full adders
FA:   entity work.FULLADD       -- connections are provided through indexed links
         port map( Op1(i), Op2(i), carry_in(i), sum(i), carry_out(i));
      carry_in(i+1) <= carry_out(i);
end generate generate_adder;

led <= carry_out(N-1) & sum;  -- the results are displayed on the onboard LEDs

end Behavioral;
```

Figure 2.9 demonstrates how the ripple adder for N = 4 has been generated. The figure also gives the user constraints file and shows how the adder can be tested.

Nested generates are also allowed and many examples of networks created with the aid of nested generates will be discussed in the next chapter. Any VHDL component may be generic and the default generic parameters can be replaced with new values by supplying a **generic map** construction. We have already explained such an opportunity when described the entity NowForNexys4Board above. For example, we can consider the following higher level component:

```
entity higher_level is
generic( New_N          : integer := 3);
   port( A              : in std_logic_vector(New_N-1 downto 0);
         B              : in std_logic_vector(New_N-1 downto 0);
         result         : out std_logic_vector(New_N downto 0));
end higher_level;

architecture Behavioral of higher_level is
begin
-- other statements
h_level: entity work.Top  -- generic map permits default generics to be replaced with new generics
      generic map( N=> New_N)          -- now N = New_N = 3
      port map(Op1=>A, Op2=>B, led=>result);
-- other statements
end Behavioral;
```

The construction **generic map** permits the default generic (N = 4 in our example for the Top entity) to be replaced with the new generic (New_N = 3 in our example).

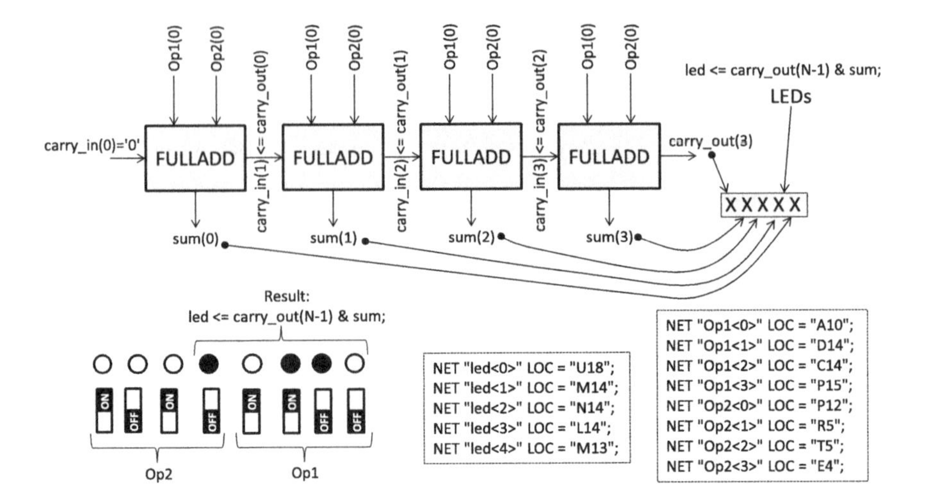

Fig. 2.9 Functionality of the ripple adder

2.6 Libraries, Packages, and Files

A *library* is a location with project's design units (entities or architectures and packages). The default library has the name *work* and contains all synthesizable source files of the project. For example, for the last project of the previous section the *work* panel displays the following five files: *Atlys.ucf*, *Full_adder.vhd*, *Half_adder.vhd*, *higher_level.vhd*, and *Top.vhd*. For the entity TestGenerics1Top, considered in the previous section, four files are displayed and one of them contains schematic: *Atlys.ucf*, *Clock_divider.vhd*, *DistTop.sch*, and *GenericsAndAssert.vhd*. If required, a user-defined library can be created, for example, with the name *MyLibrary*. In this case in the ISE the following steps can be done: (1) select *Project* → *New VHDL library* → < *specify the name MyLibrary and location (directory) of the library* >; (2) move necessary files to the *MyLibrary* (select the module and options *Source* → *Move to Library* → *MyLibrary*). Now the new library *MyLibrary* needs to be declared, for example:

```
library MyLibrary;          -- the default library work does not need to be declared
use MyLibrary.all;
```

and the library *work* in the line like: h_level: **entity** work.Top needs to be replaced with a new line: h_level: **entity** MyLibrary.Top.

A *package* permits functions, procedures, constants, types, and components to be described in a (shared) separate file. It provides a way of grouping a collection of related declarations that serve a common purpose [1]. We consider the following three groups: (1) predefined *standard packages*; (2) predefined *IEEE packages*; and (3) *user-defined packages*. The group (1), included by default, is defined in the *std* and *IEEE* standard libraries and describes the basic types: bit, bit_vector, integer, natural, real (real is frequently not fully supported by synthesis tools), and boolean. The group (2) is defined in the IEEE packages (that have to be declared) and describes common data types, functions, and procedures. We consider here the following packages supported by the XST [3]: *std_logic_1164* (describing std_logic, std_ulogic, std_logic_vector, and std_ulogic_vector types and the relevant conversion functions); *std_logic_arith* (describing unsigned and signed vectors based on the std_logic type and the relevant arithmetic operations and functions); *std_logic_unsigned* (describing unsigned arithmetic operators for the std_logic and std_logic_vector types); *std_logic_signed* (describing signed arithmetic operators for the std_logic and std_logic_vector types); and *std_logic_textio* (providing support for text-based file input/output). Note, that another available package *numeric_std* is similar to the *std_logic_arith*. The package *std.textio* (defined in the *std* standard library) provides support for a simple text-based file input/output.

A user-defined package (group 3) enables access to shared definitions from project's modules. A simplified syntax rule is given in Appendix A. A package needs to be declared and its body needs to be defined. Let us consider an example:

```
library IEEE;
use IEEE.STD_LOGIC_1164.all;
package MyPackage is                 -- declarative part of the package MyPackage
      constant limit       : integer := 10;
      type my_array is array (0 to limit-1) of std_logic_vector(1 downto 0);
      function HammingWeight (input: std_logic_vector) return integer;
      component clock_divider
        port( clk : in std_logic; divided_clk : out std_logic );
      end component;
end MyPackage;

package body MyPackage is            -- body of the package MyPackage
-- code of the function HammingWeight given above in section 2.4
end MyPackage;
```

The package is created selecting a new source in the ISE (*Project* → *New Source…*) and then *VHDL Package*. Now the package can be used in other modules something like the following:

```
library IEEE;
use IEEE.STD_LOGIC_1164.all;
use IEEE.STD_LOGIC_ARITH.all;
use work.MyPackage.all;              -- this line is required

entity UsesPackage is                -- we would like to use MyPackage from the work library
port ( clk       : in std_logic;
       sw        : in std_logic_vector(7 downto 0);
       led       : out std_logic_vector(7 downto 0));
end UsesPackage;
```

```
architecture Behavioral of UsesPackage is
  signal divided_clk       : std_logic;
begin
-- other eventual statements that might use objects declared in the MyPackage
led <= conv_std_logic_vector(HammingWeight(sw),8) when divided_clk = '1'
          else (others => '0');

my_divider : clock_divider port map (clk, divided_clk); -- positional association

end Behavioral;
```

Since the component clock_divider is declared in the MyPackage, an explicit library indication (such as my_divider : `entity` work.clock_divider) is now not needed.

In Sect. 1.7 we described an interaction of the Atlys board with a host computer using the *IOExpansion* component from Digilent [4]. The module *IOExpansion* can be taken either from a library, for example:

```
IO_interface : entity work.IOExpansion
      port map(EppAstb, EppDstb, EppWr, EppDB, EppWait, MyLed,
                  MyLBar, MySw, MyBtn, data_from_PC, data_to_PC);
```

or, alternatively, be declared in a package, for instance:

```
package InteractionWithPC is
component IOExpansion is        -- all the names have to be taken from the IOExpansion [4]
  port (EppAstb: in std_logic; EppDstb: in std_logic; EppWr : in std_logic;
        EppDB  : inout std_logic_vector(7 downto 0); EppWait: out std_logic;
        Led    : in std_logic_vector(7 downto 0);     -- 8 LEDs on the PC side
        LBar   : in std_logic_vector(23 downto 0);    -- 24 light bars on the PC side
        Sw     : out std_logic_vector(15 downto 0);   -- 16 switches on the PC side
        Btn    : out std_logic_vector(15 downto 0);   -- 16 buttons on the PC side
        dwOut : out std_logic_vector(31 downto 0);    -- 32-bit user-data from PC side
        dwIn   : in std_logic_vector(31 downto 0) );  -- 32-bit user-data to PC side
end component;
end InteractionWithPC;

package body InteractionWithPC is          -- the package body is empty
end InteractionWithPC;
```

Let us demonstrate the same interactions as shown in Fig. 1.27 (see Sect. 1.7):

```
use work.InteractionWithPC.all;

entity TestIntPC is
port ( sw         : in std_logic_vector(7 downto 0);   -- onboard switches
       led        : out std_logic_vector(7 downto 0);  -- onboard LEDs
       EppAstb    : in std_logic;                      -- signals for the IOExpansion component
       EppDstb    : in std_logic;
       EppWr      : in std_logic;
       EppDB      : inout std_logic_vector(7 downto 0);
```

```
        EppWait  : out std_logic);
end TestIntPC;

architecture Behavioral of TestIntPC is
  signal MyLed            : std_logic_vector(7 downto 0);  -- declarations of user signals
  signal MyLBar           : std_logic_vector(23 downto 0);
  signal MySw             : std_logic_vector(15 downto 0);
  signal MyBtn            : std_logic_vector(15 downto 0);
  signal data_to_PC       : std_logic_vector(31 downto 0);
  signal data_from_PC     : std_logic_vector(31 downto 0);
begin
  data_to_PC <= data_from_PC;    -- data received from the host PC are sent back to the PC
  MyLed      <= sw;      -- onboard switches are displayed on virtual LEDs (PC side)
  led       <= MySw(7 downto 0); -- 8 switches (PC side) are displayed on the board LEDs
  MyLBar     <= MySw(15 downto 8) & MyBtn;  -- 8 switches and MyBtn are displayed

  IO_interface : IOExpansion
       port map(EppAstb, EppDstb, EppWr, EppDB, EppWait, MyLed,
                MyLBar, MySw, MyBtn, data_from_PC, data_to_PC);

end Behavioral;
```

Alternatively the line **use** work.InteractionWithPC.**all** can be removed and the line IO_interface : IOExpansion needs to be replaced with: IO_interface : **entity** work.IOExpansion.

The XST (Xilinx Synthesis Technology) provides a limited support for working with files, which is described in [3]. We consider here only one example demonstrating how to read 8-bit words from a file *data.txt* and to record these words in an array my_array.

```
use std.textio.all;                 -- this package has to be used
use ieee.std_logic_textio.all;      -- this package has to be used

entity TestTextFile is              -- text file data.txt has to be recorded in the same directory
  port ( clk     : in std_logic;    -- ports can be initialized if required (see below)
         led     : out std_logic_vector(7 downto 0) := (others=>'0'));
end TestTextFile;
architecture Behavioral of TestTextFile is
 type my_array is array(0 to 15) of std_logic_vector(7 downto 0);
 impure function read_array (input_data : in string) return my_array is
   file my_file            : text is in input_data;
   variable line_name   : line;
   variable a_name        : my_array;
 begin
   for i in my_array'range loop
     readline (my_file, line_name); -- reading a line from the file my_file
     read (line_name, a_name(i));  -- reading std_logic_vector from the line line_name
   end loop;
   return a_name;
 end function;
```

```
  signal array_name :  my_array:=read_array("data.txt"); -- initializing the signal array_name
  signal divided_clk  :  std_logic;                        -- a low-frequency clock

begin

process(divided_clk)  -- changes are done with a low frequency and can be appreciated visually
  variable address :  integer range 0 to 15 := 0;
begin
  if rising_edge(divided_clk) then
        led <= array_name(address); -- displaying on the LEDs lines from the file data.txt
        address := address+1;        -- incrementing the address to get the next vector
  end if;
end process;

divider: entity work.clock_divider   port map (clk, divided_clk);

end Behavioral;
```

The file my_file is declared as follows: **file** myfile : text **is in** input_data; where input_data is a string with the file name (data.txt in our example) supplied to the function read_array as an argument (see: **signal** array_name:my_array: = read_array ("data.txt");). Two functions readline (text, line) (defined in the package std.textio) and read (line, std_logic_vector) (defined in the package std_logic_textio) are used to get data from the file data.txt, where the variables my_file and line_name have the types text and line, respectively. The variable a_name is an array of 16 vectors of type std_logic_vector(7 **downto** 0). Thus, firstly a line line_name is read: readline (my_file, line_name); and then a vector a_name(i) of type std_logic_vector(7 **downto** 0) is taken by the function read (line_name, a_name(i)); and returned from the function read_array. A similar technique is used in [3] to initialize embedded memories from files like *data.txt*. Figure 2.10 shows how the TestTextFile can be tested in the Atlys prototyping board (the file *data.txt* can be prepared in any text editor and saved in the same directory with the project). Additional examples are given in Appendix A (see *file*).

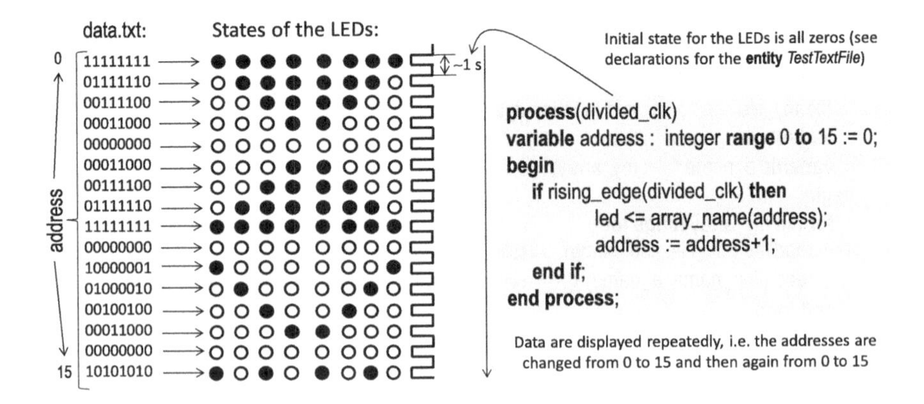

Fig. 2.10 Testing the project, which reads data from the file *data.txt*

Reading from a file can be useful to fill in a memory/array during synthesis which is similar to initialization. Writing to a file cannot be done from a working project (since it is done during synthesis). It may be used for debugging, writing specific constants or generic values [3]. Some examples can be found in [3] and one example is given in Appendix A (see ***file***).

2.7 Behavioral Simulation

This section presents a brief introduction to a behavioral (functional) simulation that can be done before an implementation of the project to verify that the logic in the project modules is correct. Additional details can be found in [5, 6]. We will use the Xilinx *ISim* simulator which is automatically installed with the ISE (and selected when needed in the *Design Properties* dialog box of the ISE).

Figure 2.11 explains how a behavioral simulation is organized for which two types of files are required: (1) the *modules* which we would like to examine (VHDL or schematic for our examples); (2) a *test bench* file created for the modules. Besides, simulation libraries for environment specific components (such as libraries for Xilinx primitives and IP cores) have to be included if the primitives/cores are used in the design. A *test bench* file is created for a particular project and supplies stimulus to the modules. The creation can be done in the ISE by adding a new source (of type *VHDL Test Bench*) and associating it with the verified module.

We consider below three examples. The first one demonstrates behavioral simulation for the full adder (FULLADD) described in VHDL in Sect. 2.1, which is a *combinational circuit*. The second example illustrates simulation of a *sequential circuit* that is an up/down binary counter with clock enable and synchronous active-high reset. The counter was taken from the ISE templates available through selection *Edit → Language Templates... → VHDL → Synthesis Constructs → Coding Examples → Counters → Binary → Up/Down Counters*. The last example enables the behavior of the circuit created in the ISE schematic editor (see Fig. 1.6 in Chap. 1) with *Xilinx library primitives* to be tested.

All the steps (a, b, and c), needed for the first example, are shown in Fig. 2.11. At the first step (a) we create a project for simulation, i.e. we add a test bench (named TestBenchFA) and associate the test bench with the FULLADD module described in Sect. 2.1. A template for the test bench is proposed by the ISE but we will change the code as it is shown at the right-hand part of Fig. 2.11. The entity FULLADD is instantiated in the architecture and the project structure is shown in Fig. 2.11 near the label a. There is one process (stim_proc) in the architecture body which generates stimulus (inputs of the FULLADD that are changed every 50 ns until the final `wait` statement is reached). At the second step b the test bench is checked for errors. In our case there is no error and we proceed to the last step c where the *Simulate Behavioral Model* option is activated. As a result, the *ISim* window with simulation waveforms is opened. For better analysis the waveforms need to be zoomed (see Fig. 2.11). A cursor permits to check values in a particular time (after 77 ns in our example depicted in Fig. 2.11).

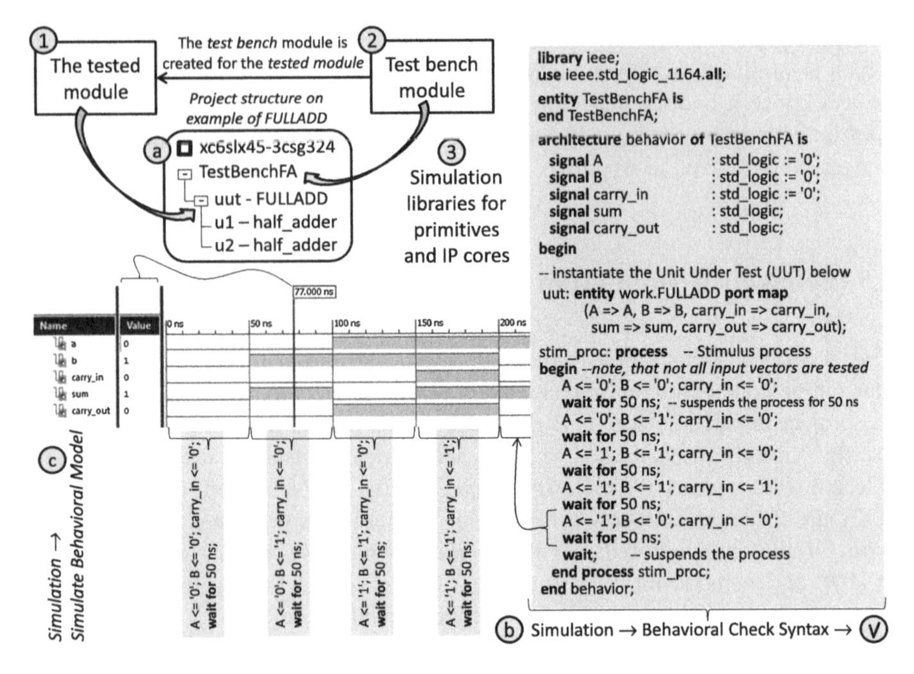

Fig. 2.11 An example of behavioral simulation for the full adder

The following module will be simulated in the second example:

```
entity Counter is
   port (        reset, clock        : in  std_logic;
                 clock_enable        : in  std_logic;
                 inc_dec             : in  std_logic;
                 outputs             : out  std_logic_vector (3 downto 0));
end Counter;

architecture Behavioral of Counter is
   signal count : std_logic_vector(3 downto 0);
begin

process (clock)
begin
  if rising_edge(clock) then
    if reset='1' then         count <= (others => '0');  -- synchronous reset
    elsif clock_enable='1' then
      if inc_dec='1' then     count <= count + 1; -- if inc_dec=1 then increment the counter
      else                    count <= count - 1;  -- if inc_dec=0 then decrement the counter
      end if;
    end if;
  end if;
end process;

outputs <= count;

end Behavioral;
```

The following test bench for_counter is added and associated with the Counter:

```
entity for_counter  is
end for_counter;

architecture behavior of for_counter  is
  signal reset           : std_logic := '0';
  signal clock           : std_logic := '0';
  signal clock_enable    : std_logic := '0';
  signal inc_dec         : std_logic := '0';
  signal outputs         : std_logic_vector(3 downto 0);
  constant clock_period  : time := 30 ns;   -- clock period definitions (valid for simulation only)
 begin

 uut: entity work.Counter port map          -- instantiate the unit under test (uut)
      (reset => reset, clock => clock, clock_enable => clock_enable,
       inc_dec  => inc_dec, outputs => outputs  );

clock_generator   : process               -- clock process definitions
begin -- the process generates clock pulses
  clock <= '0';
  wait for clock_period/2;                 -- duty cycle for the clock is 50%
  clock <= '1';
  wait for clock_period/2;
end process clock_generator   ;

stim_proc:  process                        -- stimulus process
begin
  reset <= '1';                  -- the first line **reset<='1'**
  wait for 30 ns;                -- set the reset signal to '1' and wait for 30 ns
  reset <= '0'; clock_enable <= '0'; inc_dec <= '1';
  wait for 20 ns;                -- change signals as it is indicated above and wait for 20 ns
  reset <= '0'; clock_enable <= '1'; inc_dec <= '1';
  wait for 150 ns;               -- change signals as it is indicated above and wait for 150 ns
  reset <= '0'; clock_enable <= '1'; inc_dec <= '0';
  wait for 55 0 ns;              -- change signals as it is indicated above and wait for 550 ns
end process;                     -- begin from the line **reset<='1'** after 30+20+150+550=750 ns

end behavior;
```

Since the Counter is a sequential circuit the test bench needs to supply clock signal and it is done in the clock_generator process. The simulation results with additional details are given in Fig. 2.12.

The last simulation is done for the circuit in Fig. 1.6 from which the clock_divider has been removed (see Fig. 2.13). Indeed, for simulation purposes a low frequency clock is not needed. All the required steps are exactly the same as for VHDL modules (see Fig. 2.11). The only difference is the association of the added test bench with the top-level schematic entity (DistTop.sch in our example).

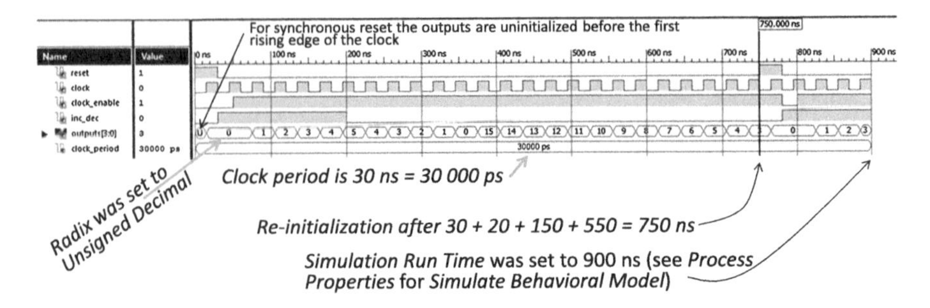

Fig. 2.12 Simulation results for the Counter with additional details

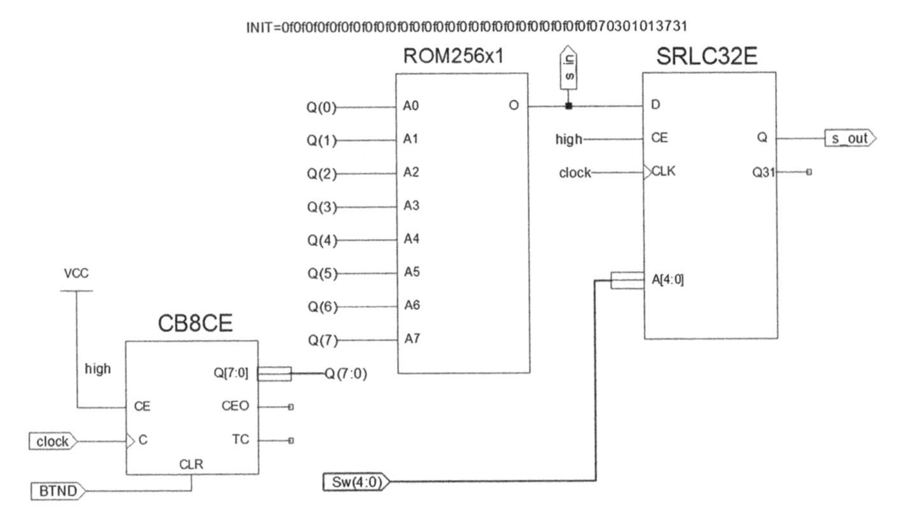

Fig. 2.13 The circuit in Fig. 1.6 without the clock_divider

The following test bench is created:

```
library unisim; -- include other libraries before this line
use unisim.Vcomponents.all; -- this package is needed for Xilinx primitives used in the schematics

entity DistTop_DistTop_sch_tb is
end DistTop_DistTop_sch_tb;

architecture behavioral of DistTop_DistTop_sch_tb is
  signal s_in, s_out       :std_logic;
  signal sw                :std_logic_vector (4 downto 0);
  signal BTND              :std_logic;
  signal clock             :std_logic;
  constant clock_period  : time := 30 ns;
begin
module_to_test: entity work.DistTop port map
          (s_in => s_in, sw => sw, s_out => s_out, BTND => BTND, clock => clock);
```

```
clock_generation: process          -- clock process definitions
begin                              -- the process generates clock pulses
  clock <= '0';
  wait for clock_period/2;
  clock <= '1';
  wait for clock_period/2;
end process clock_generation;

-- a stimulus process is not needed because we would like the values
-- of sw and BTND to be permanently assigned in the line below
sw <= (4 downto 3 => '0', others=>'1'); BTND <= '0'; -- settings are the same as in Fig. 1.7
-- if required the values of sw and BTND may be changed in the relevant stimulus process, which
-- will be used instead of the line above

end behavioral;
```

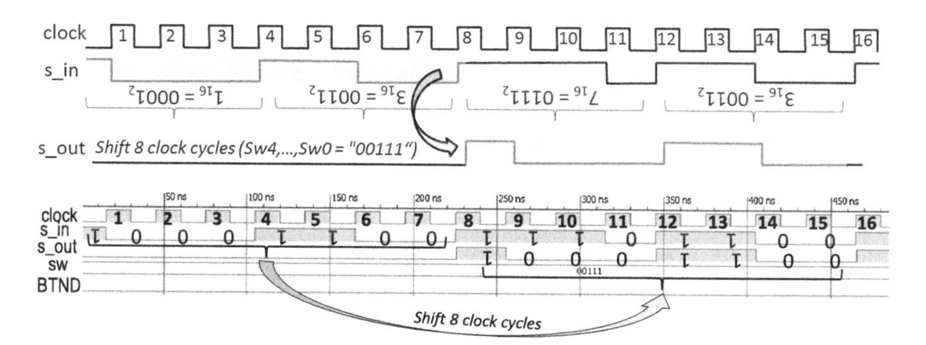

Fig. 2.14 Comparing the results of physical tests in Fig. 1.7 (*the upper part*) and behavioral simulation (*the lower part*)

The results of simulation are exactly the same as in Fig. 1.7. To show it clearer Fig. 2.14 depicts the waveforms from Fig. 1.7 and the results of behavioral simulation that uses the test bench given above.

There are many options and modes of simulation which are not described here and can be found in [5, 6].

2.8 Prototyping

The majority of the considered in this chapter examples can be implemented and tested in different prototyping boards described in Sect. 1.6. Clearly, the user constraints file (i.e. pin assignments) and the FPGA part number have to be changed properly.

The following example has been tested in the DE2-115 board (the Xilinx user constraints file has been changed to the proper Altera setting file [7]):

```
library IEEE;
use IEEE.STD_LOGIC_1164.all;
use IEEE.STD_LOGIC_ARITH.all;

entity AlteraProject is              -- all names (except clock and reset) are the same as in [7]
generic( size        : integer := 18;-- the size of vectors for the HammingWeight function
         n_LEDs      : integer := 5);-- the number of the used LEDs

port ( clock         : in std_logic; -- PIN_Y2
       reset         : in std_logic; -- PIN_M23 for key0
       sw            : in std_logic_vector(size-1 downto 0);
       ledr          : out std_logic_vector(n_LEDs-1 downto 0);
       ledg          : out std_logic_vector(n_LEDs-1 downto 0));
end AlteraProject;

architecture Behavioral of AlteraProject is
  -- code of the function HammingWeight from section 2.4 without any change
  signal count        : integer range 0 to size-1;

  signal divided_clk  : std_logic;
begin

process (divided_clk)
begin
  if rising_edge(divided_clk) then
    if not reset ='1' then count <= 0;        -- when the key0 is pressed then count is zero
    else  -- if HammingWeight(sw)>0 then count is changed from 1 to HammingWeight(sw)
      count <= (count mod HammingWeight(sw))+1; -- mod is the VHDL modulo operator
    end if;
  end if;
end process;

ledr <= conv_std_logic_vector(count, n_LEDs);

ledg <= conv_std_logic_vector(HammingWeight(sw), n_LEDs);

divider: entity work.clock_divider
        port map (clock, divided_clk);

end Behavioral;
```

If one or more switches are ON (HammingWeight(sw) > 0) then the count is changed cyclically from 1 to the HammingWeight(sw). If reset is active (*key0* button is pressed) then count = 0. The value of the HammingWeight(sw) is displayed on green LEDs (ledg) and the value of count is displayed on red LEDs (ledr). The reset signal is active *low* (that is why the **not** operation is applied to this signal).

Some projects of the book use vendor-specific libraries and technology-dependent components. The following VHDL code gives an example:

```
library IEEE;                  -- Xilinx LUT-based computation of the Hamming weight (see
use IEEE.STD_LOGIC_1164.all;   -- the simplest Hamming weight counter in section 3.9)
library UNISIM;                -- Xilinx library UNISIM for LUT primitives that are used below
```

```
use UNISIM.VComponents.all;

entity LUT_6to3 is
   port ( SixBitInput      : in  std_logic_vector (5 downto 0);   -- 6-bit input vector
          ThreeBitOutput   : out std_logic_vector (2 downto 0)); -- 3-bit Hamming weight
end LUT_6to3;
architecture Behavioral of LUT_6to3 is -- Xilinx LUTs below are configured in such a way that
begin      -- permits the Hamming weight of 6-bit input vector to be produced in a combinational circuit

LUT6_inst1 : LUT6        -- Xilinx LUT primitive LUT6
  generic map (INIT => X"fee8e880e8808000")        -- LUT Contents
  port map  (ThreeBitOutput(2), SixBitInput(0), SixBitInput(1), SixBitInput(2),
               SixBitInput(3), SixBitInput(4), SixBitInput(5));
LUT6_inst2 : LUT6        -- Xilinx LUT primitive LUT6
  generic map (INIT => X"8117177e177e7ee8")        -- LUT Contents
  port map  (ThreeBitOutput(1), SixBitInput(0), SixBitInput(1), SixBitInput(2),
               SixBitInput(3), SixBitInput(4), SixBitInput(5));

LUT6_inst3 : LUT6        -- Xilinx LUT primitive LUT6
  generic map (INIT => X"6996966996696996")        -- LUT Contents
  port map  (ThreeBitOutput(0), SixBitInput(0), SixBitInput(1), SixBitInput(2),
               SixBitInput(3), SixBitInput(4), SixBitInput(5));

end Behavioral;
```

The code above cannot be synthesized in the Quartus environment for Altera FPGAs. However, an alternative code below that uses constants instead of the Xilinx LUT6 primitive can be synthesized and works fine for both Altera and Xilinx FPGAs:

```
library IEEE;                          -- the code below is tested in the DE2-115 board
use IEEE.STD_LOGIC_1164.all;   -- with the Altera Cyclone-IVE FPGA
use IEEE.STD_LOGIC_UNSIGNED.all; -- this package is needed for type conversions below

entity LUT_6to3 is
   port ( SixBitInput      : in  std_logic_vector (5 downto 0);
          ThreeBitOutput : out  std_logic_vector (2 downto 0));
end LUT_6to3;

architecture Behavioral of LUT_6to3 is

type LUT is array (2 downto 0) of std_logic_vector(63 downto 0);
-- array below contains the same constants as used in the INIT statements in the code with LUTs above
constant conf_LUT  :  LUT := (    X"fee8e880e8808000",   -- array of constants
                                  X"8117177e177e7ee8",   -- is used here
                                  X"6996966996696996");

begin           -- Hamming weight is found in the statements below

ThreeBitOutput <=        conf_LUT(2)(conv_integer(SixBitInput)) &
                         conf_LUT(1)(conv_integer(SixBitInput)) &
                         conf_LUT(0)(conv_integer(SixBitInput));

-- alternatively the following generate statement can be used:
-- gen: for i in conf_LUT'range generate
--      ThreeBitOutput(i) <= conf_LUT(i)(conv_integer(SixBitInput));
--end generate gen;

end Behavioral; -- the same code can be used for Xilinx FPGAs without any change
```

The two given above VHDL codes describe similar functionalities and permit the Hamming weight of 6-bit input vectors to be calculated in combinational circuits. The first code explicitly configures the Xilinx LUTs and the second code implicitly configures actually the same LUTs but without the need for vendor-specific libraries. The circuit built by Altera Quartus occupies 8 logic elements and the circuit built by Xilinx ISE for the Nexys-4 board occupies 3 LUTs. Such way enables the projects of the book to be also implemented and tested in FPGAs of other companies.

Similarly the majority of other modules described in this chapter have been tested in the DE2-115 board.

Many additional examples can be found in [8, 9].

References

1. Ashenden PJ (2008) The designer's guide to VHDL, 3rd edn. Morgan Kaufmann
2. Ashenden PJ (2008) Digital design: an embedded systems approach using VHDL. Morgan Kaufmann
3. Xilinx Inc (2013) XST user guide for Virtex-6, Spartan-6, and 7 series devices. http://www.xilinx.com/support/documentation/sw_manuals/xilinx14_7/xst_v6s6.pdf. Accessed 17 Nov 2013
4. Digilent Inc (2009) Adept I/O expansion reference design. http://www.digilentinc.com/Products/Detail.cfm?NavPath=2,66,828&Prod=ADEPT2. Accessed 9 Nov 2013
5. Xilinx Inc (2011) ISE In-Depth Tutorial. http://www.xilinx.com/support/documentation/sw_manuals/xilinx13_1/ise_tutorial_ug695.pdf. Accessed 17 Nov 2013
6. Xilinx Inc (2009) Synthesis and simulation design guide. http://www.xilinx.com/support/documentation/sw_manuals/xilinx11/sim.pdf. Accessed 17 Nov 2013
7. Altera Inc (2013) Quartus II setting file with pin assignments for DE2-115. http://www.altera.com/education/univ/materials/boards/de2-115/unv-de2-115-board.html. Accessed 17 Nov 2013
8. Skliarova I, Sklyarov V, Sudnitson A (2012) Design of FPGA-based circuits using hierarchical finite state machines. TUT Press, Tallinn
9. Sklyarov V, Skliarova I (2013) Parallel processing in FPGA-based digital circuits and systems. TUT Press, Tallinn

Chapter 3
Design Techniques

Abstract This chapter begins with a brief description of widely used simple combinational and sequential circuits. Many examples are given with implementations of the circuits in FPGAs. Next, various optimization techniques are discussed with special emphasis on broad parallelism, which is very important for FPGA-based applications. More complicated digital circuits and systems are introduced, such as parallel networks for sorting and searching, Hamming weight counters/comparators, concurrent vector processing units and advanced finite state machines. The circuits are designed so that operations over multiple data items can be executed concurrently. Network-based solutions, such as sorting and counting networks in particular, and the efficient mapping of circuits to FPGA primitives (look-up tables) are examples. A number of alternative competing methods are discussed and evaluated. All the circuits and systems are described in VHDL, implemented and tested in FPGAs, and finally evaluated by applying various criteria. Many of the novel solutions proposed are parameterized, which permits very complex projects to be developed in FPGAs for solving advanced problems in several areas, such as data processing and combinatorial search.

3.1 Combinational Circuits

A combinational circuit (CC) does not have memory and, thus, output values of the circuit depend only on its current input values. This section briefly characterizes widely used (group 1) and application-specific (group 2) CCs with description of their functionality in behavioral VHDL.

The first group includes encoders, decoders, multiplexers, arithmetical circuits, and logical shifters. The second group is composed of such circuits that are synthesized from given systems of Boolean functions, such as

$$y_0 = f_0(x_{n-1}, \ldots, x_1, x_0);$$
$$y_1 = f_1(x_{n-1}, \ldots, x_1, x_0);$$
$$\ldots\ldots\ldots\ldots\ldots\ldots\ldots\ldots\ldots$$
$$y_{m-1} = f_{m-1}(x_{n-1}, \ldots, x_1, x_0);$$

V. Sklyarov et al., *Synthesis and Optimization of FPGA-Based Systems*, Lecture Notes in Electrical Engineering 294, DOI: 10.1007/978-3-319-04708-9_3,

Table 3.1 Different Boolean functions of 3 variables (n = 3, m = 1)

x_2	x_1	x_0	F_{255}	F_{254}	…….	F_{11}	F_{10}	F_9	F_8	F_7	F_6	F_5	F_4	F_3	F_2	F_1	F_0
0	0	0	1	1	… …	0	0	0	0	0	0	0	0	0	0	0	0
0	0	1	1	1	… …	0	0	0	0	0	0	0	0	0	0	0	0
0	1	0	1	1	… …	0	0	0	0	0	0	0	0	0	0	0	0
0	1	1	1	1	… …	0	0	0	0	0	0	0	0	0	0	0	0
1	0	0	1	1	… …	1	1	1	1	0	0	0	0	0	0	0	0
1	0	1	1	1	… …	0	0	0	0	1	1	1	1	0	0	0	0
1	1	0	1	1	… …	1	1	0	0	1	1	0	0	1	1	0	0
1	1	1	1	0	… …	1	0	1	0	1	0	1	0	1	0	1	0

where $y_0, y_1, \ldots, y_{m-1}$ are binary outputs of the circuit that depend on binary inputs $x_{n-1}, \ldots, x_1, x_0$ and $f_0, f_1, \ldots, f_{m-1}$ is a system F of m Boolean functions that describe how to convert input values to output values, i.e. how to construct an output vector $Y_i = (y_0, y_1, \ldots, y_{m-1})$ for any input vector $X_j = (x_{n-1}, \ldots, x_1, x_0)$: $Y_i = F(X_j)$. For $m = 1$ there are 2 in power 2^n Boolean functions of n variables and if $m > 1$, the number of different Boolean functions is rapidly increased. Table 3.1 below shows some of $2^{8=2 \text{ in power } 3} = 256$ Boolean functions F_{255}, F_{254}, …, F_2, F_1, F_0 of $n = 3$ variables.

The rightmost four functions from Table 3.1 can be described as follows:

$$y_0 = F_0(x_2, x_1, x_0) = 0;\ y_1 = F_1(x_2, x_1, x_0) = x_2 \textit{ and } x_1 \textit{ and } x_0;$$
$$y_2 = F_2(x_2, x_1, x_0) = x_2 \textit{ and } x_1 \textit{ and not } x_0;$$
$$y_3 = F_3(x_2, x_1, x_0) = (x_2 \textit{ and } x_1 \textit{ and } x_0) \textit{ or } (x_2 \textit{ and } x_1 \textit{ and not } x_0) = x_1 \textit{ and } x_2;$$

Note that the function y_3 has been simplified using the combining theorem [1]. Methods of minimization (simplification) of Boolean functions are presented in [1, 2] and they will not be considered here. The functions y_0, y_1, y_2, y_3 above or (similar functions) can be described in VHDL and available synthesizers will take care about the optimization of the relevant circuits.

Table 3.2 demonstrates Boolean functions F_{ort} and F_{int} for two frequently used operations: orthogonality and intersection that are defined in the following general form:

$$y_{ort} = \bigvee_{i=1}^{n-1} (a_i \textit{ xor } b_i);\ y_{int} = \textit{not } y_{ort};$$

where the rightmost symbol *not* requires an *inversion* operation. From Table 3.2 we can see the results for different pairs of 2-bit vectors. Bit values in the column *orthogonality* are grouped in sets of 4 bits in order to show their association with the respective hexadecimal digits that will be needed later in this section.

Table 3.2 Boolean functions for orthogonality and intersection operations

Vector A = $\{a_1,a_0\}$		Vector B = $\{b_1,b_0\}$		Orthogonality		Intersection
a_1	a_0	b_1	b_0	$y_{ort} = F_{ort}$ (A,B)		$y_{int} = F_{int}$ (A,B)
0	0	0	0	0	E	1
0	0	0	1	1		0
0	0	1	0	1		0
0	0	1	1	1		0
0	1	0	0	1	D	0
0	1	0	1	0		1
0	1	1	0	1		0
0	1	1	1	1		0
1	0	0	0	1	B	0
1	0	0	1	1		0
1	0	1	0	0		1
1	0	1	1	1		0
1	1	0	0	1	7	0
1	1	0	1	1		0
1	1	1	0	1		0
1	1	1	1	0		1

The following VHDL function describes the orthogonality operation:

```
function ort (A : std_logic_vector; B : std_logic_vector) return std_logic is
  variable result : std_logic := '0';
begin
  for i in A'range loop
      result := result or (A(i) xor B(i));
  end loop;
return result;
```

Orthogonality and/or intersection can be tested by evaluating the returned value from the function ort(A,B) and the statement not ort(A,B).

The following VHDL combinational process gives another specification:

```
process(A,B)   -- A and B are two input vectors with equal generic sizes (size)
begin
  intersected <= '1'; -- intersected and orthogonal are output ports: intersected : out std_logic;
  orthogonal <= '0';                                        -- orthogonal : out std_logic;

  for i in size-1 downto 0 loop      -- size is a generic parameter
      if A(i) /= B(i) then orthogonal <= '1'; intersected <= '0'; exit;
      end if;
  end loop;

end process;
```

Any arbitrary Boolean function can directly be described in VHDL. For example, the function y_3 from Table 3.1 can be described as y3 = x1 **and** x2. Alternatively truth tables (like Tables 3.1, 3.2) can directly be mapped to FPGA look-up tables—LUTs (see Sect. 1.2.1). For example the following VHDL code uses preliminary configured LUT (4,1) to test for orthogonality of 2-bit vectors A and B:

```
library IEEE;
use IEEE.STD_LOGIC_1164.all;
library UNISIM;
use UNISIM.vcomponents.all;      -- for using LUT primitives this library has to be included

entity TestOrt is
      generic ( size          : integer := 2);
      port (   A              : in std_logic_vector (size-1 downto 0);
               B              : in std_logic_vector (size-1 downto 0);
               orthogonal     : out std_logic );
end TestOrt;

architecture Behavioral of TestOrt is
  LUT4_inst : LUT4         -- LUT instantiation from the ISE Devise Primitive templates
  generic map (INIT => X"7BDE")  -- the initialization constant 7BDE is taken from Table 3.2
  port map (
    O => orthogonal,       -- LUT general output
    I0 => B(0),            -- LUT input
    I1 => B(1),            -- LUT input
    I2 => A(0),            -- LUT input
    I3 => A(1)             -- LUT input
  );
end Behavioral;
```

Other potential descriptions will be shown on examples of widely used circuits from the group 1. More details about these circuits can be found in [1].

3.1.1 Encoders

The following VHDL code (that can be used directly in the architecture body) presents an example of a combinational binary encoder:

```
encoder_result <= "00" when encoder_input = "0001" else
                  "01" when encoder_input = "0010" else
                  "10" when encoder_input = "0100" else
                  "11" when encoder_input = "1000" else "00";
```

Two-bit codes on the left-hand side indicate index of the value '1' in four-bit codes on the right-hand side. For example, the code "01" indicates position 1 of the value '1' in the code "0010".

Similarly, circuits that handle larger number of bits can be created. Much like the previous example an encoder (and also other circuits described below) can be mapped to FPGA LUTs.

3.1.2 Decoders

The following VHDL code presents an example of a combinational binary decoder:

```
-- the next lines can be used in architecture body
decoder_result <= "0001" when decoder_input = "00" else
                  "0010" when decoder_input = "01" else
                  "0100" when decoder_input = "10" else
                  "1000" when decoder_input = "11" else
                  "1111";
```

Four-bit codes on the left-hand side contain the value '1' in position indicated by two-bit codes on the right-hand side. For example, the code "0100" on the left-hand side is used because of the value "10" in the code on the right-hand side.

Besides of common binary decoders, other circuits may be needed. For example, to show a decimal digit on a display (see Fig. 3.1a) a *seven-segment decoder* can be designed. It has four inputs that receive binary codes and seven outputs that control individual display segments (from *a* to *g*) as illustrated in Fig. 3.1a.

Decimal digits can be written in the BCD (*Binary Coded Decimal*) code which includes 4-bit combinations from 0000 through 1001 representing decimal digits 0–9 as shown in Fig. 3.1b (combinations 1010 through 1111 are not used).

The decoder can be described in VHDL as follows:

```
with BCD select -- the segment is active when the corresponding bit in 7-bit vector below is one
  segments <= "1111110" when "0000", -- digit 0
              "0110000" when "0001", -- digit 1
              "1101101" when "0010", -- digit 2
              "1111001" when "0011", -- digit 3
              "0110011" when "0100", -- digit 4
              "1011011" when "0101", -- digit 5
              "1011111" when "0110", -- digit 6
              "1110000" when "0111", -- digit 7
              "1111111" when "1000", -- digit 8
              "1111011" when "1001", -- digit 9
              "0000000" when others; -- not valid input combinations
end Behavioral;
```

Here, the individual segments *a…g* are assumed to be active high and are grouped in a single 7-bit output vector segments (where the symbol *a* corresponds to the most significant bit in the vector and the symbol *g*—to the least significant bit).

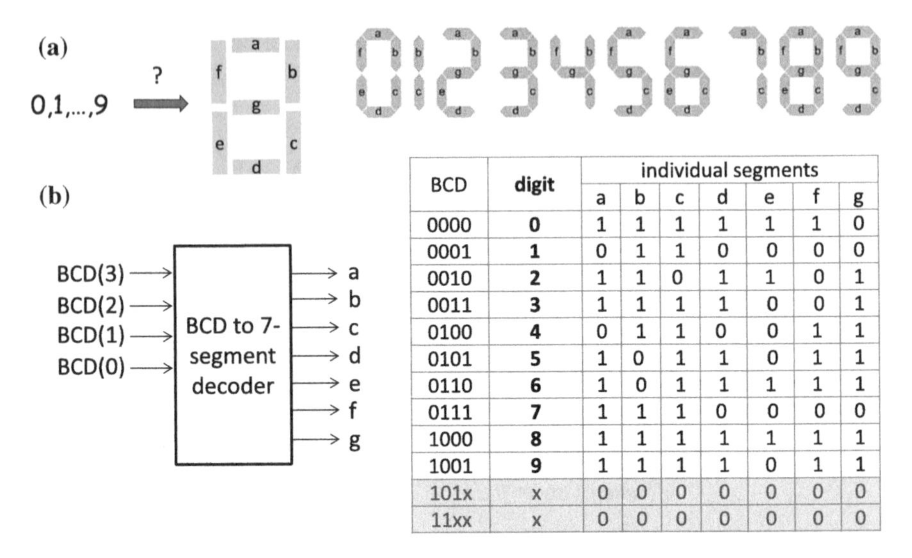

BCD	digit	individual segments						
		a	b	c	d	e	f	g
0000	**0**	1	1	1	1	1	1	0
0001	**1**	0	1	1	0	0	0	0
0010	**2**	1	1	0	1	1	0	1
0011	**3**	1	1	1	1	0	0	1
0100	**4**	0	1	1	0	0	1	1
0101	**5**	1	0	1	1	0	1	1
0110	**6**	1	0	1	1	1	1	1
0111	**7**	1	1	1	0	0	0	0
1000	**8**	1	1	1	1	1	1	1
1001	**9**	1	1	1	1	0	1	1
101x	x	0	0	0	0	0	0	0
11xx	x	0	0	0	0	0	0	0

Fig. 3.1 Identification of segments in a seven-segment display (**a**), and the truth table of a BCD to seven-segment decoder (**b**)

The following constant can be used instead of the code above:

```
type my_array is array (0 to 15) of std_logic_vector (6 downto 0);
constant converter : my_array :=  ("1111110", "0110000", "1101101", "1111001",
                                   "0110011", "1011011", "1011111", "1110000",
                                   "1111111", "1111011", "0000000", "0000000",
                                   "0000000", "0000000", "0000000", "0000000");
```

Decoding of segment is done in the following additional line:

```
segmentsP <= converter(conv_integer(BCD));
```

Since for some prototyping boards segments are active low, the following line need to be added:

```
segments      <= not segmentsP; -- segments are active low
```

Note that Appendix B contains VHDL code for the decoder that enables all hexadecimal digits (0,…, 9, A, B, C, D, E, F) to be shown on a 7-segment display.

3.1.3 Multiplexers

A combinational process below describes functionality of a 4:1 multiplexer which selects on the output O one of four inputs A, B, C, D.

```
architecture Mux of Entity_for_Mux is
begin-- 2-bit signal sel_ect permits one of four inputs (A,B,C,D) to be selected
  process (A, B, C, D, sel_ect)
  begin
    case sel_ect is
       when "00" => O <= A;           -- input A is sent to output O
       when "01" => O <= B;           -- input B is sent to output O
       when "10" => O <= C;           -- input C is sent to output O
       when "11" => O <= D;           -- input D is sent to output O
       when others => O <= A;
    end case;
  end process;
end Mux;
```

Similarly, circuits that handle larger number of bits may be created.

3.1.4 Comparators

A combinational comparator is described as follows (if the value of A is greater than or equal to the value of B then the result is '1', otherwise—'0'):

```
-- the next line can be used in architecture body
comparator_result <= '1' when A >= B else '0';
```

A similar comparator may be described in a combinational process:

```
process(A,B)
begin
  if (A >= B)  then       comparator_result <= '1';
  else                    comparator_result <= '0';
  end if;
end process;
```

Later in this chapter we will use comparators/swappers in sorting networks and they can be described similarly, for example:

```
maximum_of_A_B <= A when A >= B else B; -- signal maximum_of_A_B keeps the maximum
minimum_of_A_B <= B when A >= B else A; -- signal minimum_of_A_B keeps the minimum
```

3.1.5 Arithmetical Circuits

Arithmetical circuits have already been discussed in Sect. 2.1. We will give here one more example that demonstrates the use of operations for addition (+), subtraction (-), multiplication (*), division (/) and the rest of division (rem).

```
result <=  255 when (B = 0) and (but = "01000") else  -- "divide by 0" (only BTNC is pressed)
        A + B when but = "00001" else                 -- only BTNR is pressed
        A - B when (but = "00010") and (A>=B) else    -- only BTNL is pressed
        B - A when (but = "00010") and (A<B) else     -- only BTNL is pressed
        A * B when (but = "00100" else                -- only BTND is pressed
        A / B when but = "01000" else                 -- only BTNC is pressed
        A rem B when but = "10000" else               -- only BTNU is pressed
        0;
```

Different signals are declared as follows:

```
signal result   : integer range 0 to 255;
signal but      : std_logic_vector(4 downto 0);
signal A,B      : integer range 0 to 15;
```

Initial data can be taken from the onboard switches dip and buttons BTNU, BTNC, BTND, BTNL, and BTNR (see Figs. 1.23 and 1.24):

```
but   <= BTNU & BTNC & BTND & BTNL & BTNR;
A     <= conv_integer(dip(7 downto 4));
B     <= conv_integer(dip(3 downto 0));
```

and the results can be displayed and checked on onboard LEDs:

```
led <= conv_std_logic_vector(result, 8);
```

3.1.6 Barrel Shifters

Let us describe a 4-bit barrel shifter which has 4 data inputs—D3..D0, 4 data outputs—Y3..Y0, and two control inputs—C1C0. The output vector Y3..Y0 equals to the input vector D3..D0, rotated by a number of bit positions specified by the control inputs. For example, if the input vector is ABCD (each letter represents one bit), and the control inputs are 10, then the output vector is CDAB. The code below describes the barrel shifter functionality:

```
Y <=    D                           when C="00" else
        D(2) & D(1) & D(0) & D(3)   when C="01" else
        D(1) & D(0) & D(3) & D(2)   when C="10" else
        D(0) & D(3) & D(2) & D(1);
```

Similarly, circuits that handle larger number of bits can be created. Many additional examples of CC can be found in [3, 4].

3.2 Sequential Circuits

The majority of digital systems are sequential and they include combinational blocks as components. A fundamental notion in a sequential digital circuit (SDC) is a state that depends not only on the current inputs but also on the functionality of the SDC in the past. The state is kept in a storage allocated in the circuit and can be changed either by special signals that are clocks (in case of synchronous behavior) or by events most often on the inputs (in case of asynchronous behavior). We will discuss in the book only synchronous SDCs.

Much like combinational circuits SDCs can be divided in two groups that include widely used (group 1) and application-specific (group 2) SDCs. The latter can be further divided in numerous sub-groups many of which are not clearly identified. For example, we can point to such devices that have some common features as finite state machines (FSM), interfaces, and application-specific accelerators. One common representation of a system is called *register-transfer level* (RTL) that defines how data are transferred between registers/memories passing through combinational logic and driven by a sequential control circuit. The latter may be an FSM, dedicated asynchronous blocks, etc. Generally, it is difficult or even impossible to describe all potentially existing SDCs. So, we focus in Sect. 3.2 on simple devices from the first group, namely registers, shift registers, counters, and arithmetical devices with accumulators. SDCs from application-specific group will be discussed in the subsequent sections and chapters.

3.2.1 Registers

A SDC that is composed of R flip-flops (such as D flip-flops) with a common clock (and possibly with a common reset) input is called a register, which can be described in VHDL as follows:

```
process (clk)   -- D is an input vector and Q is an output vector
begin           -- clk is a clock and rst is a synchronous reset with active high value
   if rising_edge(clk) then
      if rst = '1' then     Q <= (others => '0');
      else                  Q <= D;
      end if;
   end if;
end process;
```

Alternatively the following code can be used:

```
Q <=  (others => '0')     when rising_edge(clk) and (rst = '1')  else
       D                   when rising_edge(clk);
```

3.2.2 *Shift Registers*

A shift register is an R-bit register which permits stored data to be shifted by one (or possibly more) bit position in each clock cycle. VHDL code below describes *parallel-in, parallel-out* shift register. The *parallel input* to_set supplies a new vector that will be written to all flip-flops of the register in parallel. Then the vector in the register can be shifted one bit position at every clock cycle (divided_clk) either to the right or to the left depending on the value of the signal shift_direction.

```
process (divided_clk)
begin
  if rising_edge(divided_clk) then
    if rst = '1' then
       reg <= (others => '0');                  -- setting all flip-flops of the register to zero
    elsif set = '1' then
       reg <= to_set;                           -- copying data to the register
    elsif clock_enable='1' then                 -- shift dependently on direction
       if shift_direction='1' then              --  if shift_direction is 1 then shift right
         reg <= reg(0) & reg(7 downto 1);       -- shifting right
       else                                     --  if shift_direction is 0 then shift left
         reg <= reg(6 downto 0) & reg(7);       -- shifting left
       end if;
    end if;
  end if;
end process;
```

The following line enables the vector from the register to be read:

```
led <= reg;     -- to display the result from the register
```

Shift can be done by more than one bit position if needed. For example, the following code enables shift to be done by three bit positions in each clock cycle:

```
if shift_direction='1' then                       --  if shift_direction is 1 then shift right
  reg <= reg(2 downto 0) & reg(7 downto 3);  -- shifting right
else                                              --  if shift_direction is 0 then shift left
  reg <= reg(4 downto 0) & reg(7 downto 5);  -- shifting left
end if;
```

3.2.3 *Counters*

A counter is a sequential circuit which iterates through a fixed cycle of states. A synchronous counter connects all of its flip-flop clock inputs to the same common clock signal, forcing in this manner all the flip-flop outputs to change at the same time. The most popular are *binary counters* which are composed of R flip-flops and basically "count" in binary from 0 to 2^R-1, returning to 0 afterwards and starting the counting process again. VHDL code below describes a counter with clock_enable and count_direction signals (all necessary explanations are given in comments):

```
process (divided_clk)
begin
  if rising_edge(divided_clk) then
    if rst='1' then
      count <= (others => '0');     -- setting all flip-flops of the counter to zero
    elsif clock_enable='1' then     -- counting dependently on direction
      if count_direction='1' then   -- if count_direction is 1 then increment the counter
        count <= count + 1;         -- incrementing the counter
      else                          -- if count_direction is 0 then decrement the counter
        count <= count - 1;         -- decrementing the counter
      end if;
    end if;
  end if;
end process;
```

The results of counting can be displayed much like in the previous example with the shift register. If required incrementing/decrementing the value in the counter can be done by more than one. For example, the following code enables an increment by 2 and a decrement by 3 to be done;

```
if count_direction='1' then      --  if count_direction is 1 then increment the counter
  count <= count + 2;            -- incrementing the counter by 2
else                             --  if count_direction is 0 then decrement the counter
  count <= count - 3;            -- decrementing the counter by 3
end if;
```

3.2.4 Arithmetical Circuits with Accumulators

The considered circuit enables any operation accu+B to be executed over an operand B and the value of accu that was saved in a special register (called accumulator and set to 0 before the first operation) holding the result of the previous operation. For example, if B = 3, then the signal accu may accumulate sequentially the following values: 3,6,9,12,.... Let us consider the following declarations:

```
signal B, accu       : integer range 0 to 255;             -- declaration of operand and accumulator
signal divided_clk   : std_logic;                          -- clocks from a clock divider
signal accu_enable : std_logic;                            -- signal enable for the accumulator
signal op_sel        : std_logic_vector(1 downto 0); -- op_sel selects an operation
```

VHDL code below describes an arithmetical circuit with accumulator accu with signals reset and accu_enable (all necessary explanations are given in the comments):

```
process (divided_clk)
begin
  if rising_edge(divided_clk) then -- low frequency clock to observe the functionality visually
    if (reset = '0') then accu <= 0; -- on active reset (zero) the accu is filled with zeros
    else
      if accu_enable = '1' then     -- arithmetical operation is allowed if accu_enable = '1'
        case op_sel is              -- op_sel selects the desired arithmetical operation
              when "00" => accu <= accu + B;  -- accumulating the results of addition
              when "01" => accu <= accu - B;  -- accumulating the results of subtraction
              when "10" => accu <= accu * B;  -- accumulating the results of multiplication
              -- if B is not zero accumulating the results of division in the next line
              when "11" => if B /= 0 then accu <= accu / B; else null; end if;
              when others => null;            -- each element of op_sel may have
        end case;                             -- any from 9 values of std_logic type
      end if;
    end if;
  end if;
end process;
```

The results from accu can be displayed using the following line:

```
led <= conv_std_logic_vector(accu, 8);
```

Descriptions for some other sequential and combinational circuits can be found in ISE/Quartus language templates.

3.3 Finite State Machines

Finite state machines (FSM) are probably the most widely used application-specific SDC in digital systems. That is why almost all the available automatic design tools that are included in industrial computer-aided design systems allow FSMs to be synthesized from their formal specifications. Since an FSM is a sequential circuit it can be characterized by a set of states $a_0,\ldots,a_{M-1}$, transitions between the states, and operations (in states and during state transitions). The number of states is finite.

Basically, FSMs are needed for two kinds of applications that are:

1. Autonomous sequential modules that are components of more complicated digital systems. For example, an FSM can read a sequence of bits and detect in the sequence two or more successive ones. Many similar examples, such as rising edge detector, debouncing circuit, etc., are given in [4, 5].
2. Control circuits. For example, an FSM-based unit for a combinatorial processor is suggested in [6]. Many other examples are given in [7, 8].

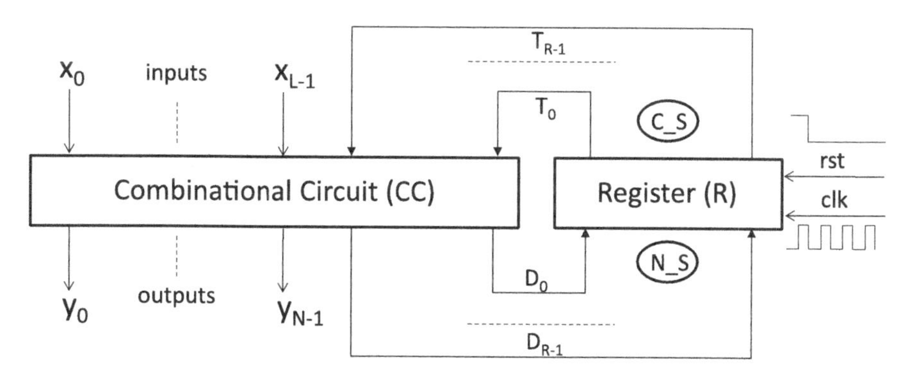

Fig. 3.2 General structure of a conventional FSM

Figure 3.2 depicts a general structure of an FSM, which is composed of a register (keeping the FSM state) and a combinational circuit (providing state transitions and generating outputs).

The most common FSM models are Mealy and Moore, which differ in the method of generating outputs. In Mealy FSM, the output signals directly depend on both the current state and the current inputs:

$D_0 = \psi_0(T_0,\ldots,T_{R-1}, x_0,\ldots,x_{L-1})$;

....

$D_{R-1} = \psi_{R-1}(T_0,\ldots,T_{R-1}, x_0,\ldots,x_{L-1})$;

$y_0 = \varphi_0(T_0,\ldots,T_{R-1}, x_0,\ldots,x_{L-1})$;

....

$y_{N-1} = \varphi_{N-1}(T_0,\ldots,T_{R-1}, x_0,\ldots,x_{L-1})$;

Here $\psi_{0,\ldots,}$ ψ_{R-1} are transition functions and $\varphi_{0,\ldots,}$ φ_{N-1} are output functions; $x_{0,\ldots,}x_{L-1}$ are input signals, and $y_{0,\ldots,}$ y_{N-1} are output signals; signals $T_0,\ldots,T_{R-1}$ represent current states (C_S) and signals $D_0,\ldots$, D_{R-1}—next states (N_S).

In Moore FSM, the output signals directly depend only on the current state:

$y_0 = \varphi_0(T_0,\ldots,T_{R-1})$;

....

$y_{N-1} = \varphi_{N-1}(T_0,\ldots,T_{R-1})$;

Both models can be structurally described in a way shown in Fig. 3.2. So, the main difference is in the representation of the combinational circuit (CC) although synchronization mechanisms can also be diverse.

Very often an FSM has a single initial state. The signal rst in Fig. 3.2 sets (resets) FSM to this state, for example, as soon as the power is switched on. The signal clk synchronizes state transitions in which the FSM changes one state to another one. Usually such transitions are executed either on a rising or on a falling edge of the signal clk.

There are many different ways to describe FSM functionality, such as state transition diagrams, state transition tables, graph-schemes, etc.

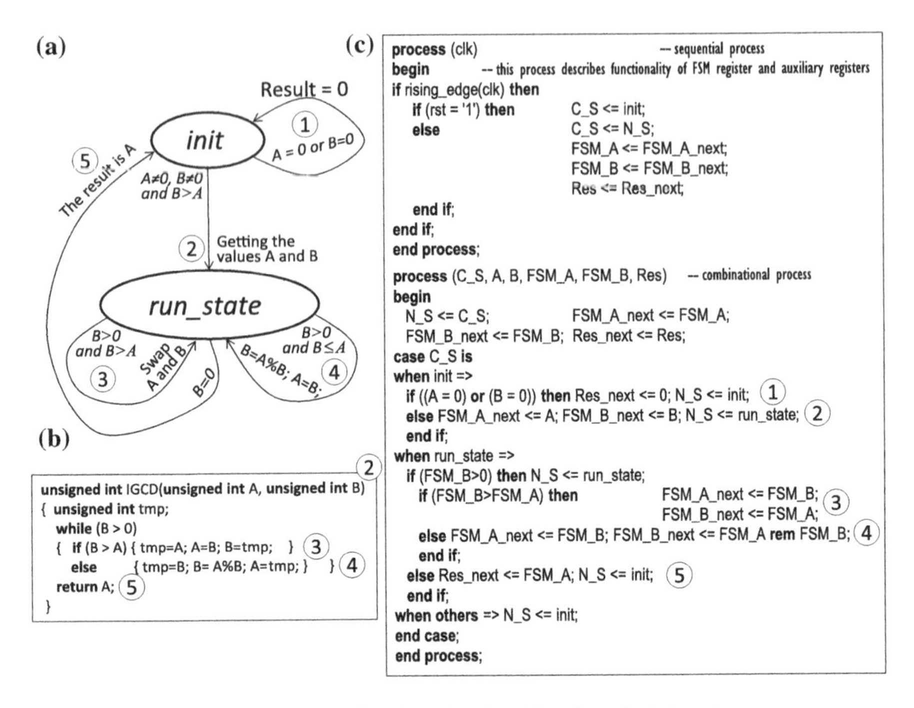

Fig. 3.3 Implementation in FPGA of an iterative algorithm for calculating the greatest common divisor: state transition diagram (**a**), C language code (**b**), FSM in VHDL code (**c**)

An FSM with datapath (FSMD) [5] combines an FSM with components of an execution unit, such as registers, counters, etc. and deals with operations at a register-transfer level (RTL). Let us consider how to design an FSMD on an example of a circuit that enables the maximum common divisor of two non-negative integer operands to be found. The following C function IGCD [4] gives a feasible iterative implementation for unsigned integers:

```
unsigned int IGCD(unsigned int A, unsigned int B)
{  unsigned int tmp;
   while (B > 0)
      {  if (B > A)        { tmp = A; A = B;    B = tmp; }
         else              { tmp = B; B = A%B; A = tmp; }          }
   return A;
}
```

Figure 3.3a depicts the FSM functionality similar to the function IGCD (see also Fig. 3.3b where a visual comparison is easier). Figure 3.3c shows VHDL code for the FSMD that calculates the greatest common divisor of two unsigned integers. There are two processes in the code. The first sequential process describes state transitions and changes states of three registers (FSM_A, FSM_B and Res) enabling data, altered in the combinational processer, to be transferred between the registers. For example, data in the registers may be swapped (FSM_A_next <= FSM_B, FSM_B_next <= FSM_A) or the rest of division is found out (FSM_B_next <=FSM_A **rem** FSM_B).

The numbers enclosed in circles indicate similar operations in the state transition diagram (Fig. 3.3a), in C code (Fig. 3.3b), and in VHDL code (Fig. 3.3c).

Note that FSM in Fig. 3.3a and 3.3b is built in accordance with the Mealy model. Operations in all state transitions depend on both the states of the FSM and on some tested values such as (FSM_B > 0 and FSM_B > FSM_A).

The following VHDL code describes a project that can easily be adapted to any prototyping board referenced in Sect. 1.6. It reads 8+8 bit data from onboard switches, calculates the greatest common divisor of the data items, and displays the result on LEDs.

```
entity FSM_OneEdge_GCD is  -- circuit with synchronization by one clock edge
port (        clk       : in std_logic;
              rst       : in std_logic;        -- BTNC button
              Ain       : in std_logic_vector(15 downto 0);  -- two 8-bit operands
              Result    : out std_logic_vector(7 downto 0)); -- 8-bit result (on LEDs)
end FSM_OneEdge_GCD;

architecture Behavioral of FSM_OneEdge_GCD is -- the circuit was tested in Nexys-4 board
  signal A, B, FSM_A, FSM_B, FSM_A_next, FSM_B_next : integer range 0 to 255;
  type state_type is (init, run_state);          -- enumeration type for the FSM states
  signal C_S, N_S          : state_type;
  signal Res, Res_next     : integer range 0 to 255;
begin
  A <= conv_integer(Ain(15 downto 8));          -- the first 8-bit operand from onboard switches
  B <= conv_integer(Ain(7 downto 0));           -- the second 8-bit operand from onboard switches
  -- copy here the FSM description from Fig. 3.3c
  Result <= conv_std_logic_vector(Res, 8);
end Behavioral;
```

Let us consider an example of FSM that is built in accordance with the Moore model (see Fig. 3.4). Numbers 1 and 2 enclosed in circles show where outputs are formed. As we can see the outputs do not depend on inputs and depend only on the state (count and final_state). Letters a, b and c enclosed in circles indicate possible transitions in the state transition diagram (see Fig. 3.4a) and in the VHDL code (see Fig. 3.4b). The following signals and types are declared for the considered example:

```
signal index, next_index          : integer range 0 to number_of_bits-1;
signal A                          : std_logic_vector(number_of_bits-1 downto 0);
signal Res, next_Res, n_o_ones, next_n_o_ones
                                  : integer range 0 to number_of_bits;
type state_type is (count, final_state);     -- enumeration type for the FSM states
signal C_S, N_S                   : state_type;
signal rst                        : std_logic;
```

and number_of_bits is a generic parameter. The number of values '1' (i.e. the Hamming weight) is counted for 16-bit vector from onboard switches of the

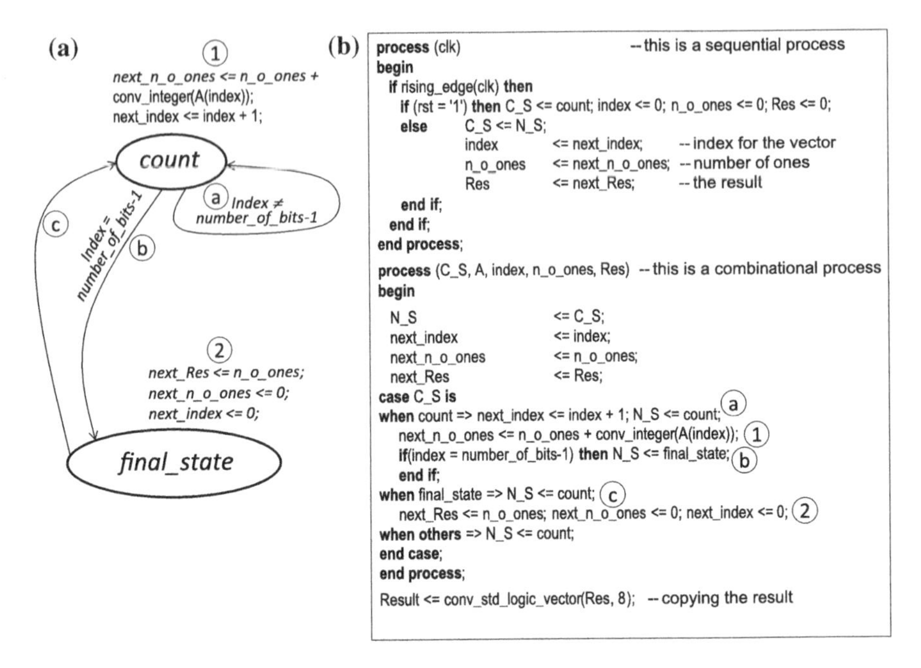

Fig. 3.4 An example of a Moore FSM that counts the number of values '1' in a given binary vector: state transition diagram (**a**), the FSM in VHDL code (**b**)

Nexys-4 board. Such circuits might be interesting to compare combinational (see Chap. 2) and sequential Hamming weight counters. Our sequential circuit occupies 8 slices of Artix-7 FPGA and has the maximum attainable clock frequency 560 MHz (these details are taken from Xilinx ISE 14.7 Design Summary/Reports). From Fig. 3.4b we can see that 16 clock cycles are needed to find out the Hamming weight of a 16-bit binary vector. Thus, the delay is about 28.6 ns.

If required an FSM that combines Mealy and Moore models can also be built [4, 7]. Many supplementary examples will be given in Chap. 5 that is dedicated to advanced FSMs.

3.4 Optimization of FPGA-Based Circuits and Systems

FPGAs still operate on a lower clock frequency than non-configurable application-specific integrated circuits and application-specific standard products and broad parallelism is evidently required to compete with potential alternatives. Many research works have been dedicated to this problem focusing on applying concurrency at different levels. We describe here several techniques that enable highly parallel circuits and systems to be designed and implemented in FPGAs. The following three major areas will be discussed:

1. Network-based solutions that are described in subsequent sections of this chapter and used in:
   - Combinational circuits with massive parallel conversions to be done simultaneously (e.g. sorting [9] and counting [10] networks).
   - Partially combinational and partially sequential circuits with highly parallel reusable segments. Such approach permits a better compromise to be found between the resources required and performance (e.g. sequential circuits with reusable combinational segments executing massive parallel conversions are described in [9]).
   - Pipelines composed of registers and highly parallel combinational circuits in between the registers (e.g. pipelined sorting and counting networks [9, 10]).
   - Highly parallel circuits that execute concurrent operations over large binary and ternary vectors (e.g. finding the maximum number of consecutive ones/zeros in binary vectors [9]).
2. On-chip systems that enable application-specific software to be run in parallel with hardware accelerators implementing solutions listed in point (1) (e.g. merging in software of sorted in hardware subsets, accelerating in hardware operations over large vectors [11]). We will consider software/hardware interaction in Chap. 4.
3. Application-specific sequential circuits executing multiple branches of implemented algorithms simultaneously (e.g. parallel hierarchical FSMs described in [12]). We will discuss such technique in Chap. 5.

Subsequent Sects. 3.4.1–3.4.3 give more details about the listed above areas and present examples of practical applications that are efficient for FPGA-based implementations.

3.4.1 Highly Parallel Network-Based Solutions

Highly parallel network-based solutions enable simultaneous operations to be executed over large sets of data items. For example, one of the fastest known parallel sorting methods is based on the *even-odd merge* and *bitonic merge* networks [13, 14]. The first type of network is shown in Fig. 3.5.

There are 6 vertical lines of comparators/swappers (they are numbered at the top) and each comparator can be described in VHDL as follows:

```
MaxValue <= A when A >= B else B; -- A and B are input data items
MinValue <= B when A >= B else A;
```

Given data (ex.: 144, 119, 150, 96, 39, 55, 17, 21) are sorted in descending order. Each vertical line is composed of some comparators/swappers and there are totally $C(N = 2^p) = (p^2 - p + 4) \times 2^{p-2} - 1$ such components [13, 15], where N is the number of items that have to be sorted. If data items are swapped they are shown in Fig. 3.5 in *italic* and underlined. Note that the decision about the result can be taken earlier than after propagation through all the vertical lines (see an

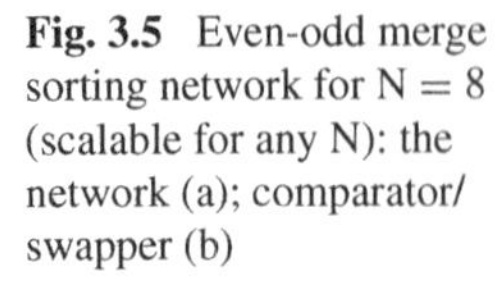

Fig. 3.5 Even-odd merge sorting network for N = 8 (scalable for any N): the network (a); comparator/swapper (b)

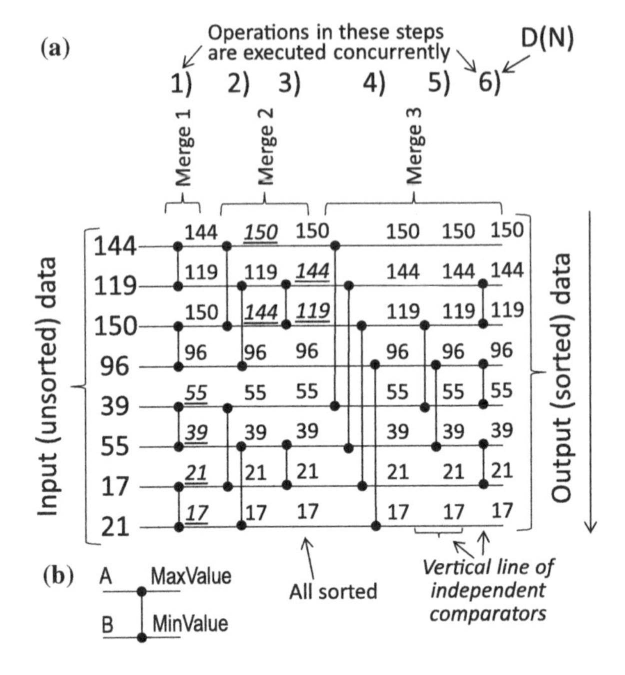

example in Fig. 3.5). Unfortunately, we cannot benefit from an earlier result (i.e. from the results that can be produced before propagating through the all vertical lines) because the network is hardwired.

Let us analyze the network in Fig. 3.5. Four parallel comparisons/swappers can be executed in parallel for the leftmost vertical line. All such operations do not have any data dependency, i.e. the result from any of them is not required for the remaining operations. The depth D(N) of a network that sorts N data items is the minimal number of data dependent steps 1, …, D(N) that have to be executed one after another. For Fig. 3.5 D(N) = 6 and the numbers n_p^s of parallel operations in steps s = 1,…,6 are: $n_p^1 = 4$; $n_p^2 = 4$; $n_p^3 = 2$; $n_p^4 = 4$; $n_p^5 = 2$; $n_p^6 = 3$. The time of sorting is equal to D(N) × t, where t is the delay of any vertical line, i.e. the delay of one comparator/swapper. It is known that for even-odd merge networks $D(N = 2^p) = p \times (p + 1)/2$ [15]. For our network in Fig. 3.5 $p = \lceil \log_2 N \rceil = 3$ and D(N) = 6. Thus, even-odd merge networks are very fast. For example, if N = 1024 then D(N) is only 55.

The following structural VHDL code demonstrates how the network in Fig. 3.5 can be described:

```
use work.set_of_data_items.all;-- the package where the type set_of_8items is declared
entity EvenOddMerge8Sort is
 generic (   M               : integer := 4;
             N               : integer := 8 );      -- cannot be changed for this project
  port (     unsorted_items  : in set_of_8items;    -- the type set_of_8items is declared in
             sorted_items    : out set_of_8items);  -- the package set_of_data_items
end EvenOddMerge8Sort;
```

```
architecture Behavioral of EvenOddMerge8Sort is
  signal out1_in2, out2_in3, out3_in4          : set_of_8items;
  signal out4_in5, out5_in6, sorted             : set_of_8items;
begin

  -- even-odd merging network
  merge1:      -- see the fragment Merge 1 in Fig. 3.5
  for i in N/2-1 downto 0 generate -- the first two parameters of the comparator are two operands
    group1merge1: entity work.Comparator
       generic map (M => M)
       port map(unsorted_items(i*2), unsorted_items(i*2+1),
                 out1_in2(i*2), out1_in2(i*2+1));
  end generate merge1; -- the last two parameters of the comparator are the maximum and the minimum

  merge2:      -- see the fragment Merge 2 in Fig. 3.5
  for i in 0 to N/4-1 generate
    incide_merge2:       -- the first data independent segment in merge 2
     for j in 0 to N/4-1 generate
       group1merge2: entity work.Comparator
         generic map (M => M)
         port map(out1_in2(i*4+j), out1_in2(i*4+j+2), out2_in3(i*4+j), out2_in3(i*4+j+2));
         out3_in4(i*4+j*3) <= out2_in3(i*4+j*3);
    end generate incide_merge2;

    group2merge2: entity work.Comparator -- the second data independent segment in merge 2
         generic map (M => M)
         port map(out2_in3(i*4+1), out2_in3(i*4+2), out3_in4(i*4+1), out3_in4(i*4+2));
  end generate merge2;

  merge3:      -- see the fragment Merge 3 in Fig. 3.5
  for i in N/2-1 downto 0 generate-- the first data independent segment in merge 3
    group1merge3: entity work.Comparator
       generic map (M => M)
       port map(out3_in4(i), out3_in4(i+4), out4_in5(i), out4_in5(i+4));

    step1merge3:         if (i >= 2 and i <= 3) generate
       group2merge3:  entity work.Comparator -- second data independent segment in merge 3
         generic map (M => M)
         port map(out4_in5(i), out4_in5(i+2), out5_in6(i), out5_in6(i+2));
    end generate;

    step2merge3: if (i < 2) generate
      out5_in6(i) <= out4_in5(i);
      out5_in6(i+6) <= out4_in5(i+6);
      sorted_items(i*7) <= out5_in6(i*7);
    end generate;

    step3merge3: if (i < N/2-1) generate  -- the third data independent segment in merge 3
      Comp2merge3       : entity work.Comparator
        generic map (M => M)
        port map(out5_in6(2*i+1), out5_in6(2*i+2), sorted_items(2*i+1),
                  sorted_items(2*i+2));
    end generate;
  end generate merge3;
end Behavioral;
```

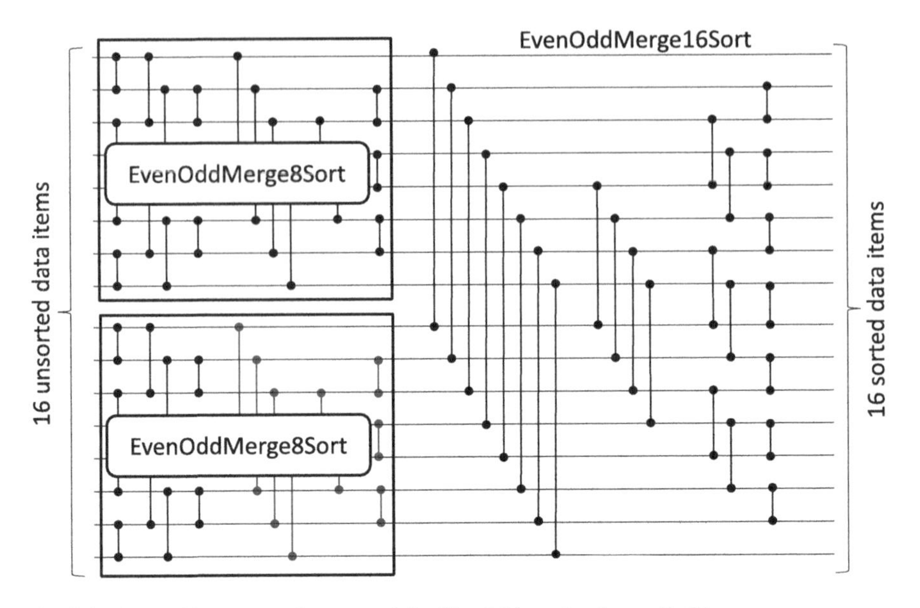

Fig. 3.6 Even-odd merge sorting network for N = 16 (see also Appendix B)

The package set_of_data_items contains the following lines:

```
constant N : integer := 8; -- cannot be changed for this project
constant M : integer := 4;
type set_of_8items is array (N-1 downto 0) of std_logic_vector (M-1 downto 0);
```

Now the component EvenOddMerge8Sort can be used in the network for N = 16 (see Fig. 3.6). An example of such circuit is given in Appendix B.

Once again the new component EvenOddMerge16Sort (see Appendix B) can be created and used in the network for N = 32 (see Fig. 3.7). Similarly, a network of any size N may be constructed. However, there is a problem. When N is increased, the complexity of the networks (the number of comparators C(N)) grows rapidly (see Fig. 3.8). Merging is executed incrementally, as shown in Fig. 3.9a for the *even-odd merge* network. Initially, each 2-item subset is merged. Then pairs of the resulting 2-item sorted subsets are merged to compose 4-item sorted subsets. Pairs of the 4-item sorted subsets are merged again and so on, until the complete sorted set of data is produced. The number of comparators/swappers required in any block is shown in the rectangles. The table in Fig. 3.9b gives the number of comparators/swappers at the last stage where the sorted subsets are merged, and the number of comparators at all stages for N varying from 8 to 2048.

Propagation delays through long combinational paths in FPGA networks are also increased and they are caused, not only by the comparators/swappers, but also by multiplexers that have to be inserted even in partially regular circuits [16], and by interconnections. Such routing overhead may be significant. The *bitonic merge* networks are also as fast as the *even-odd merge* network but the latter are slightly less resource consuming (see Fig. 3.8).

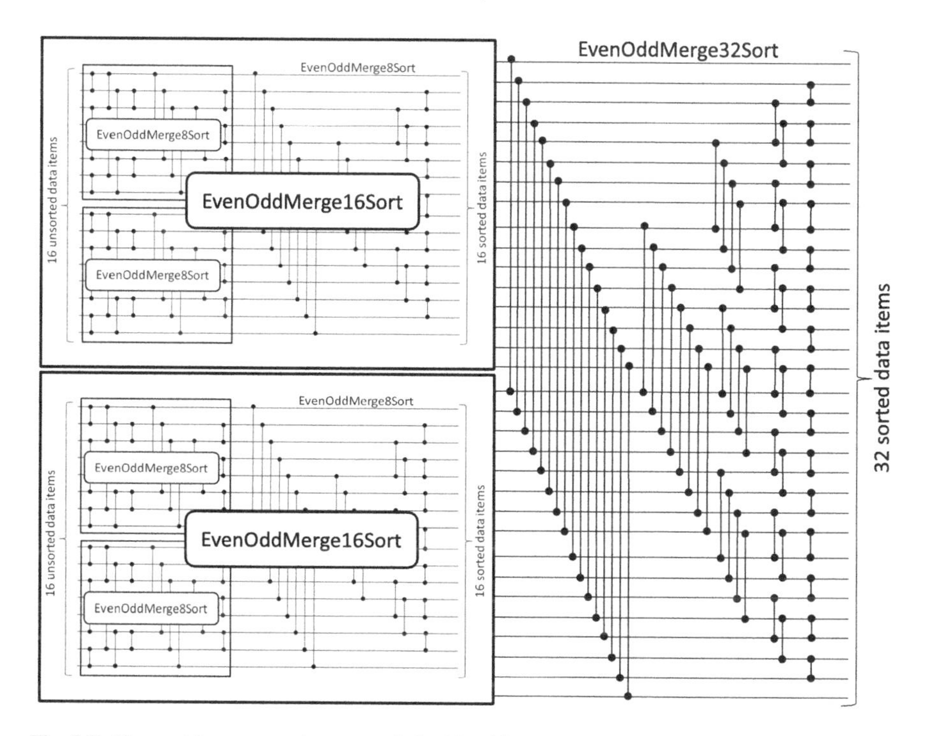

Fig. 3.7 Even-odd merge sorting network for N = 32

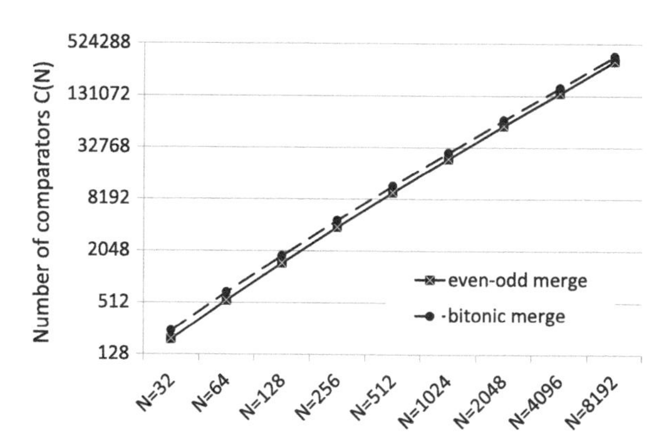

Fig. 3.8 Number of comparators for different values N of data items

Partially combinational and partially sequential circuits permit a better compromise to be found between the resources required and performance and they will be described on examples in Sects. 3.5, 3.6, where pipelined solutions will also be discussed. Similar parallel operations can be executed over large binary and ternary vectors and they will also be considered on examples in Sect. 3.10.

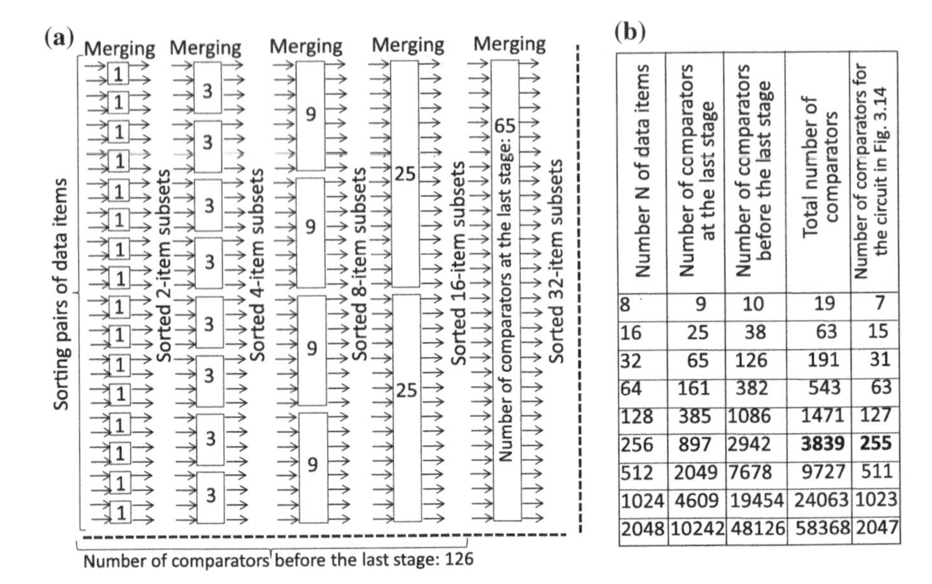

Number N of data items	Number of comparators at the last stage	Number of comparators before the last stage	Total number of comparators	Number of comparators for the circuit in Fig. 3.14
8	9	10	19	7
16	25	38	63	15
32	65	126	191	31
64	161	382	543	63
128	385	1086	1471	127
256	897	2942	**3839**	**255**
512	2049	7678	9727	511
1024	4609	19454	24063	1023
2048	10242	48126	58368	2047

Fig. 3.9 The structure of the even-odd merge network (**a**), and the number of comparators for different values of N (**b**)

3.4.2 Hardware Accelerators

From Figs. 3.8, 3.9 it is easy to conclude that sorting networks can be implemented in FPGA only for relatively small number N of items while practical applications require millions of such items to be processed. One possible way is to sort small subsets of larger sets in an FPGA and then to merge the subsets in software of a higher level system (see Fig. 3.10). The initial set of data that is to be sorted is divided into Z subsets of N items. Each subset is sorted in an FPGA using the networks discussed in the previous section. Merging is executed as shown in Fig. 3.10, in a host system/processor that interacts with the FPGA.

We will discuss in Chap. 4 two types of higher-level (host) systems (see Fig. 3.11): (1) a host PC communicating with FPGA through available ports (such as USB), and (2) a processing system (PS) in the all programmable system-on-chip (APSoC) Zynq [17] interacting with a programmable logic (PL) with the aid of on-chip high-performance interfaces.

For the considered above problem an FPGA enables sorting subsets of data to be accelerated while a host system/processor merges the sorted subsets. Thus, data need to be transferred to and from the FPGA and communication overhead may be significant especially for systems shown in Fig. 3.11a. However, such systems are also efficient in different types of experiments to support necessary data exchange. Besides systems similar to shown in Fig. 3.11b are very fast because data can be transferred through several very high speed 32/64-bit internal interfaces.

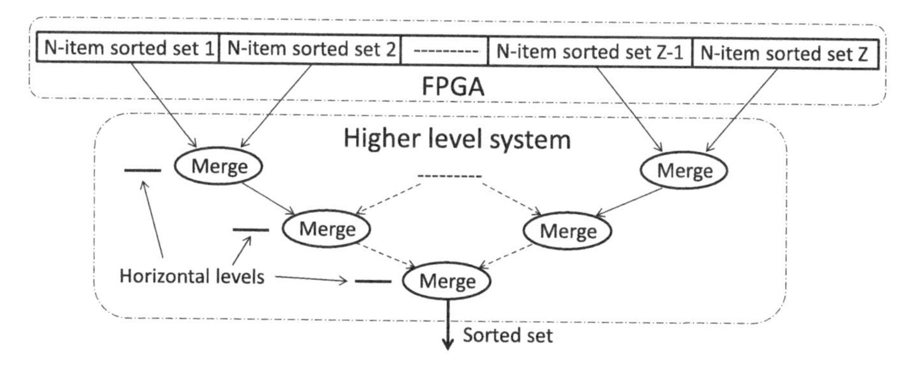

Fig. 3.10 Merging of sorted sub-sets in a software of a higher-level system

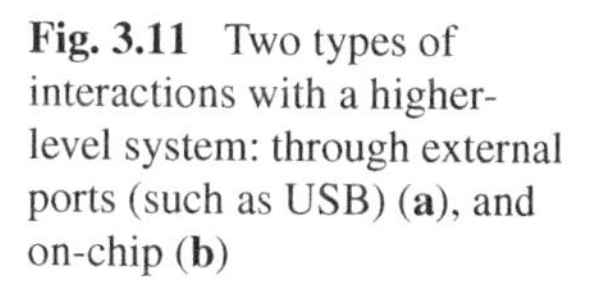

Fig. 3.11 Two types of interactions with a higher-level system: through external ports (such as USB) (**a**), and on-chip (**b**)

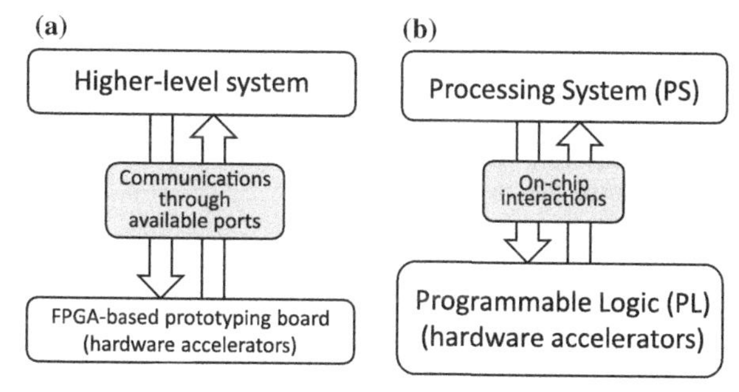

3.4.3 *Parallel Modular Algorithms Running in Hierarchical FSMs*

A hierarchical FSM (HFSM) [18] enables control algorithms composed of hierarchical modules to be implemented. A module is described by an autonomous state transition diagram that looks similar to Figs. 3.3a and 3.4a. Let us consider an example of using HFSMs for traversing $\mathcal{N}$-ary trees.

An $\mathcal{N}$-ary tree is a rooted connected graph that does not contain cycles and for which any internal node has at most $\mathcal{N}$ children [19]. Figure 3.12 depicts an example of an $\mathcal{N}$-ary tree ($\mathcal{N} = 4$) that can be seen as a graph representing operations A, B, C, D, E, …,M associated with the tree nodes a, b, c, d, e, …,m. Relationships between the operations are shown by tree edges. Alternatively, this tree can represent a set of data that are linked in accordance with given relationships. In this case the symbols A, B, C, D, E, …, M are considered to be subsets of data and edges indicate relationships between the subsets.

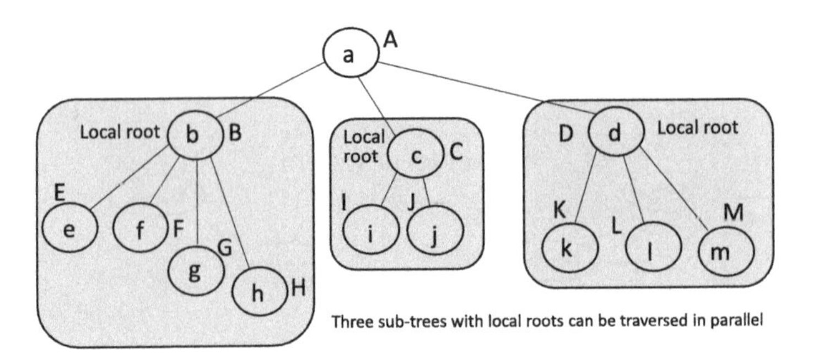

Fig. 3.12 An example of an $\mathcal{N}$-ary tree ($\mathcal{N} = 4$)

It is known that such a tree can be built, and traversed by applying either an iterative or a recursive procedure. For example, the following recursive C function from [12] does the traversal:

```
void traverse_tree(treenode* root, int depth)
{     depth++;
if (root == 0) { depth--; return; } // if root (node) does not exist it is equal to 0
if (depth == max_depth) {    executing_leaf_operation (root); depth--; return; }
for (int i = 0; i < N; i++)
   traverse_tree(root->node[i],depth);
depth--;                                    }
```

where treenode is a C structure which can be described as follows (N is a constant $\mathcal{N}$ in the C program):

```
struct treenode      {
 // other declarations   -- other declarations are collections of data or operations associated with the node
 treenode* node[N];  }; -- array of pointers to children (an element is equal to 0 if a child does not exist)
```

Similarly, an iterative function void iterative_traverse_tree(treenode* root, int depth) may be built for which the treenode structure has an additional field with a pointer to the parent node of the tree.

In Chap. 5 we will show how functions like the traverse_tree can be described as hardware modules and implemented in an HFSM. It is allowed several functions (HFSM modules) to be activated in parallel, for example, for local roots shown in Fig. 3.12 and, thus, faster tree traversing may be done. Different types of HFSMs (including parallel HFSMs) will be described in Chap. 5.

3.5 Design Examples for Parallel Sort

Sorting is a procedure that is needed in numerous computing systems [13]. For many practical applications, sorting throughput is very important. Two of the most frequently investigated parallel sorters are based on sorting [13] and linear [20]

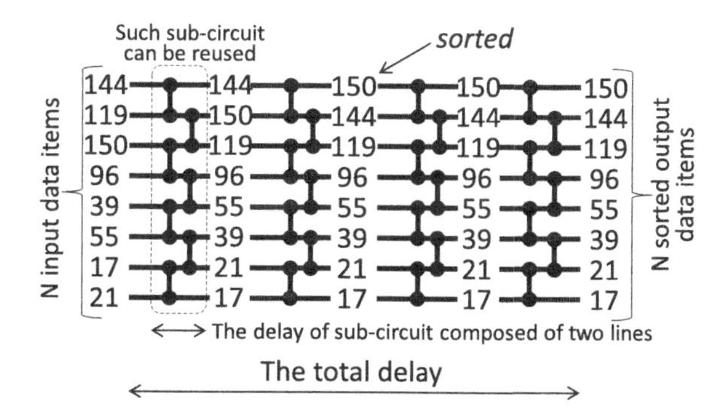

Fig. 3.13 Even-odd transition sorting network for N = 8 (scalable for any N)

networks. A sorting network is a set of vertical lines composed of comparators that can swap data to change their positions in the input multi-item vector. The data propagate through the lines from left to right to produce the sorted multi-item vector on the outputs of the rightmost vertical line. Three types of such networks have been studied: pure combinational (e.g. [3, 15, 21]), pipelined (e.g. [3, 15, 21]), and combined (partially combinational and partially sequential) [e.g. 3, 16].

We have already mentioned in Sect. 3.4.1 that the majority of sorting networks implemented in hardware use Batcher's *even-odd* and *bitonic* mergers [14]. Suppose N data items, each of size M bits, need to be sorted. The results of [15, 21] show that the referenced above sorting networks cannot be built for N > 64 (M = 32), even in the relatively advanced FPGA FX130T from the Xilinx Virtex-5 family, because the hardware resources are not sufficient. When N is increased, the complexity of the circuits (the number of comparators C(N)) grows rapidly (see Fig. 3.8). We compared the *even-odd merge* and *bitonic merge* sorting networks (which are among the fastest known) with the *even-odd transition* network [22], which is often characterized as significantly slower and more resource consuming. However it is one of the most regular networks and can be implemented very efficiently in FPGA

Figure 3.13 depicts the *even-odd transition* network for sorting the same data that are shown in Fig. 3.5a.

The network in Fig. 3.13 contains C(N) = N × (N-1)/2 comparators/swappers and the maximum depth D(N) of the network for sorting N data items is N [22]. For example, for N = 8 C(N) = 28 and D(N) = 8. Note, that for the alternative *even-odd merge* circuit (see Sect. 3.4.1) C(N) = 19 and D(N) = 6. However, for the circuit in Fig. 3.13 sub-circuits composed of two lines (even and odd) of comparators are exactly the same and can be reused iteratively in such a way that is shown in Fig. 3.14. This permits the number of comparators/swappers to be reduced by a factor of N/2 (i.e. now C(N) = 7) and the fully combinational circuit becomes sequential with two reusable lines executing highly parallel operations. Thus, N/2 iterations are required to sort data with N items but the delay of the two-line sub-circuit is smaller than the total delay (see Fig. 3.13) and, thus, clock frequency for executing iterations is high.

The circuit in Fig. 3.14 is very regular, easily scalable, and does not require any additional components when input data are written to and sorted output data

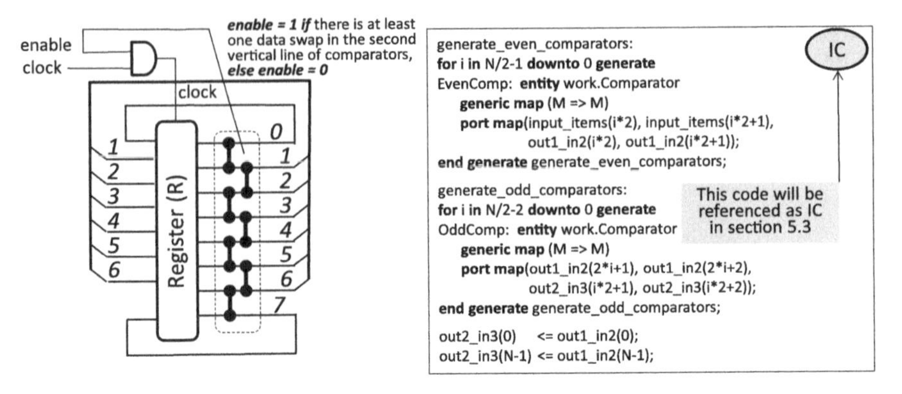

Fig. 3.14 The even-odd transition sorting network with reusable even and odd lines (VHDL code IC on the right-hand side will be referenced later on)

are read from the *register R* sequentially, applying a shift operation. To transfer data to the *register R* in parallel would require N multiplexers at the register inputs to receive data from outside before processing and from the comparators during processing. Besides, the number of clock cycles (N/2) in the *even-odd transition* network in Fig. 3.14 can be less than N/2. Let us introduce an *enable* signal which is zero at any second vertical line of the circuit in Fig. 3.14 when there is no exchange of data. As soon as *enable* = *0*, all data have been sorted. Suppose we need to sort data that are occasionally received in the sorted order, let us say: 8,7,6,5,4,3,2,1. The sequential circuit (Fig. 3.14) concludes that the data have already been sorted in time $2 \times t$, where t is the delay of a vertical line in Fig. 3.14 (i.e. the delay of one comparator/swapper). The combinational circuits in Figs. 3.5 and 3.13 still need time $D(N) \times t$ because they are hardwired. So, the simple circuit in Fig. 3.14 permits the number of steps to be reduced, which cannot be done for the circuits in Figs. 3.5 and 3.13. Chapter 5 will give an example.

A pipeline can be used for all the networks shown in Figs. 3.5 (see Fig. 3.15a), 3.13 (see Fig. 3.15b) and 3.14 (see Fig. 3.15c). In case of pipelining, the resources required are almost the same because FPGA slice flip-flops can be used without the need for additional components. The positions of pipeline registers (PLR) are shown in Fig. 3.15a–c. Figure 3.15d depicts the sequence of vectors recorded in the register R and in the PLR for the worst and the best cases in the network in Fig. 3.15c. The latter involves the *enable* signal (see Fig. 3.14) and a simple fragment of a finite state machine shown in Fig. 3.15e that tests this signal. VHDL codes for complete examples with the *enable* signal are given in Sect. 5.3.1. Clock frequency can be increased for all the circuits in Fig. 3.15a–c. Once again, the circuit in Fig. 3.15c is the least resource consuming.

At first glance, the *even-odd merge* networks seem to be faster than the circuit in Fig. 3.14. Besides, pipelining permits even better results to be obtained for these networks. However, in practice, even if the *even-odd/bitonic mergers* are faster, we cannot take advantage of such high speeds. The reasons for this conclusion are the following. Even simple experiments show that the routing overhead for the circuit

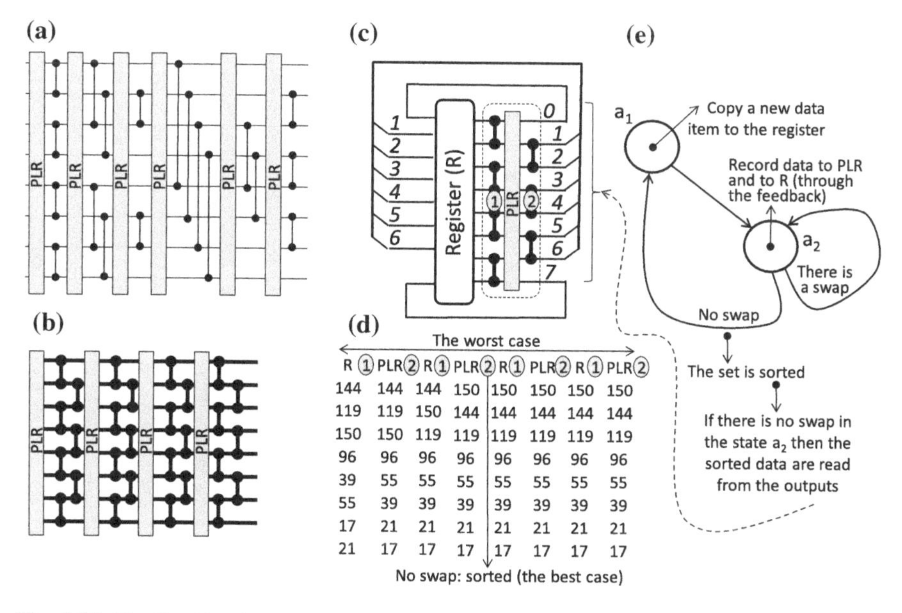

Fig. 3.15 Pipelined implementations: for even-odd merge sorter in Fig. 3.5 (**a**), even-odd transition sorter in Fig. 3.13 (**b**), the circuit in Fig. 3.14 (**c**), the sequence of vectors recorded in the register R and in the PLR (**d**), a fragment of the state transition diagram to control the circuit in Fig. 3.14 (**e**)

in Fig. 3.14 is lower. Very high throughputs cannot be achieved in practical applications because of communication overheads. Indeed, initial data need to be supplied to the sorter, the results have to be taken from the sorter, and the speed of communications is a bottleneck. The latter is more critical for networks that process small sets of data and, thus, more frequent data exchange is involved (since the number of transmitted data is actually very small it is difficult to apply full burst mode capabilities). Intensive communications between the processing system and the programmable logic in hybrid sorters that are implemented partially in software and partially in hardware do not allow the desired performance to be achieved because the processing system is frequently interrupted for necessary data exchange.

The following example gives the complete VHDL code enabling the circuit in Fig. 3.14 to be verified in the Atlys board for N = 16 and M = 4. The functionality can be tested in a host PC (in a virtual window). The details are given in Sects. 1.7 and 2.6. Additional examples with Nexys-4 board will be given in Sects. 4.1 and 4.4.2.

```
entity EvenOddTransitionIterative is              -- this code is for the Atlys board
generic (        M                                : integer := 4;
                 N                                : integer := 16 );
 port (clk                                        : in std_logic;
       BTNC, BTNU, BTND, BTNL, BTNR               : in std_logic;
       Sw                                         : in std_logic_vector(7 downto 0);
       EppAstb : in std_logic;         -- for the component IOExpansion from Digilent
```

```
        EppDstb : in std_logic;
        EppWr   : in std_logic;
        EppDB   : inout std_logic_vector(7 downto 0);
        EppWait : out std_logic);
end EvenOddTransitionIterative;
architecture Behavioral of EvenOddTransitionIterative is
  signal MyLed                  : std_logic_vector(7 downto 0);
  signal MyLBar                 : std_logic_vector(23 downto 0);
  signal MySw                   : std_logic_vector(15 downto 0);
  signal MyBtn                  : std_logic_vector(15 downto 0);
  signal data_to_PC             : std_logic_vector(31 downto 0);
  signal data_from_PC           : std_logic_vector(31 downto 0);
  signal unsortedSwBtn          : std_logic_vector(31 downto 0);
  type set_of_16items is array (N-1 downto 0) of std_logic_vector (M-1 downto 0);
  signal input_items            : set_of_16items;
  signal sorted                 : set_of_16items;
  signal out1_in2, out2_in3     : set_of_16items;
begin
  -- 32-bit signal unsortedSwBtn contains values from virtual (MySw, MyBtn) and onboard (Sw, BTN) components
  unsortedSwBtn <= MySw & Sw & BTNU & BTND & BTNL & BTNR &
                   MyBtn(3 downto 0);
  MyLBar        <= MySw & MyBtn(15 downto 8); -- these two lines are for tests only and can
  MyLed         <= MyBtn(7 downto 0);           -- be removed

process(sorted, BTNC) -- displaying the results of sorting in virtual window (signal data_to_PC)
begin
  if BTNC = '0' then       -- onboard button BTNC enables different 32-bit data (8 items) to be sent to PC
    for i in N/2-1 downto 0 loop
      data_to_PC((i+1)*M-1 downto i*M) <= sorted(i);
    end loop;
  else
    for i in N/2-1 downto 0 loop
      data_to_PC((i+1)*M-1 downto i*M) <= sorted(i+8);
    end loop;
  end if;
end process;
process(clk) -- control of iterations in the network in Fig. 3.14 without the enable signal
  variable index : integer range 0 to N := 0;
begin -- the signal input_items is used instead of the register in Fig. 3.14
  if rising_edge(clk) then
    if (index < N) then    index := index+1;
                           input_items <= out2_in3;
    else index := 0; sorted <= out2_in3;
      for i in N/2-1 downto 0 loop  -- input_items keeps 16 4-bit unsorted items
        input_items(i) <= data_from_PC((i+1)*M-1 downto i*M);
        input_items(i+N/2) <= unsortedSwBtn((i+1)*M-1 downto i*M);
      end loop;
    end if;
  end if;
end process;
```

```
IO_interface : entity work.IOExpansion      -- link with the IOExpansion component from Digilent
                 port map(EppAstb, EppDstb, EppWr, EppDB, EppWait, MyLed,
                          MyLBar, MySw, MyBtn, data_from_PC, data_to_PC);
-- even-odd transition sequential circuit shown in Fig. 3.14 (see also VHDL code IC on the right-hand side)

  generate_even_comparators:

  for i in N/2-1 downto 0 generate
    EvenComp    : entity work.Comparator
      generic map (M => M) -- the signal out1_in2 below provides connections between even and odd lines
        port map(input_items(i*2), input_items(i*2+1), out1_in2(i*2), out1_in2(i*2+1));
  end generate generate_even_comparators;

  generate_odd_comparators:

  for i in N/2-2 downto 0 generate
    OddComp     : entity work.Comparator
        generic map (M => M) -- the signal out2_in3 below provides connections with the register
        port map( out1_in2(2*i+1), out1_in2(2*i+2), out2_in3(i*2+1), out2_in3(i*2+2));
  end generate generate_odd_comparators;

  out2_in3(0)     <= out1_in2(0);     -- signals from the even line (because there are
  out2_in3(N-1)   <= out1_in2(N-1);   -- no passes through the odd line)

  end Behavioral;
```

The comparator is described as follows:

```
entity Comparator is
      generic (M : integer := 4);
      port(    Op1, Op2          : in std_logic_vector(M-1 downto 0);
               MaxValue          : out std_logic_vector(M-1 downto 0);
               MinValue          : out std_logic_vector(M-1 downto 0));
end Comparator;

architecture Behavioral of Comparator is
begin
process(Op1,Op2)
begin
  if Op1 >= Op2 then      MaxValue <= Op1; MinValue <= Op2;
  else                    MaxValue <= Op2; MinValue <= Op1;
  end if;
end process;
end Behavioral;
```

The synthesized circuit occupies 132 FPGA slices (from 6822 available slices) and the equivalent *even-odd merge* network requires 196 slices. The results of synthesis and implementation of circuits for M = 32 show that the *even-odd merge* network can be realized in the considered FPGA for only up to N = 32 (due to not sufficient resources) while the circuit in Fig. 3.14 can be customized and implemented for significantly greater number of N. Note that the generic values in the

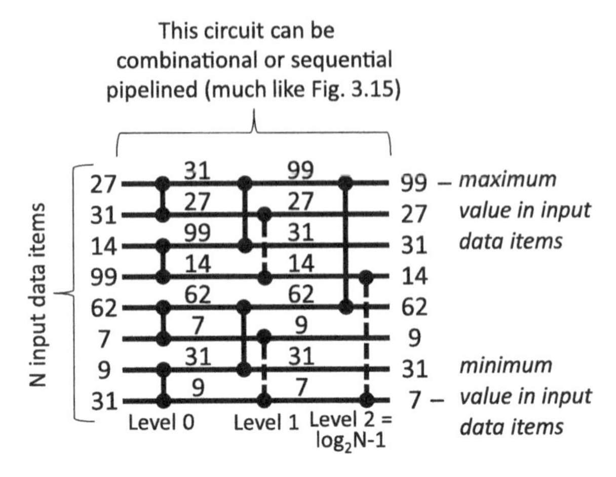

Fig. 3.16 A network for discovering the minimum and maximum values for N = 8 (scalable for any N)

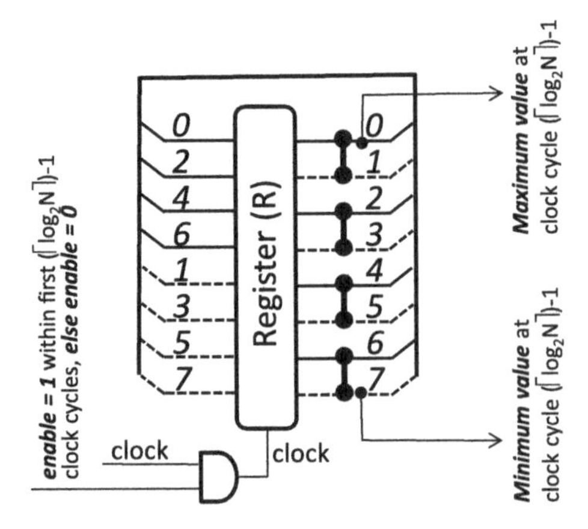

Fig. 3.17 A circuit for discovering the minimum and maximum values

entity EvenOddTransitionIterative are dependent on the available virtual and onboard peripheral devices and generally cannot be changed. However, the even-odd transition iterative network is customizable, i.e. it can be used with different values N and M as will be shown in Sect. 4.4.

3.6 Design Examples for Parallel Search

Let us look at Fig. 3.16 where the network finds the items with the maximum and minimum values for N = 8 [3, 23].

Much like the network in Fig. 3.13, the circuit in Fig. 3.16 can be implemented either combinationally or sequentially in a way shown in Fig. 3.17 [3, 23]. In the last case the hardware resources are obviously decreased. Indeed, the circuit in Fig. 3.16 requires $N + \sum_{n=1}^{(\log_2 N)-2} 2^n$ comparators/swappers whereas the circuit in

Fig. 3.17—N/2 comparators/swappers. The implementation in Fig. 3.17 is very regular, easily scalable for any N and does not involve complex interconnections. The minimum and maximum values can be found in T_f clock cycles and $T_f = \lceil \log_2 N) \rceil - 1$. Indeed, at the last iteration (T_f) the results are already on the outputs of the comparators/swappers.

It is shown in [23] that a slightly modified circuit in Fig. 3.17 can search for the maximum and minimum values in very large sets (that exceed millions of items). Besides, such circuits can also be helpful for certain types of sorting that are also discussed in [23].

VHDL code below describes the fully combinational circuit in Fig. 3.16 that enables only the maximum value to be found (generic parameters M, L, and N have default values 4, 4, and 16, respectively: L is the number of levels shown in Fig. 3.16 and they are 0, 1, 2, and 3 for the code below):

```
-- the same ports as for the entity EvenOddTransitionIterative in the example above without clk and BTNC signals
architecture Behavioral of MaxCombinational is -- the name of the entity now is MaxCombinational
  -- the same first 7 lines as in the architecture above (for the entity EvenOddTransitionIterative)
  type set_of_16items is array (N-1 downto 0) of std_logic_vector (M-1 downto 0);
  type set_of_levels is array (0 to L) of set_of_16items;
  signal to_level, from_level : set_of_levels;  -- input/output signals for each level in Fig. 3.16
begin                                            -- this code is for the Atlys board

  -- concurrent assignments for unsortedSwBtn, MyLBar and MyLed are the same as in the architecture above

  data_to_PC <= (31 downto 4 => '0') & to_level(L)(0);

process(data_from_PC, unsortedSwBtn)
begin -- preparing input data for the circuits in Fig. 3.16
  for i in N/2-1 downto 0 loop
       to_level(0)(i) <= data_from_PC((i+1)*M-1 downto i*M);
       to_level(0)(i+N/2) <= unsortedSwBtn((i+1)*M-1 downto i*M);
  end loop;
end process;

-- declaration of the component IOExpansion is the same as in the architecture above
generate_comparators:  -- generation of the circuit in Fig. 3.16 to find out the maximum value
for j in 1 to L generate
  one_level:   -- the code below is fully parameterized and can be used for any values of N and L
  for i in N/2**j-1 downto 0 generate          -- for a given L, N= 2**L
    EvenComp  : entity work.Comparator -- the comparator is generic
                 generic map (M => M)
                 port map(to_level(j-1)(i*(2**j)), to_level(j-1)(i*(2**j)+2**(j-1)),
                          from_level(j-1)(i*(2**j)), from_level(j-1)(i*(2**j)+2**(j-1)));
  end generate one_level;
  to_level(j) <= from_level(j-1); -- connects outputs of a previous level with inputs of the next level
end generate generate_comparators;

end Behavioral;
```

Since $N = 2^L$ then in the code above the generic line for N can be removed and symbol N can be replaced with 2**L. Discovering the minimum value is done trivially. It is sufficient to swap two lines in the Comparator (see Sect. 3.5), i.e. supplying to the third port MinValue and to the fourth port MaxValue (instead of MaxValue and MinValue in Sect. 3.5).

The following VHDL code describes the circuit in Fig. 3.17 that enables both the maximum and minimum values to be found. Just two generic parameters M, L with default values 4, 4 are declared and N is replaced with 2**L.

```
-- the same ports as for the entity EvenOddTransitionIterative in the example above without the BTNC signal
architecture Behavioral of MaxMinIterative is -- the name of the entity now is MaxMinIterative
  -- the same first 7 lines as in the architecture above (for the entity EvenOddTransitionIterative)
  type set_of_16items is array (2**L-1 downto 0) of std_logic_vector (M-1 downto 0);
  signal MyRegister, from_comparators       : set_of_16items;
  signal ResultMax, ResultMin               : std_logic_vector(M-1 downto 0);
begin
  -- concurrent assignments for unsortedSwBtn, MyLBar and MyLed are the same as in section 3.5
  data_to_PC <= (31 downto 8 => '0') & ResultMin & ResultMax;

process(clk)
  variable iterations : integer range 0 to L-1 := 0;
begin
  if rising_edge(clk) then
    if iterations < L-1 then
       MyRegister <= from_comparators;
       iterations := iterations+1;
    else iterations := 0;    ResultMax <= from_comparators(0);
       ResultMin <= from_comparators(2**L-1);
       for i in 2**L/2-1 downto 0 loop
                 MyRegister(i) <= data_from_PC((i+1)*M-1 downto i*M);
                 MyRegister(i+2**L/2) <= unsortedSwBtn((i+1)*M-1 downto i*M);
       end loop;
    end if;
  end if;
end process;

-- declaration of the component IOExpansion is the same as in the architecture above (in section 3.5)
single_line:      -- generating a single line of comparators shown in Fig. 3.17
for i in 2**L/2-1 downto 0 generate -- the code is parameterized and can be used for any value of L
  Comp: entity work.Comparator -- the comparator is generic
       generic map (M => M)
       port map(MyRegister(i*2), MyRegister(i*2+1),
       from_comparators(i), from_comparators(i+2**L/2));
end generate single_line;

end Behavioral;
```

All the projects in Sects. 3.5, 3.6 and also in the subsequent sections can be tested in a virtual window (the details are given in Sects. 1.7 and 2.6).

Let us look again at Fig. 3.16. Assuming that indices of the first and of the last data items are $I_{first} = 0$ and $I_{last} = N\text{-}1$, the circuit in Fig. 3.16 might be used as follows [23]: (1) discovering the maximum and the minimum values; (2) incrementing I_{first} and decrementing I_{last} and repeating the steps (1), (2) while $I_{first} < I_{last}$. Clearly, such a way permits data items to be sorted as it is shown on a simple example in Fig. 3.18.

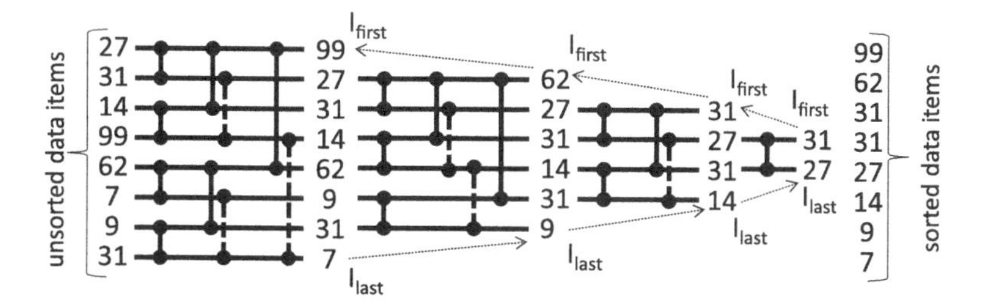

Fig. 3.18 Sorting in the network in Fig. 3.16

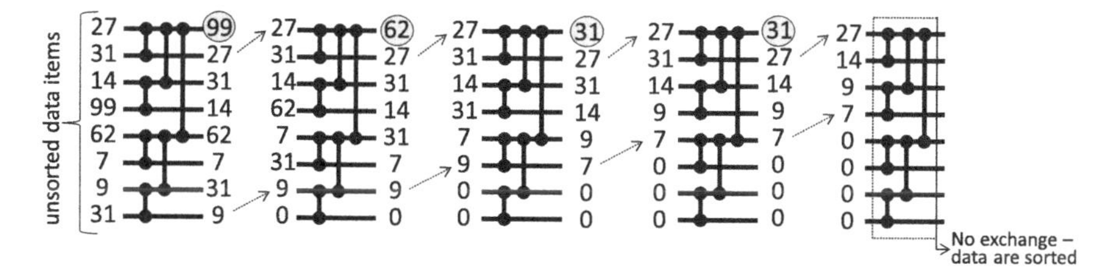

Fig. 3.19 Sorting with the aid of the network that enables the maximum value to be found

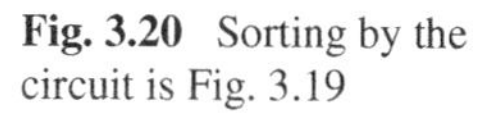

Fig. 3.20 Sorting by the circuit is Fig. 3.19

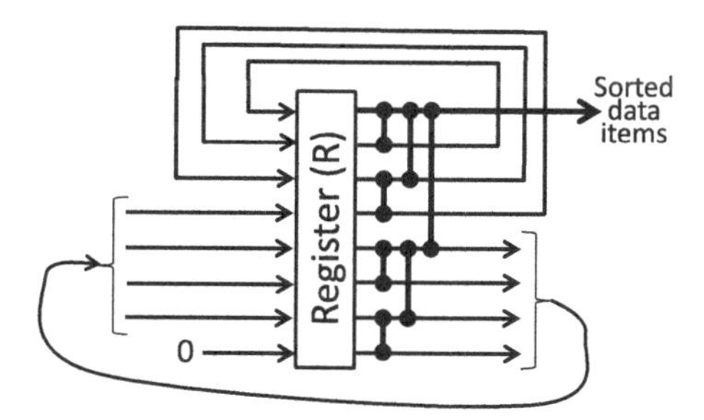

As you can see the network in Fig. 3.18 is slightly changed at each subsequent step. Figure 3.19 presents an example in which the same network for discovering the maximum value is entirely reused.

Each step (executing during one clock cycle) enables the maximum value in input data set to be found. After that data are shifted up and the next step is executed (see Fig. 3.20). The *Register R* is clocked and each new sorted data item is produced in each clock cycle. As soon as there is no data exchange in the comparators all the items are sorted. Hence, a new sorted data item is ready after one clock cycle and the depth of the network with N-1 comparators is only $\lceil \log_2 N) \rceil$ where N is the number of data items. Besides, sorting may be concluded earlier than after N clock cycles. For example, in Fig. 3.19 the results are ready after 4 clock cycles and N = 8. The experiments have demonstrated that if we need to sort all data items before outputting the result, then the previously described

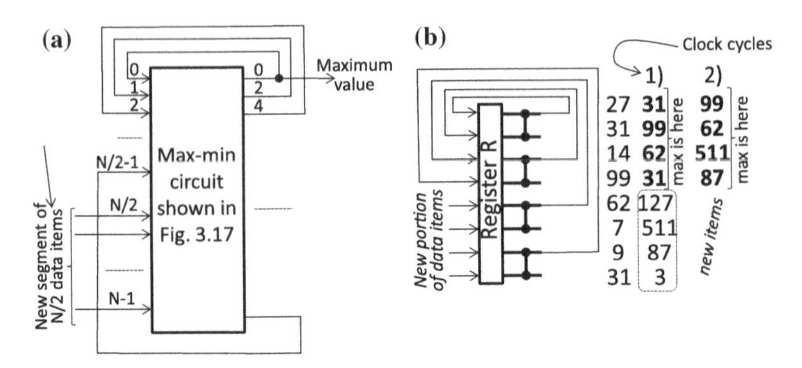

Fig. 3.21 Using the circuit in Fig. 3.17 for large scale data sets: discovering the maximum value (**a**), an example (**b**)

circuits (see Sect. 3.5) should be faster. However, if we need to output the sorted data as soon as possible, the considered here circuit is better and it is very fast. Clearly, the circuit from Fig. 3.17 may directly be used and it permits the number of comparators to be additionally reduced up to N/2. Throughput is also reduced and each new sorted item is ready after T_f clock cycles.

Let us return back again to the search problem. From Fig. 3.17 we can see that at each clock cycle N/2 data items, which include the maximum/minimum value, will be copied to the top/bottom segment of the *Register R*. Thus, the remaining (either bottom or top) part of the *Register R* can be reused to load a new portion of data. This technique enables the maximum/minimum values in very large data sets to be found even in low-cost FPGAs. Figure 3.21 gives necessary details. The circuit in Fig. 3.21a copies the even outputs of the network containing the maximum value (see Fig. 3.17) to the upper N/2 M-bit words of the *Register R*. The bottom N/2 M-bit words of the *Register R* cannot contain the maximum value and may be reused to load a new portion of data items (such as 127, 511, 87, and 3 shown in the example in Fig. 3.21b). Since a new portion can be loaded at each clock cycle, the maximum value for data sets containing Θ items can be discovered in $\tau = 2 \times (\Theta - N)/N + \lceil \log_2 N \rceil$ clock cycles. For instance, if $\Theta = 2^{20} = 1048\,576$, $N = 512$ then $\tau = 4103$. Such circuit is not resource consuming and can be implemented even in low-cost FPGAs with external memory supplying input data. In Sect. 4.5 we will describe All Programmable Systems-on-Chip. We believe that the circuit in Fig. 3.21 would allow applications implemented on the basis of APSoCs and requiring searching in large sets of data items to be accelerated significantly.

The circuit in Fig. 3.22a discovers the maximum and the minimum values in $\tau = 4 \times (\Theta - N)/N + \lceil \log_2 N \rceil$ clock cycles [23]. At the beginning, two cycles are needed to produce (in the *Register R*) the upper N/4 M-bit words with the maximum value and the bottom N/4 M-bit words with the minimum value. After that the middle N/2 M-bit words (of the *Register R*) can be reused to load a new

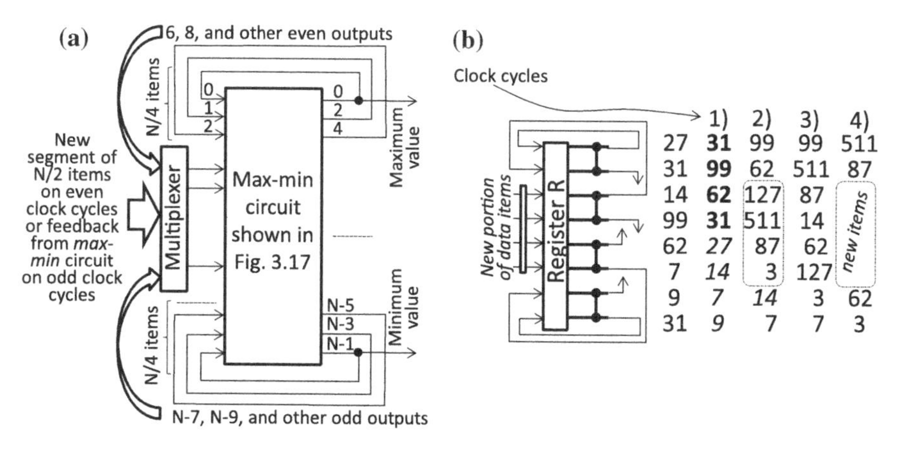

Fig. 3.22 Finding the maximum and the minimum values (**a**), an example (**b**)

portion of N/2 data items and once again the maximum and the minimum values will be transferred to the upper and to the bottom quarters of the *Register R* in 2 clock cycles. Thus, $2 \times (\Theta - N)/(N/2) = 4 \times (\Theta - N)/N$ cycles are required to process (to upload) all data and $\lceil \log_2 N \rceil$ cycles to propagate the last portion through the max–min circuit (see Fig. 3.17). If $\Theta = 2^{20} = 1\ 048\ 576$, $N = 512$ then $\tau = 8197$. Thus, the technique [23] enables large data sets to be handled.

3.7 Design Examples for Parallel Counters

Parallel computations frequently involve operations over elements of long binary and ternary vectors [3, 10, 24]. Examples include calculating the Hamming weight of a binary vector (i.e. the number of ones in the vector) [25, 26], comparing Hamming weights [25, 26], operations over ternary vectors in combinatorial search [24], and data processing [27]. In many practical applications, the execution time for operations over vectors has a significant impact on performance.

Let us consider address-based sorting [10, 27]. The basic idea is very simple. When a new data item is received, its value V is used as the memory address at which a flag (1) is recorded. We assume that all memory is zero initially and there are no duplicate input values. Once all the input data have been recorded in memory in the form of a long binary vector, the sorted sequence can be produced by sequentially reading the addresses of locations containing '1' flags. This process can be accelerated significantly if we know how many data items are recorded in each memory segment [10]. The sizes of segments can be from tens of bits to thousands of bits, or even more. Thus, we need to find a fast way to count the number of ones in a long binary vector (i.e. its Hamming weight). There

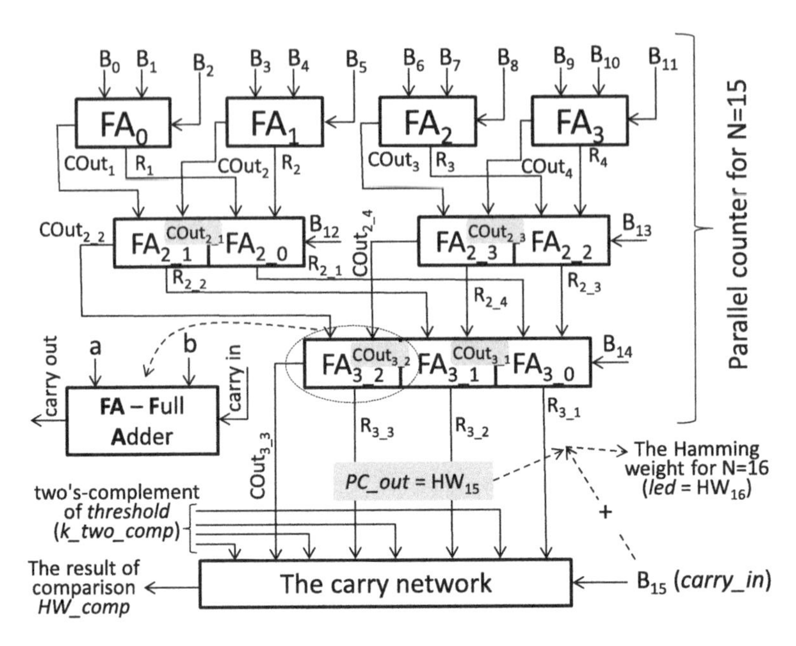

Fig. 3.23 Combinational parallel counter for N = 15 and Hamming weight comparator with 4-bit fixed threshold κ (from [25])

are several ways to solve this task. The simplest relies on sequential counting (see an example in Fig. 3.4) and is time consuming. Non-sequential circuits are frequently constructed as parallel counters [25], which are circuits based on a tree of full adders (FA). Figure 3.23 depicts a fixed-threshold Hamming weight comparator from [25], which uses a parallel counter for $N = 15$ (HW_{15}) and a carry network (CN) circuit. The result of comparison is obtained as HW_{15}-κ or the same as HW_{15} plus the 2's-complement of the threshold κ.

The circuit in Fig. 3.23 is scalable for any value of N. Formulae that determine the number of elements C(N) in Fig. 3.23 and the throughput D(N) are given in [25]: $C(N) = (N - \log_2 N - 1) \times \gamma_{FA} + \log_2 N$; $D(N) = (\log_2 N - 1) \times (\delta_{sum} + \delta_{carry}) + 1$ where γ_{FA} is the cost of a full adder (FA) relative to a gate (in [25] $\gamma_{FA} = 9$ was chosen), and $\delta_{sum}/\delta_{carry}$ are the FA delay parameters (the delay of FA relative to a gate and in [25] the value $\delta_{sum} = \delta_{carry} = 2$ were taken). It is shown in [10] that the chosen values γ_{FA}, δ_{sum}, and δ_{carry} are not appropriate for FPGAs. We found [10] that Hamming weight comparators based on parallel counters [25] are almost always less resource consuming than the networks [26, 28, 29] and they can benefit from highly optimized components supporting arithmetic operations in general-purpose FPGA slices. Thus, they are also fast.

VHDL code below describes the parallel counter for N = 16 and Hamming weight comparator with fixed 4-bit threshold κ. The relevant circuit has been designed for the Nexys-4 prototyping board (Artix-7 FPGA) and occupies 9 slices with the maximum combinational path delay 5.1 ns.

```
entity ParallelCounterComparator is
 port (        sw       : in std_logic_vector(15 downto 0);
               led      : out std_logic_vector(4 downto 0);
               ledC     : out std_logic);
end ParallelCounterComparator;

architecture Behavioral of ParallelCounterComparator is
  signal R1, R2, R3, R4, R2_1, R2_2, R2_3, R2_4, R3_1, R3_2, R3_3     : std_logic;
  signal COut1, COut2, COut3, COut4                        : std_logic;
  signal COut2_1, COut2_2, COut2_3, COut2_4                : std_logic;
  signal COut3_1, COut3_2, COut3_3                         : std_logic;
  signal B             : std_logic_vector(15 downto 0); -- represents 16-bit input vector
  signal PC_out        : std_logic_vector(3 downto 0);  -- represents 4-bit output for HW15
  signal threshold     : std_logic_vector(3 downto 0);  -- fixed threshold
  signal k_two_comp    : std_logic_vector(3 downto 0);  -- 2's-complement of the threshold
  signal HW_comp       : std_logic;                     -- the result of the comparison
begin
  B <= sw;      -- input data are taken from 16 onboard (Nexys-4) switches
  threshold <= (1 => '1', 3 => '1', others => '0'); -- threshold that is 10 is chosen as an example
  k_two_comp <= (not threshold) + 1;          -- 2's-complement of the threshold (of the value 10)

  -- structural code below allows direct mapping for the circuit in Fig. 3.23
  FA0 : entity work.FullAdder       port map(B(0), B(1), B(2), R1, COut1);
  FA1 : entity work.FullAdder       port map(B(3), B(4), B(5), R2, COut2);
  FA2 : entity work.FullAdder       port map(B(6), B(7), B(8), R3, COut3);
  FA3 : entity work.FullAdder       port map(B(9), B(10), B(11), R4, COut4);
  FA2_0:  entity work.FullAdder     port map(R1, R2, B(12), R2_1, COut2_1);
  FA2_1:  entity work.FullAdder
       port map(COut1, COut2, COut2_1, R2_2, COut2_2);
  FA2_2:  entity work.FullAdder     port map(R3, R4, B(13), R2_3, COut2_3);
  FA2_3:  entity work.FullAdder
       port map(COut3, COut4, COut2_3, R2_4, COut2_4);
  FA3_0:  entity work.FullAdder port map(R2_1, R2_3, B(14), R3_1, COut3_1);
  FA3_1:  entity work.FullAdder port map(R2_2, R2_4, COut3_1, R3_2, COut3_2);
  FA3_2:  entity work.FullAdder
       port map(COut2_2, COut2_4, COut3_2, R3_3, COut3_3);

  led <= PC_out + ("0000" & B(15));
  PC_out <= COut3_3 & R3_3 & R3_2 & R3_1;

  CN:   entity work.carry_network
       port map (PC_out, B(15), k_two_comp, HW_comp);
  ledC <= HW_comp;                -- the result of the comparison
end Behavioral;
```

Full adder is described as follows:

```
entity FullAdder is
port( A          : in std_logic;
      B          : in std_logic;
      CarryIn    : in std_logic;
      Result     : out std_logic;
      CarryOut   : out std_logic);
end FullAdder;
```

```
architecture Behavioral of FullAdder is
begin
CarryOut        <= (A and B) or (A and CarryIn) or (B and CarryIn);
Result          <= A xor B xor CarryIn;
end Behavioral;
```

VHDL code below describes the carry network from [25].

```
entity carry_network is   -- entity for 4-bit carry network from [25]
port ( PC_out           : in std_logic_vector(3 downto 0); -- see names in Fig. 3.23
       carry_in         : in std_logic;
       threshold        : in std_logic_vector(3 downto 0);  -- two's complement of threshold
       HW_comp          : out std_logic);                   -- the result of the comparison
end carry_network;

architecture Behavioral of carry_network is
  signal HW    : std_logic_vector(3 downto 0);
begin
  first_element:  entity work. CN_element
      port map(PC_out(0), threshold(0), carry_in, HW(0));

  generate_CN: for i in 1 to 3 generate
    CN_element:  entity work.CN_element
      port map(PC_out(i), threshold(i), HW(i-1), HW(i));
  end generate generate_CN;

    HW_comp <= HW(3);

  end Behavioral;
entity CN_element is    -- entity for elements of the carry network from [25]
  port ( BitFromPC              : in  std_logic;
         BitFromThreshold       : in  std_logic;
         CarryIn                : in  std_logic;
         CarryOut               : out  std_logic);
end CN_element;

architecture Behavioral of CN_element is
  signal and_out                : std_logic;
  signal or_out                 : std_logic;
  signal second_and_out: std_logic;
begin
  and_out <= BitFromThreshold and BitFromPC; -- exact mapping of the circuit from Fig. 4 in [25]
  or_out <= BitFromThreshold or BitFromPC;
  second_and_out <= or_out and  CarryIn;
  CarryOut <= second_and_out or and_out;
end Behavioral;
```

It is easy to check that the circuit in Fig. 3.23 is indeed very efficient but there are also other alternatives that permit even faster and less resource consuming solutions to be developed (see Sects. 3.8, 3.9 and 4.2).

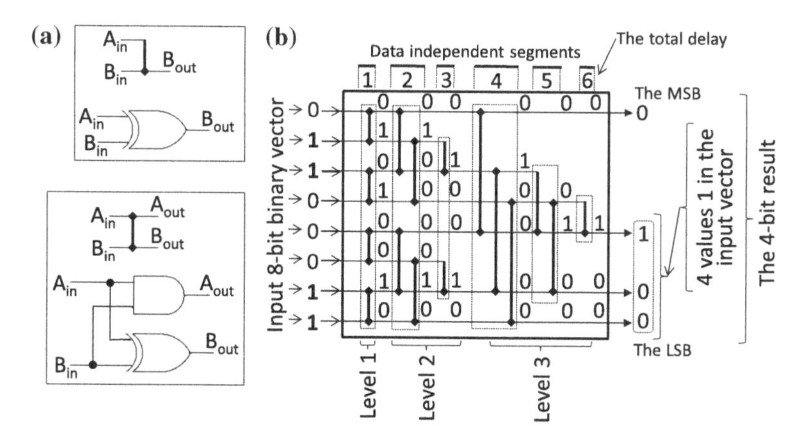

Fig. 3.24 Components of a counting network (**a**), and an example for calculating the Hamming weight of an 8-bit binary vector (**b**)

3.8 Design Examples for Counting Networks

Unlike the circuits described in Sects. 3.5, 3.6, counting networks [10] do not contain *conventional comparators*. Instead, each basic component is either a *half-adder* or an *XOR gate* (see Fig. 3.24a). To distinguish such components from conventional comparators in Fig. 3.5b, we will use *rhombs* instead of circles (see Fig. 3.24a) and we will remove such rhombs at any connection with a horizontal line if this line ends, i.e. if it does not have further connections to the right (see the upper block in Fig. 3.24a). Figure 3.24b shows an example of a counting network for $N = 2^p = 8$ inputs where p is a non-negative integer and in the example p = 3.

The *data independent segments* of the network (see Fig. 3.24b) are composed of vertical lines that do not have any data dependency between the used components; thus all necessary operations can be executed concurrently and are characterized by a single one-component delay. Hence, the total delay of the circuit in Fig. 3.24b is equal to 6. MSB is the most significant bit and LSB is the least significant bit.

The *levels* of the network (each composed of *one or more segments*) in Fig. 3.24b calculate the Hamming weights of: 2-bit (*level 1—segment 1*); 4-bit (*level 2—segments 2–3*); 8-bit (*level 3—segments 4–6*) binary vectors. An example with the input data *01100011* is shown in Fig. 3.24b. Level 1 calculates the Hamming weights in four 2-bit input vectors: *01*, *10*, *00*, and *11*; level 2 calculates the Hamming weights in two 4-bit input vectors: *0110*, and *0011*; and, finally, level 3 calculates the Hamming weight in one 8-bit input vector: *01100011* in which there are four values 1. Thus, the final result is 0100_2 (the binary code of 4_{10}).

The circuit in Fig. 3.24b is very simple and fast. It is composed of just 16 trivial components that are shown in Fig. 3.24a, which are characterized by negligible delay. General rules for designing similar circuits for very large numbers of inputs

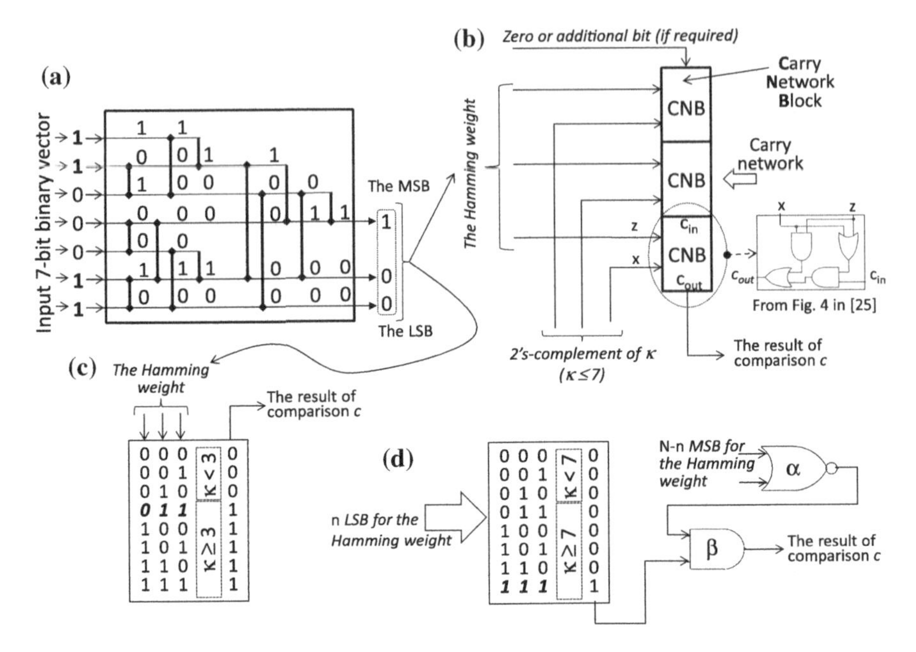

Fig. 3.25 An example of a counting network for $N \neq 2^p$ (**a**), a carry network composed of carry network blocks (CNB) from [25] (**b**), and LUT-based circuits that form the result of comparison (**c**, **d**)

N are given in [10] together with the proof that such circuits function as intended. In [10] it is shown, in particular, that counting networks for *any* N (i.e. where it is not necessary to satisfy the condition $N = 2^p$) can easily be constructed (see an example in Fig. 3.25a).

Much like the parallel counters from the previous section the counting network might be used in Hamming weight comparators (HWC) that take the result from the output of a network such as that in Fig. 3.25a and compare it with either a fixed threshold κ, or with the result of a similar to Fig. 3.25a circuit. Thus, the problem description is exactly the same as in [25]. We consider here two methods for the final comparison and these are outlined in Fig. 3.25b, c. The first method involves a carry network (CN) described in VHDL above (see the previous section and Fig. 3.25b where a CN is given for the example in Fig. 3.25a). The second method is the LUT-based circuit depicted in Fig. 3.25c for the example in Fig. 3.25a and $\kappa = 3$. Since a LUT(n,m) can implement any Boolean function of n variables, a similar circuit can easily be configured for any value of $\kappa < 2^n$. If N is greater (or even significantly greater) than n, the circuit can be built as shown in Fig. 3.25d. The NOR gate α tests if there are no '1' values in the MSBs of the Hamming weight that do not include n the LSB. The AND gate β forms the result of the comparison. The majority of currently available FPGAs contain LUTs(6,1). Thus, any value of $\kappa < 64$ can be chosen. If $\kappa \geq 64$ then the LUT can be replaced with either a set of LUTs or an embedded memory block ($9 \leq n \leq 15$ for the

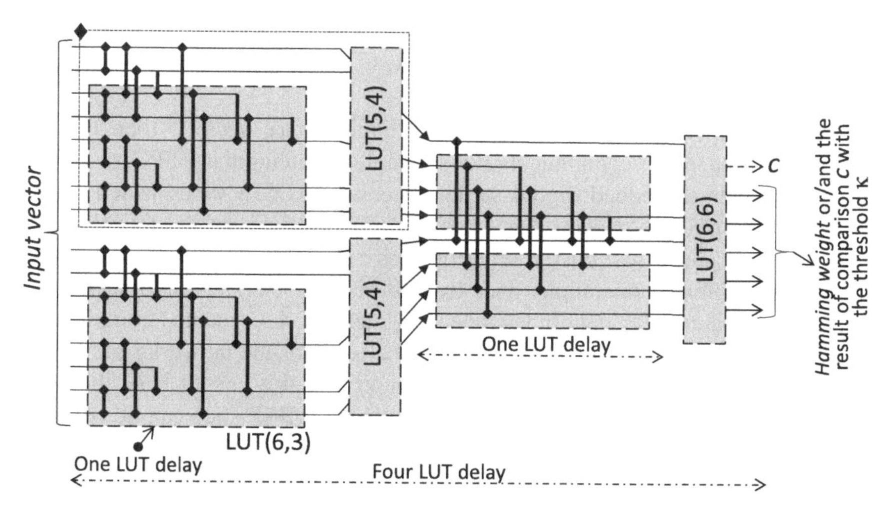

Fig. 3.26 Mapping the network fragments onto FPGA LUTs

majority of contemporary FPGAs). Since all memories (both distributed or LUT-based and embedded) are run-time configurable, the circuits in Fig. 3.25c, d are not threshold-dependent (i.e. they may be dynamically customized for any value of $\kappa < 2^n$).

Another important feature of LUT-based circuits in Fig. 3.25b, c is an opportunity to use more than one threshold. For instance, one threshold κ_l may be the lower bound and another threshold κ_u may be the upper bound. We will demonstrate how to describe such circuits in the next section.

It is shown in [10] that for Hamming weight counters/comparators based on sorting networks (such as [26]), the number of elements is significantly greater than the number of elements in the counting networks whereas both types of elements have practically equal complexity. This is because in contrast to sorting networks, the number of horizontal lines in counting networks is *incrementally reduced*. The number of *segments* is the same as the number of *segments* in the best sorting networks but due to the significantly reduced complexity, counting networks implemented on a microchip can be employed for much bigger values of N than sorting networks [26] built on the same microchip. In general, counting networks can be seen as a bridge between the circuits [26, 28, 29] and [25]. Indeed, on the one hand they look like sorting networks, and on the other hand they form a tree of levels (see Fig. 3.24b) much like parallel counters [25].

We found that if we rely on VHDL generate statements to construct a scalable counting network then the resources and delays might be worse than for the circuits from [25]. However, there is a better way. The network can be mapped to FPGA LUTs as it is shown in Fig. 3.26 for N = 16 [10].

The circuit is Fig. 3.26 was implemented and tested in the Nexys-4 prototyping board. It occupies 8 FPGA slices with the maximum combinational path delay

4.7 ns. So, it is faster and less resource consuming than the circuits from the previous section. However, there is an even better way described in the next section (see also Sect. 4.2).

If bitwise operations can be applied to large binary vectors, then the networks might be more economical. For example, contemporary FPGAs contain up to thousands embedded digital signal processing (DSP) slices which can be used in addition to the general-purpose FPGA slices (see Sect. 4.2 for additional details). For example, in Xilinx 7 series devices each slice [30] may be configured to implement bitwise operations over two 48-bit vectors. Thus, two such slices enable all operations for data independent segments to be executed combinationally for up to N = 96. This opportunity can be used for the networks and cannot be used for parallel counters. However, the latter can also benefit from the available DSP slices because the 48-bit operational unit of DSP slice can be split into smaller data segments (either *12* or *24 bit* each) where the internal carry propagation between segments is blocked to ensure independent operation for all the segments [30]. This feature is favorable for parallel counters.

Pipelining for networks (see Fig. 3.15) and parallel counters (examples are given in [10]) permits even faster solutions to be found. However, the delays in the described above circuits are so small that for the majority of practical applications an additional acceleration is not required and if nevertheless it is desirable then the results of [10] can be used.

3.9 Design Examples for LUT-Based Hamming Weight Counters/Comparators

Let us firstly implement such a simplest Hamming weight counter (SHWC) that can be optimally mapped onto FPGA LUTs and then let us take this SHWC as a base allowing Hamming weight comparators of any required complexity to be constructed. Besides, we will analyze building blocks in the known HWCs (e.g. FAs) in order to evaluate an opportunity to use some of them in the discussed solutions in case if it gives any benefit. Clearly, h LUTs(n,m) can trivially be configured to calculate the Hamming weight of $A = \{a_0,\ldots,a_{n-1}\}$, where $h = \lceil (\log_2(n+1))/m \rceil$. Thus, h LUTs may be chosen for the SHWC. The idea is to build a network from SHWCs that can compute the Hamming weight for an arbitrary vector of size N. Figure 3.27 presents a complete solution for n = 6 (i.e. for SHWC) and N = 36 as an example. All LUTs in any layer execute multiple logic operations in parallel. For example, all SHWCs of the first layer count the Hamming weights in the 6-bit vectors on their inputs concurrently. Similarly the LUTs(6,3) in the layer 2 output the results with just one LUT delay.

All LUT blocks are configured identically. Any block (i.e. SHWC) counts the Hamming weight of 6-bit vector on the input and is composed of $C_{SHWC} = \lceil (\log_2(n+1)) / m \rceil$ physical LUTs(n,m). The circuit in Fig. 3.27

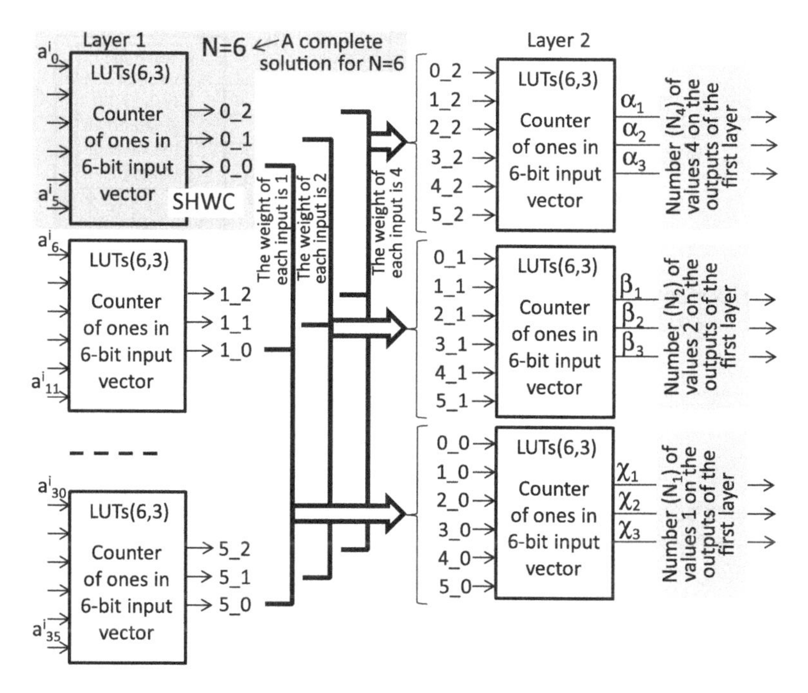

Fig. 3.27 A complete HWC for N = 36

contains $C_{SHWC} \times (\lceil N / n \rceil + \lceil (N/2)/n \rceil)$ LUTs(n, m). Even for m = 1 (the worst case when any physically implemented in FPGA LUT has just one output) and n = 6 we need only 27 LUTs for N = 36. This is negligible, because, for example, the FPGA xc7a100t (used for experiments) incorporates 15,850 slices and each slice contains 4 LUTs(6,1). However, we still need to implement the output block that forms the result of comparison. Figure 3.28 depicts a LUT-based solution.

There are two circuits in Fig. 3.28. The first one (Fig. 3.28a) takes output signals $\alpha_1\alpha_2\alpha_3\beta_1\beta_2\beta_3\chi_1\chi_2\chi_3$ from Fig. 3.27 and calculates the Hamming weight of the input vector $A = (a^i_0, \ldots, a^i_{35})$. The second circuit (Fig. 3.28b) compares the Hamming weight with a threshold κ preliminary uploaded to the LUTs (in Fig. 3.28b $\kappa = 15$: the output C is '0' when the Hamming weight of A is less than 15, otherwise C is '1'). If reconfigurable LUTs are not desirable then the outputs of the circuit in Fig. 3.28a are connected to the carry network [25]. Figure 3.28 includes INIT statements [31] to configure all the LUTs. The circuit in Fig. 3.28a is in fact a multi-bit adder (with 2-bit carry signals ρ_0 and ρ_1), which adds the following three vectors: (1) $\alpha_1\alpha_2\alpha_3$ shifted two bits left; (2) $\beta_1\beta_2\beta_3$ shifted one bit left; and (3) $\chi_1\chi_2$. This was done because the vector $\alpha_1\alpha_2\alpha_3$ contains the number N_4 of values 4, the vector $\beta_1\beta_2\beta_3$—the number N_2 of values 2, and the vector $\chi_1\chi_2\chi_3$—the number N_1 of values 1 (see Fig. 3.27). In the final Hamming weight the value $\alpha_1\alpha_2\alpha_3$ has to be multiplied by 4 (or shifted left by two bits) and the

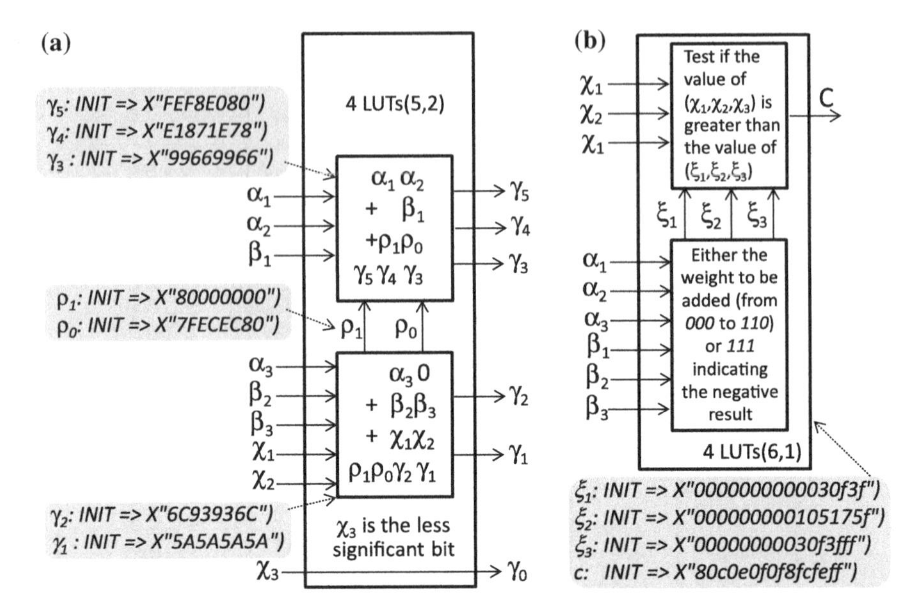

Fig. 3.28 Output circuit (see Fig. 3.27) to count the Hamming weight (**a**), and for HWC (**b**)

value $\beta_1\beta_2\beta_3$ has to be multiplied by 2 (or shifted left by one bit), which is done in the shift operations. Clearly, the value χ_3 can be taken directly.

For N < 36 some LUTs can be removed from the circuit in Fig. 3.27. For example, for N = 32 the left-hand bottom LUT can be taken off and the lines a_{30}, a_{31} will be connected directly to the LUTs of the second layer. This change reduces the number of LUTs but makes the circuit less regular. Since the number of occupied LUTs is indeed very small, the simplest way is to assign zeros to 4 unused inputs.

If for LUT(n,m) n < 6, the hierarchy is built similarly. Thus, if n = 5 we need 5 LUT groups at the layer 1 (each handling 5 signals), then 3 groups at the layer 2 (each also handling 5 signals). Finally, the same circuit as shown in Fig. 3.28 outputs the results.

A complete synthesizable VHDL specification of the circuits in Figs. 3.27 and 3.28 ready for immediate tests and assessments in many available FPGA-based prototyping boards is given in Appendix B.

For larger values N the general structure of the HWC is the same as in Fig. 3.27. An example for N = 216 is given in [3].

Our preliminary analysis has shown that the described above LUT-based circuits, parallel counters [25] and counting networks [10] are the best in resources and performance. Let us attempt to discover even better solutions combining capabilities of different designs. The circuit in Fig. 3.29 realizes such a combination for typical physical LUTs(6,1)/LUTs(5,2) from the libraries [31] and N = 18. Any LUT on the left hand side is an SHWC. Later on we will show how to build similar circuits for $N = 2^g$ (g = 5,6,…).

The full adders FA1, FA2, and FA3 in Fig. 3.29 calculate in the Hamming weight the number of values one (N_1), the number of values 2 (N_2) and the

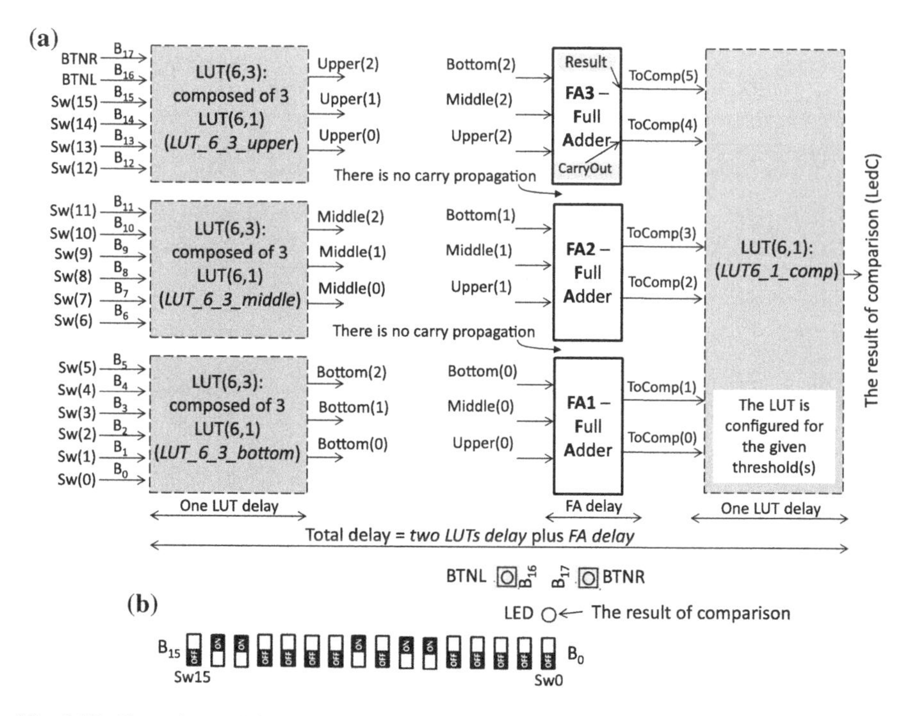

Fig. 3.29 Hamming weight comparator based on LUTs and full adders (**a**), verification of the circuit (**b**)

number of values 4 (N_4), accordingly. Let us consider an example of an input vector B = "101101_011111_110001" in which 6-bit sub-vectors are separated for better visibility. The LUTs on the left hand side output the following 3-bit vectors "100" (because there are 4 values '1' in the sub-vector "101101"), "101" (because there are 5 values '1' in the sub-vector "011111"), and "011" (because there are 3 values '1' in the sub-vector "110001"). FA1, FA2, and FA3 calculate a sum of the least significant, the middle and the most significant bits in the produced 3-bit vectors and, thus, N_1 = '0' + '1' + '1' = "10", N_2 = '0' + '0' + '1' = "01", and N_4 = '1' + '1' + '0' = "10". The final result is calculated as "10" (i.e. the vector "10" – N_1) + "10" (i.e. the vector "01" (N_2) shifted left by 1) + "1000" (i.e. the vector "10" (N_4) shifted left by 2) = "1100" giving the Hamming weight 1100_2 or 12_{10} for the input vector B. The right-hand LUT(6,1) compares inputs with preconfigured values and outputs the result of comparison. The basic idea is similar to the described above (see Fig. 3.27) but we benefit from two new features: (1) the circuits can now efficiently be scaled by 3, i.e. not by 6 as in the previous section or by 2 in [25]) and (2) the HWCs may take advantage of highly optimized arithmetic circuits (such as FA) normally offered in commercially available FPGAs. Another important characteristic is a nonexistence of carry propagation signals between the FAs (see Fig. 3.29). The total delay in the circuit is composed of only two LUTs delay and an FA delay.

The circuit in Fig. 3.29a possesses a distinctive feature. It permits more than one threshold (bound) to be used. Some examples are given in Fig. 3.30.

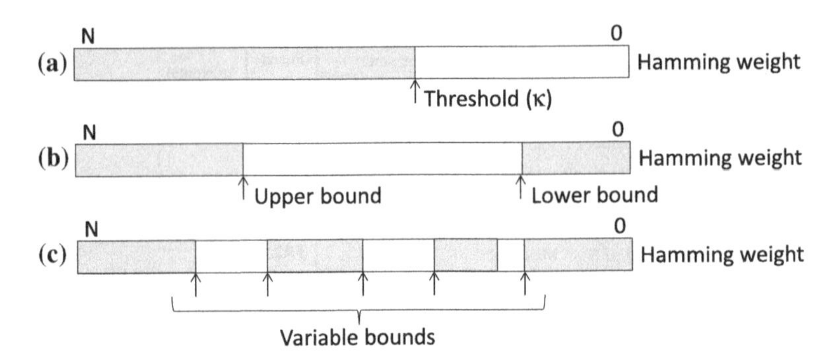

Fig. 3.30 Hamming weight comparator with a fixed threshold (**a**), lower/upper bounds (**b**), and variable bounds (**c**)

Figure 3.30a demonstrates the most frequently considered comparator with one threshold. However, we can also indicate a lower and upper bounds (see Fig. 3.30b) and even variable bounds (see Fig. 3.30c). Such feature cannot be provided for HWC [25] based on parallel counters.

VHDL code below is a complete synthesizable specification of the circuit in Fig. 3.29a.

```
entity HammingWeightComparator is              -- The code below has been tested for Nexys-4
 port ( Data_in : in std_logic_vector (17 downto 0);  -- the vector B0,...,B17 in Fig. 3.29a
        LedC    : out std_logic);                     -- the result of comparison
end HammingWeightComparator;
-- names of the signals are the same as in Fig. 3.29

architecture Behavioral of HammingWeightComparator is
signal Upper, Middle, Bottom       : std_logic_vector(2 downto 0);
signal ToComp                      : std_logic_vector(5 downto 0);
begin
  LUT_6_3_upper:  entity work.LUT_6to3
      port map(Data_in(17 downto 12), Upper);

  LUT_6_3_middle:  entity work.LUT_6to3
      port map(Data_in(11 downto 6), Middle);

  LUT_6_3_bottom:  entity work.LUT_6to3
      port map(Data_in(5 downto 0), Bottom);

  LUT6_1_comp:entity work.LUT6_1
      port map (ToComp, LedC);

  FA_generate: for i in 0 to 2 generate
    FA: entity work.FullAdder
      port map(Bottom(i), Middle(i), Upper(i), ToComp(2*i), ToComp(2*i+1));
  end generate FA_generate;

end Behavioral;
```

There are two new components (LUT_6to3 and LUT6_1) and they are described as follows:

```
library IEEE;                      -- all necessary libraries are explicitly shown here because this
use IEEE.STD_LOGIC_1164.all;       -- code has to be directly copied to an example in Appendix B
library UNISIM;                    -- (HammingWeightCounter36bits for N=36)
use UNISIM.vcomponents.all;

entity LUT_6to3 is          -- Xilinx library UNISIM [31] for LUT primitives has to be included
   port ( Data_in           : in  std_logic_vector (5 downto 0);     -- 6-bit input vector
          HW                : out  std_logic_vector (2 downto 0));   -- the Hamming weight
end LUT_6to3;

architecture Behavioral of LUT_6to3 is
begin -- for non-Xilinx FPGAs, constants can be used instead of the LUT primitives (see Sect. 2.8)

LUT6_inst1 : LUT6
  generic map (INIT => X"fee8e880e8808000")          -- LUT Contents
  port map (HW(2), Data_in(0), Data_in(1), Data_in(2), Data_in(3),
       Data_in(4), Data_in(5));

LUT6_inst2 : LUT6
  generic map (INIT => X"8117177e177e7ee8")          -- LUT Contents
  port map (HW(1), Data_in(0), Data_in(1), Data_in(2), Data_in(3),
       Data_in(4), Data_in(5));

LUT6_inst3 : LUT6
  generic map (INIT => X"6996966996696996")          -- LUT Contents
  port map (HW(0), Data_in(0), Data_in(1), Data_in(2), Data_in(3),
       Data_in(4), Data_in(5));

end Behavioral;
```

This is the code for the LUT6_1 component:

```
entity LUT6_1 is                      -- Xilinx library UNISIM [31] for LUT primitives has to be included
  port ( Data_in  : in  std_logic_vector (5 downto 0);  -- 6-bit input vector
         Comp     : out std_logic);                      -- Comp is the result of comparison
end LUT6_1;

architecture Behavioral of LUT6_1 is
begin
  LUT6_inst0 : LUT6  -- this LUT is used just for the final comparator (see the right-hand LUT in Fig. 3.29a)
  -- LUT Contents for the upper bound 10 and for the lower bound 4: (0-3: 1; 4-10: 0; 10-17: 1)
  generic map (INIT => X"fffffffcfc00003f") -- configuring such LUTs is explained in Appendix B
  port map (Comp, Data_in(0), Data_in(1), Data_in(2), Data_in(3),
       Data_in(4), Data_in(5));
end Behavioral;
```

The circuit synthesized from the code above for Artix-7 FPGA (Nexys-4) has the maximum combinational path delay equal to 2.5 ns and occupies 6 logical slices.

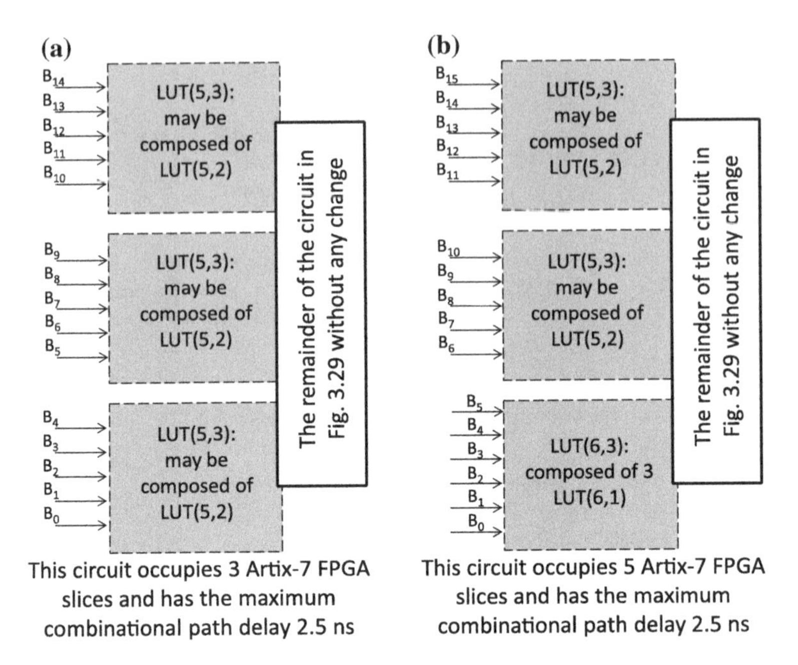

Fig. 3.31 Hamming weight comparators for N = 15 (**a**), and N = 16 (**b**)

The HWC in Fig. 3.29 is easily modifiable for other close values of N. Figure 3.31a gives an example for N = 15 and Fig. 3.31b—for N = 16. The circuit in Fig. 3.31a occupies just 3 Artix-7 FPGA slices and the complete synthesizable VHDL description for such circuit is given in Appendix B. For $6 < N<15$ some inputs can be connected to zero. Clearly for $N \leq 6$ just the SHWC (see Fig. 3.27) is sufficient.

The HWCs for larger number N of bits can be created from the HWCs shown in Figs. 3.29 and 3.31 interconnected with some others LUT-based circuits (such as SHWC) and eventually involving additional FAs. For example, Fig. 3.32 gives a solution for N = 32 based on the circuit in Fig. 3.31a. The blocks A and B are the circuits shown in Fig. 3.31a in which the right hand LUT(6,1) depicted in Fig. 3.29 is replaced with a LUT(6,4) that outputs 4-bit Hamming weight of 15-bit input vectors. The block C produces the sum of two most significant bits (MSB) of the blocks A and B and the block D produces the sum of two least significant bits (LSB) of the blocks A and B. Since the maximum value on the outputs of the block D is 6 (i.e. $11_2 + 11_2 = 110_2 = 6_{10}$) we can add one more input for the bit B_{30}. Thus, now the maximum value becomes 7 (i.e. $110_2 + 1_2 = 7_{10}$) and the number of outputs is the same (i.e. 3).

The outputs of the block C represent the number of values 4 in the Hamming weight (additional details and explanations are given in Appendix B). Finally, the block E outputs the result of comparison. Let us look at the example shown in Fig. 3.32. Suppose the blocks A and B output the values 1101 and 0110

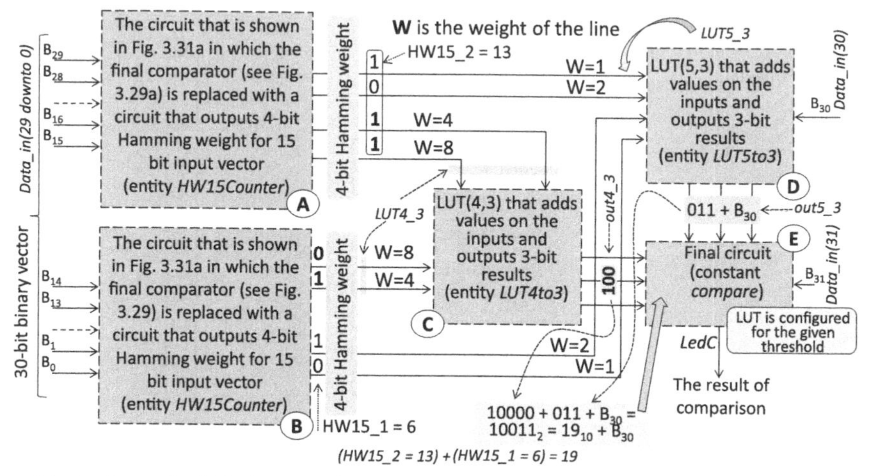

Fig. 3.32 The Hamming weight comparator for N = 32

respectively. The MSBs shown in bold font are added in the block C giving the result 100 (11 + 01 = 100) denoting four values 4. The block D outputs 011 (i.e. 01 + 10 = 011) plus the value of bit B_{30}. If $B_{30} = 0$ then the number for comparison is 100011 which corresponds to the decimal value 19 (four values 4 plus three values 1). If $B_{30} = 1$ then the number for comparison is 20. Clearly, the block E that is built on a LUT(6,1) can easily produce the results of comparison for 31-bit vector $B = \{B_0,\ldots,B_{14},B_{15},\ldots,B_{29},B_{30}\}$. The block E that is built on a LUT(6,4) enables the Hamming weight for B to be calculated. If 32-bit vectors are to be used then an additional input with a bit B_{31} is appended to the block E (see Fig. 3.32). Clearly more physical FPGA primitives will be needed for the LUT that is now LUT(7,4).

A complete synthesizable VHDL code for the circuit in Fig. 3.32 is given in Appendix B. Scaling for larger values of N can be done similarly.

3.10 Design Examples for Operations Over Vectors

Suppose we have a set of N sorted data items (produced, for example, by a sorter in Sect. 3.5) which eventually include repeated items and we need the most frequently repeated item to be found. One possible solution for this problem proposed in [9] is shown in Fig. 3.33 where N-1 comparators form a binary vector. The most frequently repeated item can be discovered if we find the maximum number of consecutive ones in the vector and take the item from any input of the comparators that forms the sub-vector with the maximum number of successive ones.

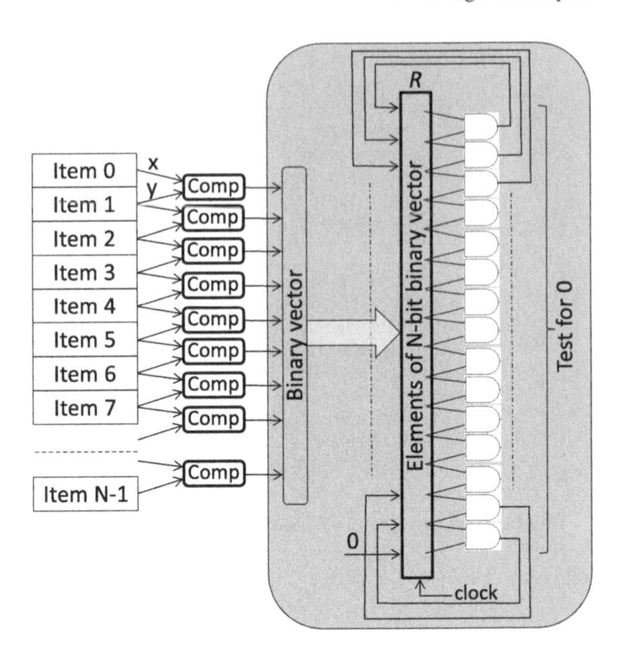

Fig. 3.33 Most frequent data item computation in a given sorted set of data

The binary vector that represents the result of comparison is saved in the feedback register *R*. The right-hand circuit in Fig. 3.33 implements the method described above (see Sects. 3.5, 3.6 and Figs. 3.14, 3.17) which enables the same combinational unit (such as that composed of AND gates in Fig. 3.33) to be reused iteratively in each subsequent clock cycle. This forces any intermediate binary vector that is formed on the outputs of the AND gates to be stored in the register *R*. Hence, any new clock cycle reduces the maximum number of consecutive ones O_{max} in the vector by one and as soon as all outputs of the AND gates are set to 0 we can conclude that $O_{max} = \xi+1$, where ξ is the number of the last clock cycle. Indeed, when there is just one value 1 in the register *R*, all the outputs of the AND gates are set to 0 and an additional clock cycle is not required to reach a conclusion. The index of the single 1 in the register is the index (position) of the first value 1 (from the top) in the set with O_{max}. The feedback from the outputs of the AND gates enables any intermediate binary vector to be stored in the register *R*. Not all the gates are entirely reused. At the first step there are N-1 active gates. In each subsequent clock the number of such gates is decremented because the lower gate is blocked by 0 to be written to the bottom bit of the register *R*. In each new clock cycle, this zero always propagates to an upper position and blocks another gate. The circuit in Fig. 3.33 is very simple and fast. It is composed of just N-1 AND gates, the register *R*, and minimal supplementary logic. Thus the maximum attainable clock frequency is high.

VHDL code below describes the part of the circuit shown in gray in Fig. 3.33. The remainder can easily be implemented using the examples in Sect. 3.4.1 (see, for instance, the entity EvenOddMerge8Sort and also the entity EvenOddMerge16Sort in Appendix B) and Sect. 3.5 (see the entity EvenOddTransitionIterative).

```
entity SucOnesEncounter is                    -- this project has been tested in the Atlys board
generic (        N         : integer := 48 );  -- number of bits in the binary vector (see Fig. 3.33)
 port (          clk       : in std_logic;
                 EppAstb : in std_logic;      -- for the component IOExpansion from Digilent
                 EppDstb : in std_logic;
                 EppWr   : in std_logic;
                 EppDB   : inout std_logic_vector(7 downto 0);
                 EppWait : out std_logic);
end SucOnesEncounter;
architecture Behavioral of SucOnesEncounter is
  signal MyLed                     : std_logic_vector(7 downto 0);
  signal MyLBar                    : std_logic_vector(23 downto 0);
  signal MySw                      : std_logic_vector(15 downto 0);
  signal MyBtn                     : std_logic_vector(15 downto 0);
  signal data_to_PC                : std_logic_vector(31 downto 0);
  signal data_from_PC              : std_logic_vector(31 downto 0);
  signal max_number_of_successive_ones : integer range 0 to N;
  signal max_number                    : integer range 0 to N;
  signal vector_with_ones, new_vector  : std_logic_vector(N-1 downto 0);
  signal Reg                           : std_logic_vector(N-1 downto 0);
begin

  MyLBar      <= MyBtn & data_from_PC(7 downto 0); -- these two lines are for tests only
  MyLed       <= data_from_PC(31 downto 24);       -- and can be removed
  data_to_PC  <= conv_std_logic_vector(max_number_of_successive_ones, 32);

process (Reg) -- this process describes AND gates and feedback to the register in Fig. 3.33
begin -- this process is combinational
  for i in 0 to N-2 loop – new_vector is formed on outputs of AND gates in Fig. 3.33
    new_vector(i) <= Reg(i) and Reg(i+1);
  end loop; -- the register changes its state in the next sequential process
  new_vector(N-1) <= '0'; -- the bottom bit is always zero (see Fig. 3.33)
end process;

process(clk)
begin -- this process is sequential
  if rising_edge(clk) then
    if ((data_from_PC = 0) and (MySw = 0)) then     -- there are no ones in input binary vector
       max_number_of_successive_ones <= 0;          -- thus, the number of ones is zero
    else Reg <= data_from_PC & MySw;                -- a vector is taken from virtual window
       max_number <= 1;  -- since the vector is not zero then there is at least one value one
       if new_vector /= 0 then
         -- if a new vector is not zero then the number of ones has to be incremented
         Reg <= new_vector;
         max_number <= max_number+1;
       else max_number_of_successive_ones <= max_number;
       end if;
    end if;
  end if;
  end process;
  IO_interface : entity work.IOExpansion
         port map(EppAstb, EppDstb, EppWr, EppDB, EppWait, MyLed, MyLBar,
                  MySw, MyBtn, data_from_PC, data_to_PC);

  end Behavioral;
```

A new vector is taken from the virtual window (from *To FPGA* field and virtual *Switches*). The result can be seen in the virtual field *From FPGA* (see Sect. 1.7 and Fig. 1.27).

The following VHDL code was tested in the Nexys-4 board.

```
entity SucOnesEncounter is
generic (      N          : integer := 16 );
 port (        clk        : in std_logic;
               sw         : in std_logic_vector(15 downto 0); -- binary vector from 16 switches
               led        : out std_logic_vector(4 downto 0)); -- the result on LEDs
end SucOnesEncounter;

architecture Behavioral of SucOnesEncounter is
signal max_number_of_successive_ones      : integer range 0 to N;
signal max_number                         : integer range 0 to N;
signal vector_with_ones, new_vector       : std_logic_vector(N-1 downto 0);
signal Reg                                : std_logic_vector(N-1 downto 0);
begin
process (Reg)
begin
      for i in 0 to N-2 loop
               new_vector(i) <= Reg(i) and Reg(i+1);
      end loop;
      new_vector(N-1) <= '0';
end process;

process(clk)
begin
 if rising_edge(clk) then
      if (sw = 0) then      max_number_of_successive_ones <= 0;
      else                  Reg <= sw;            max_number <= 1;
               if new_vector /= 0 then
                            Reg <= new_vector;  max_number <= max_number+1;
               else         max_number_of_successive_ones <= max_number;
               end if;
      end if;
 end if;
end process;
led <= conv_std_logic_vector(max_number_of_successive_ones, 5);
end Behavioral;
```

Much like the previous examples, the code above can equally be synthesized, and implemented in either ISE or Vivado (see the final part of Sect. 1.5). The only difference is in the supplied constraints file (UCF for ISE and XDC for Vivado). All the examples are available at http://sweet.ua.pt/skl/Springer2014.html for ISE and many of the examples for Vivado.

In sections above (beginning from the Sect. 3.4) we described different types of FPGA-based processing which permit broad parallelism to be supported. Many other circuits and systems can benefit from highly parallel implementations that are realizable in FPGAs and are more difficult for general-purpose and application-specific processors that impose some constraints such as the size of operands, the number of processing cores limiting parallelism and the predefined set of instructions. Note that processor-based technique still have many advantages over FPGAs particularly when complex problems need to be solved. For example, widely used systems for solving the Boolean satisfiability problem are still better implemented in general-purpose computers. However, solving some lower-level tasks that are needed for the Boolean satisfiability might be more advantageous in FPGA, which is shown, in particular, in [32]. Such frequently explored problems as stream processing are often solved partially in hardware (*e.g.* implementing sorting networks) and partially in software (*e.g.* merging pre-sorted sub-sets). Thus, it is practical to combine general-purpose or application-specific software with hardware accelerators implemented in reconfigurable logic. The latter is intended to be used in such a way that permits time-consuming operations of software (better implemented in FPGA hardware) to be speeded up. In the next chapter we will discuss such issues.

References

1. Wakerly JF (2006) Digital design. Principles and practices. Pearson Prentice Hall, Upper Saddle River
2. De Micheli G (1994) Synthesis and optimization of digital circuits. McGraw-Hill, Inc, New York
3. Sklyarov V, Skliarova I (2013) Parallel processing in FPGA-based digital circuits and systems. TUT Press, Tallinn
4. Skliarova I, Sklyarov V, Sudnitson A (2012) Design of FPGA-based circuits using hierarchical finite state machines. TUT Press, Tallinn
5. Chu PP (2008) FPGA prototyping using VHDL examples: Xilinx Spartan-3 version. John Willey & Sons Inc, New Jersey
6. Skliarova I, Ferrari A (2001) Design and implementation of reconfigurable processor for problems of combinatorial computations. In: Proceedings of the Euromicro symposium on digital system design, Warsaw, 2001
7. Baranov S (1994) Logic synthesis for control automata. Kluwer Academic Publishers, Dordrecht
8. Baranov S (2008) Logic and system design of digital systems. TUT Press, Tallinn
9. Sklyarov V, Skliarova I (2013) Digital hamming weight and distance analyzers for binary vectors and matrices. Int J Innovative Comput Inf Control 9(12):4825–4849
10. Sklyarov V, Skliarova I (2013) Design and implementation of counting networks. Computing. doi: 10.1007/s00607-013-0360-y
11. Sklyarov V, Skliarova I, Rjabov A, Sudnitson A (2013) Implementation of parallel operations over streams in extensible processing platforms. In: Proceedings of the IEEE 56th international Midwest symposium on circuits & systems, Columbus, Ohio, 2013
12. Sklyarov V, Skliarova I (2013) Hardware implementations of software programs based on HFSM models. Comput Electr Eng 39(7):2145–2160
13. Knuth DE (2011) The art of computer programming. Sorting and searching, vol 3. Addison-Wesley, New York
14. Batcher KE (1968) Sorting networks and their applications. In: Proceedings of AFIPS spring joint computer conference, USA, 1968
15. Mueller R, Teubner J, Alonso G (2012) Sorting networks on FPGAs. Int J Very Large Data Bases 21(1):1–23
16. Zuluada M, Milder P, Puschel M (2012) Computer generation of streaming sorting networks. In: Proceedings of the 49th design automation conference, San Francisco, 2012
17. Xilinx Inc. (2013) Zynq-7000 All Programmable SoC Overview. http://www.xilinx.com/support/documentation/data_sheets/ds190-Zynq-7000-Overview.pdf. Accessed 21 Nov 2013

18. Sklyarov V (1999) Hierarchical finite-state machines and their use for digital control. IEEE Trans VLSI Syst 7(2):222–228
19. Rosen KH, Michaels JG, Gross JL, Grossman JW, Shier DR (eds) (2000) Handbook of discrete and combinatorial mathematics. CRC Press, Florida
20. Ortiz J, Andrews D (2010) A configurable high-throughput linear sorter system. In: Proceedings of IEEE international symposium on parallel & distributed processing, Phoenix, 2010
21. Mueller R (2010) Data stream processing on embedded devices. Ph.D. dissertation, Swiss Federal Institute of Technology
22. Kipfer P, Westermann R (2005) Improved GPU sorting. In: Pharr M, Fernando R (eds) GPU gems 2: programming techniques for high-performance graphics and general-purpose computation. Addison-Wesley. http://developer.nvidia.com/GPUGems2/gpugems2_chapter46.html. Accessed 21 Nov 2013
23. Sklyarov V, Skliarova I (2013) Fast regular circuits for network-based parallel data processing. Adv Electr Comput Eng 13(4):47–50
24. Zakrevskij A, Pottosin Y, Cheremisiniva L (2008) Combinatorial algorithms of discrete mathematics. TUT Press, Tallinn
25. Parhami B (2009) Efficient hamming weight comparators for binary Vectors based on accumulative and up/down parallel counters. IEEE Trans Circuits Syst II: Express Briefs 56(2):167–171
26. Piestrak SJ (2007) Efficient hamming weight comparators of binary vectors. Electron Lett 43(11):611–612
27. Sklyarov V, Skliarova I, Mihhailov D, Sudnitson A (2011) Implementation in FPGA of address-based data sorting. In: Proceedings of the 21st international conference on field-programmable logic and applications, Crete, 2011
28. Pedroni VA (2003) Compact fixed-threshold and two-vector Hamming comparators. Electron Lett 39(24):1705–1706
29. Pedroni VA (2004) Compact Hamming-comparator-based rank order filter for digital VLSI and FPGA implementations. In: Proceedings of the IEEE international symposium on circuits and systems, Vancouver, 2004
30. Xilinx Inc. (2013) 7 Series DSP48E1 Slice User Guide. http://www.xilinx.com/support/documentation/user_guides/ug479_7Series_DSP48E1.pdf. Accessed 16 Nov 2013
31. Xilinx Inc. (2011) Xilinx 7 series FPGA libraries guide for HDL designs. http://www.xilinx.com/support/documentation/sw_manuals/xilinx13_3/7series_hdl.pdf. Accessed 21 Nov 2013
32. Davis JD, Tan Z, Yu F, Zhang L (2008) A practical reconfigurable hardware accelerator for Boolean satisfiability solvers. In: Proceedings of the 45th ACM/IEEE design automation conference, Anaheim, California, June 2008

Chapter 4
Embedded Blocks and System-Level Design

Abstract This chapter begins with examples that demonstrate how commercially available intellectual property cores can be embedded in different designs. In particular, arithmetical circuits constructed from digital signal processing slices, and parameterized memory blocks that provide support for data buffering (such as FIFO—first input first output), are described. More details on digital signal processing slices are then given and it is shown how these may be used efficiently in practical circuits such as Hamming weight counters/comparators. The major part of this chapter is dedicated to interactions between a host computer and FPGA-based prototyping boards through the Digilent enhanced parallel port and the UART (Universal Asynchronous Receiver and Transmitter) interfaces. Complete details of the communication modules are described, including both software for general-purpose computers that was developed in the C++ language, and hardware for FPGAs. The next section makes use of the designed modules for projects that involve such interactions for different purposes. A more complicated design for a network-based iterative data sorter from Chap. 3 is implemented and tested in this way as a complete fully functioning example. The chapter concludes with a brief description of programmable systems-on-chip (PSoC) that combine an embedded processing system with a reconfigurable logic which can lead to more efficient implementations of the applications. Proposals for mapping the designs from Chap. 3 to the PSoC are given and discussed.

4.1 Using IP Cores

Intellectual Property (IP) cores are preconfigured blocks that can be included in the design. For example, Xilinx ISE supplies a wide selection of IPs for memories (both embedded and distributed), digital signal processing, math functions, bus interfaces, clock distribution, etc. A block can be chosen, customized and attached to the design with the aid of the Xilinx CORE Generator™ tools. We consider in

V. Sklyarov et al., *Synthesis and Optimization of FPGA-Based Systems*,
Lecture Notes in Electrical Engineering 294, DOI: 10.1007/978-3-319-04708-9_4,

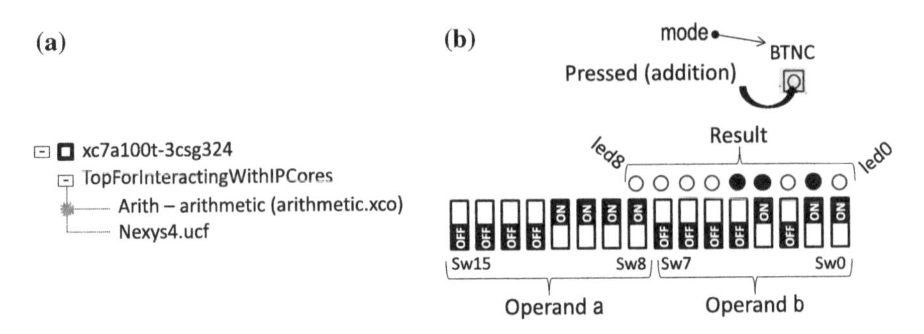

Fig. 4.1 The structure of the project with IP core (**a**) and demonstration (**b**)

this section a few examples and show how to use IP cores for our previous designs from Chap. 3.

The first example demonstrates how to include a DSP-based adder-subtractor to the project. The project has been created in Xilinx ISE 14.7 and a *New Source* named *arithmetic* with the type *IP Core* has been added from the *Math Functions* and *Adders & Subtractors groups*. We requested to use the DSP48 slice (available for Xilinx FPGAs), 8-bit unsigned operands, 9-bit result, *Add Subtract* mode, *Latency* 0, and active high *carry in* signal. After that the core was generated and included in the project as follows (mapping is done in accordance with details given in the ISE *HDL Instantiation Template*):

```
entity TopForInteractingWitIPCores is
  port ( Sw     : in std_logic_vector (15 downto 0);
         mode : in std_logic; -- the BTNC button is used (for addition it has to be pressed)
         led    : out std_logic_vector (8 downto 0) );
end TopForInteractingWitIPCores;

architecture Behavioral of TopForInteractingWitIPCores is
begin
Arith: entity work.arithmetic
      port map (a => Sw(15 downto 8), b => Sw(7 downto 0), add => mode,
                c_in => '0', s => led);
end Behavioral;
```

Figure 4.1a depicts the structure of the project. The file *arithmetic.xco* was created by the Xilinx CORE Generator. Figure 4.1b shows how to test the project in the Nexys-4 board. Two operands are taken from the 8 leftmost (a) and 8 rightmost (b) switches. The result is displayed on 9 LEDs. The mode is chosen by the button *BTNC* (addition if the button is pressed and subtraction if the button is released). In Fig. 4.1b a $= 00001111_2 = 15_{10}$, b $= 00001011_2 = 11_{10}$. Thus, if *BTNC* is pressed then a + b $= 26_{10} = 000011010_2$ (see Fig. 4.1b). If *BTNC* is released then a-b $= 4_{10} = 000000100_2$ which can be easily examined in the Nexys-4 board. The circuit occupies 0 logical slices and just 1 DSP48E1 slice (from 240 available slices). Similar projects can be created and tested in the Atlys prototyping board.

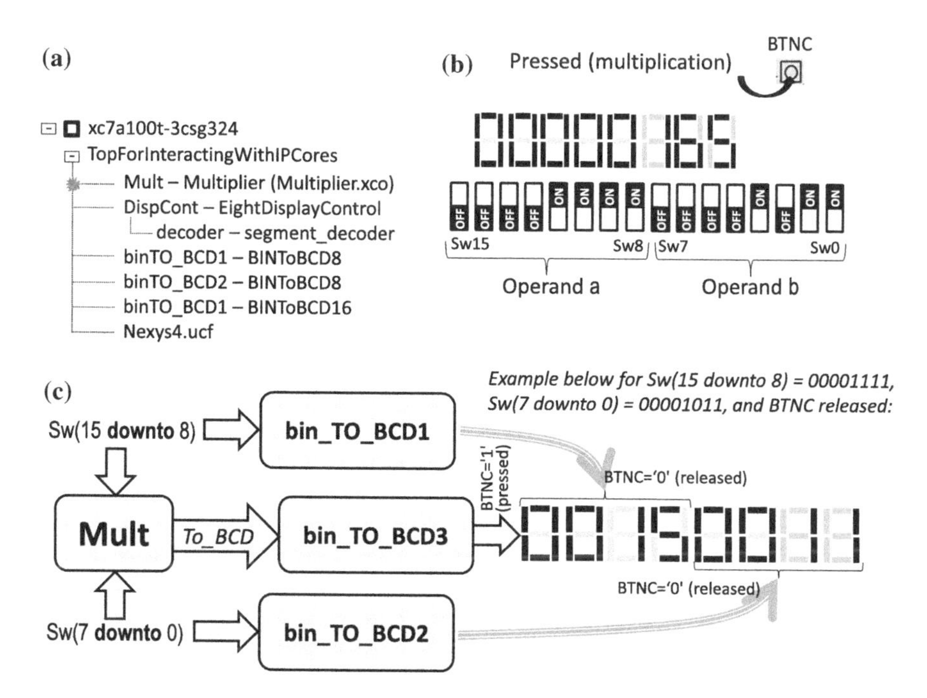

Fig. 4.2 The structure of the project with IP core (**a**), demonstration (**b**), and a simplified component diagram (**c**)

The next example will use IP core for multiplication operation which can be taken from *Multipliers* in the ISE group of *Math functions*. Figure 4.2a depicts the structure of the project. The file *Multiplier.xco* was created by the Xilinx CORE Generator. From this project we will use available in the Nexys-4 eight onboard segment displays. The displays are managed by two components EightDisplayControl and segment_decoder (their VHDL codes are given in Appendix B). The project (see Fig. 4.2a) also uses components BinToBCD8 and BinToBCD16 which convert 8-bit and 16-bit binary vectors to their BCD representation that is needed to display decimal numbers on segment displays. VHDL code for these components is also given in Appendix B.

The project has been created in Xilinx ISE 14.7 and a *New Source* named Multiplier with the type *IP Core* has been added from the *Math Functions* and *Multipliers* groups. We requested to use the DSP48 slice, 8-bit unsigned operands and 16-bit result. After that the core was generated and included in the project as follows (as before, mapping is done in accordance with the details given in the ISE *HDL Instantiation Template*):

```
entity TopForInteractingWitIPCores is          -- this project is for the Nexys-4 board
  port (  clk           : in  std_logic;
          seg           : out std_logic_vector(6 downto 0);    -- segments
          sel_disp      : out std_logic_vector(7 downto 0);    -- display selections
          Sw            : in  std_logic_vector (15 downto 0);  -- onboard switches
          BTNC          : in  std_logic;                       -- onboard BTNC button
```

```
        reset            : in std_logic);                  -- onboard BTND button
end TopForInteractingWitIPCores;

architecture Behavioral of TopForInteractingWitIPCores is

signal BCD4, BCD3, BCD2, BCD1, BCD0          : std_logic_vector(3 downto 0);
signal BCD2_L, BCD1_L, BCD0_L                : std_logic_vector(3 downto 0);
signal BCD2_R, BCD1_R, BCD0_R                : std_logic_vector(3 downto 0);
signal BCD3_D, BCD2_D, BCD1_D, BCD0_D        : std_logic_vector(3 downto 0);
signal BCD7_D, BCD6_D, BCD5_D, BCD4_D        : std_logic_vector(3 downto 0);
signal To_BCD                                : std_logic_vector(15 downto 0);

begin -- see the simplified components diagram in Fig. 4.2c

Mult: entity work.Multiplier                 -- DSP-based multiplier
    port map (a=>Sw(15 downto 8), b=>Sw(7 downto 0), p=>To_BCD);

DispCont: entity work.EightDisplayControl   -- display controller (see Appendix B)
    port map(clk, BCD7_D, BCD6_D, BCD5_D, BCD4_D, BCD3_D, BCD2_D,
             BCD1_D, BCD0_D, sel_disp, seg);

binTO_BCD1: entity work.BinToBCD8        -- binary to BCD converter      (see Appendix B)
    port map (clk, reset, open, Sw(15 downto 8), BCD2_L, BCD1_L, BCD0_L);

binTO_BCD2: entity work.BinToBCD8        -- binary to BCD converter      (see Appendix B)
    port map (clk, reset, open, Sw(7 downto 0), BCD2_R, BCD1_R, BCD0_R);

binTO_BCD3: entity work.BinToBCD16       -- binary to BCD converter      (see Appendix B)
    port map (clk, reset, open, To_BCD, BCD4, BCD3, BCD2, BCD1, BCD0);

process(BTNC, BCD4, BCD3, BCD2, BCD1, BCD0,   -- combinational process
        BCD2_L, BCD1_L, BCD0_L, BCD2_R, BCD1_R, BCD0_R)
begin -- this process selects either operands (if BTNC=0) or the result (if BTNC=1) to display
  BCD7_D <= (others => '0'); BCD6_D <= (others => '0');
  BCD5_D <= (others => '0'); BCD4_D <= (others => '0');

  BCD3_D <= (others => '0'); BCD2_D <= (others => '0');
  BCD1_D <= (others => '0'); BCD0_D <= (others => '0');

if (BTNC = '0') then                   -- display mode selection
    BCD7_D <= (others => '0'); BCD6_D <= BCD2_L;
    BCD5_D <= BCD1_L;          BCD4_D <= BCD0_L;
    BCD3_D <= (others => '0'); BCD2_D <= BCD2_R;
    BCD1_D <= BCD1_R;          BCD0_D <= BCD0_R;
else BCD7_D <= (others => '0'); BCD6_D <= (others => '0');
    BCD5_D <= (others => '0'); BCD4_D <= BCD4;
    BCD3_D <= BCD3;            BCD2_D <= BCD2;
    BCD1_D <= BCD1;            BCD0_D <= BCD0;
end if;
  end process;
end Behavioral;
```

Figure 4.2b shows how to test the project in the Nexys-4 board. Two operands are taken from the 8 leftmost (a) and 8 rightmost (b) switches. The result is converted to a BCD representation and then is shown on the segment displays (see Fig. 4.2b). If the button *BTNC* is released then a decimal value of a is shown on 4 leftmost displays and a decimal value of b—on 4 rightmost displays (see

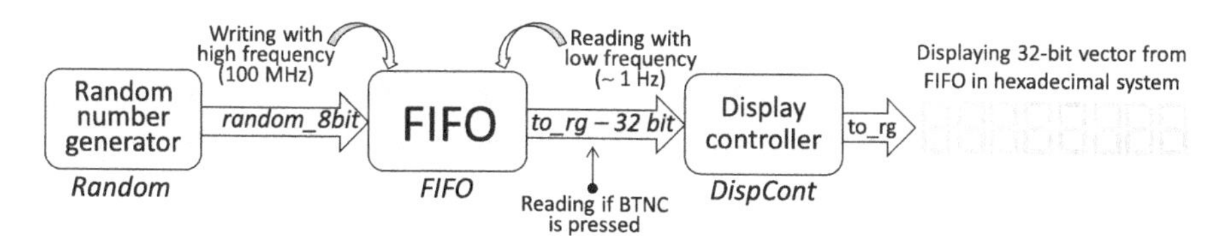

Fig. 4.3 Testing FIFO memory

Fig. 4.2c). If the button *BTNC* is pressed then a decimal value of the result is shown on 8 displays (see Fig. 4.2b). The circuit occupies 63 logical slices (from 15,850 available slices) and just 1 DSP48E1 slice. Similar projects can be created and tested in the Atlys prototyping board.

We did not discuss above how to control segment displays, which is described in detail in [1]. The binary to BCD converters implement the algorithm explained in [2] and they are based on VHDL code from [3]. The only difference is immediate conversion without the need for additional signals that have to be used in [2, 3]. This means that as soon as inputs of the converter are changed, the result is ready after a few clock cycles delay. Since we examine the results visually, such delay cannot give rise to any problem and interface with the converter becomes very simple. Let us look at Fig. 4.2. The result of multiplication $00001111_2 \times 00001011_2 = 10100101_2$ in binary format is converted (by the component *binTO_BCD3* in Fig. 4.2c) to three least significant BCD digits 0001_2, 0110_2, and 0101_2. The latter can directly be decoded to decimal format giving the result 165. Five most significant digits are 0000_2 giving five decimal zeros on the leftmost segment displays (see Fig. 4.2b). Complete VHDL codes for all the used components are given in Appendix B.

The next example demonstrates how to construct FIFO (First Input First Output) block based on embedded memory. A simplified component diagram for the first project is shown in Fig. 4.3.

The project has been created in Xilinx ISE 14.7 and a *New Source* named FIFO_mem with the type *IP Core* has been added from the *Memories & Storage Elements* and *FIFOs* groups. We requested to use independent read/write clocks for block RAM, write width 8, write depth 64, read width 32, read depth 16, and write/read data count slice. Note that the width for the input (8 bit) differs from the width for the outputs (32 bits). Finally, the core has been generated and included in the project as follows:

```
entity TestFIFO is          -- Fig. 4.4 demonstrates how to test this project
generic (data_in_size              : integer := 8;     -- width of FIFO input data
         data_out_size             : integer := 32);   -- width of FIFO output data
 port (  clk                       : in std_logic;     -- clock 100 MHz
         led_full                  : out std_logic;    -- ON if FIFO is full
         led_empty                 : out std_logic;    -- ON if FIFO is empty
         led_rd_data_count         : out std_logic_vector (3 downto 0);
         led_wr_data_count         : out std_logic_vector (5 downto 0);
         led_div_clk               : out std_logic;    -- low frequency clock (~1  Hz)
```

```
        seg              : out std_logic_vector(6 downto 0);
        sel_disp         : out std_logic_vector(7 downto 0);
        BTNC             : in std_logic );     -- to read data from FIFO
end TestFIFO;

architecture Behavioral of TestFIFO is

signal divided_clk     : std_logic;          -- low frequency clock (~1 Hz)
signal random_8bit     : std_logic_vector(data_in_size-1 downto 0);
signal wr_en           : std_logic;          -- write enable to FIFO
signal rd_en           : std_logic;          -- read enable from FIFO
signal to_rg           : std_logic_vector(data_out_size-1 downto 0);
signal full            : std_logic;          -- FIFO is full
begin
  led_div_clk    <= divided_clk;
  led_full       <= full;

enables_gen: process(full, BTNC) -- support for FIFO write/read modes
begin
  if (full /= '1') then wr_en <= '1';   -- if FIFO is not full then new data can be written
  else wr_en <= '0';        end if;
  if (BTNC = '1') then rd_en <= '1'; -- data are read from FIFO when BTNC is pressed
  else rd_en <= '0';        end if;
end process enables_gen;

FIFO: entity work.FIFO_mem                    -- see Fig. 4.5
      port map (wr_clk => clk, rd_clk => divided_clk, din => random_8bit,
                wr_en => wr_en, rd_en => rd_en, dout => to_rg, full => full,
                empty => led_empty, rd_data_count => led_rd_data_count,
                wr_data_count => led_wr_data_count );

Random: entity work.RanGen        -- the code is available in Appendix B
      generic map(width => data_in_size)
      port map (clk, random_8bit);

DispCont: entity work.EightDisplayControl   -- the code is available in Appendix B
      port map(clk, to_rg(31 downto 28), to_rg(27 downto 24), to_rg(23 downto 20),
                to_rg(19 downto 16), to_rg(15 downto 12), to_rg(11 downto 8),
                to_rg(7 downto 4), to_rg(3 downto 0), sel_disp, seg);

div:  entity work.clock_divider               -- the code is available in Appendix B
      port map( clk, '0', divided_clk);

end Behavioral;
```

The project can be tested in the Nexys-4 board as it is shown in Fig. 4.4. The process enables_gen above enables writing when the FIFO is not full and reading when the onboard button *BTNC* is pressed. Data are written to the FIFO from a random number generator (see VHDL code in Appendix B) and are read from the FIFO and shown on eight 7-segment displays when the button *BTNC* is pressed (see Fig. 4.5). The output vector is split in 8 4-bit segments and a hexadecimal code corresponding to each segment is shown on the associated display (the most significant segment is associated with the leftmost display).

Two additional projects in this section demonstrate how to link the FIFO memory with inputs of the 32-bit Hamming weight counter (see Sect. 3.9) and the

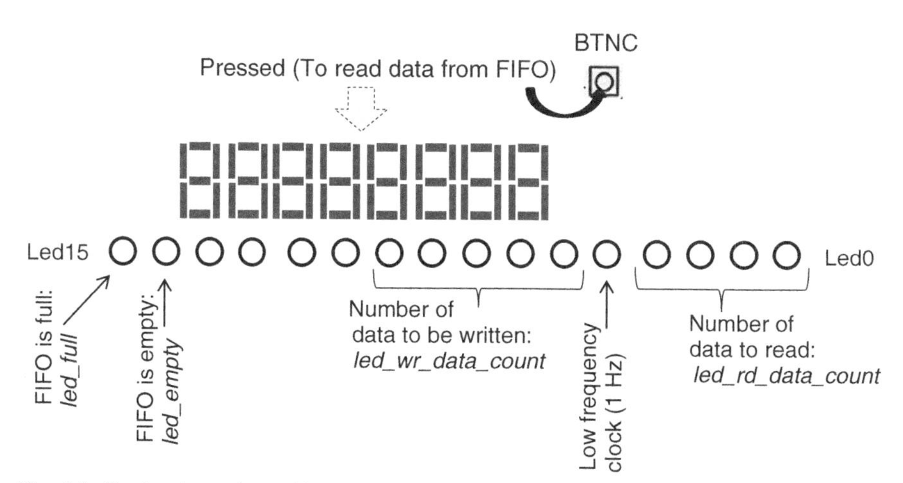

Fig. 4.4 Testing the project with FIFO memory

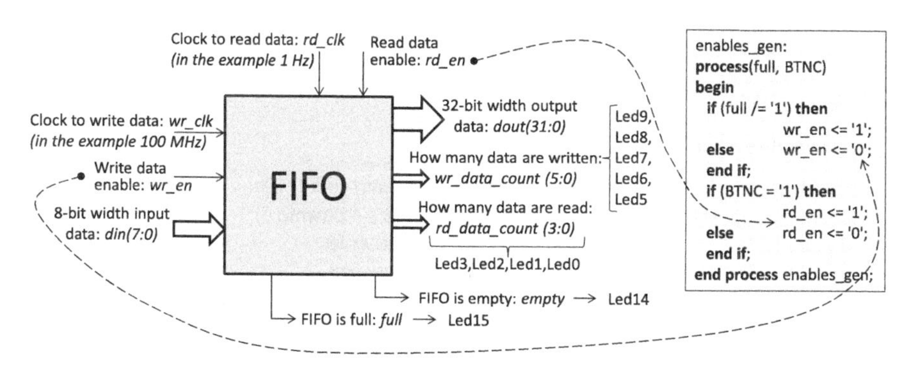

Fig. 4.5 Interface with FIFO memory

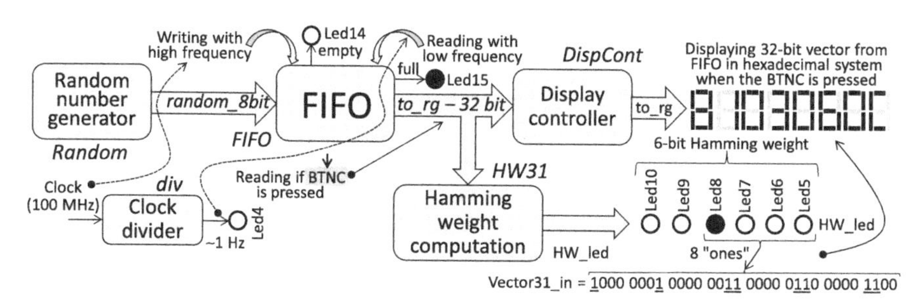

Fig. 4.6 A simplified component diagram and an example for the project that links the Hamming weight counter (Fig. 3.24a) with the FIFO memory (Fig. 4.3)

sorting network shown in Fig. 3.5. A simplified component diagram for the first project is given in Fig. 4.6. As soon as a 32-bit vector is available on the outputs of the FIFO, the entity HW31_HWC32 computes the Hamming weight for 31-bit vector that is shown on 5 LEDs (LED9,..., LED5). The most significant bit of the input vector is added in the line (see appendix B for details):

```
HW_led <= ("00000" & to_rg(data_out_size-1)) + ('0' & bits4_0);
```

Figure 4.6 shows an example where the random number $8103060C_{16}$ (which contains 8 "ones") is displayed on the LEDs as $001000_2 = 8_{10}$.

The following VHDL code describes the functionality of the circuit shown in Fig. 4.6.

```
entity TestFIFO is          -- the project was tested in the Nexys-4 board
generic (data_in_size       : integer := 8;
         data_out_size      : integer := 32);
 port ( clk                 : in std_logic;        -- system clock is 100 MHz
        led_full            : out std_logic;       -- ON if FIFO is full
        led_empty           : out std_logic;       -- ON if FIFO is empty
        HW_led              : out std_logic_vector (5 downto 0); -- 6-bit Hamming weight
        LedC                : out std_logic;       -- the result of comparison (see appendix B)
        led_div_clk         : out std_logic;       -- low frequency clock (~1 Hz)
        seg                 : out std_logic_vector(6 downto 0);     -- see appendix B
        sel_disp            : out std_logic_vector(7 downto 0);     -- see appendix B
        BTNC                : in std_logic);                        -- onboard button BTNC
end TestFIFO;

architecture Behavioral of TestFIFO is

signal divided_clk          : std_logic;           -- low frequency clock (~1 Hz)
signal random_8bit          : std_logic_vector(data_in_size-1 downto 0);
signal wr_en                : std_logic;           -- FIFO write enable
signal rd_en                : std_logic;           -- FIFO read enable
signal to_rg                : std_logic_vector(data_out_size-1 downto 0);
signal full                 : std_logic;           -- '1' if FIFO is full
signal bits4_0              : std_logic_vector(4 downto 0); -- bits 4...0 for Hamming weight

begin

HW_led <= ("00000" & to_rg(data_out_size-1)) + ('0' & bits4_0); -- handling an additional bit
led_div_clk     <= divided_clk;
led_full        <= full;
  -- insert here the process enables_gen from the previous project

FIFO: entity work.FIFO_mem        -- FIFO memory component
      port map ( wr_clk => clk, rd_clk => divided_clk, din => random_8bit,
                 wr_en => wr_en, rd_en => rd_en, dout => to_rg, full => full,
                 empty => led_empty, rd_data_count => open,
                 wr_data_count => open );

Random: entity work.RanGen        -- random number generator (see appendix B)
      generic map(width => data_in_size )
      port map (clk, random_8bit);

DispCont: entity work.EightDisplayControl  -- display controller (see appendix B)
port map(clk, to_rg(31 downto 28), to_rg(27 downto 24), to_rg(23 downto 20),
to_rg(19 downto 16), to_rg(15 downto 12), to_rg(11 downto 8),
to_rg(7 downto 4), to_rg(3 downto 0), sel_disp, seg);
```

```
div: entity work.clock_divider                       -- clock divider (see appendix B)
        port map (clk, '0', divided_clk);            -- reset is always deasserted ('0')

HW31: entity work.HW31_HWC32  -- the code of this component is given in appendix B
        port map (Data_in => to_rg, led => bits4_0, LedC => LedC);

end Behavioral;
```

The project above can be tested in the Nexys-4 board as, for example, it is shown in Fig. 4.6. The circuit occupies 50 logical slices (from 15,850 available slices) and 1 block RAMB36E1 (from 135 available blocks).

A simplified component diagram for the second project is given in Fig. 4.7. As soon as a 32-bit vector is available on the outputs of the FIFO, the block sorter sorts eight 4-bit data items in the vector. Eight hexadecimal digits are read from FIFO and shown on the segment displays when the button *BTNC* is pressed for a few seconds (until zeros shown on the segment displays are replaced with non-zero values) and then released. If the button *BTND* is pressed then the data previously shown on the segment displays are put in the sorted (ascending) order. For example, data from Fig. 4.6 will be displayed as shown in Fig. 4.7. If the button *BTNC* is pressed once again then new random items will be displayed and they will be sorted as soon as the button *BTND* is pressed.

VHDL code for the project is almost the same as shown above. The ports HW_led and LedC are no longer needed and they are removed. The component HW31_HWC32 is replaced with the sorter:

```
sorter : entity work.EvenOddMerge8Sort     -- see the code of the sorter in section 3.4.1
        port map (input_data => to_rg, sorted_data => sorted_data);
```

The sorted data (sorted_data) are displayed when the button *BTND* is pressed:

```
-- displaying either randomly generated data to_rg or sorted_data
data_to_display <= to_rg when BTND = '0' else sorted_data;
```

This button has to be added to the ports:

```
BTND : in std_logic;
```

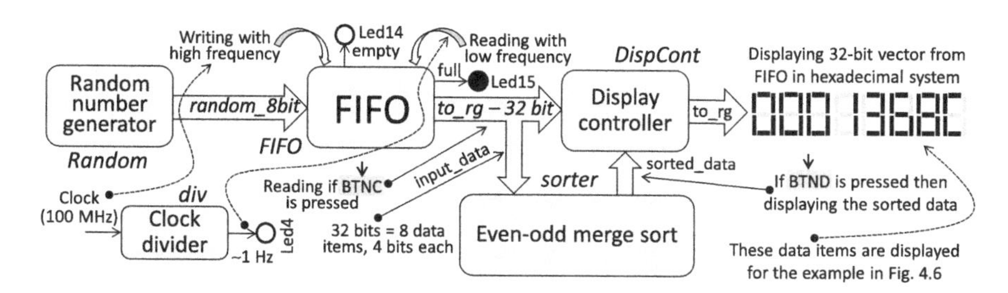

Fig. 4.7 A simplified component diagram and an example for the project that links the sorting network (Fig. 3.5) with the FIFO memory (Fig. 4.3)

All other lines in VHDL code are the same as in the previous project. The circuit can be tested in the Nexys-4 prototyping board. It occupies 84 logical slices and 1 block RAMB36E1. Note that there are a number of additional examples in subsequent Sects. 4.3 and 4.4 that involve embedded memory blocks using IP cores.

The last example here demonstrates the use of clocking circuit. The project has been created in Xilinx ISE 14.7 and a *New Source* named clock_mult with the type *IP Core* has been added from the *FPGA Features & Design* and *Clocking* groups. We requested to use input frequency 100 MHz and 6 output clocks with frequencies 150, 200, 300, 400, 50, and 25 MHz. Finally, the core has been generated and included in the project as follows:

```
entity TopForClockGenerator is
  port (clk                 : in std_logic;
        clock_sel           : in std_logic;     -- switch 1 for the Nexys-4 board
        led25, led50, led100, led150, led200, led300, led400
                            : out std_logic;    -- LEDs 0,1,2,3,4,5,6 for the Nexys-4 board
       variable_clock       : out std_logic);   -- LED 15 for the Nexys-4 board
end TopForClockGenerator;

architecture Behavioral of TopForClockGenerator is

signal clk25, clk50, clk100, clk150, clk200, clk300, clk400, var_clk : std_logic;

begin

clk_man: entity work.clock_mult -- this core has been generated by the Xilinx IP core generator
      port map(CLK_IN1=>clk, CLK_OUT1=>clk100, CLK_OUT2=>clk150,
               CLK_OUT3=>clk200, CLK_OUT4=>clk300, CLK_OUT5=>clk400,
               CLK_OUT6=>clk50, CLK_OUT7=>clk25);

 div100 : entity work.clock_divider        -- generic parameter how_fast in the clock_
        port map(clk100, '0', led100);    -- parameter how_fast is set to 28 (see appendix B for details)

-- similar to div100 clock dividers for signals clk150, clk200, clk300, clk400, clk 50, clk25
div_var: entity work.clock_divider
      port map(var_clk, '0', variable_clock); -- reset is always deasserted ('0')

var_clk <= clk100 when clock_sel = '1' else clk400;

end Behavioral;
```

The respective project can be tested in the Nexys-4 board. Clock frequencies from the component clock_mult are additionally divided by a factor of $2^{how\_fast+1} = 2^{29} = 536{,}870{,}912$ (see the code of clock divider in Appendix B). Thus, the LED 0 changes the states from ON to OFF and vice versa every 21.5 s (i.e. 536,870,912/25,000,000), the LED 1 changes the states two times faster, etc. Switching frequency for the LEDs 0,…,6 is increased from LED 0 to LED 6. The frequency of LED 15 is controlled by the switch 0 and can be increased or decreased with a factor of 4.

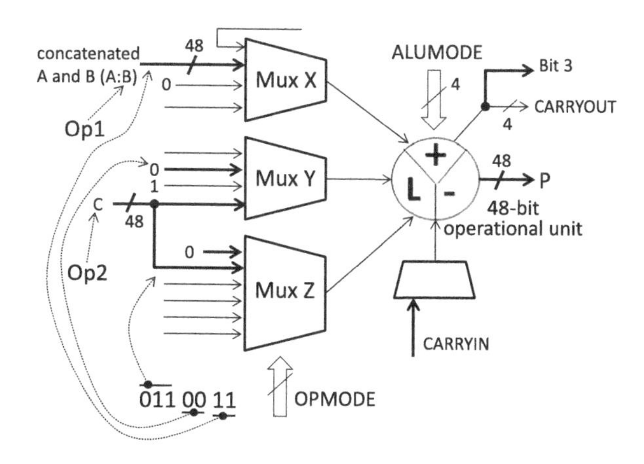

Fig. 4.8 Part of DSP48E1 [4] that will be used in examples below

4.2 Design with Embedded DSP Slices

The majority of contemporary FPGAs have embedded DSP slices (ex. DSP48E1 slice for Xilinx FPGAs [4]) and they can be employed to implement arithmetical and logical operations. For example, segments of counting networks described in Sect. 3.8 and in [5] may benefit from available bitwise operations. Indeed, the operations "*and*" and "*xor*" (see Fig. 3.24a) can be executed concurrently in two DSP slices (the "*and*" operation in the first slice and the "*xor*" operation in the second slice) for $N \leq 2 \times \xi$, where ξ is the size of the operands for a DSP slice (for the DSP48E1 $\xi = 48$ and the most advanced FPGAs contain thousands of such slices). For $N > 2 \times \xi$ the network can be decomposed in fragments implemented in different slices.

We consider in this section only a few examples that illustrate using DSP slices to solve some problems discussed earlier in the book, such as the design of Hamming weight counters/comparators. Exhaustive material about DSPs can be found in guides of the relevant companies, such as [4]. Let us consider at the beginning a part of DSP48E1 slice including only components that will be used for the examples below (see Fig. 4.8). Inputs and outputs that will be needed for our circuits are shown in Fig. 4.8 by thick lines.

The following simple example demonstrates how to test different bitwise operations. The language template for 48-bit multi-functional arithmetic block DSP48E1 may be chosen (through the path in Xilinx ISE: *VHDL → Device Primitive Instantiation → Artix-7 → Arithmetic Functions*) and customized. The DSP multiplier that is not shown in Fig. 4.8 is bypassed by setting in the DSP48E1 generic map USE_MULT attribute to NONE (i.e. USE_MULT => "NONE") [4]. Now bitwise logic operations (indicated by letter L in Fig. 4.8) over two 48-bit binary vectors can be executed and they may be controlled dynamically changing the ALUMODE control signals (see Fig. 4.8). Figure 4.9 shows how the indicated above template for the block DSP48E1 has been customized, which was done in accordance with the Xilinx guide [4].

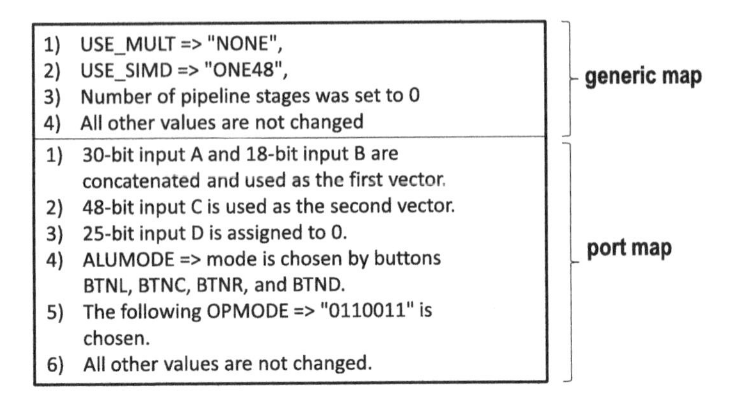

1) USE_MULT => "NONE",
2) USE_SIMD => "ONE48",
3) Number of pipeline stages was set to 0
4) All other values are not changed
generic map
1) 30-bit input A and 18-bit input B are concatenated and used as the first vector.
2) 48-bit input C is used as the second vector.
3) 25-bit input D is assigned to 0.
4) ALUMODE => mode is chosen by buttons BTNL, BTNC, BTNR, and BTND.
5) The following OPMODE => "0110011" is chosen.
6) All other values are not changed.
port map

Fig. 4.9 Changes in the template for DSP48E1

Now the following VHDL code can be examined:

```
entity Test_bitwise_with_DSP is
  port ( Sw     : in  std_logic_vector (15 downto 0);      -- 8+8 bits for two vectors
         mode   : in std_logic_vector(3 downto 0);         -- this is ALUMODE for DSP48E1
         led    : out  std_logic_vector (15 downto 0));  -- the result of bitwise operations
end Test_bitwise_with_DSP;

architecture Behavioral of Test_bitwise_with_DSP is
  signal Op1, Op2, Y     : std_logic_vector(47 downto 0);
begin
Op1 <= (47 downto 8 => '0') & Sw(15 downto 8);          -- the first vector
Op2 <= (47 downto 8 => '0') & Sw(7 downto 0);           -- the second vector
led <= Y(15 downto 0);                                  -- the result
DSP: entity work.TesDSP48E1_bitwise          -- link with the template DSP48E1 component
       port map (Op1, Op2, mode, Y);         -- the library UNISIM is included in the template
end Behavioral;
```

Bitwise operations are selected by OPMODE bits 3 and 2 and by ALUMODE. Examples of such operations are: "*xor*" for OPMODE(3:2) = "00" and ALUMODE = "0100"; "*and*" for OPMODE(3:2) = "00" and ALUMODE = "1100"; and "*or*" for ALUMODE = "1100" and OPMODE(3:2) = "10" (further details can be found in [4]).

In Single Instruction, Multiple Data (SIMD) mode the 48-bit adder/subtractor/accumulator can be split into either 4 independent 12-bit or 2 independent 24-bit adders/subtractors/accumulators performing the same function specified by the ALUMODE. USE_SIMD option in the DSP48E1 generic map has to be changed appropriately to either USE_SIMD => "FOUR12" (for four operations over 12-bit operands) or USE_SIMD => "TWO24" (for two operations over 24-bit operands). OPMODE controls multiplexer outputs in the DSP48E1 and enables different operands to be selected (see Fig. 4.8). For the considered above VHDL code *OPMODE* is the same

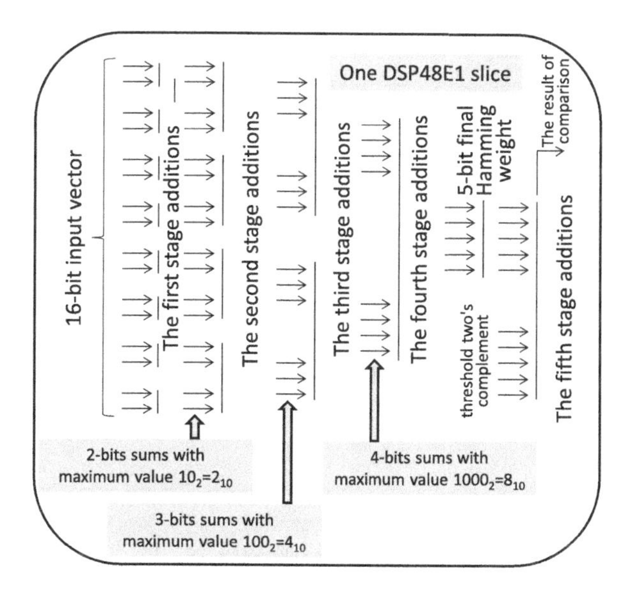

Fig. 4.10 Computing the Hamming weight and the result of comparison with a fixed threshold for a 16-bit binary vector

("0110011") where the first three bits ("011") select our second operand (Op2) on the outputs of Z-multiplexer [4], the second two bits ("00") set zeros on the outputs of Y-multiplexer [4] and the last two bits ("11") select our first operand (Op1) on the outputs of X-multiplexer [4]. All necessary details can be found in [4]. For the examples below we will need addition operations with carry in and carry out signals. Let us implement the Hamming weight counter and comparator considered in Sects. 3.7–3.9 in a way shown in Fig. 4.10.

At the beginning all eight pairs of bits in 16-bit binary vectors are added. The size of sum is 2 bits with possible values 00, 01 and 10. The produced pairs of sums are again added giving 3-bit results with possible values 000, 001, 010, 011, 100. Any similar operations in subsequent steps give results with the size n + 1, where n is the size of used operands.

This feature was taken into account in the DSP-based implementation shown in Fig. 4.11. Only one DSP48E1 slice is required. The outputs of 48-bit adder are taken as inputs for the next stage selecting the proper number of bits that have to be used for operands and results at each stage. For example, for additions at the first stage even bits of 48-operands (A_0,A_2,…,B_0,B_2,…,) are set to zero. Two-bit results (n = 2) were taken from associated even and odd bits and they are used again as inputs with indices 16,17; 19,20, 22,23, 25,26. The most significant bits in each group of inputs with n + 1 bits (i.e. 18, 21, 24, 27) are again not used and set to zero. The result of comparison is formed similarly to [6] (see also Sect. 3.7). As in Sect. 3.9, a LUT-based comparator can also be used to support features shown in Fig. 3.30.

We found that the considered method permits quite complex Hamming weight counters/comparators to be implemented with very small resources. For example, the circuit in Fig. 4.11 requires just one DSP48E1 slice. For N = 32 the number

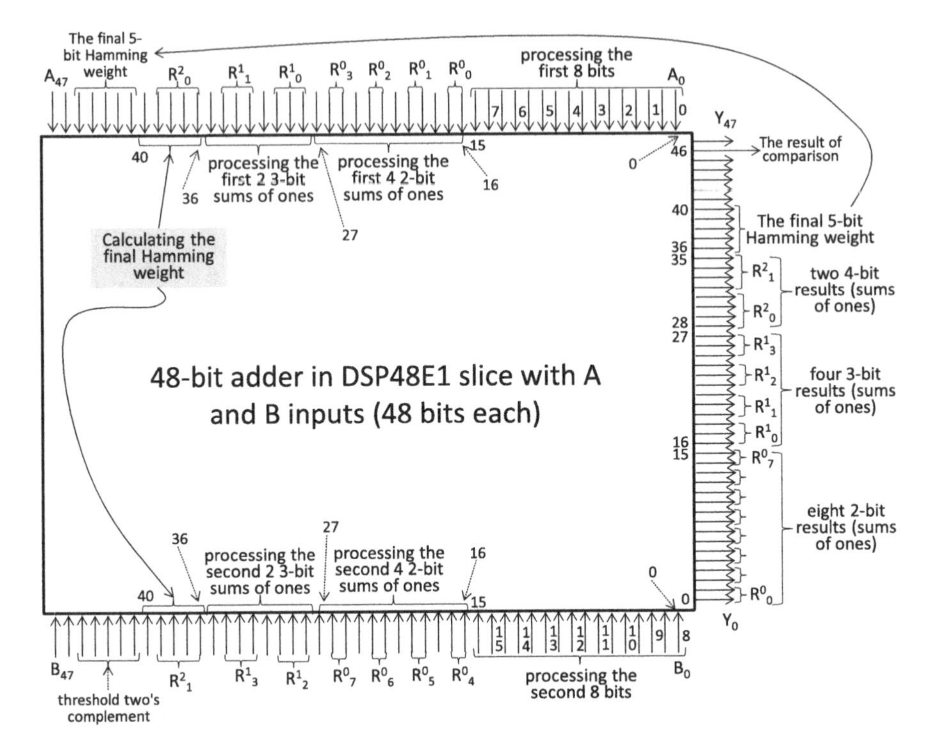

Fig. 4.11 Implementation of Hamming weight counter/comparator for 16-bit binary vectors in one DSP48E1 slice

of such slices is 2 and for N = 64 the number is 4. Complete synthesizable VHDL codes for DSP-based counters/comparators are given in Appendix B. Similar circuits can also be built for the DSP48A1 slice in previous families of Xilinx FPGAs (such as that are available in Spartan-6 FPGA of the Atlys board).

Let us consider now the synthesizable VHDL code for Hamming weight counter/comparator shown in Figs. 4.10 and 4.11.

```
entity Test_HW16 is -- the circuit occupiers 1 DSP slice and 0 logical slices
  port ( Sw : in std_logic_vector (15 downto 0);       -- 16-bit input vector
         led : out std_logic_vector ( 4 downto 0);     -- the Hamming weight and the result
         led_comp: out std_logic );                    -- of comparison with fixed threshold
end Test_HW16;

architecture Behavioral of Test_HW16 is
  signal A, B, Y: std_logic_vector(47 downto 0); -- DSP operands (A,B) and the result (Y)
  signal threshold : std_logic_vector(4 downto 0);
begin
threshold <= not "01010" + 1;                          -- threshold two's complement

process(Sw, Y, threshold)
begin
  A <= (others => '0');                -- the first 48-bit DSP operand
```

```
  B <= (others => '0');                    -- the second 48-bit DSP operand
  for i in 7 downto 0 loop                 -- the first stage in Fig. 4.10
     A(2*i) <= Sw(i);
     B(2*i) <= Sw(i+8);
  end loop;
  for i in 3 downto 0 loop                 -- the second stage in Fig. 4.10
     A(16+3*i+1 downto 16+3*i) <= Y(2*i+1 downto 2*i);
     B(16+3*i+1 downto 16+3*i) <= Y(2*i+1+8 downto 2*i+8);
  end loop;
  for i in 1 downto 0 loop                 -- the third stage in Fig. 4.10
     A(28+4*i+2 downto 28+4*i) <= Y(16+3*i+2 downto 16+3*i);
     B(28+4*i+2 downto 28+4*i) <= Y(16+3*i+2+6 downto 16+3*i+6);
  end loop;
  A(39 downto 36) <= Y(31 downto 28); -- the fourth stage in Fig. 4.10
  B(39 downto 36) <= Y(35 downto 32);
  A(45 downto 41) <= Y(40 downto 36);  -- Hamming weight comparison
  B(45 downto 41) <= threshold;
end process;

led <= Y(40 downto 36);            -- the resulting Hamming weight
led_comp  <= Y(46);                            -- the result of Hamming weight comparison
DSP: entity work.TesDSP48E1_HW16
      port map (A, B, "0000", Y);
end Behavioral;
```

The entity TesDSP48E1_HW16 is described as follows:

```
entity TesDSP48E1_HW16 is
   port ( A_conc_B        : in  std_logic_vector (47 downto 0);
          C               : in  std_logic_vector (47 downto 0);
          mode            : in  std_logic_vector(3 downto 0);
          Result          : out  std_logic_vector (47 downto 0));
end TesDSP48E1_HW16;
```

The component TesDSP48E1_HW16 has the shown below changes comparing with the template:

1. The following two lines are used in the architecture body

```
A <= A_conc_B(47 downto 18);    -- A is 30-bit operand
B <= A_conc_B(17 downto 0);     -- B is 18-bit operand
```

2. The following lines have been changed in the port map:

```
P => Result,                        -- see Fig. 4.8
A => A,                     -- 30-bit input: A data input
B => B,                     -- 18-bit input: B data input
C => C,                     -- 48-bit input: C data input
OPMODE => "0001111",  -- Mux X and Mux Y are used (see Fig. 4.8)
ALUMODE => mode,        -- mode = "0000": addition operation is chosen
```

3. The other changes are shown in Fig. 4.9

The following VHDL code shows changes needed to implement Hamming weight counter for N = 19 in one DSP slice:

```
entity Test_HW19 is     -- the circuit occupiers 1 DSP slice and 0 logical slices
  port ( Sw : in STD_LOGIC_VECTOR (15 downto 0);          -- 16 bit input vector
         led: out STD_LOGIC_VECTOR (15 downto 0);         -- The Hamming weight
         BTNL, BTNR, BTNC : in std_logic); -- additional 3 bits for input vector
  end Test_HW19;

-- below only changes comparing to the previous project are shown

process(Sw,Y,BTNL,BTNR)
  A(41) <= BTNL;
  B(41) <= BTNR;
  A(44 downto 43) <= Y(42 downto 41);
  B(47 downto 43) <= Y(40 downto 36);
end process;

led <= (15 downto 5 => '0') & Y(47 downto 43);
DSP: entity work.TesDSP48E1_HW19
       port map (A, B, "0000", Y, BTNC);
```

The following additional change is done in the DSP template (in the port map):
CARRYIN => BTNC,

Both projects were tested in the Nexys-4 board. Projects for N = 32 and N = 64 are given in Appendix B, where an additional signal CarryOutBit is used which is the most significant bit (bit 3) of 4-bit CARRYOUT signal shown in Fig. 4.8. It keeps the carry out signal for 48-bit operational unit in DSP48E1.

4.3 Interaction with FPGA

Up to now we interacted with all the developed circuits with the aid of peripheral components (such as push buttons, switches, LEDs and 7-segment displays) available on prototyping boards. This type of interaction permits to test simple projects but is inappropriate for more complicated designs. Moreover capabilities of onboard components to supply (large) input data for processing in FPGA are limited. That is why a random number generator was used in some of the examples above. So, when considerable amounts of data have to be transferred from software running on a host computer to a circuit implemented in an FPGA and back, we need to provide support for interaction between the computer and the FPGA.

We will explore two types of interaction: Digilent parallel port interface and UART (Universal Asynchronous Receiver and Transmitter) and illustrate how to develop communicating modules in both PC software and FPGA hardware.

4.3.1 Digilent Parallel Port Interface

The Digilent Parallel port interface follows the EPP (Enhanced Parallel Port) mode of parallel communications. This interface can only be used in Digilent prototyping boards equipped with Digilent EPP data transferring capability (such as Nexys-2, Nexys-3, and Atlys; please note that the Nexys-4 board does not support this interface).

EPP is a half-duplex bi-directional interface which means that a receiver and a transmitter share a single parallel data bus but not at the same time. Data bus is 8-bit wide and there are 6 handshaking lines to control the data transfer (only 4 handshaking lines are used by Digilent EPP). The logic in FPGA has to include a single 8-bit address register and up to 256 8-bit data registers which can be read and written by the host PC. Individual data registers are addressable through the value specified in the address register. The functions of all the Digilent EPP interface lines are the following:

- EppDB—8-bit bi-directional data bus.
- EppAstb—address strobe driven by the host PC which causes data to be read from or written to the address register.
- EppDstb—data strobe driven by the host PC which causes data to be read from or written to a data register.
- EppWait—synchronization signal driven by the FPGA, which is used to indicate when the FPGA is ready to transmit or receive data.
- EppWr—direction of data transfer chosen by the host PC (when *high*—PC reads data from an FPGA register, when *low*—PC writes data to an FPGA register).

Access to the registers is done through transfer cycles and four types of transactions are possible: read the address register, write the address register, read a data register, and write a data register. The direction of the data transfer is controlled by the host PC through the EppWr signal. Timing diagrams from Fig. 4.12 illustrate read and write transfer cycles. More details are available in [7].

4.3.1.1 Digilent EPP Communication Module

Let us support three data registers in the EPP communication module:

- Register 0x00—will hold an address for memory transactions (for example when we would like to fill in a memory block with data supplied from the PC it would be helpful to specify memory locations to which these data are to be written; the same technique can be used during read cycles to enable the host to read from a specific memory address).
- Register 0x01—holds 8-bit user data received from the PC.
- Register 0x05—holds 8-bit data to be sent to the PC.

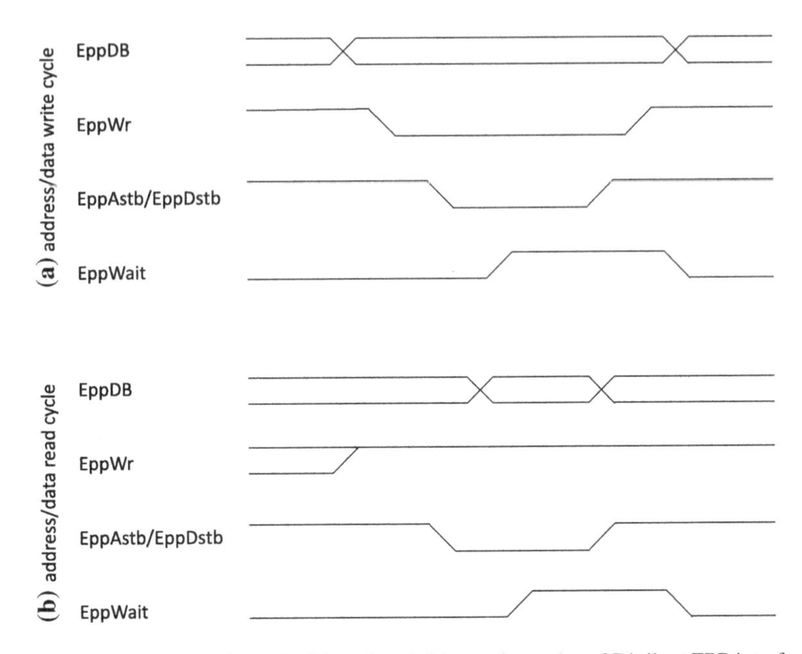

Fig. 4.12 Timing diagrams for write (**a**) and read (**b**) transfer cycles of Digilent EPP interface

Please note that the data register addresses (i.e. 0x00, 0x01, and 0x05) were chosen arbitrary and other addresses could be equally applied. If needed, more registers (up to 256) can easily be added to the communication module (see an example in [8] where 16 data registers are involved).

The following entity EPP_interface implements the address register and the listed above data registers and interacts with the parallel port bus according to the timing diagrams in Fig. 4.12. Besides of communication signals, the block includes three additional output ports and one input port:

- data_to_PC is an 8-bit value supplied by a circuit implemented in the FPGA which holds data to be sent to the host PC over the Digilent EPP interface.
- data_from_PC is an 8-bit value received from the host PC over the Digilent EPP interface, which is further transferred to other logic circuits implemented in the FPGA.
- data_ready is a 1-bit signal which indicates that data have just been received from the PC and are ready for further processing.
- address—is an 8-bit signal which holds an address for memory transactions, this address is set by the host PC.

```
library ieee;
use ieee.std_logic_1164.all;
entity EPP_interface is
  port (-- EEP handshaking signals and data bus
      EppAstb: in std_logic;                    -- address strobe
      EppDstb: in std_logic;                    -- data strobe
```

```
        EppWr  : in std_logic;                                  -- direction of data transfer
        EppDB  : inout std_logic_vector(7 downto 0);            -- parallel data bus
        EppWait: out std_logic;                                 -- synchronization wait signal
        -- user extended signals
        -- address for memory access operations (stored in the data register 0x00)
        address : out std_logic_vector (7 downto 0);
        -- signal which indicates that data are ready to be used in other design blocks
        data_ready : out std_logic;
        -- 8-bit user data received from the PC (stored in the data register 0x01)
        data_from_PC: out std_logic_vector(7 downto 0);
        -- 8-bit data to send to the PC (held in the data register 0x05)
        data_to_PC : in std_logic_vector(7 downto 0));
end EPP_interface;

architecture Behavioral of EPP_interface is
  signal EppAddressRegister: std_logic_vector (7 downto 0);   -- Epp address register
  signal EppInternalBus: std_logic_vector(7 downto 0);        -- internal bus
begin

--activate EppWait when either address strobe or data strobe is asserted
EppWait <= '1' when EppAstb = '0' or EppDstb = '0' else '0';
--write to the data bus during PC read cycles
EppDB <= EppInternalBus when (EppWr = '1') else (others => 'Z');
--write address or data to the bus
EppInternalBus <= EppAddressRegister when (EppAstb = '0') else data_to_PC;

address_register: process (EppAstb)
begin
  if rising_edge(EppAstb) then                --end of address access cycle
    if EppWr = '0' then                       --this is address write cycle
       EppAddressRegister <= EppDB;           --update the address register
    end if;
  end if;
end process address_register;

data_registers: process (EppDstb)
begin
  if rising_edge(EppDstb) then                --end of data access cycle
    if EppWr = '0' then                       --this is data write cycle
       data_ready <= '0';
       if EppAddressRegister = x"00" then     --memory address register
         address <= EppDB;
       elsif EppAddressRegister = X"01" then--register holding user data received from PC
         data_from_PC <= EppDB;
         data_ready <= '1';
       end if;
    end if;
  end if;
end process data_registers;

end Behavioral;
```

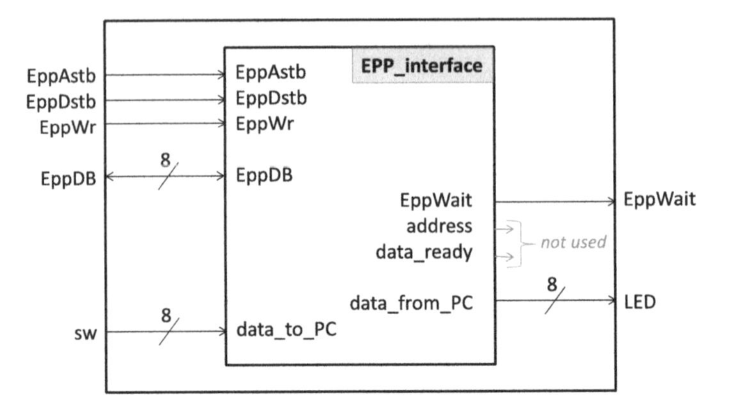

Fig. 4.13 Structure of a circuit which receives an 8-bit value from the host PC via the Digilent EPP interface, displays this value on the LEDs of Atlys board, and sends the 8-bit value selected on board's switches to the host PC

Let us use the developed communication module to design a simple circuit which receives from the host PC a randomly generated 8-bit value and displays it on LEDs of the Atlys board. Similarly, the 8-bit value selected on the board's switches is sent to the host PC. The circuit is organized as shown in Fig. 4.13.

This circuit can be described in VHDL as follows:

```
library ieee;
use ieee.std_logic_1164.all;

entity main is
  port (EppAstb : in  std_logic;
        EppDstb : in  std_logic;
        EppWr : in  std_logic;
        EppDB : inout  std_logic_vector(7 downto 0);
        EppWait : out  std_logic;
        sw : in std_logic_vector(7 downto 0);
        LED : out std_logic_vector(7 downto 0));
end main;

architecture Behavioral of main is
  signal data_from_PC, data_to_PC : std_logic_vector(7 downto 0);
begin
EPP: entity work.EPP_interface port map (EppAstb => EppAstb,
      EppDstb => EppDstb, EppWr => EppWr, EppDB => EppDB,
      EppWait => EppWait, address => open, data_ready => open,
      data_to_PC => data_to_PC, data_from_PC => data_from_PC);
LED <= data_from_PC;
data_to_PC <= sw;
end Behavioral;
```

Note that data transfers in this example are trivial, that is why, for now, signals address and data_ready, which are generated in the communication module, are not required and were left unconnected. In more complicated examples from Sect. 4.4 these signals will be used intensively.

All ports of the main entity have to be connected to appropriate FPGA pins. The respective pin locations can be found in the master user constraints file (UCF) available in the board's documentation.

To test the designed circuit, the second communication module has to be developed in PC software. The following section gives the necessary explanations and an example.

4.3.1.2 Application Software

Once the required hardware modules are developed, to test them and to appreciate the data transfer facilities we need to design application software that would interact with the FPGA logic. We will perform data transfer over USB with the aid of *Adept* SDK (software development kit) which provides an Application Programming Interface (API) DPCUTIL allowing EPP-equipped Digilent prototyping boards to communicate with application software running under Microsoft Windows on a host computer [9]. The API requires a parallel port interface to be implemented in FPGA in a way described in Sect. 4.3.1.1. The API is composed of a number of C functions and can be used with programs written in C/C++ [9]. The available API functions allow for accessing (both reading and writing) a single register or a set of registers. The detailed description of the API can be found in the reference manual [9].

In order to use DPCUTIL API functions, a program must be linked with dpcutil library (available at [9]) and include the following header files:

```
#include <windows.h>
#include "dpcdefs.h"
#include "dpcutil.h"
```

C++ code below illustrates the interaction between a software program running on a host PC and an EPP-equipped Digilent prototyping board:

```
//The following header files are required to use the DPCUTIL API
//The program must be linked with the dpcutil.lib library.
#include <windows.h>
#include "dpcdefs.h"
#include "dpcutil.h"
#include <iostream>
#include <ctime>

const int INITIALIZATION_FAILED = 1;
const int NO_DEFAULT_DEVICE = 2;
const int INTERNAL_ERROR = 3305; //internal error in DPCUTIL
const int devNameLength = 16;
char nameDevice[devNameLength+1];
```

```
void SendDataToFPGA(unsigned data);
bool WriteData(HANDLE hif, unsigned data);
void ReceiveResultFromFPGA(unsigned& data);
bool ReadData(HANDLE hif, unsigned& data);

using namespace std;

int main(int argc, char* argv[])
{ ERC error_code;
  if ( !DpcInit(&error_code) ) //before using DPCUTIL API functions, call DpcInit
     return INITIALIZATION_FAILED; //error occurred while initializing
  //obtain the index of the default device in the Device Table
  int idDevice = DvmgGetDefaultDev(&error_code);
  if (idDevice == -1) //no devices in the Device Table
  {    cerr << "No default device"<< endl;
       cerr << "Run Digilent Adept and modify the Device Table" <<
                  " (Settings tab, Device Manager option)" << endl;
       return NO_DEFAULT_DEVICE;
  }
  else //get the default device name
       DvmgGetDevName(idDevice, nameDevice, &error_code);

  unsigned data, result;
  const int range_min = 0, range_max = 0xff;
  srand (static_cast<unsigned>(time(0)));
  char operation;

  do
  {    cout << "Select an operation (r - read switches, s - send a value, e - exit)\n";
       cin >> operation;
       switch (operation)
       { case 'r': ReceiveResultFromFPGA(result);
                   cout << "The result from FPGA is: " << hex << result << endl;
                   break;
         case 's': //randomly generate an 8-bit number
                   data = static_cast<unsigned>((double)rand() / ( RAND_MAX + 1) *
                             (range_max - range_min)  + range_min);
                   SendDataToFPGA(data); //send data to the FPGA
                   break;
            case 'e' : break;
            default: cout << "Wrong parameter" << endl;
         }
    }
    while (operation != 'e');
    return 0;
}

void SendDataToFPGA(unsigned data) //sends an 8-bit data item to the FPGA
{ ERC error_code;
   HANDLE hif;
   TRID trid; //transaction ID type
   //before using data transfer functions, connect to a communication device
```

```
    if (!DpcOpenData(&hif, nameDevice, &error_code, &trid))
    {    cerr << "DpcOpenData failed." << endl;
         return;
    }
    //wait for the last (trid) transaction to be completed
    if (!DpcWaitForTransaction(hif, trid, &error_code))
    {    DpcCloseData(hif, &error_code); // close the communications module
         cerr << "DpcOpenData failed." << endl;
         return;
    }
    if (!WriteData(hif, data)) return; //data transfer
    error_code = DpcGetFirstError(hif); //search for the first transaction with an error
    if ((error_code == ercNoError) || (error_code == INTERNAL_ERROR) )
    {    DpcCloseData(hif, &error_code); //close the communications module
         cout << "Value " << hex << data << " successfully written to the FPGA." << en(
    }
    else
    {    DpcCloseData(hif, &error_code); //close the communications module
         cerr << "An error occurred while setting the register" << endl;
  }
}
bool WriteData(HANDLE hif, unsigned data)
{ ERC error_code;
  unsigned char idData;
  unsigned idReg;
  idReg = 0x01;
  idData = data;
  //send a single data byte (idData) to the register idReg
  if (!DpcPutReg(hif, idReg, idData, &error_code, 0))
  {    DpcCloseData(hif, &error_code); //close the communications module
       cerr << "DpcPutReg failed." << endl;
       return false;
  }
  return true;
}
void ReceiveResultFromFPGA(unsigned& data)
{ ERC error_code;
  HANDLE      hif;
  //before using data transfer functions, connect to a communication device
  if (!DpcOpenData(&hif, nameDevice, &error_code, 0))
  {    cerr << "DpcOpenData failed." << endl;
       return;
  }
  if (!ReadData(hif, data)) return; //data transfer
  error_code = DpcGetFirstError(hif); //search for the first transaction with an error
  if ((error_code == ercNoError) || (error_code == INTERNAL_ERROR) )
  {    DpcCloseData(hif, &error_code); //close the communications module
       cout << "Values successfully received from the FPGA." << endl;
  }
```

```
    else
    {   DpcCloseData(hif, &error_code); //close the communications module
        cerr << "An error occurred while reading the register" << endl;
    }
}
bool ReadData(HANDLE hif, unsigned& data)
{ ERC error_code;
  unsigned char idData;
  unsigned idReg;
  data = 0;
  idReg = 0x05;
  //get a single data byte (idData) from the register idReg
  if (!DpcGetReg(hif, idReg, &idData, &error_code, 0))
  {   DpcCloseData(hif, &error_code); //close the communications module
      cerr << "DpcGetReg failed." << endl;
      return false;
  }
  data = idData;
  return true;
}
```

The program starts by initializing dpcutil and obtaining the index of the default device in the *Device Table*. The *Device Table* is managed by the Digilent *Adept* utility (*Settings* tab, *Device Manager* option). If there is a problem during the initialization phase, ensure that the *dpcutil* is visible to the linker and that the connected board appears in the *Device Table*. Afterwards the program iteratively suggests the user to choose one of 3 available options: "r"—read the value of the board's switches and print it on the screen, "s"—send a randomly generated 8-bit value to the board and show it on LEDs, and "e"—exit. The functions SendDataToFPGA and WriteData send an 8-bit data item to the FPGA (by writing to the data register 0x01). The function ReceiveResultFromFPGA and ReadData get an 8-bit value from the communication module implemented in the FPGA (by reading the data register 0x05).

The following dpcutil functions are used in the code above:

- DpcInit—must be called before using any dpcutil API function. The function returns true value if initialization was successful, otherwise false value is returned.
- DvmgGetDefaultDev—permits to get the index of the default device in the Digilent *Device Table*. The function returns -1 if there is no default device. In this case the user has to set a default device in the *Device Table* managed through the Digilent *Adept* utility (*Settings* tab, *Device Manager* option).
- DvmgGetDevName—gets the name of the selected default device.
- DpcOpenData—establishes connection with the default device. This function creates a handle to be used in subsequent data transfer cycles.
- DpcWaitForTransaction—waits for a transaction to be complete.
- DpcCloseData—closes the device.
- DpcGetFirstError—searches for the first transaction with error and returns the respective error code. Error codes are detailed in [9].

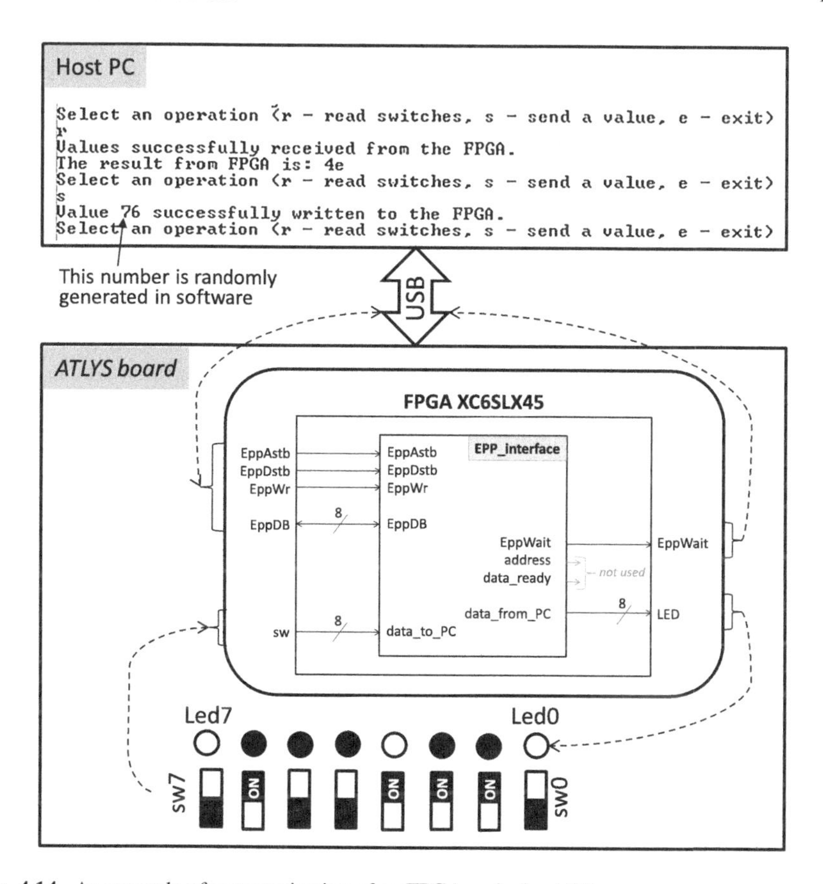

Fig. 4.14 An example of communication of an FPGA and a host PC

- DpcPutReg—sends one byte to a specified data register. In the code above the only register that is written is 0x01.
- DpcGetReg—gets one byte from a specified data register. In the code above the only register that is read is 0x05.

Finally, the code above and the circuit described in Sect. 4.3.1.1 permit an FPGA and a host PC to communicate as depicted in Fig. 4.14. The explored communication scenario is very simple but serves as a base to develop more complex interaction models as will be shown in Sect. 4.4.

4.3.2 UART Interface

For boards that do not provide support for Digilent EPP data transferring capability (such as the Nexys-4) other communication interfaces need to be explored. Let

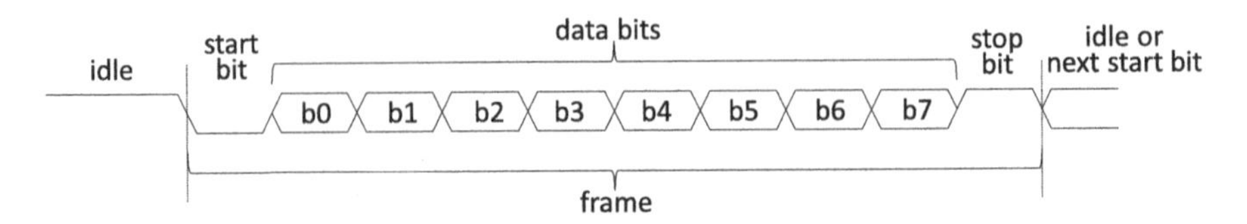

Fig. 4.15 An example of a frame containing one start bit, 8 data bits, and one stop bit

us consider the UART interface which implements a very simple serial communication protocol.

The Nexys-4 board includes a USB-UART bridge that permits a host PC to communicate with the board using standard Windows COM (Communication port) commands [1]. To establish a communication, USB-COM port drivers [10] are required to convert USB packets to UART data. Four pins of the onboard FPGA are connected to four lines of the USB controller: RXD (pin C4)—for data transmission from a host PC to the FPGA, TXD (pin D4)—for data transmission from the FPGA to a host PC, and RTS/CTS (pins E5/D3) are the handshaking control signals. In this way the board supports a full duplex bi-directional interface, i.e. both the FPGA and the PC can transmit data at the same time using separate lines (RXD and TXD).

Since the FPGA has to transmit data to a host PC as well as to receive data from the host PC, both a transmitter and a receiver circuits have to be implemented. Basically, a transmitter gets parallel data and shifts them to the communication line bit by bit at a special rate. The receiver executes the opposite task, i.e. it extracts data bit by bit from the serial line (also at a special rate) and converts the result to a parallel representation. The rate of data (bit) sampling is called a baud rate and, for UART, this is essentially the number of bits transmitted per second over a serial communication line. For the interface to work properly both the receiver and the transmitter must function at the same baud rate. The serial line is always '1' when idle. The transmission begins with a start bit, which is always '0', followed by several data bits and an optional parity bit, and ends with one or several stop bits, which are '1'. The number of data bits transmitted is typically equal to 8 (i.e. one byte). The parity bit might be used by the receiver to detect errors in the transmitted data. Only odd number of errors can be detected. Once again both the receiver and the transmitter must agree on the presence of the parity bit and the number of stop bits.

In this section we will consider a simple UART communication module which transmits 8 data bits, without parity bits and with one stop bit. The respective frame structure is shown in Fig. 4.15.

4.3.2.1 UART Communication Module

We will design a UART communication module which operates at the baud rate 9600 and whose frame structure is depicted in Fig. 4.15. We will target our design to the Nexys-4 prototyping board, which includes a 100 MHz crystal oscillator. The external interface of the communication module will be the following:

```
entity UART_comm is
  port (clk : in std_logic;                          -- board's clock (100 MHz)
        WR : in std_logic;                           -- write strobe (to send data to PC)
        DIN : in std_logic_vector (7 downto 0);      -- data to send
        DOUT : out std_logic_vector (7 downto 0);    -- data received
        TX_ready : out std_logic;                    -- ready to transmit
        RX_ready : out std_logic;                    -- received data are available
        TXD : out std_logic;                         -- transmission line
        RXD : in std_logic);                         -- reception line
end UART_comm;
```

Here, clk is the board's clock signal. The communication lines TXD and RXD are used for serial data transmission in both directions. DIN is the byte to be transmitted from the FPGA to the host PC (or other device). DOUT is the data byte received from the host PC (or other device). WR is an input signaling that data on DIN bus are ready and have to be transmitted. The outputs TX_ready and RX_ready are asserted when the module has finished the last communication cycle and is ready to transmit more data (TX_ready) or that data have just been received and are ready for processing (RX_ready). The module will not provide any buffering of data, so other FPGA logic has to guarantee that new data are not supplied if TX_ready is not asserted.

To support the selected baud rate (9600 bits per second), the input clock signal (100 MHz) has to be divided by $10^8/9600 \approx 10416$ and the communication lines have to be sampled/written at the middle of a bit period. The operation of the module will be controlled by the original 100 MHz clock signal.

Let us start with designing a transmitter, which is simpler than a receiver. The flowchart, which describes the control operations, is shown in Fig. 4.16 (outputs are specified in VHDL). The control finite state machine (FSM) includes 3 states: READY, LOAD_BIT, and SEND_BIT. The FSM is in the state READY when it is ready to receive more data from other FPGA logic and to transmit them over UART. Essentially, this is a waiting state which constantly drives the TXD communication line with logic value '1' (i.e. the line is idle). The signal TX_bitIndex stores the index of the next bit to be transmitted over the TXD line; it ranges from 0 (for the start bit) to 9 (for the stop bit). The signal TX_div is a counter which permits bits to be sent over TXD line with the established baud rate; it counts from 0 to 10416 (the maximum value is stored in a constant CLK_DIV). In the READY state the counter is reset to 0. Once the transmission request comes through the WR signal, the FSM composes the frame to be transmitted and changes its state to the LOAD_BIT. The frame is stored in a 10-bit signal TX_frame and includes 1 start bit ('0'), 8 data bits from DIN input and 1 stop bit ('1'), according to Fig. 4.15. In the LOAD_BIT state the FSM deasserts the TX_ready signal and drives the TXD communication line with one bit from the frame. The bit to send is indicated by TX_bitIndex. At the same time, TX_bitIndex is incremented to point to the next bit in the frame to be transmitted. Finally, the state SEND_BIT continues to drive the TXD line with the selected in the

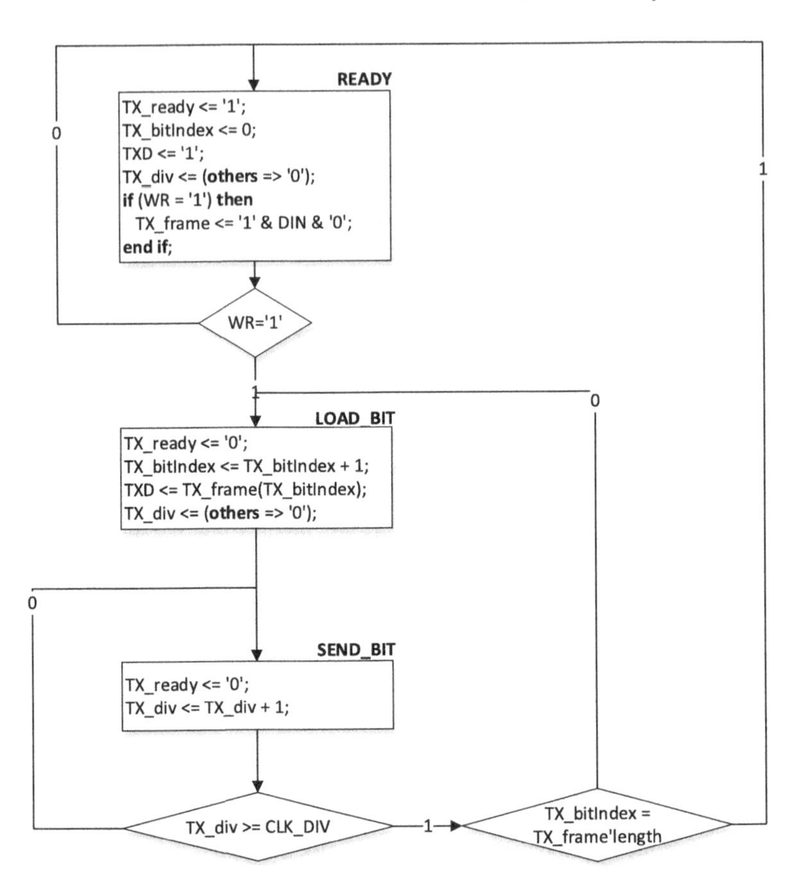

Fig. 4.16 Flowchart for the UART transmitter

previous state bit until CLK_DIV clock cycles (with the frequency 100 MHz) have passed. By this time, the next bit might be extracted from the frame and driven to the TXD line in the state LOAD_BIT. Once the last (stop) bit from the frame is transmitted, the FSM returns to the READY state and waits for new data to arrive from other FPGA circuits.

The receiver is a bit more complex. Essentially, it has to permanently monitor the RXD line until a transition from '1' (idle) to '0' (start bit) is detected. Once the start of a new arriving frame is identified, the receiver has to wait for half a baud rate period and then sample the first (start) bit. Afterwards the remaining 9 bits (8 data bits and a stop bit) have to be read from the RXD line with intervals equal to the period for the selected baud rate. The flowchart that describes behavior of the receiver is shown in Fig. 4.17.

The receiver control FSM includes six states: READY, DETECT_START_BIT, PUT_BIT, READ_BIT, DATA_READY, and DONE. The FSM is in the state READY when it is waiting for new data to arrive on the RXD line. The signal RX_ready is deasserted indicating that no data have been received so far. The signal RX_div is a counter,

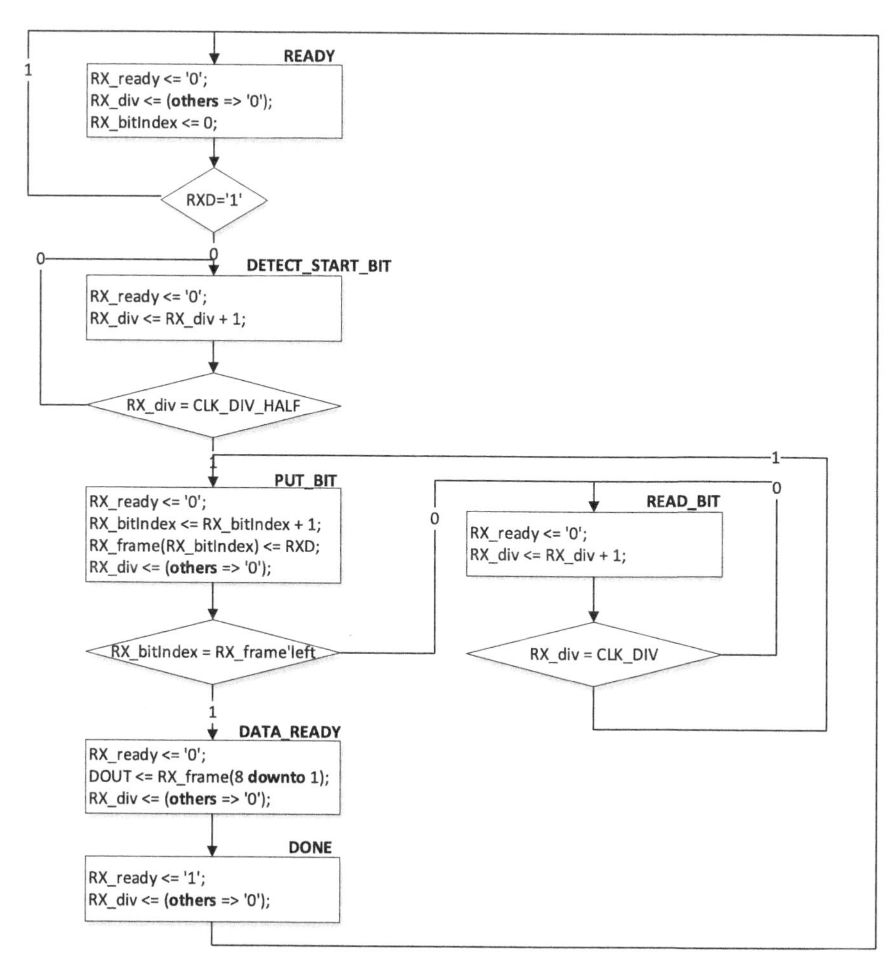

Fig. 4.17 Flowchart for the UART receiver

similar to TX_div described above, which permits bits to be received over the RXD line with the established baud rate; it counts either from 0 to 5208 (10416/2, half the baud rate period, stored in a constant CLK_DIV_HALF) or from 0 to 10416 (the baud rate period stored in a constant CLK_DIV). In the READY state the counter is reset to 0. The signal RX_bitIndex stores the frame index of the next bit to be received from the RXD line; it ranges from 0 (for the start bit) to 9 (for the stop bit). In the state READY the RXD signal is constantly sampled until a transition from '1' to '0' is detected.

Once the start bit is detected, the FSM changes its state to the DETECT_START_BIT. In this state RX_div counter is incremented once every (100 MHz) clock cycle until it reaches value CLK_DIV_HALF (which means that half a baud rate period has elapsed, the RXD line is to be sampled for the start bit) and the FSM changes its state to the PUT_BIT. In the PUT_BIT state the FSM samples the RXD communication line and stores the extracted bit in the receiver frame RX_frame at the position indicated by RX_bitIndex. At the same time, RX_bitIndex is incremented to point to the next bit in the frame to be received. The next state is the READ_BIT which is a waiting state counting until CLK_DIV (100 MHz) clock cycles have passed after the last bit had been sampled from the RXD line. By this time, the next bit might be sampled from the RXD line in the state PUT_BIT. Once the last (stop) bit is extracted, the FSM changes to the DATA_READY state, where information bits from the frame are driven to the output DOUT, which is to be used by other FPGA circuits. The start and stop bits (with indices 0 and 9, respectively) are discarded. The last FSM state is DONE where the signal RX_ready is asserted indicating that new data have been received and are available on DOUT output. Afterwards the FSM returns to the state READY and waits for new data to arrive over the RXD line.

VHDL code describing both the transmitter and the receiver is shown below.

```
library IEEE;
use IEEE.std_logic_1164.all;
use IEEE.std_logic_unsigned.all;

--9600 baud rate, 8 data bits, no parity, 1 stop bit
entity UART_comm is
  port (clk : in std_logic;                        -- board's clock (100 MHz)
        WR : in  std_logic;                        -- write strobe (to send data to PC)
        DIN : in  std_logic_vector (7 downto 0);   -- data to send
        DOUT : out  std_logic_vector (7 downto 0); -- data received
        TX_ready : out std_logic;                  -- ready to transmit
        RX_ready : out std_logic;                  -- received data are available
        TXD : out std_logic;                       -- transmission line
        RXD : in std_logic);                       -- reception line
end UART_comm;

architecture Behavioral of UART_comm is

--CLOCK
--100 MHz/9600 = 10416 = 0x28B0
constant CLK_DIV : std_logic_vector(13 downto 0) := "10" & x"8B0";
--100 MHz/9600/2 = 5208 = 0x1458
constant CLK_DIV_HALF : std_logic_vector(12 downto 0) := "1" & x"458";

--TRANSMISSION
signal TX_div : std_logic_vector(13 downto 0) := (others => '0');
--frame = 1 start + 8 data + 1 stop = 10 bits
signal TX_frame : std_logic_vector(9 downto 0) := '1' & x"ff" & '0';
type TX_TYPE is (READY, LOAD_BIT, SEND_BIT);
signal TX_state, TX_next_state : TX_TYPE := READY;
signal TX_bitIndex : natural; -- index of the next bit in TX_frame to be transferred

--RECEPTION
signal RX_div : std_logic_vector(13 downto 0) := (others => '0');
signal RX_frame : std_logic_vector(9 downto 0); --1 start + 8 data + 1 stop = 10 bits
type RX_TYPE is (READY, DETECT_START_BIT, READ_BIT, PUT_BIT,
                 DATA_READY, DONE);
signal RX_state, RX_next_state : RX_TYPE := READY;
```

```
signal RX_bitIndex : natural; -- index of the next bit in the RX_frame to be received

begin
------------------------------------------------------
---     TRANSMISSION
------------------------------------------------------
TX_state_transition: process (clk)
begin
  if (rising_edge(clk)) then
    TX_state <= TX_next_state;
  end if;
end process TX_state_transition;

TX_output_logic: process (clk)
begin
  if (rising_edge(clk)) then
    case TX_state is
      when READY =>
        TX_ready <= '1';
        TX_bitIndex <= 0;
        TXD <= '1'; -- idle
        TX_div <= (others => '0');
        if (WR = '1') then
                TX_frame <= '1' & DIN & '0';
        end if;
      when LOAD_BIT =>
        TX_ready <= '0';
        TX_div <= (others => '0');
        TX_bitIndex <= TX_bitIndex + 1;
        TXD <= TX_frame(TX_bitIndex);
      when SEND_BIT =>
        TX_ready <= '0';
        TX_div <= TX_div + 1;
    end case;
  end if;
end process TX_output_logic;

TX_next_state_logic: process (TX_state, WR, TX_div, TX_bitIndex)
begin
  case TX_state is
    when READY =>
      if (WR = '1') then
        TX_next_state <= LOAD_BIT;
      else
        TX_next_state <= READY;
      end if;
    when LOAD_BIT =>
      TX_next_state <= SEND_BIT;
    when SEND_BIT =>
      if (TX_div >= CLK_DIV) then
        if (TX_bitIndex = TX_frame'length) then
                TX_next_state <= READY;
        else
                TX_next_state <= LOAD_BIT;
        end if;
```

```
        else
        TX_next_state <= SEND_BIT;
        end if;
       when others => -- should never be reached
         TX_next_state <= READY;
  end case;
end process TX_next_state_logic;
------------------------------------------------------
---     RECEPTION
------------------------------------------------------
RX_state_transition: process (clk)
begin
  if (rising_edge(clk)) then
    RX_state <= RX_next_state;
  end if;
end process RX_state_transition;

RX_output_logic: process (clk)
begin
  if (rising_edge(clk)) then
    case RX_state is
      when READY =>
        RX_ready <= '0';
        RX_div <= (others => '0');
        RX_bitIndex <= 0;
      when DETECT_START_BIT =>
        RX_ready <= '0';
        RX_div <= RX_div + 1;
      when PUT_BIT =>
        RX_ready <= '0';
        RX_bitIndex <= RX_bitIndex + 1;
        RX_frame(RX_bitIndex) <= RXD;
        RX_div <= (others => '0');
      when READ_BIT =>
        RX_ready <= '0';
        RX_div <= RX_div + 1;
      when DATA_READY =>
        RX_ready <= '0';
        DOUT <= RX_frame(8 downto 1); --extract only data bits
        RX_div <= (others => '0');
      when DONE =>
        RX_ready <= '1';
        RX_div <= (others => '0');
    end case;
  end if;
end process RX_output_logic;

RX_next_state_logic: process (RX_state, RXD, RX_div, RX_bitIndex)
begin
  case RX_state is
```

```
case RX_state is
  when READY =>
    if (RXD = '1') then --idle
      RX_next_state <= READY;
    else
      RX_next_state <= DETECT_START_BIT; --start bit detected
    end if;
  when DETECT_START_BIT =>
    if (RX_div = CLK_DIV_HALF) then
      RX_next_state <= PUT_BIT;
    else
      RX_next_state <= DETECT_START_BIT;
    end if;
  when PUT_BIT =>
    if (RX_bitIndex = RX_frame'left) then
      RX_next_state <= DATA_READY;
    else
      RX_next_state <= READ_BIT;
    end if;
  when READ_BIT =>
    if (RX_div = CLK_DIV) then
      RX_next_state <= PUT_BIT;

    else
      RX_next_state <= READ_BIT;
    end if;
    when DATA_READY =>
      RX_next_state <= DONE;
    when DONE =>
      RX_next_state <= READY;
    when others => -- should never be reached
      RX_next_state <= READY;
  end case;
end process RX_next_state_logic;

end Behavioral;
```

Let us show how the developed UART transmitter/receiver can be used in a circuit which receives an 8-bit value from the host PC and displays this value on the Nexys-4 board rightmost LEDs. Similarly, the 8-bit value selected on the board's rightmost switches is sent to the host PC. The circuit is organized as shown in Fig. 4.18.

This circuit can be described in VHDL as follows:

```
library ieee;
use ieee.std_logic_1164.all;
use ieee.std_logic_unsigned.all;

entity main is
  port ( clk : in std_logic;
         TXD : out std_logic;
         RXD : in std_logic;
         sw : in std_logic_vector(7 downto 0);
         LED : out std_logic_vector(7 downto 0));
end main;
```

```
architecture Behavioral of main is
signal data_from_PC, data_to_PC : std_logic_vector(7 downto 0);
signal WR : std_logic;
constant CLK_DIV : std_logic_vector(23 downto 0) := x"98967F"; --10 times per second
signal div : std_logic_vector(23 downto 0) := (others => '0');

begin

UART: entity work. UART_comm port map (clk => clk, WR => WR,
        DIN => data_to_PC, DOUT => data_from_PC, TX_ready => open,
        RX_ready => open, TXD => TXD, RXD => RXD);

process (clk)
begin
  if (rising_edge(clk)) then
    if (div = CLK_DIV) then        div <= (others => '0');       WR <= '1';
    else                           div <= div + 1;               WR <= '0';
    end if;
  end if;
end process;

LED <= data_from_PC;
data_to_PC <= sw;

end Behavioral;
```

In this example the status of the board's switches is sent over UART with the frequency 10 times per second. The timing is controlled by a simple counter which counts from 0 to the maximum value (equal to 1/10th of a second) defined in the CLK_DIV constant; once the maximum value is reached the counter is reset and the WR signal is asserted starting the transfer cycle. As soon as data are received from the host PC they are immediately shown on the board's LEDs. Since the involved processing and data transfers in this example are trivial, the signals RX_ready

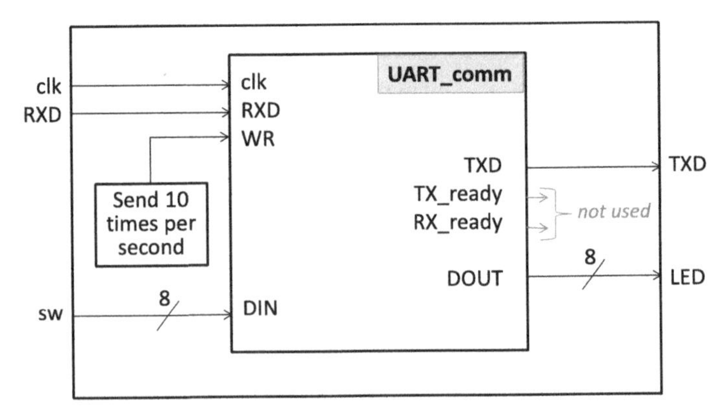

Fig. 4.18 Structure of a circuit which receives an 8-bit value from the host PC via the UART interface, displays this value on the Nexys-4 board rightmost LEDs and sends the 8-bit value selected on the board's rightmost switches to the host PC. The value from the switches is sent to the host PC 10 times per second

and TX_ready, which are generated in the UART communication module, are not required and were left unconnected. In the more complicated examples from Sect. 4.4 these signals will be used intensively.

All ports of the main entity have to be connected to appropriate FPGA pins. The respective pin locations can be found in the master UCF file available in the board's documentation.

4.3.2.2 Application Software

Once the required hardware modules are developed, we have to design software functions that would connect to the host PC serial port (to which the board is attached) and interact with the FPGA through UART interface. The following C++ code gives an example for Windows system:

```
#include <windows.h>
#include <iostream>
#include <ctime>

void set_up_serial_port(HANDLE& h);
bool get_data_from_serial_port(unsigned& data);
bool write_data_to_serial_port(unsigned data);

const int NO_DEFAULT_DEVICE = 2;

using namespace std;

int main(int argc, char* argv[])
{  unsigned data, result;
   const int range_min = 0, range_max = 0xff;
   srand (static_cast<unsigned>(time(0)));

   char operation;
   do { cout << "Select operation (r - read switches, s - send a value, e - exit)" << endl;
       cin >> operation;
       switch (operation)
       {  case 'r':
                 if (get_data_from_serial_port(result))
                   cout << "The result from FPGA is: " << hex << result << endl;
                 break;
          case 's':
                 //randomly generate an 8-bit number
                 data = static_cast<unsigned>((double)rand() / (RAND_MAX + 1) *
                             (range_max - range_min)  + range_min);
                 //send data to the FPGA
                 if (write_data_to_serial_port(data))
                   cout << "The data " << hex << data <<
                             " have been successfully transmitted to the FPGA" << endl;
                 break;
          case 'e' :
                 break;
```

```
            default:
                    cout << "Wrong parameter" << endl;
        }
      } while (operation != 'e');
    return 0;
}
void set_up_serial_port(HANDLE& serial_port)
{  const long baud_rate = 9600;      //baud rate
   char port_name[] = "COM9:";       //name of serial port (consult the Device Manager)
    //open up a handle to the serial port
   serial_port = CreateFile(port_name, GENERIC_READ | GENERIC_WRITE, 0, 0,
        OPEN_EXISTING, 0, 0);
   if (serial_port == INVALID_HANDLE_VALUE) //make sure the port was opened
   {    cerr << "Error opening port" << endl;
        CloseHandle(serial_port);
   }
   //set up the serial port
   DCB properties;                              //properties of serial port
   GetCommState(serial_port, &properties);  //get the properties
   properties.BaudRate = baud_rate;             //set the baud rate
   //set the other properties
   properties.Parity = NOPARITY;
   properties.ByteSize = 8;
   properties.StopBits = ONESTOPBIT;
   SetCommState(serial_port, &properties);
}
bool get_data_from_serial_port(unsigned& data)
{  unsigned long bytes_to_receive = 1;        //number of bytes to receive from COM
   unsigned long bytes_received;               //number of bytes actually received from COM
   HANDLE serial_port = 0;          set_up_serial_port(serial_port);
   //receive data from the serial port
  ReadFile(serial_port,static_cast<void *>(&data),bytes_to_receive,&bytes_received, 0);
   if (bytes_received != bytes_to_receive)
   {    cerr << "Error reading file" << endl;
        CloseHandle(serial_port);
        return false;
   }
   CloseHandle(serial_port);
   data = *data & 0xff;
   return true;
}
bool write_data_to_serial_port(unsigned data)
{  unsigned long bytes_to_send = 1;           //number of bytes to send to COM
   unsigned long bytes_sent;                   //number of bytes actually sent to COM
   data = data & 0xff;
   HANDLE serial_port = 0;          set_up_serial_port(serial_port);
   //send data to the serial port
   WriteFile(serial_port, static_cast<void *>(&data), bytes_to_send, &bytes_sent, 0);
   if (bytes_sent != bytes_to_send)
```

```
    {   cerr << "Error writing file" << endl;
        CloseHandle(serial_port);
        return false;
    }
    CloseHandle(serial_port);
    return true;
}
```

The program starts by printing a menu of options: "r"—to read the value of the board's switches and display it on the screen, "s"—to send a randomly generated 8-bit value to the board and show it on the LEDs, and "e"—to exit. The functions get_data_from_serial_port and write_data_to_serial_port provide for read/write interface to the serial port. Both functions start by setting up the port with the aid of set_up_serial_port function. The latter creates a handle to the serial port (whose name port_name can be found in the *Device Manager*) with the OpenFile function and sets all the required communication parameters, such as the baud rate, the number of data bits, stop bits, and parity bits. Once the communication is set up, the attached input/output device can be read and written through ReadFile and WriteFile functions and, when finished, the handle is closed. The user interface is the same as in Fig. 4.14. We will show in Sect. 4.4 a more complex example where many data (i.e. not just a single byte) are transferred to and received from the FPGA through UART.

4.4 Software/Hardware Co-design and Co-simulation

Developing efficient and reliable digital systems demands hardware/software co-design and co-simulation. There are different aspects that motivate co-design and co-simulation. First of all, the majority of methods of designing digital (and especially embedded) systems rely on a separation of software and hardware parts of the future system at early stages (usually during the specification phase). Once the separation is done, software and hardware are developed independently and, typically, by different people/teams. Such a priori separation has a number of limitations (e.g. time to market, suboptimal designs) which are better addressed if the interrelated software and hardware are developed simultaneously [11]. Second, co-simulation permits different design strategies to be explored more easily to detect most critical system parts that have to be assigned to hardware for acceleration, while the sequential control-oriented parts are more efficiently implemented in software.

Let us illustrate how co-design can be done with a PC computer running a software program and a standalone FPGA-based prototyping board executing a computationally intensive algorithm. We will explore communication through the Digilent EPP and UART on the example of a sorting system depicted in Fig. 4.19. The system executes the following actions:

1. Randomly generates in software 16 32-bit data items.
2. Sends them to the FPGA.

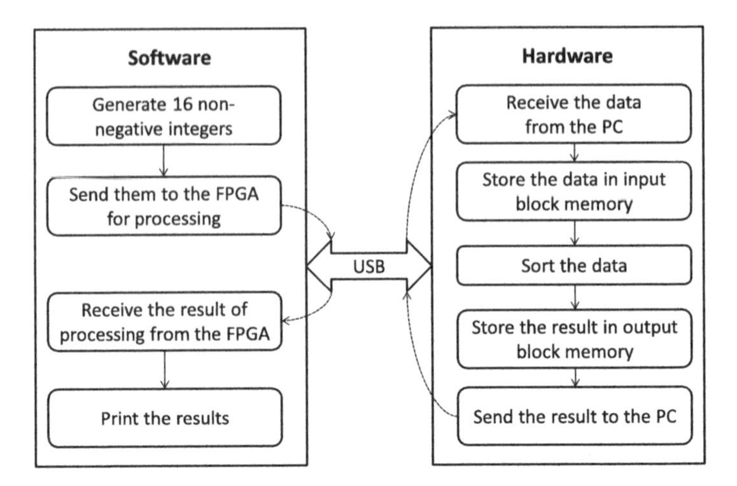

Fig. 4.19 Structure of a co-design project for data sort

3. Sorts the data in the FPGA with an iterative even-odd transition network (see Sect. 3.5).
4. Sends the results back to the host PC.
5. Displays the sorted data.

We examine a scenario where the FPGA receives data from the PC with the aid of either Digilent EPP or UART communication module and stores them in the input memory module. Then the received data are sorted by the EvenOddTransitionIterative block described in Sect. 3.5. The block EvenOddTransitionIterative is parameterizable with the number of data items (N) and the width of each item in bits (M). For our example we set $N = 16$ and $M = 32$. The block requires the input data to be supplied in a single $N \times M$-bit register input_data and generates the result of sorting also in a single $N \times M(16 \times 32 = 512)$-bit register sorted_data. To begin sorting the input signal *sort_en* must be asserted and once sorting is finished the output signal ready goes *high*.

The input memory module is a simple dual-port RAM where the port *A* permits 8-bit data to be written (by the host PC) and port *B* allows 512-bit data to be read (by the block EvenOddTransitionIterative). The input memory module has been created with the aid of Xilinx CORE Generator, where the port A width is set to 8 (since we are only able to transfer 8-bit values from the PC), port A depth is set to 64, and port B width is set to 256 (the maximum allowed). Since we need to read 512 bits at once, two input memories are required.

The 512-bit result of sorting is written to an output memory. The output memory module is also a simple dual-port RAM where the port A permits 512-bit data to be written (by the sorting block) and the port B allows 8-bit data to be read (by the host PC). The output memory module can be created with the aid of Xilinx CORE Generator, where the port A width is set to 256, port A depth is set to 2 (the minimum allowed), and the port B width is set to 8 (because we are only able to transfer 8-bit values to the PC). Since we need to write 512 bits at once, two output memories are required.

If we target our designs to Spartan-6 or Artix-7 FPGA (either Atlys or Nexys-4 prototyping boards) then both input and output memory modules require 16 embedded Block RAMs each. This is because in the CORE Generator we used the algorithm optimized for minimum area which leads to the instantiation of 16 Block RAMs (that allow at most 32 bits to be read/written) for each type of memory, while we need 512 bits to be processed. Other available options for Artix-7 FPGA permit fixed primitives with data widths of up to 72 bits.

4.4.1 Software-Hardware Co-design with Digilent Parallel Port Interface

Let us consider software-hardware co-design on the example of data sort described above and in Chap. 3 assuming the following functionality. The main hardware module communicates with the host PC through USB port according to the Digilent EPP protocol, fills in the input memory, starts the sorting module, and, when sorting is finished, writes the results to the output memory to be further read by the PC. The flowchart that describes behavior of the main module is shown in Fig. 4.20 (inputs and outputs are specified in VHDL).

In the WAIT_FOR_DATA state the control FSM waits for new data to arrive through the EPP data bus. Once new 8-bit data are received (which is indicated by the asserted data_ready signal) the FSM changes its state to the WRITE_INPUT where either the signal write_enable_in1 or the signal write_enable_in2 is asserted depending on the value of the bit 5 in the PC_address. The PC_address signal is controlled by the address output of the EPP communication module, which, as described in Sect. 4.3.1.1, holds an address for memory transactions, set by the host PC. When the bit 5 is equal to '0', the circuit is processing the first 32 bytes from 64 bytes that have to be received from the PC (since there are 16 data items 4-bytes each). In this case all the received data bytes are stored in the first input memory by asserting the signal write_enable_in1. If the bit 5 of PC_address is '1', the circuit is processing the last 32 data bytes (i.e. the remaining 8 data items) and these have to be stored in the second input memory by asserting the signal write_enable_in2. At the same time in the WRITE_INPUT state the FSM checks whether the last data item (whose address is 0x3f = 63) has been received. If so, the FSM changes its state to the START_PROC. Otherwise the state WAIT_NEXT_ADDRESS is activated where the FSM simply waits for the end of the current EPP data transfer cycle and then returns to the WAIT_FOR_DATA state to look for the next data item to be received over the EPP data bus.

In the START_PROC state the start_processing signal is asserted which activates the sorting module EvenOddTransitionIterative. In this state the FSM first checks if processing has been started (i.e. the ready signal from the block EvenOddTransitionIterative

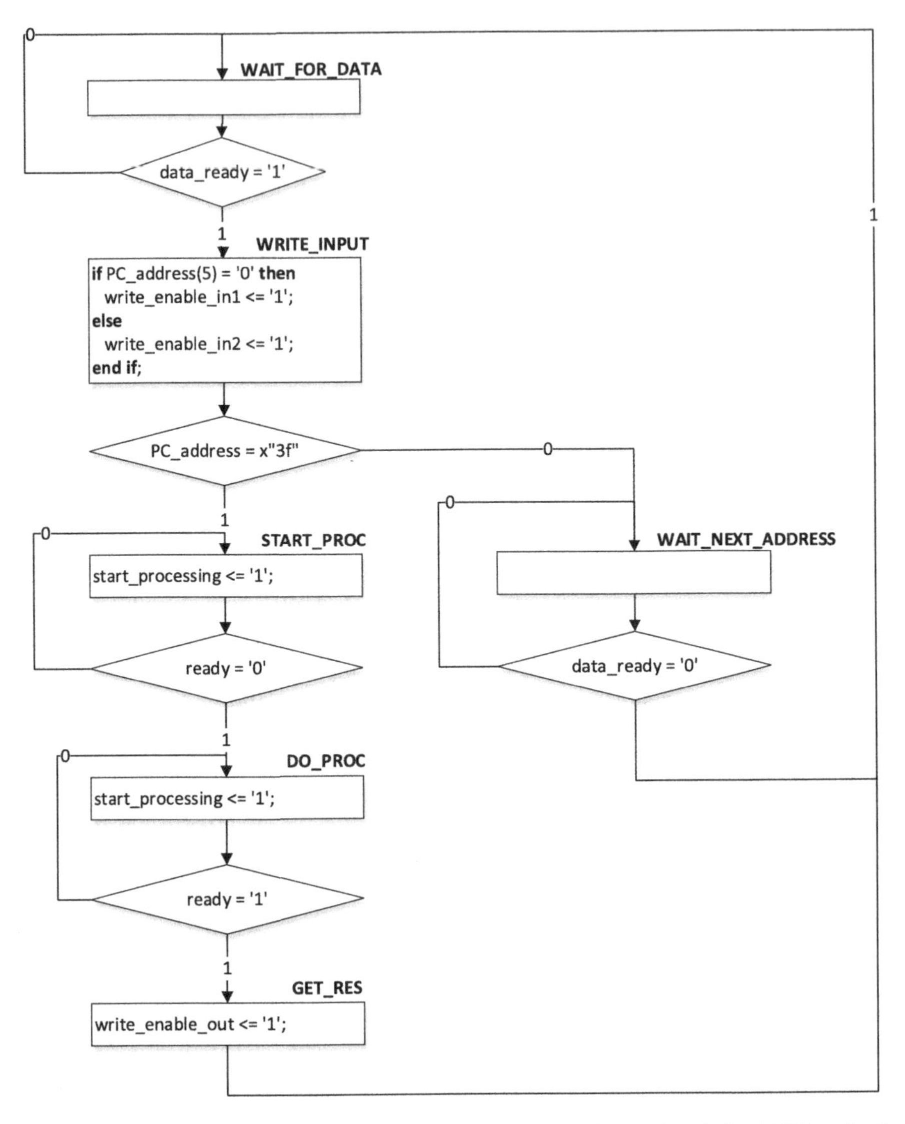

Fig. 4.20 Flowchart for the main module in the data sort co-design project (using EPP interface)

has been deasserted) and if so, advances to the DO_PROC state. In this state the output ready of the module EvenOddTransitionIterative is monitored and, once asserted, which indicates that sorting has finished, the FSM changes its state to the GET_RES. By this moment the results of sorting are stored in the output memory (by asserting the write_enable_out signal) and the FSM returns to the WAIT_FOR_DATA state to be able to receive a new set of data for sorting. Please note that sorting starts immediately, once data are written to the last memory position (0x63).

The following VHDL code shows how the hardware part of the system has been specified.

```
library ieee;
use ieee.std_logic_1164.all;

entity main is
   port ( clk : in std_logic;
          EppAstb : in  std_logic;
          EppDstb : in  std_logic;
          EppWr : in  std_logic;
          EppDB : inout  std_logic_vector(7 downto 0);
          EppWait : out  std_logic);
end main;

architecture Behavioral of main is

--communication signals
signal data_from_PC, data_to_PC1, data_to_PC2, data_to_PC :
      std_logic_vector(7 downto 0);
signal data_ready : std_logic;

--processing signals
signal start_processing : std_logic := '0'; --signal that starts the processing block
signal ready : std_logic := '0';             --signal that reports that processing block has finished
type PROC_TYPE is (WAIT_FOR_DATA, WRITE_INPUT, WAIT_NEXT_ADDRESS,
                   START_PROC, DO_PROC, WRITE_RES, GET_RES);
signal next_proc_state, proc_state : PROC_TYPE := WAIT_FOR_DATA;

--memory signals
signal PC_address: std_logic_vector(7 downto 0);
signal write_enable_in1, write_enable_in2, write_enable_out : std_logic;
signal FPGA_write_address, FPGA_read_address: std_logic;
signal PC_read_write_address: std_logic_vector(5 downto 0);
signal write_item, read_item: std_logic_vector(511 downto 0);

begin

------------------------------------------------------------
---Interface with dual-port memory
------------------------------------------------------------
memory_from_PC_1: entity work.INPUT_MEMORY port map(CLKA => clk,
      WEA(0) => write_enable_in1, ADDRA => PC_read_write_address,
      DINA => data_from_PC, CLKB => clk, ADDRB(0) => FPGA_read_address,
      DOUTB => read_item(255 downto 0));

memory_from_PC_2: entity work.INPUT_MEMORY port map(CLKA => clk,
      WEA(0) => write_enable_in2, ADDRA => PC_read_write_address,
      DINA => data_from_PC, CLKB => clk, ADDRB(0) => FPGA_read_address,
      DOUTB => read_item(511 downto 256));

memory_to_PC_1 : entity work.OUTPUT_MEMORY port map (CLKA => clk,
      WEA(0) => write_enable_out, ADDRA(0) => FPGA_write_address,
      DINA => write_item(255 downto 0), CLKB => clk,
      ADDRB => PC_read_write_address, DOUTB => data_to_PC1);
```

```
memory_to_PC_2 : entity work.OUTPUT_MEMORY port map (CLKA => clk,
        WEA(0) => write_enable_out, ADDRA(0) => FPGA_write_address,
        DINA => write_item(511 downto 256), CLKB => clk,
        ADDRB => PC_read_write_address, DOUTB => data_to_PC2);

data_to_PC <= data_to_PC1 when PC_address(5) = '0' else data_to_PC2;
PC_read_write_address <= '0' & PC_address(4 downto 0);

FPGA_read_address <= '0';
FPGA_write_address <= '0';

------------------------------------------------------------
---Interface with PC
------------------------------------------------------------
EPP: entity work.EPP_interface port map (EppAstb => EppAstb, EppDstb => EppDstb,
        EppWr => EppWr, EppDB => EppDB, EppWait => EppWait,
        address => PC_address, data_ready => data_ready,
        data_to_PC => data_to_PC, data_from_PC => data_from_PC);

------------------------------------------------------------
---Processing block (the control FSM)
------------------------------------------------------------
state_transition: process (clk)
begin
  if (rising_edge(clk)) then
        proc_state <= next_proc_state;
  end if;
end process state_transition;

output_logic: process (clk)
begin
  if (rising_edge(clk)) then
    start_processing <= '0';
    write_enable_in1 <= '0';
    write_enable_in2 <= '0';
    write_enable_out <= '0';
    case proc_state is
      when WAIT_FOR_DATA =>
      when WRITE_INPUT =>
                if PC_address(5) = '0' then
                        write_enable_in1 <= '1';
                else
                        write_enable_in2 <= '1';
                end if;
      when WAIT_NEXT_ADDRESS =>
      when START_PROC =>
                start_processing <= '1';
      when DO_PROC =>
                start_processing <= '1';
      when GET_RES =>
                write_enable_out <= '1';
      when others =>    --should never be reached
    end case;
  end if;
end process output_logic;
```

```
next_state_logic: process (proc_state, data_ready, PC_address, ready)
begin
  case proc_state is
    when WAIT_FOR_DATA =>
      if (data_ready = '1') then
                next_proc_state <= WRITE_INPUT;
      else      next_proc_state <= WAIT_FOR_DATA;
      end if;
    when WRITE_INPUT =>
      if (PC_address = x"3f") then --last address, input data transfer completed
                next_proc_state <= START_PROC;
      else      next_proc_state <= WAIT_NEXT_ADDRESS;
      end if;
    when WAIT_NEXT_ADDRESS =>
      if data_ready = '0' then
                next_proc_state <= WAIT_FOR_DATA;
      else      next_proc_state <= WAIT_NEXT_ADDRESS;
      end if;
    when START_PROC =>
      if (ready = '0') then --processing started
                next_proc_state <= DO_PROC;
      else      next_proc_state <= START_PROC;
      end if;
    when DO_PROC =>
      if (ready = '1') then--processing finished
                next_proc_state <= GET_RES;
      else      next_proc_state <= DO_PROC;
      end if;
    when GET_RES =>
      next_proc_state <= WAIT_FOR_DATA; --write the result to the output memory
    when others =>        --should never be reached
      next_proc_state <= WAIT_FOR_DATA;
  end case;
end process next_state_logic;

sort: entity work.EvenOddTransitionIterative
  generic map (M => 32, N => 16)
  port map (clk => clk, sort_en => start_processing, ready => ready,
       input_data => read_item, sorted_data => write_item);

end Behavioral;
```

From the software side, the main function gives a user two options: (1) to randomly generate 16 32-bit unsigned data, send them to the FPGA, sort, receive the result and display, or (2) to exit. C++ code of the main function is shown below.

```
#include <windows.h>
#include "dpcdefs.h"
#include "dpcutil.h"
#include <iostream>
#include <ctime>

const int INITIALIZATION_FAILED = 1;
const int NO_DEFAULT_DEVICE = 2;
```

```
const int INTERNAL_ERROR = 3305; //internal error in DPCUTIL
const int devNameLength = 16;
char nameDevice[devNameLength+1];

void SendDataToFPGA(unsigned* data, unsigned size);
bool WriteData(HANDLE hif, unsigned address, unsigned data);
void ReceiveResultFromFPGA(unsigned* data, unsigned size);
bool ReadData(HANDLE hif, unsigned address, unsigned& data);

const unsigned N = 16;
const unsigned M = 32;

using namespace std;

int main(int argc, char* argv[])
{ ERC error_code;
  if ( !DpcInit(&error_code) )              //before using DPCUTIL API functions, call DpcInit
      return INITIALIZATION_FAILED;        //error occurred while initializing
  //obtain the index of the default device in the Device Table
  int idDevice = DvmgGetDefaultDev(&error_code);
  if (idDevice == -1) //no devices in the Device Table
  {   cerr << "No default device"<< endl;
      cerr << "Run Digilent Adept and modify the Device Table (Settings tab, "<<
              "Device Manager option)"<< endl;
      return NO_DEFAULT_DEVICE;
  }
  else //get the default device name
      DvmgGetDevName(idDevice, nameDevice, &error_code);

  unsigned* data = new unsigned[N];
  unsigned* result = new unsigned[N];
  const unsigned range_min = 0, range_max = RAND_MAX;
  srand (static_cast<unsigned>(time(0)));
  char operation;
  do
  {   cout << "Select an operation (s - sort data in FPGA, e - exit)" << endl;
      cin >> operation;
      switch (operation)
      { case 's':
          for(int j = 0; j < N; j++) //randomly generate N M-bit numbers
          {     data[j] = static_cast<unsigned>((double)rand() / (RAND_MAX ) *
                         (range_max - range_min) + range_min);
                data[j] = data[j] << M/2 | static_cast<unsigned>((double)rand() /
                         (RAND_MAX) * (range_max - range_min) + range_min);
          }
          SendDataToFPGA(data, N); //send data to the FPGA
          cout << "Original data: " << endl;
          for(unsigned j = 0; j < N; j++)
          {     cout.width(8);      cout << hex << data[j] << endl;
          }
          ReceiveResultFromFPGA(result, N);
          cout << "The result in FPGA is: " << endl;
          for(unsigned j = 0; j < N; j++)
          {     cout.width(8);      cout << hex  <<  result[j] << endl;
          }
```

```
            break;
        case 'e' : break;
        default  : cout << "Wrong parameter" << endl;
      }
  } while (operation != 'e');
  delete [] data;            delete [] result;
  return 0;
}
```

The function SendDataToFPGA was modified to include 2 parameters: array of input 32-bit data and the size of the array. The FPGA circuit is designed in such a way that permits sorting to be started when the last data item is written to the address 0x3f in the input memory (i.e. when 64th 8-bit word is received from the PC). That is why, if less than 16 32-bit data items are to be processed, the software simply fills the remaining memory positions with 0. As a result, the function SendDataToFPGA is almost the same as shown above, but the line marked as "data transfer" in Sect. 4.3.1.2 is changed to the following code (i.e. instead of a single byte an array of size 32-bit values are transferred to the FPGA):

```
for (unsigned n = 0; n < size; n++)
      if (!WriteData(hif, n, data[n])) return;
if (size*4*8 < N*M) //fill in the remaining memory positions with 0 and finish writing
      for (unsigned n = size; n < N*M/8/4; n++)
            if (!WriteData(hif, n, 0)) return;
```

The function WriteData sends a 32-bit data item to FPGA in 8-bit fractions. This function firstly specifies an address to which a data item will be sent and saved. The memory address is stored in the register 0x00 in the communication module. Then one byte of a data item is written to the register 0x01. Note that 4 address/data write cycles are needed to transfer a 32-bit data item. The code is the following:

```
bool WriteData(HANDLE hif, unsigned address, unsigned data)
{ ERC error_code;        unsigned char idData;        unsigned idReg;
  for (int b = 0; b < M/8; b++) //M/8 transactions are needed to send an M-bit data item
  {   idData = address * M/8 + b; //specify address to which to write to
      idReg = 0x00; //send address
      //send a single data byte (idData) to the register idReg
      if (!DpcPutReg(hif, idReg, idData, &error_code, 0))
      {        DpcCloseData(hif, &error_code); // close the communications module
               cerr << "DpcPutReg failed." << endl;
               return false;
      }
      idReg = 0x01;      idData = (data >> b*8) & 0xff;
      //send a single data byte (idData) to the register idReg
      if (!DpcPutReg(hif, idReg, idData, &error_code, 0))
      {        DpcCloseData(hif, &error_code); //close the communications module
               cerr << "DpcPutReg failed." << endl;
```

```
                        return false;
            }
    }
    return true;
}
```

The ReceiveResultFromFPGA function is the same as shown in Sect. 4.3.1.2 except two changes: the list of parameters was modified to be able to receive an array (data) of size sorted data items and the data transfer is done with the following loop:

```
for (unsigned n = 0; n < size; n++)
        if (!ReadData(hif, n, data[n])) return;
```

The code for ReadData function is given below. Once again several data read cycles are required to get a 32-bit data item back from the FPGA.

```
bool ReadData(HANDLE hif, unsigned address, unsigned& data)
{ ERC error_code;        unsigned char idData;        unsigned idReg;
  data = 0;
  //M/8 transactions are needed to receive an M-bit data item
  for (int b = 0; b < M/8; b++)
  {   //specify address which to read from
      idData = address * M/8 + b;
      idReg = 0x00; //send address
      //send a single data byte (idData) to the register idReg
      if (!DpcPutReg(hif, idReg, idData, &error_code, 0))
      {         DpcCloseData(hif, &error_code); //close the communications module
                cerr << "DpcPutReg failed." << endl;
                return false;
      }
      idReg = 0x05;
      //get a single data byte (idData) from the register idReg
      if (!DpcGetReg(hif, idReg, &idData, &error_code, 0))
      {         DpcCloseData(hif, &error_code); //close the communications module
                cerr << "DpcGetReg failed." << endl;
                return false;
      } data = data | (idData << b*8);
  }
  return true;
}
```

Once both software and hardware are developed, the project can be tested. An example of user interface is given in Fig. 4.21 where the randomly generated input data are sorted in descending order.

4.4.2 Software-Hardware Co-design with UART Interface

The functionality of the top hardware module is very similar to that described in the previous section. The only difference is that instead of the Digilent EPP

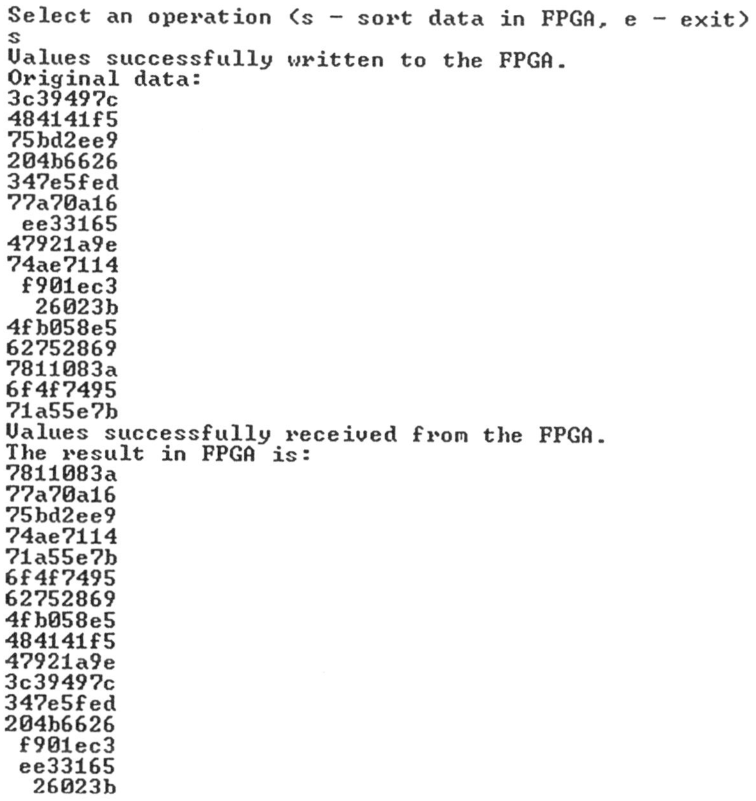

```
Select an operation (s - sort data in FPGA, e - exit)
s
Values successfully written to the FPGA.
Original data:
3c39497c
484141f5
75bd2ee9
204b6626
347e5fed
77a70a16
 ee33165
47921a9e
74ae7114
 f901ec3
  26023b
4fb058e5
62752869
7811083a
6f4f7495
71a55e7b
Values successfully received from the FPGA.
The result in FPGA is:
7811083a
77a70a16
75bd2ee9
74ae7114
71a55e7b
6f4f7495
62752869
4fb058e5
484141f5
47921a9e
3c39497c
347e5fed
204b6626
 f901ec3
 ee33165
  26023b
```

Fig. 4.21 The result of sorting in the FPGA communicating with the host PC through Digilent EPP

communication module the UART communication module will be used. The flow-chart that describes behavior of the main module is shown in Fig. 4.22 (inputs and outputs are specified in VHDL).

The sequence of steps required to fill in the input memory modules as well as to send the result of sorting back to the PC is more complicated because there is no dedicated address register as in the case of the EPP communication module. Therefore all the received over UART data bytes are written to input memory sequentially, starting with the address 0x00, until the last address 0x3f (64th data byte) is reached.

In the RESET state the module initializes the signal PC_address to 0x00 and advances to the WAIT_FOR_DATA state. In the WAIT_FOR_DATA state the module waits for new data to arrive through the RXD line. Once a new 8-bit data item is received (which is indicated by the asserted RX_ready signal) the control FSM changes its state to the WRITE_INPUT where either the signal write_enable_in1 or the signal write_enable_in2 is asserted depending on the value of the bit 5 in the PC_address. When the bit 5 is '0' the circuit is processing the first 32 bytes from 64 bytes that have to be received from the PC (since there are 16 data items 4-bytes each). In this

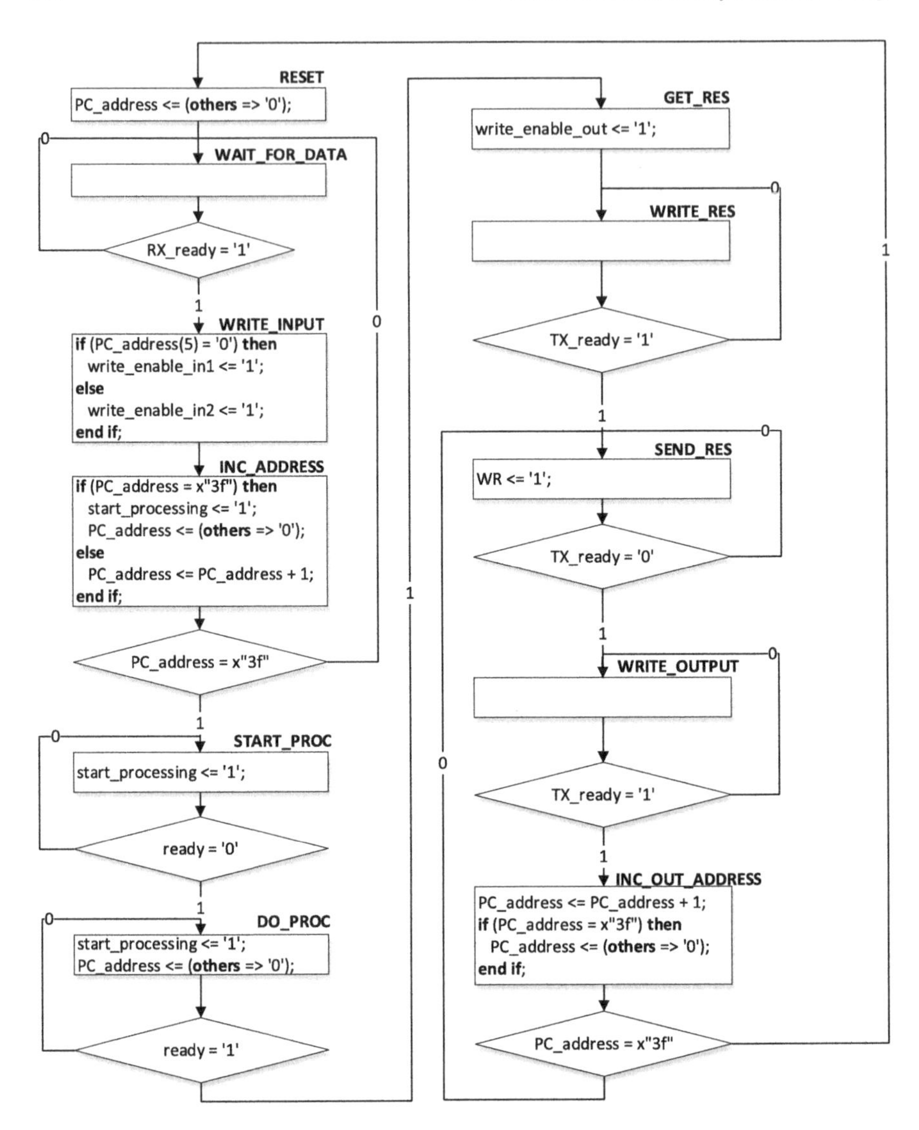

Fig. 4.22 Flowchart for the main module in the data sort co-design project (using the UART interface)

case all the received data bytes are stored in the first input memory by asserting the signal write_enable_in1. If the bit 5 of PC_address is '1', the circuit is processing the last 32 data bytes and these are stored in the second input memory by asserting the signal write_enable_in2. The next state is INC_ADDRESS where the PC_address signal is incremented and a test is done whether the last data item (whose address is 0x3f $=$ 63) has already been received. If so, the FSM changes to the START_PROC state, otherwise the state WAIT_FOR_DATA is activated once again to look for the next data item to be received.

In the START_PROC state the start_processing signal is asserted which activates the sorting module EvenOddTransitionIterative. In this state the FSM first checks if sorting has been started (i.e. the ready signal from the block EvenOddTransitionIterative has been deasserted) and if so, advances to the DO_PROC state. In this state the output ready of EvenOddTransitionIterative is monitored and, once asserted, which indicates that sorting has finished, the FSM state is changed to the GET_RES which activates the write_enable_out signal allowing the results of sorting to be stored in the output memory. The FSM changes to the WRITE_RES state which monitors the TXD_ready signal to check whether the transmission line TXD is available. If so, the FSM state is changed to the SEND_RES which activates the signal WR indicating that a new data byte is ready to be sent. Once sending of a data byte over the TXD line has started, the FSM advances to the WRITE_OUTPUT state, deasserts the WR signal and waits for the transmission to be finished. Once the transmission is finished, the PC_address signal is incremented in the INC_OUT_ADDRESS state. If the last result from the address 0x3f has been transmitted to the PC, the FSM returns to the RESET state, otherwise the state SEND_RES is activated and the next result data item is dispatched over the UART TXD line.

The following VHDL code shows specification of the top module which connects all the hardware components of the system and controls the flow of operations in accordance with Fig. 4.22.

```
library ieee;
use ieee.std_logic_1164.all;
use IEEE.std_logic_unsigned.all;

entity main is
  port ( clk : in std_logic;
         TXD : out std_logic;
         RXD : in std_logic);
end main;

architecture Behavioral of main is

--communication signals
signal data_from_PC, data_to_PC1, data_to_PC2, data_to_PC :
      std_logic_vector(7 downto 0);
signal WR : std_logic;
signal TX_ready, RX_ready_prev, RX_ready : std_logic := '1';

--processing signals
signal start_processing : std_logic := '0'; --signal that starts the processing block
signal ready : std_logic := '0';            --signal that reports that processing block has finished
type PROC_TYPE is (RESET, WAIT_FOR_DATA, WRITE_INPUT, INC_ADDRESS,
      START_PROC, DO_PROC, WRITE_RES, GET_RES,
      SEND_RES, WRITE_OUTPUT, INC_OUT_ADDRESS);
signal next_proc_state, proc_state : PROC_TYPE := RESET;

--memory signals
signal PC_address: std_logic_vector(7 downto 0);
signal write_enable_in1, write_enable_in2, write_enable_out : std_logic;
signal FPGA_write_address, FPGA_read_address: std_logic;
```

```
signal PC_read_write_address: std_logic_vector(5 downto 0);
signal write_item, read_item: std_logic_vector(511 downto 0);

begin

-----------------------------------------------------------
---Interface with dual-port memory
-----------------------------------------------------------
memory_from_PC_1: entity work.INPUT_MEMORY port map(CLKA => clk,
        WEA(0) => write_enable_in1, ADDRA => PC_read_write_address,
        DINA => data_from_PC, CLKB => clk, ADDRB(0) => FPGA_read_address,
        DOUTB => read_item(255 downto 0));

memory_from_PC_2: entity work.INPUT_MEMORY port map(CLKA => clk,
        WEA(0) => write_enable_in2, ADDRA => PC_read_write_address,
        DINA => data_from_PC, CLKB => clk, ADDRB(0) => FPGA_read_address,
        DOUTB => read_item(511 downto 256));

memory_to_PC_1 : entity work.OUTPUT_MEMORY port map (CLKA => clk,
        WEA(0) => write_enable_out, ADDRA(0) => FPGA_write_address,
        DINA => write_item(255 downto 0), CLKB => clk,
        ADDRB => PC_read_write_address, DOUTB => data_to_PC1);

memory_to_PC_2 : entity work.OUTPUT_MEMORY port map (CLKA => clk,
        WEA(0) => write_enable_out, ADDRA(0) => FPGA_write_address,
        DINA => write_item(511 downto 256), CLKB => clk,
        ADDRB => PC_read_write_address, DOUTB => data_to_PC2);

data_to_PC <= data_to_PC1 when PC_address(5) = '0' else data_to_PC2;
PC_read_write_address <= '0' & PC_address(4 downto 0);

-----------------------------------------------------------
---Interface with PC
-----------------------------------------------------------
UART: entity work.UART_comm port map (clk => clk, WR => WR,
        DIN => data_to_PC, DOUT => data_from_PC, TX_ready => TX_ready,
        RX_ready => RX_ready, TXD => TXD, RXD => RXD);

FPGA_read_address <= '0';
FPGA_write_address <= '0';

-----------------------------------------------------------
---Processing block (the control FSM)
-----------------------------------------------------------
state_transition: process (clk)
begin
  if (rising_edge(clk)) then
        proc_state <= next_proc_state;
  end if;
end process state_transition;

output_logic: process (clk)
begin
  if (rising_edge(clk)) then
     start_processing <= '0';
     write_enable_in1 <= '0';
```

```
      write_enable_in2 <= '0';
      write_enable_out <= '0';
      WR <= '0';
      case proc_state is
      --START FILL IN INPUT MEMORY
        when RESET =>
                PC_address <= (others => '0');
        when WAIT_FOR_DATA =>
        when WRITE_INPUT =>
                if (PC_address(5) = '0') then
                        write_enable_in1 <= '1';
                else
                        write_enable_in2 <= '1';
                end if;
        when INC_ADDRESS =>
                if (PC_address = x"3f") then --last position, input data transfer completed
                        start_processing <= '1';
                        PC_address <= (others => '0');
                else
                        PC_address <= PC_address + 1;
                end if;
      --FINISH FILL IN INPUT MEMORY
      --START PROCESSING
        when START_PROC =>
                start_processing <= '1';
        when DO_PROC =>
                start_processing <= '1';
                PC_address <= (others => '0');
        when GET_RES =>
                write_enable_out <= '1';
        when WRITE_RES =>
      --FINISH PROCESSING
      --START SEND THE RESULT TO PC
        when SEND_RES =>
                WR <= '1';
        when WRITE_OUTPUT =>
        when INC_OUT_ADDRESS =>
                PC_address <= PC_address + 1;
                if (PC_address = x"3f") then --last position, input data transfer completed
                        PC_address <= (others => '0');
                end if;
      --FINISH SEND THE RESULT TO PC
        when others =>   --should never be reached
      end case;
    end if;
  end process output_logic;

  next_state_logic: process (proc_state, RX_ready, PC_address, ready, TX_ready)
  begin
    case proc_state is
     --START FILL IN INPUT MEMORY
      when RESET =>
```

```
        next_proc_state <= WAIT_FOR_DATA;
      when WAIT_FOR_DATA =>
        if (RX_ready = '1') then
                    next_proc_state <= WRITE_INPUT;
        else        next_proc_state <= WAIT_FOR_DATA;
        end if;
      when WRITE_INPUT =>
        next_proc_state <= INC_ADDRESS;
      when INC_ADDRESS =>
        if (PC_address = x"3f") then --last position, input data transfer completed
                    next_proc_state <= START_PROC;
        else        next_proc_state <= WAIT_FOR_DATA;
        end if;
--FINISH FILL IN INPUT MEMORY
--START PROCESSING
      when START_PROC =>
        if (ready = '0') then --processing started
                    next_proc_state <= DO_PROC;
        else        next_proc_state <= START_PROC;
        end if;
      when DO_PROC =>
        if (ready = '1') then --processing finished
                    next_proc_state <= GET_RES;
        else        next_proc_state <= DO_PROC;
        end if;
      when GET_RES =>
        next_proc_state <= WRITE_RES; --write the result to output memory
      when WRITE_RES =>
        if (TX_ready = '1') then
                    next_proc_state <= SEND_RES ; --ready to transmit
        else        next_proc_state <= WRITE_RES;
        end if;
--FINISH PROCESSING
--START SEND THE RESULT TO PC
      when SEND_RES =>
        if (TX_ready = '0') then --transmission started
                    next_proc_state <= WRITE_OUTPUT;
        else        next_proc_state <= SEND_RES;
        end if;
      when WRITE_OUTPUT =>
        if (TX_ready = '1') then
                    next_proc_state <= INC_OUT_ADDRESS;
        else        next_proc_state <= WRITE_OUTPUT;
        end if;
      when INC_OUT_ADDRESS =>
        if (PC_address = x"3f") then --last position, input data transfer completed
                    next_proc_state <= RESET;
        else        next_proc_state <= SEND_RES;
        end if;
--FINISH SEND THE RESULT TO PC
      when others =>       --should never be reached
```

```
        next_proc_state <= RESET;
    end case;
  end process next_state_logic;

  sort: entity work.EvenOddTransitionIterative
        generic map (M => 32, N => 16)
        port map (clk => clk, sort_en => start_processing, ready => ready,
                  input_data => read_item, sorted_data => write_item);
end Behavioral;
```

The software part of the projects includes the main function which provides for interaction with the user, randomly generates unsigned integers to sort, creates a handle to the serial port, sends data to the FPGA with the aid the write_data_to_serial_port function, receives the result of sorting back with the get_data_from_serial_port function, prints the original and sorted data, and finally closes the handle. The main function can be described in C++ as follows:

```
#include <windows.h>
#include <iostream>
#include <ctime>

void set_up_serial_port(HANDLE& h);
bool get_data_from_serial_port(HANDLE h, unsigned* data, unsigned long data_size);
bool write_data_to_serial_port(HANDLE serial_port, unsigned* data,
                               unsigned long size);

const int NO_DEFAULT_DEVICE = 2;

const unsigned N = 16; //number of data items
const unsigned M = 32; //size of each data item in bits

using namespace std;

int main(int argc, char* argv[])
{ using namespace std;
  HANDLE serial_port = 0;
  set_up_serial_port(serial_port);

  unsigned* data = new unsigned[N];
  unsigned* result = new unsigned[N];

  const unsigned range_min = 0, range_max = RAND_MAX;
  srand (static_cast<unsigned>(time(0)));
  char operation;

  do
  {   cout << endl << "Select an operation (s - sort data in FPGA, e - exit)" << endl;
      cin >> operation;

      switch (operation)
      {
        case 's':
              for(int j = 0; j < N; j++) //randomly generate N M-bit numbers
              { data[j] = static_cast<unsigned>((double)rand() / (RAND_MAX ) *
                         (range_max - range_min) + range_min);
```

```
                    data[j] = data[j] << M/2 | static_cast<unsigned>((double)rand() /
                              (RAND_MAX) * (range_max - range_min) + range_min);
                }
                write_data_to_serial_port(serial_port, data, N); //send data to the FPGA
                cout << "Original data: " << endl;
                for(unsigned j = 0; j < N; j++)
                {  cout.width(8);
                   cout << hex << data[j] << endl;
                }
                get_data_from_serial_port(serial_port, result, N); //get the result of sort
                cout << "The result in FPGA is: " << endl;
                for(unsigned j = 0; j < N; j++)
                {  cout.width(8);
                   cout << hex << result[j] << endl;
                }
                break;
        case 'e' :
                break;
        default:
                cout << "Wrong parameter" << endl;
      }
   } while (operation != 'e');

   delete [] data;
   delete [] result;

   CloseHandle(serial_port); //close handle
   return 0;
}
```

The function set_up_serial_port is identical to the function presented in Sect. 4.3.2.2. The remaining two functions get_data_from_serial_port and write_data_to_serial_port were modified to receive three parameters: a handle to the serial port, an array of 32-bit data (to send to the FPGA or to fill in from the FPGA) and the size of the array.

The function write_data_to_serial_port sends the array data of 16 32-bit unsigned numbers to the serial port. For each data item a 4-byte buffer is filled in and then the contents of the buffer is written to the attached COM port. The function has the following C++ code:

```
bool write_data_to_serial_port (HANDLE h, unsigned* data, unsigned long data_size)
{  unsigned long bytes_sent = 0;              //number of bytes actually sent to COM
   const unsigned BUF_SIZE = 4;
   char buffer[BUF_SIZE];                     //buffer to store a data item to send
   unsigned new_data;

   for (unsigned i = 0; i < data_size; i++)
   {
      new_data = data[i];
      for (unsigned j = 0; j < BUF_SIZE; j++)
              buffer[j] = (new_data & (0xff << j*8) ) >> (j*8);
```

```
        WriteFile(h, static_cast<void *>(buffer), BUF_SIZE, &bytes_sent, 0);
        if (bytes_sent != BUF_SIZE)
        {           cerr << "Error writing file" << endl;
                    CloseHandle(h);
                    return false;
        }
    }
    return true;
}
```

The function get_data_from_serial_port receives the sorted data from the FPGA and stores them in the array data of size 32-bit unsigned numbers. For each data item a 4-byte buffer is filled in by reading from the attached COM port.

```
bool get_data_from_serial_port(HANDLE h, unsigned* data, unsigned long data_size)
{ unsigned long bytes_received = 0;//number of bytes actually received from COM
   const unsigned BUF_SIZE = 4;
   char buffer[BUF_SIZE];              //buffer to store 4 bytes to read from the FPGA
   unsigned new_data;
   for (unsigned i = 0; i < data_size; i++)
   {   //receive data from the serial port
      ReadFile(h, static_cast<void *>(buffer), BUF_SIZE, &bytes_received, 0);
      if (bytes_received != BUF_SIZE)
      {           cerr << "Error reading file" << endl;
                  CloseHandle(h);
                  return false;
      }
      new_data = 0;
      for (unsigned j = BUF_SIZE; j > 0; j--)
                  new_data = (new_data << 8) | (buffer[j-1] & 0xff);
      data[i] = new_data;
   }
   return true;
}
```

Now the developed software/hardware co-design project can be tested. The user interface is identical to the interface depicted in Fig. 4.21.

4.5 Programmable Systems-on-Chip

This section presents very brief introduction to the Xilinx all programmable systems-on-chip (APSoC) and suggests several APSoC-based designs. The APSoCs of the Xilinx Zynq-7000 family contain an industry-standard ARM® dual-core Cortex™-A9 MPCore™ processing system (PS) and 7 series FPGA-based programmable logic (PL) combining logical slices, DSP blocks, memories and other embedded components. The ARM® dual-core Cortex™-A9 MPCore™ PS can be

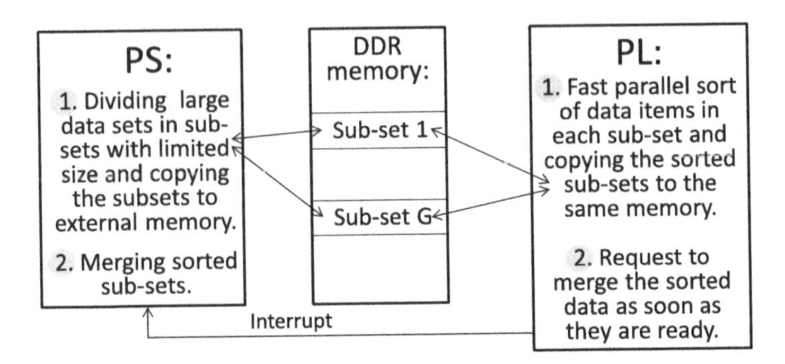

Fig. 4.23 Using fast sorting networks in hardware and merging in software

used autonomously or interact with circuits implemented in the PL through high performance interfaces. The PL can generate up to 16 interrupts that are handled by the PS and may be used as signals from the circuits indicating that the processing task is completed or for some other needs. The interaction between the PS and the PL can be organized through the following interfaces [12]:

- High-performance Advanced eXtensible Interface (AXI) optimized for high bandwidth access from the PL to external DDR memory and to dual-port on-chip memory [12, 13]. There are totally four 32/64-bit ports available connecting the PL to the memory through FIFOs [13]. The multi-protocol DDR memory controller supports speed of up to 1333 Mb/s for DDR3 and allows shared access to a common memory from the PS and the PL [13].
- Four (two slave and two master) General-Purpose Interfaces (GPI) optimized for access from the PL to the PS peripheral devices and from the PS to the PL registers [12].
- Accelerator Coherency Port (ACP) permits a coherent access from the PL (where hardware accelerators might be implemented) to the PS memory cache enabling a low latency path between the PS and the PL [13].

Design with APSoCs is supported by the Xilinx ISE and Vivado computer-aided design systems and permits configuration of hardware in the PL, linking the PL with the PS, and development of software for the PS that interacts with hardware in the PL. All such steps require comprehensive knowledge of different topics and they cannot be presented here with sufficient details. Many of such details can be found in [14]. In this section we mainly focus on potential applications of Xilinx APSoCs for solving problems described in the book and for co-design that enables the developed hardware circuits and systems to be linked with software running in the PS (or possibly in general-purpose computers interacting with APSoCs through widely available interfaces [13]). We will start with data processing that involves high-performance parallel circuits that may efficiently be implemented in the PL such as network-based sorters and searchers (see Sects. 3.4–3.6).

Figure 4.23 demonstrates one potential application which enables large sets of data to be sorted in software with the aid of hardware accelerators. Suppose the PS has to sort a set of data stored in the external DDR memory (these data either are

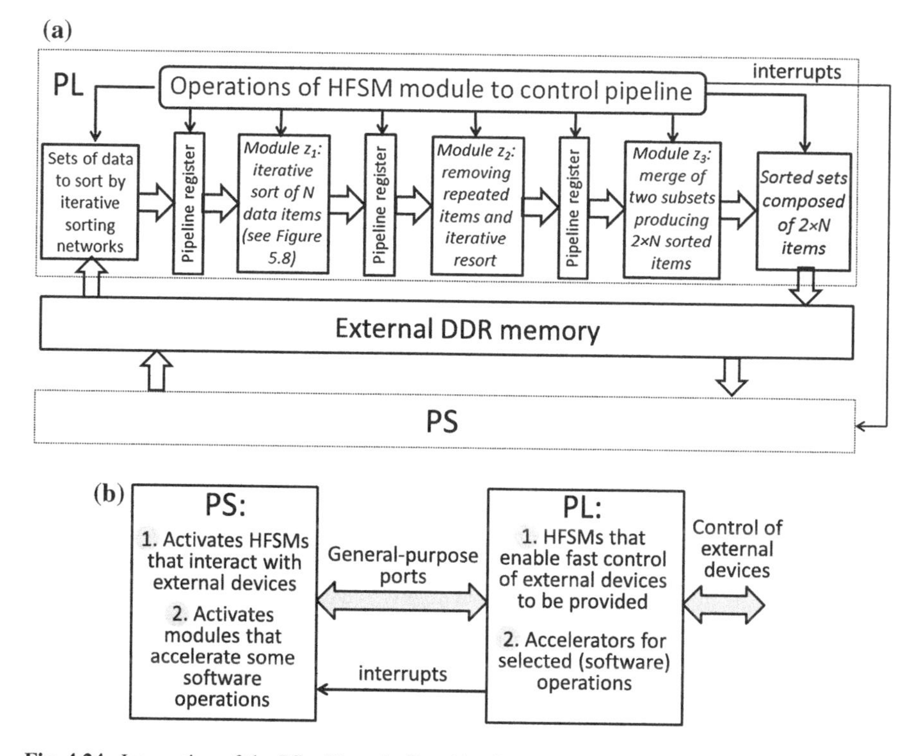

Fig. 4.24 Interaction of the PS with a pipeline (details are given in Sect. 5.8) implemented in the PL (**a**) and with circuits that autonomously control external devices (**b**)

received from outside or may be preliminary created by the PS and copied to the memory). There are ν items in the set and they cannot be sorted in the PL because the PL resources are not sufficient. Let us consider the following steps:

- The PS divides the given set in such subsets that can be sorted in the PL. Suppose there are G such subsets with N items in each one.
- The numbers G, N and DDR addresses are transferred to the PL and the latter reads subsets from the DDR memory through the internal high-performance interfaces, sorts data in each subset and copies the sorted subsets to the DDR.
- As soon as some subsets (with a predefined number) are sorted, the PL generates a dedicated interrupt that informs the PS that certain sorted subsets are already available in the memory and ready for further processing.
- The PS processes the sorted subsets (e.g. merges them) and completes sorting.

Figure 4.24 demonstrates other potential applications. The first system (see Fig. 4.24a) enables an interaction between a pipeline (implemented in the PL) and the PS that uses the results of processing in the pipeline. There are three modules z_1, z_2, z_3 between pipeline registers and they solve problems that will be described with details in Sect. 5.8. Input data are received from the external DDR memory and the results are copied to the memory. The PS supplies initial data and handles the results of processing. Eventually FIFO memories are used in the PL on inputs

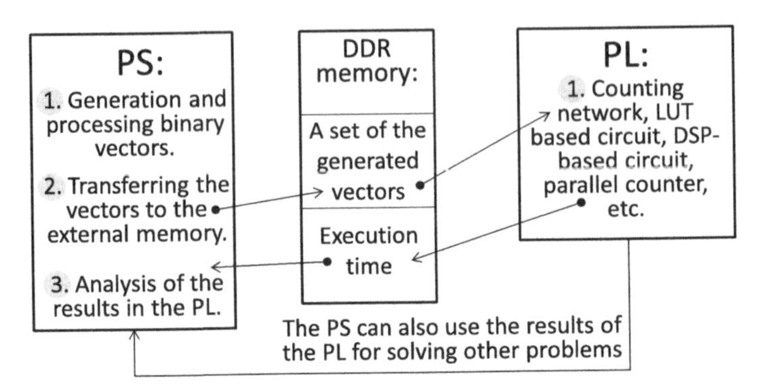

Fig. 4.25 A system that provides support for experiments and comparisons with alternative circuits

and outputs of the pipeline. Such high-speed processing permits many practical algorithms to be accelerated.

Figure 4.24b demonstrates another potential interaction between the PS and the PL through general purpose ports. The PL may provide support for control of external devices through APSoCs pins. In the next chapter we will show that advanced FSMs enable dedicated modules that execute application-specific functions to be implemented. Such modules can be triggered from the PS and the latter continues its functionality in parallel with the hardware modules. This way permits, in particular, concurrent hardware accelerators (for software operations) to be activated/deactivated. For example, in Chap. 5 we will describe accelerators that permit the greatest common divisor of many integers to be found in parallel.

APSoCs are also very efficient for experiments and comparisons. For example, often the most appropriate algorithm needs to be chosen from a large number of available alternatives (e.g. to compute and compare Hamming weights). It should be noted that a reliable evaluation of competitive algorithms needs to be done in the same hardware and under similar conditions. For such purposes the following technique can be used:

- Competitive devices are implemented in different areas of the PL and they can receive the same sets of data from a shared window in the DDR memory.
- Initial data are prepared and stored in memory segments $S_1,\ldots,S_K$ in such a way that a segment S_k is then owned by the device k.
- The same data sets are processed by competitive devices in parallel.
- The results are supplied to the PS, which makes the conclusion.

Figure 4.25 gives an example of a system that provides support for such experiments. The PL contains different circuits for Hamming weight counters and comparators described in the book. The following operations are executed:

- The PS sends a request to the PL to activate all the components which begin execution in parallel. The PS and/or the PL measure timing intervals for each component before the requested task is completed.

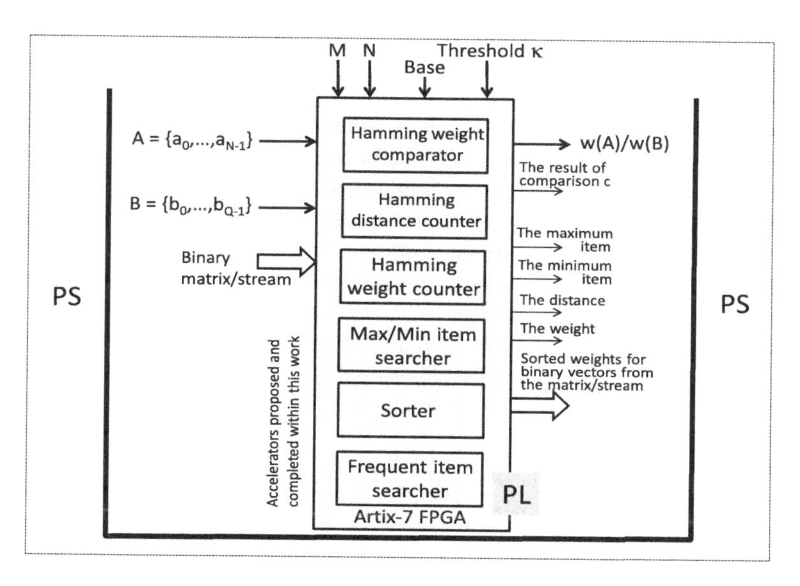

Fig. 4.26 Architecture of the analyzer from [15]

- As soon as any device solves the problem, the results are stored in the respective segment of a shared window in the memory and an associated interrupt is generated requesting the relevant PS measuring circuit to be stopped.
- As soon as all the results are produced by all the devices, the PS validates the correctness of these results, measures time slots, carries out the final analysis and makes the conclusion.

The selected from the experiments circuits (see the right-hand part of Fig. 4.25) can be used as accelerators of software programs (e.g. for digital filtering).

Figure 4.26 presents architecture of an analyzer for binary vectors and matrices described in [15] which includes: a number of accelerators in the PL and a subsystem implemented in software of the PS. Software applications were developed in C language and they execute the following tasks: (1) getting data from a host PC; (2) partitioning the data and transmitting them to the PL when required; (3) getting an application-specific analysis of the results from the PL; (4) supporting experiments with the developed hardware in the PL.

Similarly other problems requiring high-performance computations may be solved. For example, many designs described in [16] can be implemented faster and more efficiently because of availability of on-chip high-performance interfaces. In [17] an FPGA-based accelerator for algorithms that solve the Boolean satisfiability problem is proposed. The idea was to use an FPGA for the Boolean constraints propagation algorithm while other required tasks are executed in software. Such partitioning can directly be transformed to APSoCs. Many other applications from the scope of video processing, driver assistance, communications and control can also benefit from APSoCs [13].

Comprehensive design examples based on APSoCs can be found in [14] where the following methods and tools are described:

- Design flow and design steps for APSoCs in the Xilinx Vivado environment.
- Tutorials demonstrating all necessary steps for simple projects involving the PS, the PL, various interactions between the PS and the PL as well as interaction of APSoCs with peripheral devices such as switches, buttons, LEDs, displays and a number of supplementary modules [18].
- Many details about design techniques and prototyping with ZedBoard [19] and Xilinx evaluation kit [20].
- Implementation of the majority of the projects described here (Chaps. 3–5) in the PL of Xilinx APSoCs, development of software for the PS (that uses the results of the projects) with demonstration of different modes for interactions between the PS and the PL.
- Data exchange between the PS and the PL through the external DDR memory and the results of measurements for different projects.
- Software/hardware co-design for APSoCs with demonstration of more complicated projects from such areas as data processing, combinatorial optimization, interaction of software programs with hardware accelerators, using advanced controllers based on hierarchical and parallel finite state machines, and others.
- Using interrupts for different projects.
- Support for experiments that involve software of general-purpose computers and the PS interacting with different circuits implemented in the PL.

References

1. Digilent Inc. (2013) Nexys-4 reference manual. http://www.digilentinc.com/Data/Products/NEXYS4/Nexys4_RM_VB1_Final_3.pdf. Accessed 9 Nov 2013
2. Chu PP (2008) FPGA prototyping using VHDL examples: Xilinx Spartan-3 version. Willey, New Jersey
3. Sklyarov V, Skliarova I (2013) Parallel processing in FPGA-based digital circuits and systems. TUT Press, Tallinn
4. Xilinx Inc. (2013) 7 series DSP48E1 slice user guide. http://www.xilinx.com/support/documentation/user_guides/ug479_7Series_DSP48E1.pdf. Accessed 16 Nov 2013
5. Sklyarov V, Skliarova I (2013) Design and implementation of counting networks. Computing. doi:10.1007/s00607-013-0360-y
6. Parhami B (2009) Efficient Hamming weight comparators for binary vectors based on accumulative and up/down parallel counters. IEEE Trans Circuits Syst—II: Express Briefs 56(2):167–171
7. Digilent Inc. (2004) Digilent parallel interface model reference manual. http://www.digilentinc.com/Data/Products/ADEPT/DpimRef%20programmers%20manual.pdf. Accessed 9 Nov 2013
8. Digilent Inc. (2009) Adept I/O expansion reference design. http://www.digilentinc.com/Products/Detail.cfm?NavPath=2,66,828&Prod=ADEPT2. Accessed 9 Nov 2013
9. Digilent Inc. (2007) Adept SDK API. http://www.digilentinc.com/Products/Detail.cfm?NavPath=2,66,828&Prod=ADEPT2. Accessed 9 Nov 2013
10. Future Technology Devices International Ltd. (2013) Virtual COM port drivers. http://www.ftdichip.com/Drivers/VCP.htm. Accessed 9 Nov 2013

11. Teich J (2012) Hardware/software codesign: the past, the present, and predicting the future. Proc IEEE 100:1411–1430
12. Neuendorffer S, Martinez-Vallina F (2013) Building Zynq® accelerators with Vivado® high level synthesis. Tutorial. In: Proceedings of the 21st ACM/SIGDA international symposium on field-programmable gate arrays, Monterey, California, 2013. http://tcfpga.org/fpga2013/VivadoHLS_Tutorial.pdf. Accessed 25 Nov 2013
13. Xilinx Inc. (2013) Zynq-7000 all programmable SoC. http://www.xilinx.com/products/silicon-devices/soc/zynq-7000/. Accessed 16 Nov 2013
14. Sklyarov V, Skliarova I, Rjabov A, Silva J, Sudnitson A, Cardoso C (2014) Hardware/software co-design for programmable systems-on-chip. TUT Press, Tallinn
15. Sklyarov V, Skliarova I (2013) Digital Hamming weight and distance analyzers for binary vectors and matrices. Int J Innovative Comput Inf Control 9(12):4825–4849
16. Mueller R (2010) Data stream processing on embedded devices. Ph.D. dissertation, Swiss Federal Institute of Technology
17. Davis JD, Tan Z, Yu F, Zhang L (2008) A practical reconfigurable hardware accelerator for Boolean satisfiability solvers. In: Proceedings of the 45th ACM/IEEE design automation conference, Anaheim, California, 2008
18. Digilent Inc. (2013) Peripheral modules. http://www.digilentinc.com/pmods/. Accessed 16 Nov 2013
19. Digilent Inc. (2013) ZedBoard Zynq™-7000 Development Board. http://www.digilentinc.com/Products/Detail.cfm?NavPath=2,400,1028&Prod=ZEDBOARD. Accessed 16 Nov 2013
20. Xilinx Inc. (2013) Xilinx Zynq-7000 All Programmable SoC ZC702 Evaluation Kit. http://www.xilinx.com/products/boards-and-kits/EK-Z7-ZC702-G.htm. Accessed 16 Nov 2013

Chapter 5
Design Technique Based on Hierarchical and Parallel Specifications

Abstract This chapter gives an overview of the design techniques based on hierarchical and parallel specifications. First, hierarchical graph-schemes (HGSs) are introduced that enable complex digital control algorithms to be decomposed and described efficiently. A module, described by an HGS, is the fundamental entity that provides the basis for the technique, and is an autonomous, complete, and potentially reusable component. A module has to be designed such that: (1) it can be verified independently of other modules; (2) it possesses a well-defined external interface so it can be reused in different specifications. It is shown that a set of HGSs (modules) can be implemented in a hierarchical finite state machine (HFSM) with stack memory. Many VHDL examples are given that demonstrate that HFSMs permit the execution of hierarchical algorithms and provide support for recursion if required. Various types of HFSMs are described and synthesizable VHDL templates for these are given that can be customized for particular problems. Parallel specifications and parallel HFSMs are also discussed. Many fully functioning VHDL examples for all the types of HFSMs above are presented and evaluated. It is also shown how software programs can be mapped to hardware with the aid of HFSM models. Finally, a variety of HFSM optimization techniques are proposed.

5.1 Modular Hierarchical Specifications

Nowadays, the development of software and hardware becomes more and more interrelated. The emphasis has significantly shifted from general-purpose to application-specific products in the form of embedded processing modules in various areas such as communications, industrial automation, automotive computers, and home electronics [1]. To support application-specific computations, a number of new engineering solutions and technological innovations have been proposed. There is a tendency to integrate components on a chip that not so long ago were separated and implemented as autonomous Application-Specific Integrated

V. Sklyarov et al., *Synthesis and Optimization of FPGA-Based Systems*,
Lecture Notes in Electrical Engineering 294, DOI: 10.1007/978-3-319-04708-9_5,

Circuits (ASICs) or Application-Specific Standard Products (ASSP). In the past individual ASICs/ASSPs were assembled together with the surrounding logic, often implemented in autonomous FPGAs; today all these components are coupled within the same micro-chip. For example, the Zynq-7000 all programmable system-on-chip (APSoC) briefly described in Sect. 4.5 incorporates a processing system (PS) and programmable logic (PL) on the same microchip and they are linked through advanced interfaces.

APSoCs can run software that interacts with parallel processing elements (PE) that have been mapped to hardware. The main objective of any PE is to provide greater performance than an equivalent software component with similar functionality that is typically composed of a set of functions in C, or methods in Java. The relative effectiveness (e.g. performance) of software modules that have been mapped to hardware PEs needs to be tested, analyzed and compared. Thus, it is important to be able to create the functionality of typical software constructions directly in hardware circuits. This chapter addresses the provision of modularity, hierarchy (including recursion), and parallelism in hardware. Modularity and hierarchy are very widely used techniques in general-purpose programming [2, 3]. They are supported by the majority of application-specific development systems for the design of software in single/multi-core autonomous and built-in microcontrollers, mainly originating from specifications in C. In many practical cases, there is a need for hardware accelerators to achieve higher performance by parallelizing the most critical parts of the programs in hardware circuits. Thus, mapping such processor-intensive software fragments to hardware by applying potential parallelism becomes very important. There are many known methods that allow modularity, hierarchy and parallelism to be realized in hardware and a survey of some of these is presented in [4]. The described here technique is based on a hierarchical finite state machine (HFSM) model, which is less constrained than potential alternatives [4], can easily be implemented in hardware, and is very consistent with the corresponding software technique. The model is also supported by known templates that are synthesizable [5] in commercial computer-aided design (CAD) systems such as ISE of Xilinx.

The main objective of this chapter is to develop an approach to the synthesis of digital circuits and systems whose functionality can be expressed hierarchically (also allowing recursive invocations) represented in form of hierarchical graph-schemes (HGSs) with the following formal description (see Fig. 5.1).

An HGS is a directed connected graph containing rectangular (Fig. 5.1a), rhomboidal (Fig. 5.1b), and triangular (Fig. 5.1c) nodes. Each HGS has one entry point, which is a rectangular node named *Begin* (Fig. 5.1d) and one exit point, which is a rectangular node named *End* (Fig. 5.1e). Other rectangular nodes contain either *micro instructions* (Fig. 5.1f) or *macro instructions* (Fig. 5.1g) or both (Fig. 5.1h). We will also allow *micro instructions* to be assigned to the nodes *Begin* and *End* if required. Any *micro instruction* Y_j (Fig. 5.1f) includes a subset of *micro operations* from the set $Y = \{y_1,\ldots, y_N\}$. A *micro operation* is an output binary signal. Any *macro instruction* Z_k incorporates a subset of *macro operations* from the set $Z = \{z_1,\ldots, z_Q\}$ (Fig. 5.1g). Each *macro operation* is described by another HGS of

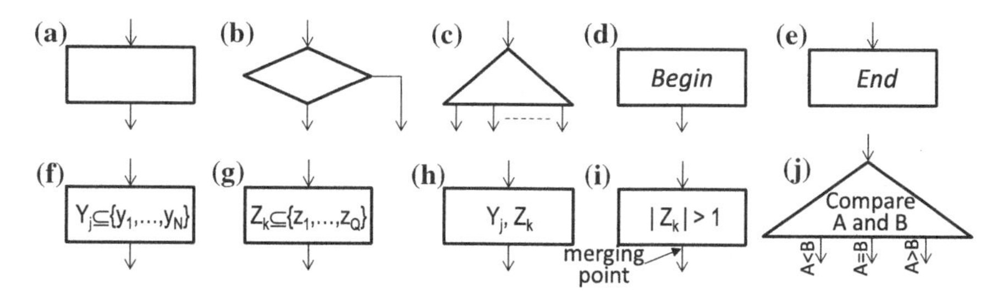

Fig. 5.1 Nodes of a hierarchical graph-scheme: rectangular (**a**); rhomboidal (**b**); triangular (**c**); entry point—node *Begin* (**d**); exit point—node *End* (**e**); with micro instructions (**f**); with macro instructions (**g**); with micro and macro instructions (**h**); with parallel macro operations (**i**); triangular node with expression producing a set of one-hot values (**j**)

a lower level called a *module*. If a macro instruction includes more than one macro operation then these macro operations have to be executed in parallel (Fig. 5.1i). Each rhomboidal node contains one element from the set $X \cup \Theta$, where $X = \{x_1,\ldots, x_L\}$ is the set of *logic conditions*, and $\Theta = \{\theta_1,\ldots, \theta_I\}$ is the set of *logic functions*. A *logic condition* is an input signal, which communicates the result of a test. Each *logic function* is calculated by performing a predefined set of sequential steps that are described by an HGS (a module) of a lower level. Directed lines (arcs) connect the inputs and outputs of the nodes in the same manner as for an ordinary graph-scheme [6]. Each triangular node contains an expression which can produce a set of one-hot values associated with the outputs of this node. As soon as the control flow passes a triangular node, exactly one output must be selected enabling the control flow to proceed (see examples in Fig. 5.1c and j). The output of a rectangular node k with more than one element $z_i, z_j,\ldots$ from the set Z is called a merging point (Fig. 5.1i). Control flow passes the merging point if and only if all the elements $z_i, z_j,\ldots$ have been completed. This means that a node following the node k is only activated after terminating all the macro operations $z_i, z_j,\ldots$.

Using HGSs enables complex control algorithms to be developed step by step, concentrating the efforts at each stage on a specified level of abstraction [7]. Each separate HGS (i.e. module) is usually simple, and can be tested independently. Besides, a module may be easily updated if required.

Figure 5.2 demonstrates an example of describing the function IGCD from Sect. 3.3 in form of an HGS.

Conversion from the C function in Fig. 5.2a to the HGS in Fig. 5.2b is easier and better understandable than in Fig. 3.3a. Since the HGS in Fig. 5.2b does not involve any hierarchy, FSM states can be assigned to the HGS much like it is done for an ordinary graph-scheme using the methods [6] where the rules are different for Mealy and Moore models. For example, the states init and run_state for Mealy FSM can be assigned as it is shown in Fig. 5.2b. They are associated with the input of the node following the node *Begin* (init), the input of the node *End* (the same state init), and the input of the triangular node (run_state). Transition and output signals can also be identified with the aid of the methods [6]. Finally, VHDL code that describes FSM state transitions with elements of datapath can be built

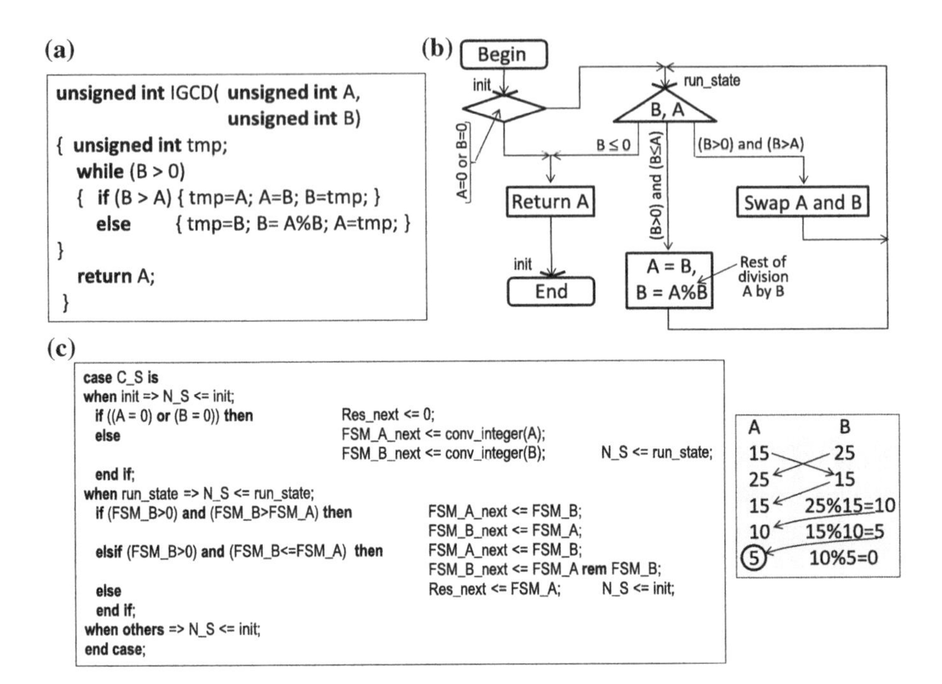

Fig. 5.2 C function IGCD from Sect. 3.3 (**a**); description of the function in form of HGS (**b**); VHDL description with example (**c**)

as it is shown in Fig. 5.2c (Fig. 5.2c also depicts an example of FSM functionality for A = 15 and B = 25). All necessary details needed for the considered above conversions are given in [6, 8]. Coding of HGSs in VHDL is discussed in [5].

The complete parameterizable VHDL module for the IGCD is given below:

```
entity FSM_OneEdge_GCD is  -- circuit with synchronization by one clock edge
generic(        data_size: integer := 8);
port (          clk       : in  std_logic;
                rst       : in  std_logic;
                A         : in std_logic_vector(data_size-1 downto 0);
                B         : in std_logic_vector(data_size-1 downto 0);
                Result    : out std_logic_vector(data_size-1 downto 0));
end FSM_OneEdge_GCD;

architecture Behavioral of FSM_OneEdge_GCD is
  signal FSM_A, FSM_B, FSM_A_next, FSM_B_next
       : integer range 0 to 2**data_size-1;
  type state_type is (init, run_state);
  signal C_S, N_S        : state_type;
  signal Res, Res_next   : integer range 0 to 2**data_size-1;
begin

process (clk)            -- this process describes functionality of the FSM state register and
begin                    -- registers of datapath
  if rising_edge(clk) then
    if (rst = '1') then       C_S <= init;
```

```
                        FSM_A  <= conv_integer(A);
                        FSM_B  <= conv_integer(B);
                        Res    <= 0;
      else        C_S <= N_S;
                  FSM_A <= FSM_A_next;     FSM_B <= FSM_B_next;
                  Res <= Res_next;
      end if;
    end if;
  end process;

  process (C_S, A, B, FSM_A, FSM_B, Res) -- this is a combinational process
  begin
    N_S <= C_S;
    FSM_A_next <= FSM_A;
    FSM_B_next <= FSM_B;
    Res_next <= Res;
    case C_S is
      when init =>
        if ((A = 0) or (B = 0)) then   Res_next <= 0; N_S <= init;
        else                           FSM_A_next <= conv_integer(A);
                                       FSM_B_next <= conv_integer(B);
                                       N_S <= run_state;
        end if;
      when run_state => N_S <= run_state;
        if (FSM_B>0) and (FSM_B>FSM_A) then          FSM_A_next <= FSM_B;
                                                     FSM_B_next <= FSM_A;
        elsif (FSM_B>0) and (FSM_B<=FSM_A) then  FSM_A_next <= FSM_B;
                                           FSM_B_next <= FSM_A rem FSM_B;
        else                               Res_next <= FSM_A; N_S <= init;
        end if;
      when others => N_S <= init;
    end case;
  end process;

  Result <= conv_std_logic_vector(Res, data_size);

  end Behavioral;
```

The code above can be used as a component in more complicated projects. For example, it can be linked with modules from Appendix B in order to show the results on 7-segment displays of the Nexys-4 board in decimal format:

```
entity TestGCD is
  generic(    data_size: integer := 8);
      port (  clk       : in  std_logic;          -- clock signal
              rst       : in  std_logic;          -- reset signal (active high)
              A         : in std_logic_vector(data_size-1 downto 0);
              B         : in std_logic_vector(data_size-1 downto 0);
              sel_disp  : out std_logic_vector(7 downto 0);
              seg       : out std_logic_vector(6 downto 0));
end TestGCD;
```

```
architecture Behavioral of TestGCD is
  signal BCD2,BCD1,BCD0        : std_logic_vector(3 downto 0);
  signal Result                : std_logic_vector(data_size-1 downto 0);
begin
BCD:          entity work.BinToBCD8
              port map (clk, rst, open, Result, BCD2, BCD1, BCD0);
DispCont:     entity work.EightDisplayControl
              port map(clk, "0000", "0000", "0000", "0000", "0000",
              BCD2, BCD1, BCD0, sel_disp, seg);
GCD:          entity work.FSM_OneEdge_GCD
              port map(clk, rst, A, B, Result);
end Behavioral;
```

5.2 Hierarchical Finite State Machines

It is known [5, 7, 9, 10] that a set of HGSs can be implemented in an HFSM with stack memory, which permits the execution of hierarchical algorithms. The HFSM model was proposed in [9] and further elaborated in [10]. The model was realized in hardware and successfully tested in a number of industrial products. Further improvements were made in [5, 7, 11, 12] and consequently new practical applications have been implemented, tested and evaluated. Theoretical and practical issues of HFSMs have been analyzed and used for further extensions and improvements in numerous publications [13–24]. Statecharts [25] specifications are also applicable to HFSMs, and they were adapted for object-oriented programming and used as a part of the unified modeling language. Hierarchical and concurrent finite state machines of other types are discussed extensively and their applicability to embedded systems is demonstrated in [26]. It is important to point out that HFSMs permit not only an abstract conversion, but also physical implementations because they are synthesizable.

We will skip here the formal mathematical definition and will describe the HFSM model informally. Let $x_1,\ldots, x_L/y_1,\ldots, y_N$ be sets of input/output signals. Structurally, an HFSM contains one or two stacks. In case of two stacks one of them (*FSM_stack*) keeps states and the other (*M_stack*) enables transitions between modules to be done. Any module is considered to be either an FSM or an HFSM. The stacks are managed by a circuit (C) that is responsible for new module invocations and state transitions in active modules that are designated by the outputs of the *M_stack*. Since each particular module has a unique identification code, the same HFSM states can be repeated in different modules. Any non-hierarchical (conventional) transition is performed through the change of a code only on the top register of the *FSM_stack* (see Fig. 5.3 and the mark ●). Any hierarchical call activates a push operation and alters the states of the both stacks in such a way that the *M_stack* will store the code for the new (called) module and the *FSM_stack* will be set to an initial state of the called module (see Fig. 5.3 and the mark ■).

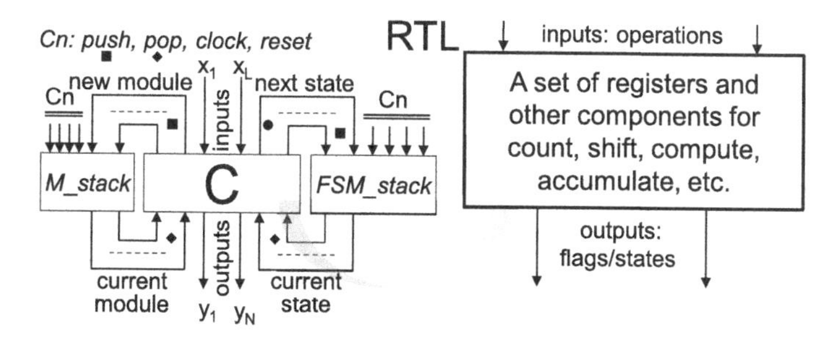

Fig. 5.3 HFSM which provides support for hierarchy and recursive calls

Any hierarchical return just activates a pop operation without any change in the stacks (see Fig. 5.3 and the mark ◆). As a result, a transition to the state following the state where the terminated module was called will be executed. The stack pointer is common to the both stacks. In the explored here HFSM with datapath the circuit C has RTL structure (see Fig. 5.3) enabling operations of high-level languages to be either mapped directly or in a slightly altered manner and consequently to be executed in hardware.

The model depicted in Fig. 5.3 possesses the following advantages:

- It does not have the limitations that exist for processing cores, such as the constrained size of operands, a predefined set of instructions, limited parallelism, the impossibility of fast combinational operations;
- It is entirely synthesizable, which is demonstrated in [5];
- It implements hierarchy (including potential recursion) faster than in software [27, 28], i.e. a smaller number of clock cycles is required.

We can distinguish two types of HFSMs [5, 11]: HFSMs with explicit and with implicit modules. HFSMs with explicit modules (see Fig. 5.3) include two stacks (the *FSM_stack* and the *M_stack*) and the circuit C, which is responsible for state transitions within any active module selected by the stack of modules (*M_stack*). An HFSM with implicit modules includes just one stack that keeps track of returns from a currently active module.

Design of HFSM-based circuits can be done from templates that are described below. The stacks in the templates are entirely reusable. The design method just requires state transitions to be specified in the templates that describe functionality of the circuit C.

5.2.1 HDL Template for HFSM with Explicit Modules

Figure 5.3 depicts the structure of an HFSM with explicit modules [5, 11]. *M_stack* and *FSM_stack* enable the currently executing module and the current state of the module to be explicitly indicated. The top register of the *M_stack*

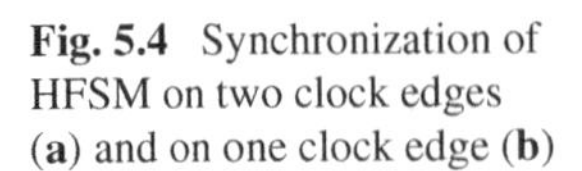

Fig. 5.4 Synchronization of HFSM on two clock edges (**a**) and on one clock edge (**b**)

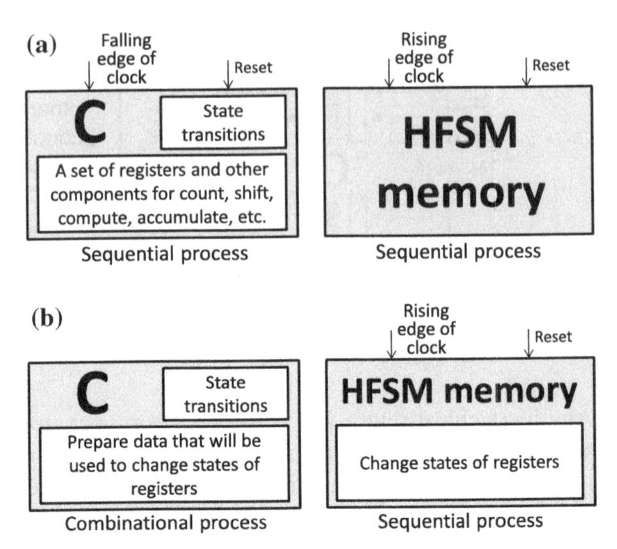

contains the code of the currently executing module. The top register of the *FSM_stack* is used as a register for the currently executing module, i.e. it supplies states (codes of states) for any state transition required within the currently executing module. At the beginning, the top registers of both stacks are set to the initial state a_0 of the initial module (z_0), which must be activated first according to the given algorithm. After that the following three allowed types of state transitions can be executed:

1. Transitions between states that belong to the same module. In this case the HFSM operates like an ordinary FSM.
2. A transition to the first state of a next module z_p. In this case the operation *push("the code of* z_p*")* is applied to the *M_stack* and the operation *push("the first state of* z_p*")* is applied to the *FSM_stack*. This transition is known as a *hierarchical call*.
3. A transition from a currently executing module z_p to a module z_q from which z_p was activated. In this case the operation *pop* is applied to the *M_stack* (thus, the top register of the *M_stack* will contain the code of z_q) and the operation *"pop + state transition"* is applied to the *FSM_stack*. This third type of transition is known as a *hierarchical return*.

HFSMs with explicit modules have the following features. There are two stack memories that keep vectors of $\lceil \log_2 Q \rceil$ bits for modules and $\lceil \log_2 R \rceil$ bits for states, where Q is the number of modules and R is the maximum number of states in a module. States in different modules can be assigned the same codes.

We discuss here two types of HFSMs. The first one is exactly the same as it was described in [5]. Synchronization is done with two clock edges: rising and falling (see Fig. 5.4a). In the second type (see Fig. 5.4b) synchronization is done with one clock edge in such a way that combinational process is responsible for preparation

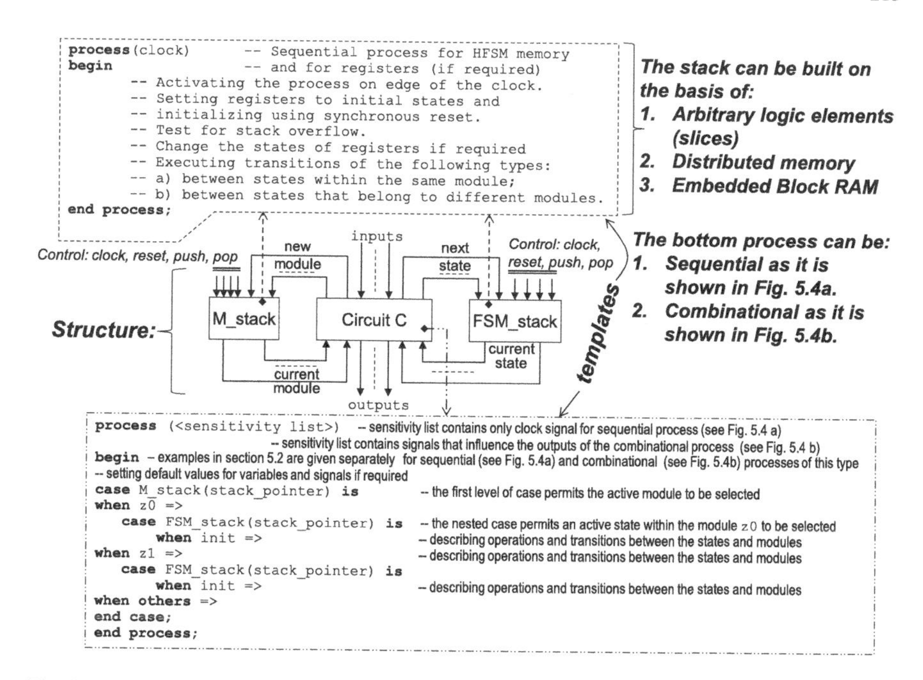

Fig. 5.5 A template for HFSM with explicit modules

of the next HFSM state and the next contents of registers in the datapath. The states of HFSM and the registers are changed on the selected edge of the clock (such as rising edge in Fig. 5.4b). Implementation of the circuit in Fig. 5.4b is usually smaller and faster while the first circuit in Fig. 5.4a has clearer and simpler description and, thus, potential errors can be detected more easily and avoided.

The HFSM with explicit modules template for synthesis from VHDL [7, 11] is shown in Fig. 5.5 and includes two processes describing: (1) reusable stacks for modules (*M_stack*) and states (*FSM_stack*); (2) a structure of the circuit C allowing transitions at the level of modules and states to be executed. Here, stack_pointer is a stack pointer common to both *M_stack* and *FSM_stack*; signals push and pop increment and decrement the stack_pointer.

Let us consider an example. The following C language code (where the function IGCD with two arguments was described above in Fig. 5.2a) finds the greatest common divisor of four non-negative integers A, B, C, and D:

```
unsigned int IGCD(unsigned int A, unsigned int B, unsigned int C, unsigned int D)
{  return IGCD (IGCD(A,B), IGCD(C,D)); }
```

The function IGCD with four arguments A, B, C, D can be described by an HGS depicted in Fig. 5.6. You can see that it looks similar to a flowchart. Different states (*initAB*, *initCD*, *c1_z1*, *c2_z1*, *c3_z1*, *init1_2*, *final_state*) and modules (z_0, z_1) are written in *italic*. In Sect. 5.3 we will show how to associate HFSM states with nodes of HGSs. From comments given in Fig. 5.6a the functionality of the

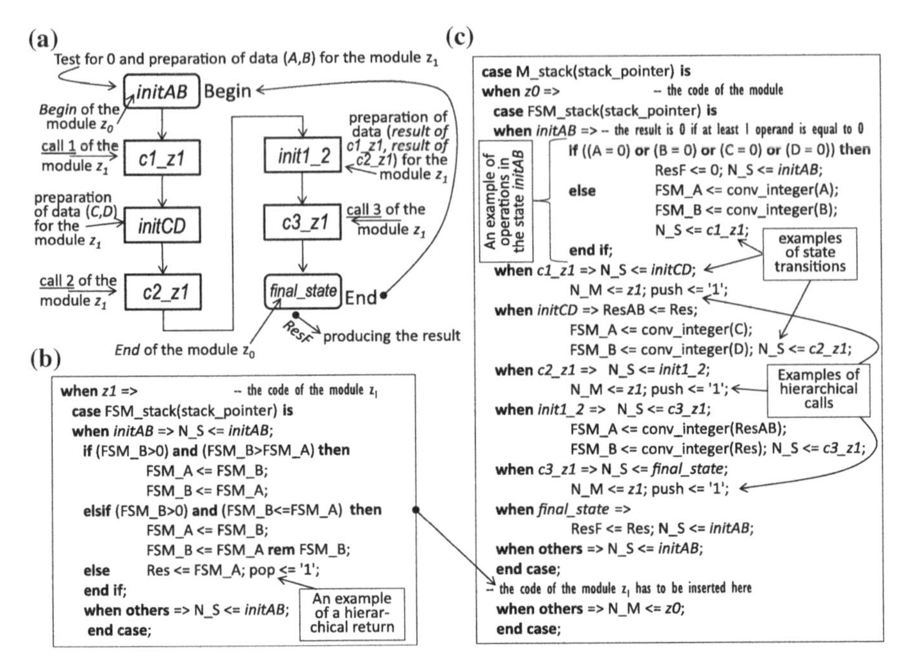

Fig. 5.6 Description of IGCD function with 4 arguments A, B, C, D (**a**); VHDL code for the module z_1 (**b**); state transitions and hierarchical calls (**c**)

module z_0 should be understandable. The module z_1 (that finds the greatest common divisor of two non-negative integers) has already been described above. VHDL code for this module is depicted in Fig. 5.6b. VHDL code in Fig. 5.6c gives a general idea of the intended functionality of the module z_0.

Any hierarchically called module (such as z_1) can be replaced or modified without any influence on upper modules (such as z_0). For example, the module of greatest common divisor in [1, 5] was described in VHDL on the basis of the following recursive C function *RGCD*:

```
unsigned int RGCD (unsigned int A, unsigned int B)
{      if (B > A) return RGCD (B,A);
       else if (B<=0) return A;
       else return RGCD(B, A%B);            }
```

As we mentioned above recursive module invocations are supported by the considered HFSM modules, although recursive implementations of cyclic functions (like *RGCD* above) are not efficient at all [27]. Later on we will discuss algorithms over trees for which recursion might be profitable [28].

VHDL code below gives a complete synthesizable specification of HFSM with datapath for implementation of the HGS in Fig. 5.6a and applying synchronization with rising and falling clock edges (see Fig. 5.4a).

```
entity IGCD is -- the function IGCD with four arguments A, B, C, and D
generic ( stack_size       : integer := 1; data_size        : integer := 4);
port ( clk                 : in  std_logic;       -- clock signal
       rst                 : in  std_logic;       -- reset signal (active high)
       A                   : in std_logic_vector(data_size-1 downto 0);
       B                   : in std_logic_vector(data_size-1 downto 0);
       C                   : in std_logic_vector(data_size-1 downto 0);
       D                   : in std_logic_vector(data_size-1 downto 0);
       stack_overflow      : out std_logic;       -- indicates HFSM stack overflow
       Result              : out std_logic_vector(data_size-1 downto 0));
end IGCD;

architecture Behavioral of IGCD is
  signal FSM_A, FSM_B : integer range 0 to 2**data_size-1; -- data_size is the size of data
  signal Res, ResF, ResAB : integer range 0 to 2**data_size-1;
  type state_type is (initAB, initCD, c1_z1, c2_z1, c3_z1,init1_2,final_state); --states ***
  signal N_S               : state_type;
  type MODULE_TYPE is (z0, z1);            -- HFSM modules                   -- ***
  signal N_M               : MODULE_TYPE;
  type stack is array(0 to stack_size) of STATE_TYPE;                        -- ***
  signal FSM_stack         : stack;        -- FSM_stack for HFSM             -- ***
  signal stack_pointer : integer range 0 to stack_size+1; -- +1 to allow test for overflow ***
  signal push              : std_logic;    -- forces to increment the stack_pointer -- ***
  signal pop               : std_logic;    -- forces to decrement the stack_pointer -- ***
  type Mstack is array(0 to stack_size) of MODULE_TYPE;                      -- ***
  signal M_stack           : Mstack;       -- M_stack for HFSM               -- ***
begin

process(clk)               -- beginning of the M_stack and FSM_stack
begin                      -- this is a sequential process
  if rising_edge(clk) then stack_overflow <= '0';
    if rst = '1' then  stack_pointer <= 0;        FSM_stack(0)      <= initAB;
                    M_stack(0)      <= z0;
    else
      if push = '1' then
        if stack_pointer = stack_size+1 then stack_overflow <= '1';
        else    stack_pointer                    <= stack_pointer + 1;
                FSM_stack(stack_pointer+1)<= initAB;
                FSM_stack(stack_pointer)    <= N_S;
                M_stack(stack_pointer+1)    <= N_M;
        end if;
      elsif pop = '1' then          stack_pointer              <= stack_pointer - 1;
      else                          FSM_stack(stack_pointer)   <= N_S;
      end if;
    end if;
  end if;
end process;  -- description of the M_stack and FSM_stack ends here

process (clk)  -- description of the left-hand circuit in Fig. 5.4a
begin                               -- this is a sequential process
  if falling_edge(clk) then         push <= '0'; pop <= '0'; N_M <= z0;
    if rst = '1' then               FSM_A <= 0; FSM_B <= 0;   Res <= 0;
```

```
else
   case M_stack(stack_pointer) is
     when z0 =>        -- the code is the same as in Fig. 5.6c
       case FSM_stack(stack_pointer) is
             when initAB =>
               if ((A = 0) or (B = 0) or (C = 0) or (D = 0)) then
                 ResF <= 0; N_S <= initAB;
               else FSM_A <= conv_integer(A);
                 FSM_B <= conv_integer(B); N_S <= c1_z1;      end if;
             when c1_z1 => N_S <= initCD; N_M <= z1; push <= '1';
             when initCD => ResAB <= Res; FSM_A <= conv_integer(C);
               FSM_B <= conv_integer(D); N_S <= c2_z1;
             when c2_z1 =>  N_S <= init1_2; N_M <= z1; push <= '1';
             when init1_2 =>  N_S <= c3_z1; FSM_A <= conv_integer(ResAB);
               FSM_B <= conv_integer(Res); N_S <= c3_z1;
             when c3_z1 => N_S <= final_state; N_M <= z1; push <= '1';
             when final_state => ResF <= Res; N_S <= initAB;
             when others => N_S <= initAB;
       end case;
     when z1 =>        -- the code is the same as in Fig. 5.6b
       case FSM_stack(stack_pointer) is
             when initAB => N_S <= initAB;
               if (FSM_B>0) and (FSM_B>FSM_A) then FSM_A <= FSM_B;
                 FSM_B <= FSM_A;
               elsif (FSM_B>0) and (FSM_B<=FSM_A)  then FSM_A <= FSM_B;
                 FSM_B <= FSM_A rem FSM_B;
               else    Res <= FSM_A; pop <= '1';
               end if;
             when others => N_S <= initAB;
       end case;
     when others => N_M <= z0;
   end case;
 end if;
end if;
end process;

Result <= conv_std_logic_vector(ResF, data_size);

end Behavioral;
```

VHDL code below gives a complete synthesizable specification of HFSM with datapath for implementation of the HGS in Fig. 5.6a and applying synchronization with the only one rising clock edge (see Fig. 5.4b).

```
entity Hierarchical_IGCD is
-- this entity is described exactly the same as the entity IGCD above
end Hierarchical_IGCD;

architecture Behavioral of Hierarchical_IGCD is --some lines have to be copied from the IGCD
  --the same declarations as in the IGCD above are not shown (they are marked with *** in the IGCD)
 signal FSM_A, FSM_B, FSM_A_next, FSM_B_next  :
      integer range 0 to 2**data_size-1;
```

```
  signal Res, Res_next, ResAB, ResAB_next, ResF, ResF_next:
      integer range 0 to 2**data_size-1;
  signal N_S, C_S        : state_type;
  signal N_M, C_M        : MODULE_TYPE;
begin

process(clk)   -- beginning of the M_stack and FSM_stack
begin                    -- this is a sequential process
  if rising_edge(clk) then -- stack memory is described differently compared to the previous example
    if rst = '1' then
      if pop /= '1' then stack_pointer <= 0; -- to avoid warnings because of line +++ below
      end if;
      C_M <= z0; C_S <= initAB; M_stack(0) <= z0; FSM_stack(0) <= initAB;
    else FSM_A <= FSM_A_next; FSM_B   <= FSM_B_next;
      Res      <= Res_next;        ResAB   <= ResAB_next;
      ResF     <= ResF_next;       C_M     <= N_M; C_S <= N_S;
      FSM_stack(stack_pointer) <= N_S; M_stack(stack_pointer) <= C_M;
      if push = '1' then
        if stack_pointer = stack_size+1 then           stack_overflow <= '1';
        else  stack_pointer <= stack_pointer + 1;       stack_overflow <= '0';
        end if;
      elsif pop = '1' then stack_pointer <= stack_pointer - 1; -- +++ (see comment above)
        C_S <= FSM_stack(stack_pointer-1);
        C_M <= M_stack(stack_pointer-1);
      end if;
    end if;
  end if;
end process;  -- description of the M_stack and FSM_stack ends here

process (A, B, C, D, C_M, C_S, FSM_A, FSM_B, Res, ResAB, ResF)
begin           -- this combinational process describes the left-hand circuit in Fig. 5.4b
N_S <= C_S;
FSM_A_next <= FSM_A; FSM_B_next <= FSM_B; Res_next   <= Res;
N_M <= C_M; push <= '0'; pop <= '0';
ResAB_next <= ResAB; ResF_next <= ResF;
case C_M is
  when z0 =>
    case C_S is
      when initAB =>
        if ((A = 0) or (B = 0) or (C = 0) or (D = 0)) then
              ResF_next <= 0; N_S <= initAB;
        else  FSM_A_next <= conv_integer(A);
              FSM_B_next <= conv_integer(B); N_S <= c1_z1;
        end if;
      when c1_z1 => N_S <= initCD; N_M <= z1; push <= '1';
      when initCD => ResAB_next <= Res; FSM_A_next <= conv_integer(C);
              FSM_B_next <= conv_integer(D); N_S <= c2_z1;
      when c2_z1 => N_S <= init1_2; N_M <= z1; push <= '1';
      when init1_2 => N_S <= init1_2; FSM_A_next <= conv_integer(ResAB);
              FSM_B_next <= conv_integer(Res); N_S <= c3_z1;
      when c3_z1 => N_S <= final_state; N_M <= z1; push <= '1';
      when final_state => N_S <= initAB; ResF_next <= Res;
      when others => N_S <= initAB;
    end case;
```

```
      when z1 =>
         case C_S is
           when initAB => N_S <= initAB;
             if (FSM_B>0) and (FSM_B>FSM_A) then FSM_A_next <= FSM_B;
                    FSM_B_next <= FSM_A;
             elsif (FSM_B>0) and (FSM_B<=FSM_A)  then FSM_A_next <= FSM_B;
                    FSM_B_next <= FSM_A rem FSM_B;
             else  N_S <= initAB; Res_next <= FSM_A; pop <= '1';
             end if;
           when others => N_S <= initAB;
         end case;
      when others => N_M <= z0;
   end case;
end process;

Result <= conv_std_logic_vector(ResF, data_size);

end Behavioral;
```

The projects for the entities IGCD and Hierarchical_IGCD above have been tested in the Nexys-4 and Atlys prototyping boards. Only onboard switches, buttons and LEDs have been used for the Nexys-4 board. Since there are 16 switches available, 4-bit operands A, B, C, D were supplied. The following VHDL code has been used:

```
entity Top_GCD_4items is          -- this project has been tested in the Nexys-4 board
generic (stack_size : integer := 1; data_size            : integer := 4  );
   port ( A,B,C,D          : in  std_logic_vector (data_size-1 downto 0);
          Result           : out  std_logic_vector (data_size-1 downto 0);
          stack_overflow : out std_logic;
          clk, rst         : in  std_logic);
end Top_GCD_4items;

-- IGCD: synchronized by two clock edges;  Hierarchical_IGCD: synchronized by one clock edge
architecture Behavioral of Top_GCD_4items is
begin          -- either the first or the second entity has to be uncommented below

HIGCD: entity work.Hierarchical_IGCD      -- the number of the occupied slices (Ns) is 29
     generic map (stack_size, data_size)  -- the maximum attainable clock frequency (Fmax)
     port map (clk, rst, A, B, C, D, stack_overflow, Result);      -- is 365.5 MHz

--HIGCD: entity work.IGCD                            -- Ns = 29, Fmax = 241.5 MHz
--                generic map (stack_size, data_size)
--                port map (clk, rst, A, B, C, D, stack_overflow, Result);

end Behavioral;
```

Note that both VHDL codes for the entities IGCD and Hierarchical_IGCD are easily parameterizable for any size of data (data_size). The following VHDL code has been used for testing these entities in the Atlys board communicating with a host computer with the aid of the Digilent IOExpansion component (see Sect. 1.7). The size of data was set to 8 and the 8-bit values A, B, C, D were taken from 32-bit value (data_from_PC) received from a host computer. The result is displayed in the virtual window of a host computer (signal data_to_PC).

```
entity Iterative_GCD is
generic (stack_size : integer := 1; data_size : integer := 8);
  port ( clk            : in std_logic;
         EppAstb        : in std_logic;
         EppDstb        : in std_logic;
         EppWr          : in std_logic;
         EppDB          : inout std_logic_vector(7 downto 0);
         EppWait        : out std_logic);
end Iterative_GCD;

architecture Behavioral of Iterative_GCD is -- see interaction with PC in Sects. 1.7 and 2.6
  signal MyLed          : std_logic_vector(7 downto 0);
  signal MyLBar         : std_logic_vector(23 downto 0);
  signal MySw           : std_logic_vector(15 downto 0);
  signal MyBtn          : std_logic_vector(15 downto 0);
  signal data_to_PC     : std_logic_vector(31 downto 0);
  signal data_from_PC   : std_logic_vector(31 downto 0);
  signal A,B,C,D        : std_logic_vector(data_size-1 downto 0);
  signal rst, stack_overflow       : std_logic;       -- reset and HFSM stack overflow signals
  signal Result         : std_logic_vector(data_size-1 downto 0);
begin --IGCD: synchronization by two clock edges; Hierarchical_IGCD: synchronization by one clock edge
  MyLed(0) <= stack_overflow;                          -- HFSM stack overflow
  MyLed(7 downto 1) <= MyBtn(7downto 1);               -- for tests only
  MyLBar <= MySw & MyBtn(15 downto 8);                 -- for tests only
  rst <= MyBtn(0);                                     -- HFSM reset
  A <= data_from_PC(31 downto 24);       B <= data_from_PC(23 downto 16);
  C <= data_from_PC(15 downto 8);        D <= data_from_PC(7 downto 0);

  -- either the first or the second entity has to be uncommented below
  HIGCD: entity work.Hierarchical_IGCD -- Ns = 122, Fmax = 61.5 MHz
      generic map (stack_size, data_size)   -- note, that Ns and Fmax above are
      port map (clk, rst, A, B, C, D, stack_overflow, Result); -- for the entire project

  --HIGCD: entity work.IGCD                    -- Ns = 136, Fmax = 61.5 MHz
  --    generic map (stack_size, data_size)   -- note, that Ns and Fmax above are
  --    port map (clk, rst, A, B, C, D, stack_overflow, Result); -- for the entire project

  data_to_PC <= (31 downto 8 => '0') & Result;

  IO_interface: entity work.IOExpansion
      port map(EppAstb, EppDstb, EppWr, EppDB, EppWait, MyLed,
               MyLBar, MySw, MyBtn, data_from_PC, data_to_PC);

end Behavioral;
```

As we can see from the previous examples, HFSMs enable executable algorithms to be built from preliminary tested modules. Thus, we can benefit from the following:

- Any module can be debugged and verified independently of a more complex algorithm which can potentially use this module.
- Any module becomes reusable and can be included in different algorithms. Reusability assumes that for any new call all variables and signals that might be potentially changed during previous calls have to be set to initial values.

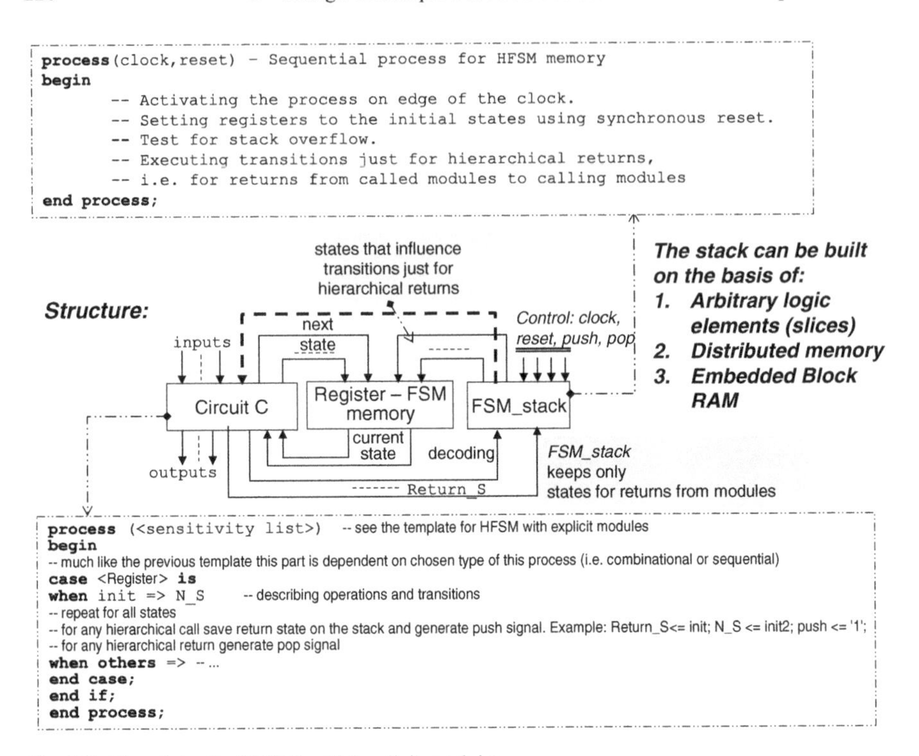
process(clock,reset) - Sequential process for HFSM memory
begin
-- Activating the process on edge of the clock.
-- Setting registers to the initial states using synchronous reset.
-- Test for stack overflow.
-- Executing transitions just for hierarchical returns,
-- i.e. for returns from called modules to calling modules
end process;
states that influence transitions just for hierarchical returns
The stack can be built on the basis of:
1. Arbitrary logic elements (slices)
2. Distributed memory
3. Embedded Block RAM
Structure:
next state
inputs
Control: clock, reset, push, pop
Circuit C
Register – FSM memory
FSM_stack
current state
decoding
outputs
Return_S
FSM_stack keeps only states for returns from modules
process (<sensitivity list>) -- see the template for HFSM with explicit modules
begin
-- much like the previous template this part is dependent on chosen type of this process (i.e. combinational or sequential)
case <Register> is
when init => N_S -- describing operations and transitions
-- repeat for all states
-- for any hierarchical call save return state on the stack and generate push signal. Example: Return_S<= init; N_S <= init2; push <= '1';
-- for any hierarchical return generate pop signal
when others => -- ...
end case;
end if;
end process;

Fig. 5.7 Templates for HFSM with implicit modules

- Any module can be optimized, generally without requiring changes in the rest of the algorithms.
- Alternative and competitive modules (such as *IGCD* and *RGCD*) can easily be examined and compared.
- Complexity of algorithms mapped to hardware can be significantly increased.

5.2.2 HDL Template for HFSM with Implicit Modules

Figure 5.7 depicts the structure of an HFSM with implicit modules [29]. The HFSM behaves like an ordinary FSM and a single stack of states is used just for returns from the called modules.

There are three basic blocks in Fig. 5.7: a *Register*, an *FSM_stack*, and a *circuit C* that calculates next states for state transitions and generates the required outputs. Now states in different modules have to be assigned different codes [29]. The *FSM_stack* is needed just to know which state has to be the target of the transition when a called module is terminated. All state transitions are executed with the register, much like it is done in a conventional FSM. Here, *Return_S* is a code of the return state.

Suppose that a new module z_p has to be called in a state a_m. In this case the following operations are executed at the same time: (1) the state a_n that follows a_m is saved in the *FSM_stack*; (2) the *stack_pointer* is incremented; and (3) the transition from a_m to a_0 (the first state of z_p) is performed in the register.

When the called module z_p is terminated, the stack pointer is decremented and the *stack_pointer* points to the register of the stack with the state a_n that has to be selected for the next state transition. We consider *two modes* of returns [5]. In the first mode, transition from the state a_m does not depend on the execution of the called module (z_p). Thus, we can explicitly save in the stack the target state (such as a_n) for transition after the return from the called module (such as z_p). In the second (more complicated) case the transition from a_m can be changed in the called module and a method based on the use of a special return flag [28] has to be applied. All necessary details can be found in [5, 28].

Note that identical module calls might appear in different states, and, thus, the returns might also be done to different states from the same called module. That is why the returned state has to be chosen in the calling modules (but not in the called modules). The stack is needed just to know which state has to be the target of the transition when a called module is terminated. The number of states is increased comparing to HFSMs with explicit modules. However, the number of stacks and the size of the stack registers are reduced [5]. Another feature of this model is that it is directly applicable to all known optimization techniques that have been proposed for conventional FSMs. An example of complete synthesizable VHDL code for HFSM with implicit modules is given in [5].

5.3 Synthesis of HFSMs

Synthesis of an HFSM with the structure shown in Figs. 5.5 and 5.7 includes the following steps:

1. Marking the given HGSs with labels that will be considered as the HFSM states. For example, the labels initAB, initCD, c1_z1, c2_z1, c3_z1, init1_2, final_state in Fig. 5.6a are HFSM states.
2. Customizing the proposed HDL templates (VHDL templates in Figs. 5.5 and 5.7).
3. Synthesis of HFSM circuits from the customized VHDL templates using commercially available computer-aided design tools, such as ISE of Xilinx or Quartus of Altera.

Various types of HGS marking (labeling) have been proposed and these types depend on the selected HFSM model (Mealy, Moore or combined Mealy and Moore with either explicit or implicit modules). In the example in Fig. 5.6a an HFSM with explicit modules is based on Moore model. In consequent sections we will present methods of synthesis for different types of HFSM.

5.3.1 Synthesis of HFSMs with Explicit Modules

Synthesis can be done for Moore, Mealy, and mixed (Moore and Mealy) models [8]. For Moore machine it includes the following steps, which are very similar to the methods for conventional FSMs [6, 8]:

- The label a_0 is assigned to the node *Begin* of all HGSs.
- The labels $a_1, a_2, \ldots, a_{M-1}$ are assigned to unmarked rectangular nodes (including *End* node) in each HGS.
- The labels can be repeated in different HGSs but cannot be repeated within the same HGS (except the label a_0 in the main HGS z_0, which can also be assigned to the node *End*).
- All rectangular nodes have to be labeled.

The considered type of labeling allows HGS to be executed from the node *Begin* of the main module z_0 and to be terminated in the node *End* of the main module z_0. If z_0 has to be executed cyclically then the node *End* of z_0 has to be assigned the same label as the node *Begin* of z_0, i.e. the label a_0. Alternatively transition from the node *End* can be performed explicitly to the node *Begin* (see example in Fig. 5.6a). Now the labels $a_0, .., a_{M-1}$ are considered to be HFSM states. State transitions are formed using the same rules as in [6, 8]. Each state transition is used to customize the proposed template in Fig. 5.5. All other details will be shown on a simple practical example in Fig. 5.8.

Let us design an HFSM that provides support for full functionality of the iterative sorter in Fig. 3.14. The HFSM analyzes the *enable* signal in Fig. 3.14 and permits sorting to be concluded in less than N/2 clock cycles (see Sect. 3.5 for additional details). Now each comparator in the right-hand line of Fig. 5.8c has the following VHDL code:

```
entity ComparatorOdd is
      generic (M          : integer := 4 );
      port( Op1           : in std_logic_vector(M-1 downto 0);
              Op2         : in std_logic_vector(M-1 downto 0);
            MaxValue      : out std_logic_vector(M-1 downto 0);
            MinValue      : out std_logic_vector(M-1 downto 0);
            test_sorted   : out std_logic);    -- test_sorted =0 if data are not swapped
end ComparatorOdd;

architecture Behavioral of ComparatorOdd is
begin
process(Op1,Op2)
begin
  if Op1 >= Op2 then MaxValue <= Op1; MinValue <= Op2; test_sorted <= '0';
  else MaxValue <= Op2; MinValue <= Op1; test_sorted <= '1';
  end if;
end process;

end Behavioral;
```

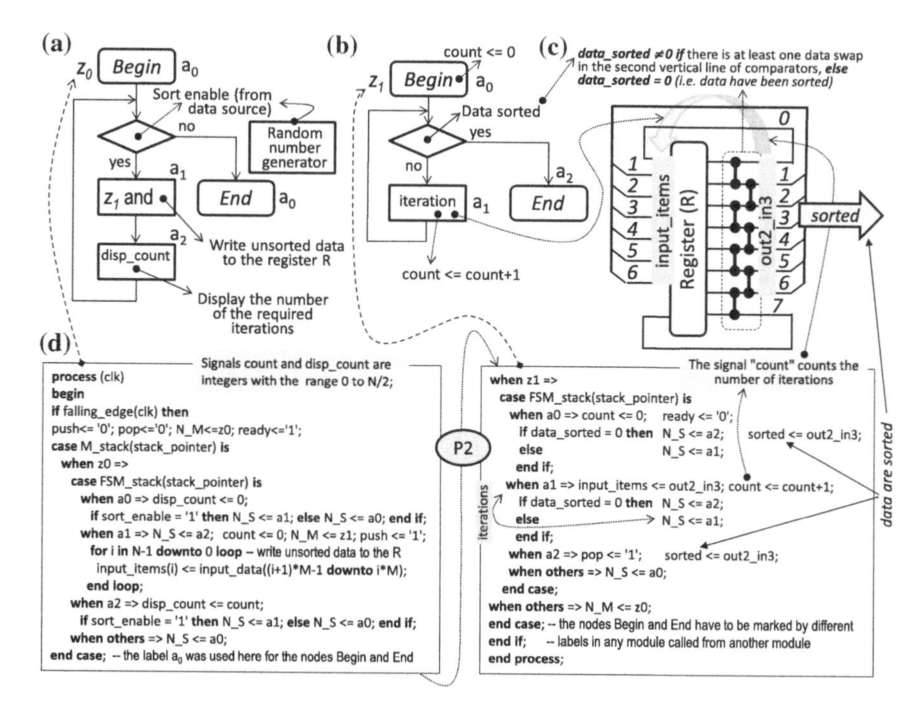

Fig. 5.8 HGS module z_0 (**a**), HGS module z_1 (**b**), iterative sorter (**c**), part of VHDL code for the modules z_0 and z_1 (**d**)

The first HFSM module z_0 checks if a new set of data is available from a source. The latter, for example, could be a random number generator from Appendix B (see Fig. 5.9a), a host PC (see Sects. 4.3 and 4.4), or a host processor (see Sect. 4.5). As soon as a new set of data is available the data are copied to the register R of the sorter and the module z_1 (see Fig. 5.8b) is called. The latter sorts data controlling the sorter shown in Fig. 5.8c. The main function of z_1 is an execution of iterations in the sorter until there is no data swap in the second line of comparators, which indicates that the data have already been sorted and can be copied from the outputs of the sorter (see Fig. 5.8c). Hence, the number of iterations may now be less than N/2 (see Sect. 3.5 for additional details) and the sorting may be accelerated. The statement count <=0; can be associated with the node *Begin* of z_1 or with the calling node a_1 in the module z_0.

The HGSs in Fig. 5.8a and b are marked with the labels a_0, a_1, a_2 which are considered to be HFSM states. All state transitions and operations in the states are shown in Fig. 5.8d. For the sake of simplicity we used synchronization with two clock edges (see Fig. 5.4a). Stack memory in the HFSM is exactly the same as in the entity IGCD (see Sect. 5.2.1). Functionality of the circuits can be tested in more complicated examples given in Sect. 4.1 and one of them is shown in Fig. 5.9.

As distinct from Fig. 4.7 the iterative sorter from Fig. 5.8c for N items is used and it is controlled by the HFSM a part of which is shown in Fig. 5.9 (see reusable

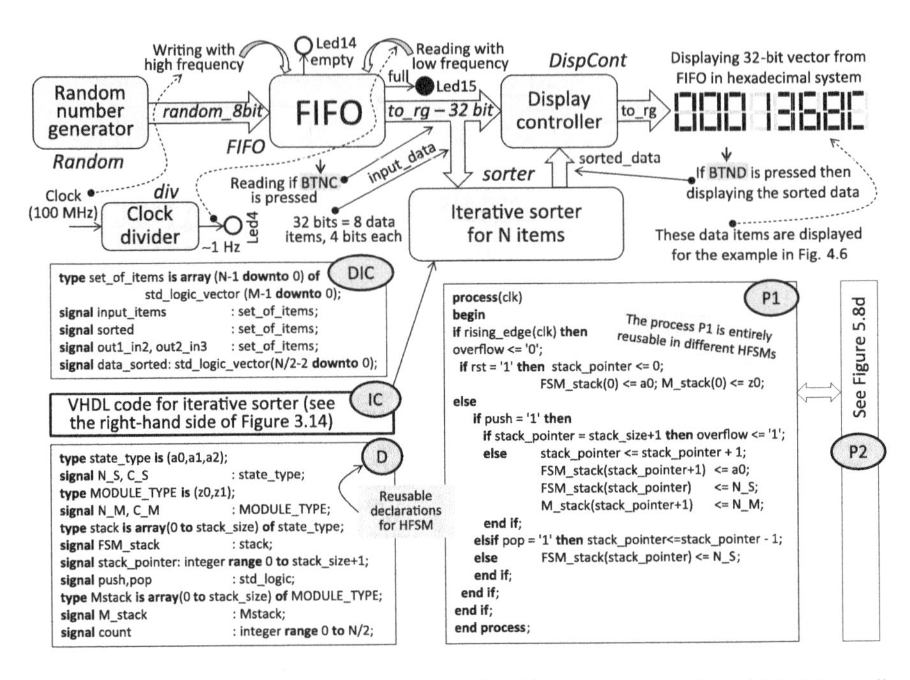

Fig. 5.9 The use of HFSM and the sorter from Fig. 5.8 in a more complicated (slightly modified) example from Sect. 4.1 (see also Fig. 4.7)

declarations D and the reusable process P1) and another part—in Fig. 5.8d (the process P2). VHDL code for the iterative sorter (see the block IC in Fig. 5.9) can almost completely be reused from Fig. 3.14. The only difference is the replacement of the Comparator for the OddComp component with the shown above ComparatorOdd, the signals test_sorted from which are individual signals of type std_logic in the following vector data_sorted:

```
signal data_sorted: std_logic_vector(N/2-2 downto 0); -- the bottom line in the block DIC
```

Thus, if data_sorted = 0 then we can conclude that sorting is completed (see the states a_0 and a_1 in the module z_1 in Fig. 5.8d). Signal declarations for the iterative sorter (DIC) shown in Fig. 5.9 have to be also provided. The entire circuit in Fig. 5.9 occupies 117 slices and 1 embedded block RAM (Nexys-4 board with Artix-7 FPGA) and it has been tested in hardware. The maximum attainable clock frequency is 299 MHz. Note that such frequency is increased if synchronization with one clock edge (see Fig. 5.4b and Sect. 5.2.1) is applied.

For Mealy machine synthesis includes the following steps, which are very similar to the methods for conventional FSMs [6, 8]:

1. The label a_0 is assigned to the input of node following the node *Begin* of all HGSs.
2. The labels $a_1, a_2, \ldots, a_{M-1}$ are assigned to unmarked inputs of nodes that follow rectangular nodes and to inputs of *End* node in each HGS.

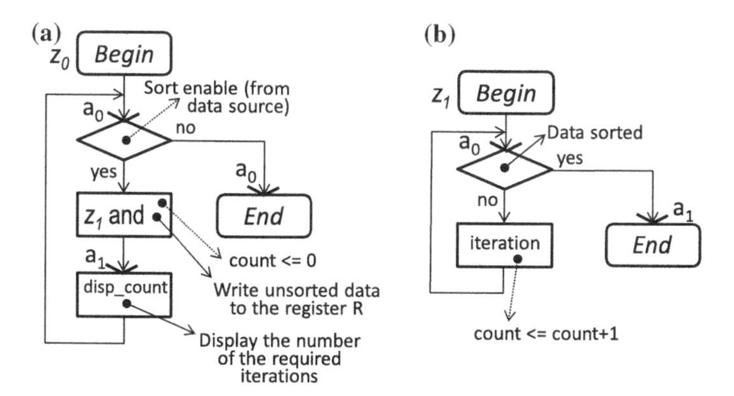

Fig. 5.10 Labeling the modules z_0 (**a**) and z_1 (**b**) from Fig. 5.8 a and b for Mealy HFSM

3. The labels can be repeated in different HGSs but cannot be repeated within the same HGS (except the label a_0 in the main HGS z_0, which can also be assigned to the input of the node *End*).
4. Any input is labeled only once.

The considered type of labeling allows HGS to be executed from the node *Begin* of the main module z_0 and to be terminated in the node *End* of the main module z_0. If z_0 has to be executed cyclically than the same label a_0 can be used after the node *Begin* and at the input of the node *End* of z_0 much like it was done for Moore HFSM. Now the labels a_0,.., a_{M-1} are considered to be HFSM states. State transitions are formed applying the same rules as in [6, 8]. Each state transition is used to customize the proposed template in Fig. 5.5. Figure 5.10 demonstrates an example of marking for the same HGSs that were used for the Moore machine in Fig. 5.8a and b.

VHDL code below gives complete specification of the circuit in Fig. 5.9 in which Mealy HFSM is used instead of Moore HFSM. Blocks with exactly the same code as in Fig. 5.9 are indicated in comments and VHDL statements are not shown explicitly. The sorter is described in the following component:

```
sorter : entity work.EvenOddTransitionIterative -- this specification has to be used below
    port map (clk=>clk, ready=>ready, input_data=>to_rg, sorted_data=>sorted_data,
              overflow =>overflow, disp_count=>disp_count_int, rst=>rst,
              sort_enable=>sort_en);
```

where the following signals are used:

- clk, rst—clock and reset;
- ready—indicates that the sorter is ready to sort a new set of data items;
- input_data, sorted_data—unsorted and sorted sets of data items;
- overflow—HFSM stack overflow;
- disp_count—the number of iterations in the iterative sorter;
- sort_enable—enables a new set of items to be sorted in the iterative sorter.

The top level module has the following VHDL code:

```
entity TestFIFO_withMealyHFSM_Component is
generic (data_in_size      : integer := 8;       -- the width of the input for FIFO
         data_out_size     : integer := 32;      -- the width of the output for FIFO
         N                 : integer := 8 );     -- we consider here eight 4-bit items to sort
 port (  clk               : in std_logic;       -- system clock 100 MHz
         led_full          : out std_logic;      -- LED 15 of the Nexys-4 was used
         led_empty         : out std_logic;      -- LED 14 of the Nexys-4 was used
         led_div_clk       : out std_logic;      -- LED 4 of the Nexys-4 was used
         seg               : out std_logic_vector(6 downto 0);  -- display segments
         sel_disp          : out std_logic_vector(7 downto 0);  -- display selections
         disp_data         : in std_logic;       -- switch 0 of the Nexys-4 was used
         disp_sorted_data  : in std_logic;       -- switch 1 of the Nexys-4 was used
         overflow          : out std_logic;      -- LED 13 of the Nexys-4 was used
         rst               : in std_logic;       -- BTNL of the Nexys-4 was used
         disp_count        : out std_logic_vector(2 downto 0)); -- LEDs 2,1,0 of the Nexys-4
end TestFIFO_withMealyHFSM_Component;

architecture Behavioral of TestFIFO_MealyHFSM_Component is
  signal divided_clk       : std_logic;  -- see also section 4.1
  signal random_8bit       : std_logic_vector(data_in_size-1 downto 0);
  signal wr_en             : std_logic;
  signal rd_en             : std_logic;
  signal to_rg             : std_logic_vector(data_out_size-1 downto 0);
  signal sorted_data       : std_logic_vector(data_out_size-1 downto 0);
  signal data_to_display   : std_logic_vector(data_out_size-1 downto 0);
  signal wr_ack            : std_logic;
  signal rd_valid          : std_logic;
  signal full              : std_logic;
  signal ready             : std_logic;
  signal disp_count_int    : integer range 0 to N/2;
  signal sort_en           : std_logic;
  signal to_HFSM_to_count: std_logic;
begin

  disp_count   <= conv_std_logic_vector(disp_count_int, 3);
  led_div_clk  <= divided_clk;
  led_full     <= full;

 process(full, rd_valid, disp_data)
 begin
   if (full /= '1') then        wr_en <= '1';
   else                         wr_en <= '0';
   end if;
   if (disp_data = '1') then rd_en <= '1';
   else                         rd_en <= '0';
   end if;
 end process;
```

```
process(clk)
begin
  if rising_edge(clk) then
    if ((rd_en = '1') and (ready = '1')) then      sort_en <= '1';
    else                                            sort_en <= '0';
    end if;
  end if;
end process;
data_to_display <= to_rg when disp_sorted_data = '0' else
                   sorted_data when ready = '1'
                   else (others => '0');
FIFO  : entity work.FIFO_mem       -- see section 4.1
      port map (wr_clk=>clk, rd_clk=>divided_clk, din=>random_8bit, wr_en=>wr_en,
      rd_en=>rd_en, dout=>to_rg, full=>full, empty=>led_empty,
      rd_data_count=>open, wr_data_count=>open);
Random: entity work.RanGen       -- see VHDL code in Appendix B
      generic map(width => data_in_size)
      port map (clk, random_8bit);
DispCont: entity work.EightDisplayControl   -- see VHDL code in Appendix B
      port map(clk, data_to_display(31 downto 28), data_to_display(27 downto 24),
      data_to_display(23 downto 20), data_to_display(19 downto 16),
      data_to_display(15 downto 12), data_to_display(11 downto 8),
      data_to_display(7 downto 4), data_to_display(3 downto 0), sel_disp, seg);
div: entity work.clock_divider                   -- see VHDL code in Appendix B
      port map(clk, '0', divided_clk);
sorter: entity work.EvenOddTransitionIterative  -- see the specification above
end Behavioral;
```

The sorter has the following VHDL code:

```
entity EvenOddTransitionIterative is
generic (M              : integer := 4;   -- M is the size of any data item
        stack_size      : integer := 1;   -- there are two registers in the HFSM stack 0 and 1
        N               : integer := 8 ); -- N is the number of data items
 port (clk              : in std_logic;
       ready            : out std_logic;
       input_data       : in std_logic_vector(N*M-1 downto 0);
       sorted_data      : out std_logic_vector(N*M-1 downto 0);
       overflow         : out std_logic;
       disp_count       : out integer range 0 to N/2;
       rst              : in std_logic;
       sort_enable      : in std_logic);
end EvenOddTransitionIterative;
architecture Behavioral of EvenOddTransitionIterative is
-- Declarations needed for the HFSM: insert here all lines from the block D in Fig. 5.9
-- (the state a2 is not needed for the Mealy HFSM and is removed from the state_type)
-- Declarations needed for the iterative sorter: insert here all lines from the block DIC in Fig. 5.9
```

```
begin
process(sorted) -- this combinational process converts a set of N data items to a single vector
begin
  for i in N-1 downto 0 loop
    sorted_data((i+1)*M-1 downto i*M) <= sorted(i);
  end loop;
end process;

generate_even_comparators: -- even-odd transition iterative circuit is given below in its entirety
  for i in N/2-1 downto 0 generate-- see also the block IC in Fig. 5.9 and Fig. 3.14
    EvenComp: entity work.Comparator -- this is exactly the same comparator as in Fig. 3.14
       generic map (M => M)
       port map(input_items(i*2), input_items(i*2+1), out1_in2(i*2), out1_in2(i*2+1));
end generate generate_even_comparators;

generate_odd_comparators: -- the code below is slightly different compared to Fig. 3.14
  for i in N/2-2 downto 0 generate
    OddComp: entity work.ComparatorOdd -- see the code at the beginning of section 5.3.1
       generic map (M => M)
       port map(out1_in2(2*i+1), out1_in2(2*i+2), out2_in3(i*2+1),
                   out2_in3(i*2+2), data_sorted(i));
end generate generate_odd_comparators;

out2_in3(0)      <= out1_in2(0);      -- these two lines are exactly the same as in Fig. 3.14
out2_in3(N-1)    <= out1_in2(N-1);

-- The process for HFSM stacks: insert here all lines from the block P1 in Fig. 5.9

process (clk)  -- Description of transitions and operations in the Mealy HFSM (see Fig. 5.10)
begin
  if falling_edge(clk) then
    push<='0'; pop<='0'; N_M<=z0; ready<='1';
    case M_stack(stack_pointer) is
       when z0 =>
         case FSM_stack(stack_pointer) is
           when a0 => disp_count <= 0;
             if sort_enable = '1' then          N_S <= a1;          count <= 0;
                 for i in N-1 downto 0 loop -- copying unsorted data items to the register R
                   input_items(i) <= input_data((i+1)*M-1 downto i*M);
                 end loop;
                 N_M <= z1; push <= '1';
             else N_S <= a0;
             end if;
           when a1 => disp_count <= count;
             if sort_enable = '1' then          N_S <= a1;          count <= 0;
               for i in N-1 downto 0 loop       -- copying unsorted data items to the register R
                 input_items(i) <= input_data((i+1)*M-1 downto i*M);
               end loop;
               N_M <= z1;  push <= '1';
             else N_S <= a0;
             end if;
           when others => N_S <= a0;
```

```
          end case;
        when z1 =>
          case FSM_stack(stack_pointer) is
            when a0 => ready <= '0'; input_items <= out2_in3;
              if data_sorted = 0 then -- test if there is no swap in the second line of Fig. 5.8c
                  N_S <= a1; sorted <= out2_in3; -- sorted data are ready
              else N_S <= a0; count <= count+1;
              end if;
            when a1 => pop <= '1';
            when others => N_S <= a0;
          end case;
        when others => N_M <= z0;
      end case;
    end if;
  end process;

  end Behavioral;
```

The described circuit occupies 109 slices and 1 embedded block RAM (Nexys-4 board with Artix-7 FPGA) and it has been tested in hardware. The maximum attainable clock frequency is 266.7 MHz. Note that such frequency is increased if synchronization with one clock edge (see Fig. 5.4b and Sect. 5.2.1) is applied.

5.3.2 Synthesis of HFSMs with Implicit Modules

Much like HFSMs with explicit modules, synthesis of HFSMs with implicit modules can be done for Moore, Mealy, and mixed (Moore and Mealy) models. For Moore HFSM synthesis includes the following steps:

1. The label a_0 is assigned to the node *Begin* of the main HGS usually designated z_0.
2. The labels $a_1, a_2, \ldots, a_{M-1}$ are assigned to unmarked rectangular nodes (including *End* node) in each HGS.
3. The labels cannot be repeated in different HGSs and within the same HGS (except the label a_0 in the main HGS z_0, which can also be assigned to the node *End*).
4. All rectangular nodes have to be labeled.
5. All other details are the same as for HFSM with explicit modules.

For Mealy HFSM synthesis includes the following steps:

1. The label a_0 is assigned to the input of the node following the node *Begin* of all HGS.
2. The labels $a_1, a_2, \ldots, a_{M-1}$ are assigned to unmarked inputs of nodes that follow rectangular nodes and inputs of *End* nodes in each HGS.
3. The labels cannot be repeated in different and within the same HGSs (except the label a_0 in the main HGS z_0, which can also be assigned to the input of the node *End*).
4. Any input is labeled only once.

All other details are the same as for HFSM with explicit modules. A mixed HFSM permits Mealy and Moore models to be combined and such model is the most preferable in many practical applications. Indeed, the circuit C may use the most appropriate signals that depend on either only the states or the states and inputs.

Many different examples for synthesis of HFSMs with implicit modules (including mixed HFSMs) are given in [5].

5.4 Parallel Specifications and Parallel HFSMs

Some modules (such as *c1_z1* and *c2_z1* in Fig. 5.6a) can be executed in parallel (see Fig. 5.11). Let us take for further study only HGS rectangular nodes with more than one *macro operation* making up sets $Z_1, Z_2, \ldots, \ldots$ [1]. Thus, parallel execution of *macro operations* assigned to each set has to be provided. For example in Fig. 5.11b there are three sets: $Z_1 = \{z_1, z_2, z_3\}$, $Z_2 = \{z_1, z_4\}$, and $Z_3 = \{z_2, z_3, z_4\}$. The main module $Z_0 = \{z_0\}$ also needs to be implemented and up to three modules (see the sets Z_1 and Z_3) need to be executed in parallel. According to the proposal in [1], a parallel HFSM (PHFSM) can be designed by applying the following rules:

1. *Macro operations* from each set Z_i are assigned to different HFSMs running in parallel. The HFSM implementing the calling module is responsible for the parallel activation of the called modules and for verification that all the called modules from the same set have been completed (i.e. execution can proceed after the relevant merging point such as that is shown in Fig. 5.1i). For our example in Fig. 5.11b, the assignment can be done as follows: $HFSM_1 \leftarrow z_0, z_1, z_2$; $HFSM_2 \leftarrow z_2, z_3, z_4$; $HFSM_3 \leftarrow z_3, z_4$. For the example in Fig. 5.11a: $HFSM_1 \leftarrow Z_1^4, Z_1^2(A, B), Z_1^2(R1, R2)$; $HFSM_2 \leftarrow Z_1^2(C, D)$.
2. Each $HFSM_p$ is described as a VHDL component with three additional signals that are introduced in the next point.
3. If a calling ($z_q \rightarrow$) and a called ($\rightarrow z_p$) module ($z_q \rightarrow z_p$) belong to the same HFSM component, then functionality is exactly the same as for

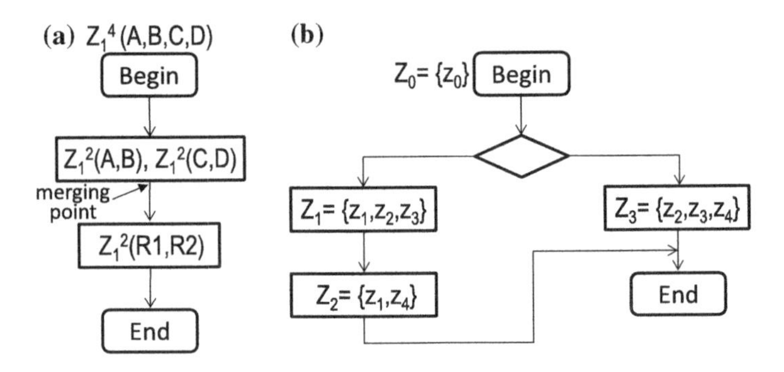

Fig. 5.11 Examples of parallel operations

a non-parallel HFSM (see sections above). Suppose now that $z_q \rightarrow z_p$ and the modules $z_q \rightarrow$, $\rightarrow z_p$ belong to different components $HFSM_q$ and $HFSM_p$. To trigger a *macro operations* $\rightarrow z_p$ from $z_q \rightarrow$, the following three additional signals are involved: (1) $start_p$ to activate the $HFSM_p$ ($HFSM_q \rightarrow HFSM_p$); (2) z_p to choose the module $\rightarrow z_p$ in the $HFSM_p$ from the $HFSM_q$; (3) $finish_p$ to indicate that the module $\rightarrow z_p$ is completed. The signals $start_p$ and z_p are formed (assigned) in $z_q \rightarrow$ and used (tested) in $\rightarrow z_p$. The signal $finish_p$ is assigned in $\rightarrow z_p$ and tested in $z_q \rightarrow$.

4. Finally parallel execution of *macro operations* in the sets Z_1, Z_2, Z_3 in Fig. 5.11b will be provided in the following three HFSM components: $Z_1 \rightarrow \{HFSM_1(z_1), HFSM_2(z_2), HFSM_3(z_3)\}$; $Z_2 \rightarrow \{HFSM_1(z_1), HFSM_2(z_4)\}$; $Z_3 \rightarrow \{HFSM_1(z_2), HFSM_2(z_3), HFSM_3(z_4)\}$. Parallel execution of *macro operations* from the set $Z_1 = \{Z_1^2(A, B), Z_1^2(C, D)\}$ in Fig. 5.11a will be provided in two HFSM components: $\{HFSM_1(Z_1^2(A, B)), HFSM_2(Z_1^2(C, D))\}$.

The technique described above enables any reasonable number of HFSMs mapped to VHDL components to be executed at the same time. All the HFSM features discussed in the previous sections are entirely supported. Concurrent execution of VHDL components is combined with modularity and recursion within individual HFSMs. However, parallel calls from recursively activated modules are not allowed [1]. The maximum number of concurrent HFSMs has to be known in advance to provide the necessary mapping to VHDL components. The graph of *parallel* invocations (such as $Z_1 \rightarrow \{z_1, z_2, z_3\}$; $Z_2 \rightarrow \{z_1, z_4\}$; $Z_3 \rightarrow \{z_2, z_3, z_4\}$) has to be a tree (i.e. cycles are not allowed for *parallel* invocations but they are allowed for *sequential* invocations). Thus, any called module cannot call any of its predecessors with parallel calls.

Let us consider an example of a PHFSM that has to be synthesized from specification shown in Fig. 5.12. Let us assume that four pairs of operands (A, B), (C, D), (E, F), and (G, H) need to be processed in parallel. One of the operations, for instance (A, B), can be executed in the main module called PHFSM. For each pair of the remaining operations the main module PHFSM activates parallel branches (PB) supplying to each branch the signal *start* (activating the branch), the name of macro-operation z_p that has to be executed, and the relevant operands (see Fig. 5.12). As soon as any branch completes the required operation, it generates the signal *finish*. Signals *finish* from all the parallel branches are verified in the main module, which decides whether it can continue execution after the parallel calls.

Suppose, parallel operations have to be executed over individual operands, such as A, B, C, D. The method is exactly the same: one operand (let us say A) is associated with the main module and the remaining operands (B, C, D) are processed in parallel branches, activated from the main module.

Let us implement the following C function *gcd* with 8 arguments in FPGA:

```
unsigned int gcd(unsigned int A, unsigned int B, unsigned int C,
  unsigned int D, unsigned int E, unsigned int F,
  unsigned int G, unsigned int H)
{  return gcd(gcd(gcd(A,B), gcd(C,D)), gcd(gcd(E,F), gcd(G,H))); }
```

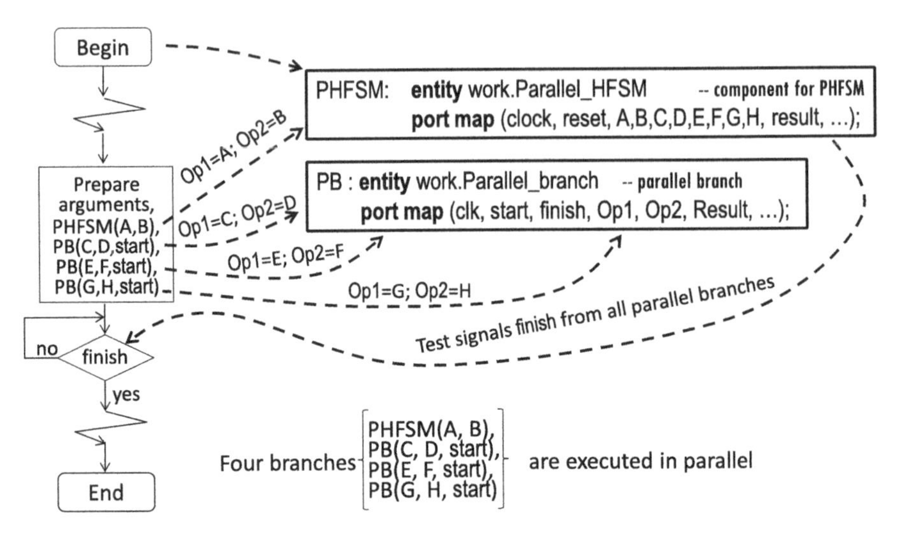

Fig. 5.12 An example of a parallel specification

This function permits the greatest common divisor of eight operands A, B, C, D, E, F, G, H to be found and calls another function *gcd* with two operands:

```
unsigned int gcd(unsigned int A, unsigned int B)
{ int tmp;
  while (B > 0)
  { if (B > A) { tmp=A; A=B; B=tmp; }
    else       { tmp=B; B= A%B; A=tmp; }    }
  return A;                                       }
```

Clearly, four functions gcd(A,B), gcd(C,D), gcd(E,F), gcd(G,H) can be executed in parallel at the first step giving the results Result_A_B, Result_C_D, Result_E_F, and Result_G_H. At the second step these results will be used as the arguments of the functions: gcd(Result_A_B, Result_C_D), and gcd(Result_E_F, Result_G_H), which can also be executed in parallel giving the result Result_A_B_C_D, and Result_E_F_G_H. At the next (last) step the function gcd(Result_A_B_C_D, Result_E_F_G_H) computes the final result which is the greatest common divisor of 8 unsigned integers A, B, C, D, E, F, G, H. All the discussed above functions can be implemented in the PHFSM with the functionality described by parallel HGSs depicted in Fig. 5.13. At the beginning the operands A, B, C, D, E, F, G, H are tested and if there is at least one zero operand then the subsequent steps are not executed and the result is assigned to 0. If all the operands are not equal to zero then 4 modules z_1 with different arguments are active at the same time. As soon as all of them terminate, the results of these modules are used as operands for two new invocations of z_1 also running in parallel. The final result is produced in the bottom module z_1.

We present below two complete synthesizable VHDL specifications that allow the hardware circuit S that implements the algorithm in Fig. 5.13 to be designed.

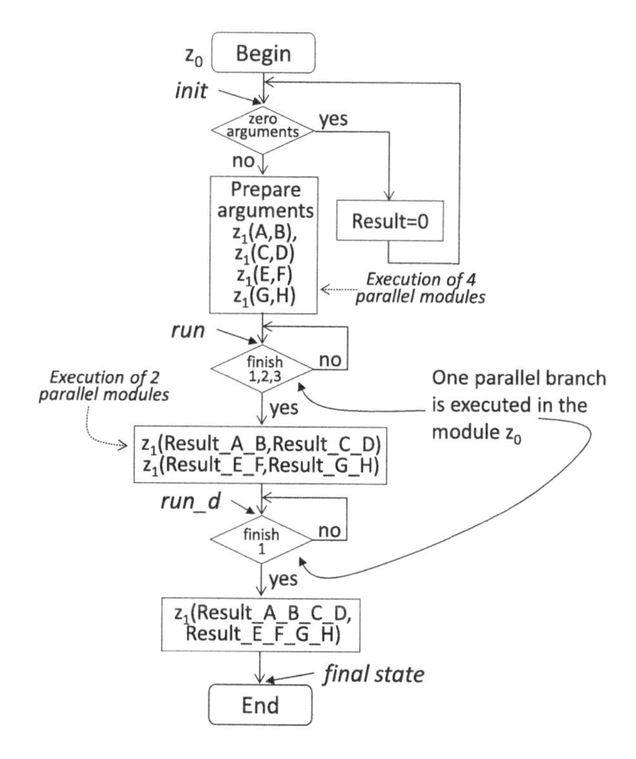

Fig. 5.13 Parallel HGS that permits the greatest common divisor of eight non-negative integers to be found

The first specification (an entity Parallel_HFSM_iterative) corresponds to the C function discussed above. The second specification (an entity Parallel_HFSM_recursive) is based on the recursive C function *RGCD* given in Sect. 5.2.1. Thus, there are recursive calls in all modules z_1 running in parallel. Figure 5.14 demonstrates the general interface. Interactions are organized with the aid of two additional signals: *enable* and *ready*. The latter is generated by the circuit S and indicates that it is ready to process a new set of operands A, B, C, D, E, F, G, H. The signal *enable* is formed by a system that interacts with the circuit S and indicates that a new set of data A, B, C, D, E, F, G, H is available for further processing.

Note that in the previous examples we used the VHDL *rem* operation to find the rest of division of two operands. From preliminary experiences we found that if the size of operands is increased, the maximum attainable clock frequency is rapidly decreased and, besides the maximum size of integers (that are the requested type for the operation *rem*) is limited. So, we decided to implement a similar operation using an additional HFSM module z_2. A simple algorithm was adopted from [30]. Figure 5.15 demonstrates the basic organization of different HGSs (modules) z_0, z_1, and z_2.

The first HGS z_0 tests the operands for zeros and activates the iterative module z_1. As soon as a remainder needs to be found, necessary data are copied to the variables local_divisor and local_remainder and the module z_2 is activated. The latter executes a cyclic algorithm [30] requiring totally M cycles controlled by a variable index, where M is the size of operands. Parallel call of several modules (see Fig. 5.13) is

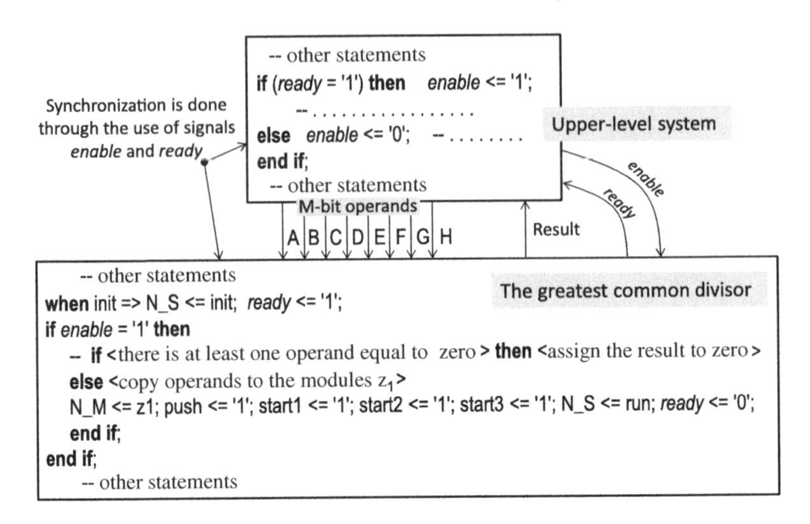

Fig. 5.14 Interface of an upper-level system with the greatest common divisor

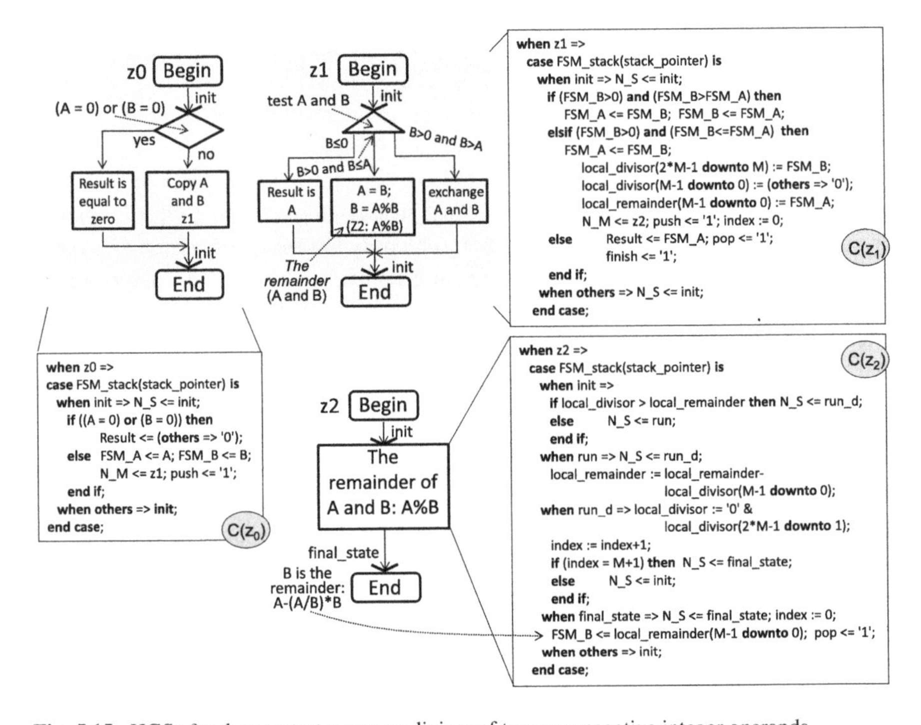

Fig. 5.15 HGSs for the greatest common divisor of two non-negative integer operands

done in a separate module (see Fig. 5.16) that executes its own operation (discovering the greatest common divisor of two operands) and activates parallel modules that are described as VHDL components. Activations are done by the signals *start* and terminations of modules are checked examining the values of the signals *finish* (see Fig. 5.12).

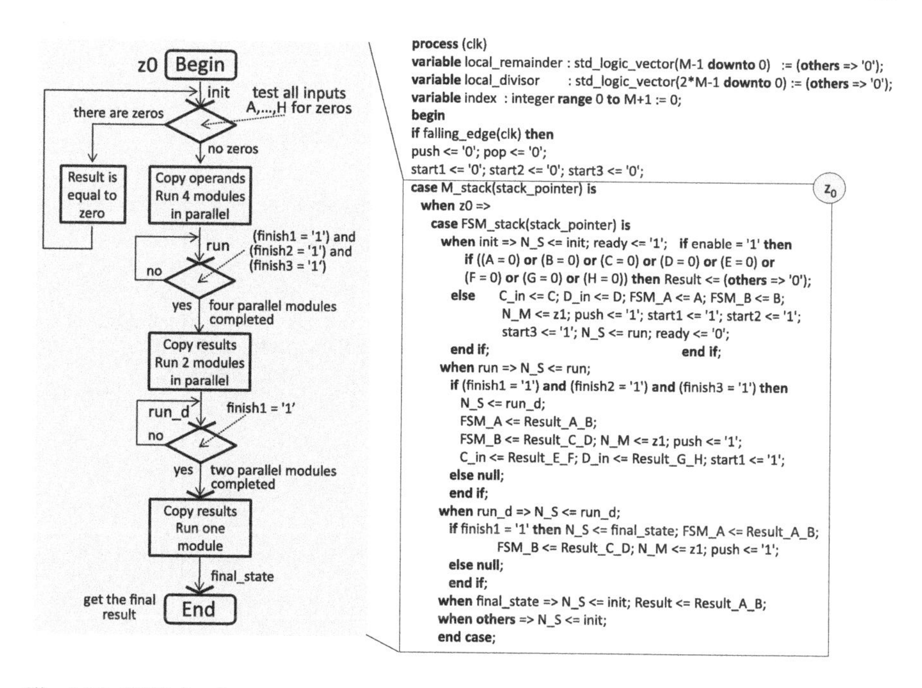

Fig. 5.16 HGSs for the greatest common divisor of eight non-negative integer operands

The following VHDL code gives the complete specification of the greatest common divisor for 8 non-negative integer operands using an iterative algorithm:

```
entity Parallel_HFSM is
generic (stack_size       : integer;             -- stack_size is the size of the HFSM stack
         M                : integer);            -- M is the size (the number of bits) of operands
port (   clk              : in  std_logic;       -- system clock (100 MHz for Nexys-4)
         rst              : in  std_logic;       -- reset signal (the button BTNC was used)
         A,B,C,D,E,F,G,H  : in std_logic_vector(M-1 downto 0);   -- M-bit operands
         Result           : out std_logic_vector(M-1 downto 0);  -- M-bit result
         overflow         : out std_logic;       -- HFSM stack overflow signal
         enable           : in std_logic;        -- enable signal
         ready            : out std_logic);      -- ready signal
end Parallel_HFSM;

architecture Behavioral of Parallel_HFSM is
  signal FSM_A, FSM_B : std_logic_vector(M-1 downto 0);
  type state_type is (init, run, run_d, final_state);                 -- *
  signal N_S               : state_type;                              -- *
  type MODULE_TYPE is (z0, z1, z2);                                    -- *
   -- the lines marked with *** in the entity IGCD in section 5.2.1
   -- except for state_type and MODULE_TYPE
  signal C_in, D_in        : std_logic_vector(M-1 downto 0);
  signal Result_A_B        : std_logic_vector(M-1 downto 0);
  signal Result_C_D        : std_logic_vector(M-1 downto 0);
```

```
  signal Result_E_F      : std_logic_vector(M-1 downto 0);
  signal Result_G_H      : std_logic_vector(M-1 downto 0);
  signal overflow1, overflow2, overflow3, overflow4     : std_logic;
  signal start1, start2, start3, finish1, finish2, finish3     : std_logic;
begin
  overflow <= overflow1 or overflow2 or overflow3 or overflow4;

  -- the process from the entity IGCD in section 5.2.1 in which the state initAB is
  -- replaced with the state init

process (clk)
  variable local_remainder : std_logic_vector(M-1 downto 0)     := (others => '0');
  variable local_divisor : std_logic_vector(2*M-1 downto 0)     := (others => '0');
  variable index : integer range 0 to M+1                       := 0;
begin
  if falling_edge(clk) then
    push <= '0'; pop <= '0'; start1 <= '0'; start2 <= '0'; start3 <= '0';
    -- the module z0 from Fig. 5.16
    -- the module C(z1) from Fig. 5.15 without the signal finish
    -- the module C(z2) from Fig. 5.15
    when others => null;
    end case;
  end if;
end process;

C_D:  entity work.Parallel_branch
        generic map(stack_size => stack_size, M => M)
        port map (clk, start1, finish1, C_in, D_in, Result_C_D, overflow2);

E_F:  entity work.Parallel_branch
        generic map(stack_size => stack_size, M => M)
        port map (clk, start2, finish2, E, F, Result_E_F, overflow3);

G_H:  entity work.Parallel_branch
        generic map(stack_size => stack_size, M => M)
        port map (clk, start3, finish3, G, H, Result_G_H, overflow4);
end Behavioral;
```

The components C_D, E_F, and G_H have the following VHDL code:

```
entity Parallel_branch is
generic (  stack_size     : integer;
           M              : integer   );

port ( clk                : in  std_logic;
       reset              : in  STD_LOGIC;
       finish             : out  STD_LOGIC;
       A, B               : in std_logic_vector(M-1 downto 0);
       Result             : out std_logic_vector(M-1 downto 0);
       overflow           : out std_logic);
end Parallel_branch;

architecture Behavioral of Parallel_branch is
  signal FSM_A, FSM_B : std_logic_vector(M-1 downto 0);
```

```
   -- the lines marked with * from the code above (entity Parallel_HFSM)
   -- the lines marked with *** in the entity IGCD in section 5.2.1
   -- except for state_type and MODULE_TYPE
begin

   -- the process from the entity IGCD in section 5.2.1 in which the state initAB
   -- is replaced with the state init

process (clk)
  variable local_remainder : std_logic_vector(M-1 downto 0)     := (others => '0');
  variable local_divisor     : std_logic_vector(2*M-1 downto 0) := (others => '0');
  variable index               : integer range 0 to M+1            := 0;
begin
  if falling_edge(clk) then    push <= '0'; pop <= '0'; finish <= '0';
    case M_stack(stack_pointer) is
      -- the module C(z0) from Fig. 5.15
      -- the module C(z1) from Fig. 5.15
      -- the module C(z2) from Fig. 5.15
      when others => null;
    end case;
  end if;
end process;

end Behavioral;
```

The entity Parallel_HFSM can be used as a component of an upper-level system like in the following code:

```
entity test_4_parallel_HFSM_iterative is
generic (stack_size : integer := 2; M : integer := 32 );  -- the size of operands is 32 bits
port ( clk        : in std_logic;         -- system clock 100 MHz for the Nexys-4 board
       rst        : in std_logic;         -- BTNC button for the Nexys-4 board was used
       rec        : in std_logic;                              -- switch15 (Nexys-4)
       sel        : in std_logic_vector(2 downto 0);           -- switches 2-0 (Nexys-4)
       use_sw     : in std_logic;                              -- switch14 (Nexys-4)
       sw         : in std_logic_vector(10 downto 0);          -- switches 13-3 (Nexys-4)
       overflow   : out std_logic;         -- stack overflow in at least one HFSM
       Result1    : out std_logic_vector(M-1 downto 15);-- bits of the result on PMod pins
       led        : out std_logic_vector(14 downto 0) ); -- bits of the result on LEDs
end test_4_parallel_HFSM_iterative;

architecture Behavioral of test_4_parallel_HFSM_iterative is
  signal A,B,C,D,E,F,G,H: std_logic_vector(M-1 downto 0) := (others => '0');
  signal Result          : std_logic_vector(M-1 downto 0););-- M-bit result
  signal enable, ready   : std_logic;
begin

  Result1 <= Result(M-1 downto 15);          -- bits of the result to PMod pins
process (clk)
begin
  if rising_edge(clk) then
    if (ready = '1') then   enable <= '1';
      if (rec = '1') then   -- use one of fixed or generated data sets
        case (sel) is       -- fixed (or generated) data sets selected by onboard switches 2,1,0
```

```
            when "000" => A<=A+1; B<=B+1; C<=C+1; D<=D+1; E<=E+1; F<=F+1;
            G<=G+1; H<=H+1;         -- change values of operands somehow
            when "001" =>A<=conv_std_logic_vector(152,M); -- the first fixed set
                  B<=conv_std_logic_vector(38, M); C<=conv_std_logic_vector(209, M);
                  D<=conv_std_logic_vector(133, M); E<=conv_std_logic_vector(95, M);
                  F<=conv_std_logic_vector(57, M); G<=conv_std_logic_vector(247, M);
                  H<=conv_std_logic_vector (171, M);   -- the result is 19:    10011
                  -- other fixed sets
            when others =>
                  A <= conv_std_logic_vector (3303375, M); -- the last fixed set
                  B<=conv_std_logic_vector(20809539, M);
                  C<=conv_std_logic_vector(127666539, M);
                  D<=conv_std_logic_vector(19533, M);
                  E<=conv_std_logic_vector(1147851, M);
                  F<= conv_std_logic_vector(1320201, M);
                  G<=conv_std_logic_vector(20980740, M);
                  H<= conv_std_logic_vector(688479651, M);
                  -- the result is 1149:  10001111101
          end case;
          if use_sw = '1' then
            H <= (31 downto 11 => '0') & sw; -- onboard switches can be used
          end if; -- onboard switches 13,...,3 can be used to change 10 least significant bits of H
        else       -- a default set, the result is 3: 11
          A<= conv_std_logic_vector(33, M);
          B<= conv_std_logic_vector(60, M);
          C<= conv_std_logic_vector(1200, M);
          D<= conv_std_logic_vector(57, M);
          E<= conv_std_logic_vector(6, M);
          F<= conv_std_logic_vector(399, M);
          G<= conv_std_logic_vector(63, M); H<=conv_std_logic_vector (24, M);
        end if;
      else enable <= '0';
      end if;
    end if;
  end process;

  PHFSM: entity work.Parallel_HFSM_iterative
          generic map(stack_size => stack_size, M => M)
          port map (clk, rst, A, B, C, D, E, F, G, H, Result, overflow, enable, ready);

  led <= Result(14 downto 0);          -- 15 onboard LEDs were used for indicating the binary result

  end Behavioral;
```

In the second (recursive) specification (an entity Parallel_HFSM_recursive) only one module z_1 is changed and now it has the following VHDL code:

```
when z1 =>
  case FSM_stack(stack_pointer) is
    when init => N_S <= final_state;
      if (FSM_B>0) then
        if (FSM_B>FSM_A) then   FSM_A <= FSM_B; FSM_B <= FSM_A;
                N_M <= z1; push <= '1';       -- recursive call of z1
```

```
        else FSM_A <= FSM_B;   N_S <= run;
                local_divisor(2*M-1 downto M)        := FSM_B;
                local_divisor(M-1 downto 0) := (others => '0');
                local_remainder(M-1 downto 0)        := FSM_A;
                N_M <= z2;  push <= '1'; index := 0; -- non-recursive call of z2
        end if;
      else Result <= FSM_A;
      end if;
   when run => N_M <= z1; push <= '1'; N_S <= final_state; -- recursive call of z1
   when final_state => N_S <= final_state;  finish <= '1'; pop <= '1';  -- note that the
   when others => init; -- statement finish<='1' has to be removed in the entity Parallel_HFSM
 end case;
```

The remaining code is the same and tests can be done in the entity test_4_parallel_HFSM_iterative above where the Parallel_HFSM_iterative needs to be replaced with a new component that calls the recursive module z_1.

The results of synthesis, implementation, and tests have shown the following. The project based on the iterative algorithm requires 645 slices (from 15,850 slices available in the FPGA of the Nexys-4 board) and permits the maximum attainable clock frequency 133.1 MHz. The project based on the recursive algorithm requires 683 slices and permits the maximum attainable clock frequency 124.2 MHz. Clearly for cyclic algorithms recursive calls do not give any advantage, however for algorithm that use tree-based structures recursive modules might be more beneficial than iterative modules [2, 3, 5, 27, 28].

5.5 Hardware Implementations of Software Programs Based on HFSM Models

It is known [27, 28] that iteration in general and recursion in particular can be implemented more efficiently in hardware than in software. This is because any activation of a module can be combined with the execution of operations (micro operations) that are required by the algorithm. The same event takes place when any module is being terminated, i.e. when control has to be returned to the point after the last recursive call and an operation of the executing algorithm that follows the last recursive call has to be activated. The number of states required for the execution of recursion in hardware can be reduced comparing with software. Moreover, we will show later (in Sect. 5.6) that states can be accommodated on stacks that are implemented on built-in memory blocks. Besides, broad parallelism can directly be supported (see the Sect.5.4). The results obtained for known methods, such as those reviewed in [4], that enable hierarchical calls to be executed in hardware have shown that hardware circuits may be faster than alternative software programs implementing similar functionality. Enhanced models of HFSMs that allow different types of arguments to be passed to hardware modules and to be returned from the modules, much as in software programs, are described in [1].

Let us look at Fig. 5.13 where several modules z_1 can run in parallel. Each module has two arguments and returns a value. To provide similar functionality in hardware, we need to be able to: (a) *pass arguments by values*; (b) *return values*.

Let us consider an example from [1]. The following C code (where the function treesort is called recursively) constructs and returns a sorted list from a given binary tree (such as that studied in [28]):

```
ValueAndCounter* treesort(treenode* node) { // node is a pointer to the root of the tree
  ValueAndCounter* tmp;                     // tmp is a temporary pointer to a list item
  static ValueAndCounter* ttmp=0;           // at the beginning the list is empty
  if(node!=0)
  { // if the node exists
      treesort(node->lnode);          // sort left sub-tree
      tmp = new ValueAndCounter; // allocate memory for a new list item tmp
      tmp->next=ttmp;                 // store pointer to the previous list item
      tmp->val = node->val;           // save the value
      tmp->count = node->count;       // save the number of repetitions of the value node->val
      ttmp = tmp;                     // extend the list
      treesort(node->rnode);          // now sort right sub-tree
      return ttmp;
  }
```

Any tree node has the following structure:

```
struct treenode {
  int val;                 // value of an item of type int
  int count;               // number of items with the value val
  treenode* lnode;         // pointer to left sub-tree
  treenode* rnode; };      // pointer to right sub-tree
```

Any list item has the following structure:

```
struct ValueAndCounter {
  int val;                        // value of an item of type int
  int count;                      // number of items with the value val
  ValueAndCounter* next; };       // pointer to the next item of type ValueAndCounter
```

We assume here that the tree has already been built (using, for example, the method [28]). The nodes of the tree contain four fields: a pointer to the right child node, a pointer to the left child node, a counter, and a value (an integer in our case). The nodes are maintained so that at any node, the left sub-tree contains only values that are less than the value at the node, and the right sub-tree contains only values that are greater. The counter indicates the number of occurrences of the value associated with the respective node.

If we call the function with the statement beginning = treesort(root);, it returns a pointer to the list of sorted data items. To provide similar functionality in hardware, we need to be able to: (a) *pass arguments through pointers*; (b) *return pointers*.

To support the described above features the third stack memory (called *AR_stack*) has been introduced for arguments together with an additional register for the returned value [1] which is shown in Fig. 5.17.

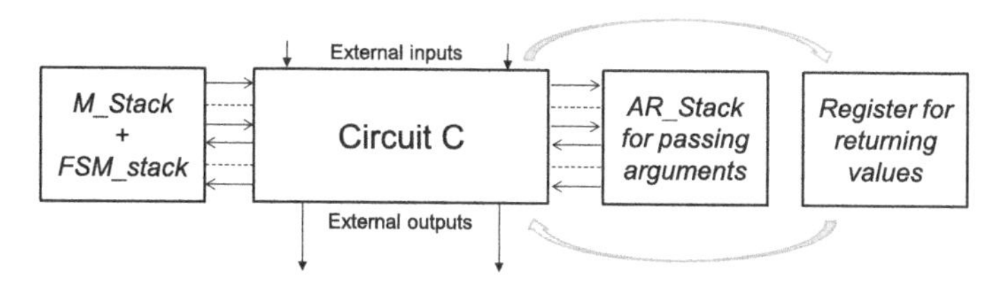

Fig. 5.17 Using additional elements for passing arguments and returning values/pointers

Now C functions can be converted to an HFSM as follows:

1. Stacks are specified in VHDL using the considered above templates.
2. Other blocks are described based on the VHDL templates and using the following additional rules:
   - Arguments passed by value are stored in the *AR_stack* when a module (created for the respective C function) is being activated.
   - Different numbers of arguments passed to the same function are recognized by specifying a different HDL module depending on the actual number of arguments. This can be seen as a hardware technique for replicating function overloading in software.
   - For each argument that is a pointer, the address is stored in the *AR_stack* when a module (created for the respective C function) is being activated.
   - A single returned value/pointer is copied to a specially allocated register when a module is terminated and all arguments previously passed to this module are destroyed.

All three stacks (*FSM_stack*, *M_stack*, and *AR_stack*) are described in the following VHDL process:

```
process(clock)
begin                  -- a0 is an initial state; z0 is a top-level module
  if rising_edge(clock) then          stack_overflow <= '0';
    if reset = '1' then               stack_pointer <= 0; FSM_stack(0) <= a0;
      M_stack(0) <= z0;  stack_overflow <= '0'; AR_stack(0) <= (others => '0');
    else
      if push = '1' then
        if stack_pointer = stack_size then -- handling stack overflow
        else  stack_pointer <= stack_pointer + 1;
          FSM_stack(stack_pointer+1) <= a0; -- initial state is a0
          FSM_stack(stack_pointer) <= N_S;  -- N_S is the next state in the calling module
          M_stack(stack_pointer+1) <= NextModule;  -- NextModule is the next module
          AR_stack(stack_pointer+1) <= pass_arguments; -- passing arguments
        end if;
      elsif pop = '1' then
        stack_pointer <= stack_pointer - 1;  -- decrementing the stack_pointer when the
      else                                   -- module is terminated
        FSM_stack(stack_pointer) <= N_S;  -- conventional state transition to N_S
      end if;
    end if;
  end if;
end process;
```

Since there is just a single value returned, it is kept in a signal that is declared as:

```
signal return_value : std_logic_vector(size_of_operands-1 downto 0);
```

where size_of operands is a generic constant.

The arguments are prepared in the calling module like the following:

```
when stateWhereTheCalledModuleActivated => push <= '1'; NextModule <= <name>;
      pass_arguments(<index range>) <= <arguments>; -- preparing arguments
```

The returned value is produced as follows:

```
when stateWhereTheResultIsProduced  => N_S <= indicatingTheNextState;
      return_value <= signalThatKeepsTheResult;
```

5.6 Using Stacks Based on Embedded or Distributed Memories

Note that HFSM stack memory might require excessive hardware resources when it is built as a logic block. However, it can also be constructed from embedded to FPGA or distributed memories. Since the signals *push*, *pop*, *clock*, *reset*, *stack_pointer* are common to all the stacks, the memory can be organized as shown in Fig. 5.18. VHDL code for the stacks constructed from block/distributed RAM (see RAM_block in Fig. 5.18) looks like this:

```
process(clock)
begin -- states and modules are represented by binary codes
  if rising_edge(clock) then  stack_overflow <= '0';
    if reset = '1' then
       stack_pointer  <= 0; stack_overflow <= '0'; -- see Fig. 5.18a
       FSM_Register <= (others => '0');          -- see Fig. 5.18c
    else
       if push = '1' then                                  -- hierarchical call
         if stack_pointer = 2**ram_addr_bits-1 then stack_overflow <= '1';
         else     stack_pointer <= stack_pointer + 1;
            -- the arguments are passed through the signal to_AR
            FSM_Register <= to_AR & N_M &
                 (size_of_FSM_stack_words-1 downto 0 => '0');
           RAM_block(stack_pointer) <= to_AR & C_M & N_S;
         end if;
       elsif pop = '1' then                                -- hierarchical return
         stack_pointer <= stack_pointer - 1;
         FSM_Register <= RAM_block(stack_pointer-1);
       else                                                -- conventional transition
         FSM_Register(size_of_FSM_stack_words-1 downto 0) <= N_S; end if;
    end if;
  end if;
end process;
```

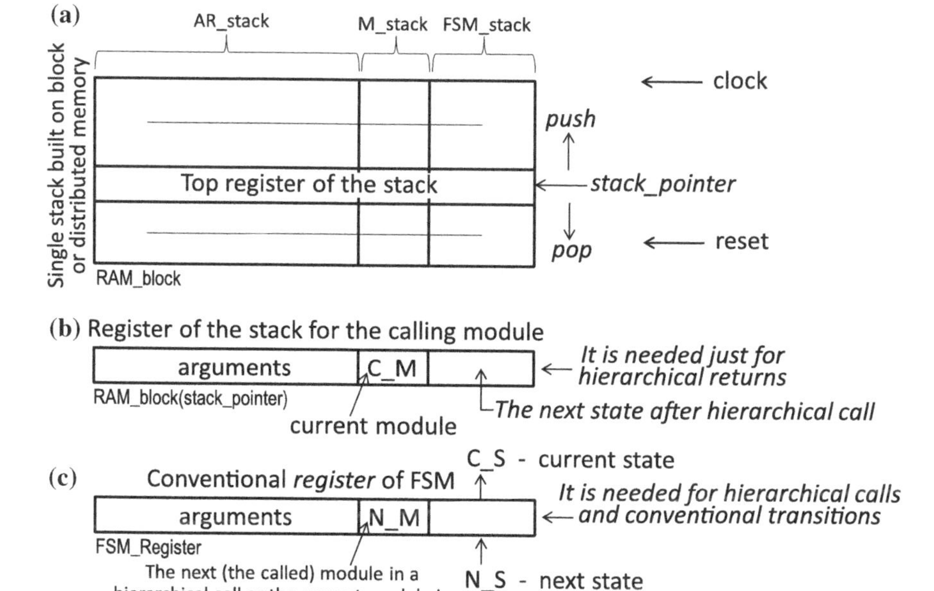

Fig. 5.18 Single block of embedded/distributed RAM for three stacks in Fig. 5.17 (**a**), active stack register (**b**), state transitions/hierarchical calls through the FSM conventional register (**c**)

RAM_block is declared as an array:

```
constant ram_width       : integer := <size of words for the stack shown in Fig. 5.18a,b>
constant ram_addr_bits : integer :=  <size of RAM addresses>
type DistributedRAM is array (2**ram_addr_bits-1 downto 0) of
        std_logic_vector (ram_width-1 downto 0);
signal RAM_block: DistributedRAM; --  Block RAM is declared similarly to distributed RAM
```

Figure 5.19 illustrates different types of transitions in the HFSM for *hierarchical calls* (Fig. 5.19a), *conventional state transitions* (Fig. 5.19b), and *hierarchical returns* (Fig. 5.19c). Note, that the stack is passive in a *hierarchically called module* (the stack is needed just for a *hierarchical return* from the *called module*). Thus, just a *register* (FSM_Register) can be used for passing arguments and executing state and module transitions. As soon as a transition to the next module has to be done (in the case of a *hierarchical call*), a binary vector (BVc = to_AR & N_M & <first state with all zeros>) with the arguments (to_AR) and the codes of the called module (N_M) with its initial state (all zeros) is copied to the *register* as shown in Fig. 5.19a.

Conventional state transitions are executed similarly to an ordinary FSM using the register FSM_Register (see Fig. 5.19b). The arguments are taken directly from the register (FSM_Register).

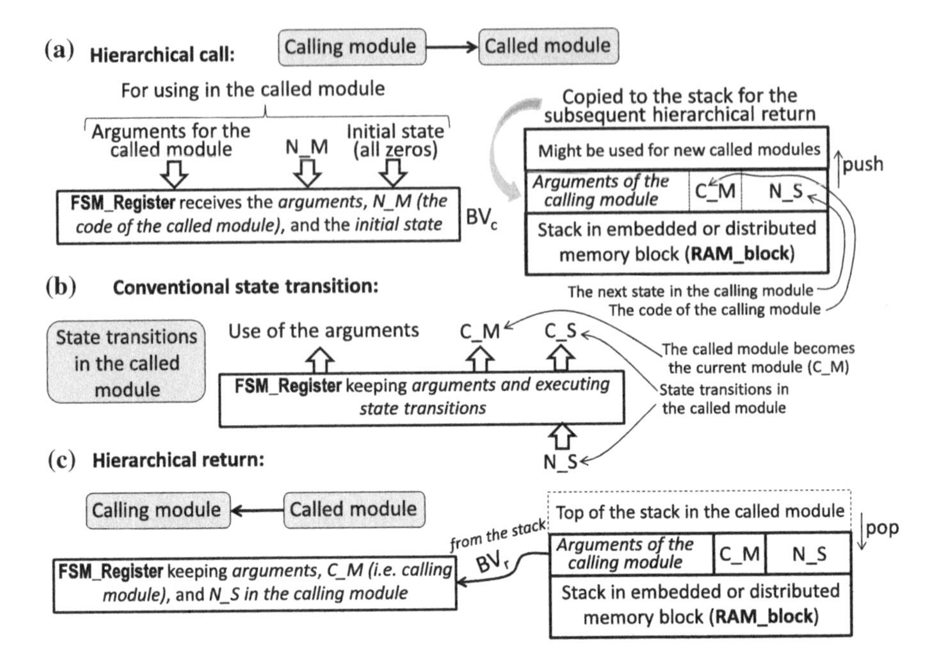

Fig. 5.19 Three types of stack transitions in HFSM: hierarchical calls (**a**); conventional transitions (**b**), and hierarchical returns (**c**)

As soon as a *hierarchical return* has to be done, the binary vector (BV_r) *from the stack* shown in Fig. 5.19c (containing the arguments, the code of the calling module and the code of the next state in the calling module after termination of the called module) is copied to the FSM_Register (FSM_Register <= RAM_block(stack_pointer-1);). Thus, the calling module will continue its execution.

The line RAM_block(stack_pointer) <= to_AR & C_M & N_S; in the process that describes embedded or distributed memory sets the code of the next state N_S that is needed after the termination of the called module. As a result, after the corresponding hierarchical return, the transition to the proper HFSM state occurs (FSM_Register <= RAM_block(stack_pointer-1);). Since the next state is determined before the invocation of a module, the called module cannot change the predetermined state transition. For the majority of practical applications this does not create a problem. However, in some cases it is a problem, which must be resolved. This can be done by replacing the line above with the statement: RAM_block(stack_pointer) <= to_AR & C_M & C_S; where C_S (the current state in the calling module) has to be further replaced with such N_S in the calling module that is found taking into account potentially changed conditions in the called module(s). Methods for such a replacement are discussed in [31] and in the next section (see Sect. 5.7.1).

Note that complete synthesizable VHDL projects for HFSMs with stacks based on embedded and distributed memories can be found in [5].

5.7 Optimization Techniques

This section presents optimization techniques [5] for synthesis of HFSMs, namely, execution of hierarchical returns, using multiple entry points to sub-algorithms (HGSs), and fast stack unwinding.

5.7.1 Execution of Hierarchical Returns

The line FSM_stack(stack_ pointer)<=N_S; sets the code of the next state N_S during a hierarchical call. As a result, after a hierarchical return the top register of the *FSM_stack* contains the code of the proper HFSM state. Since the next state is determined before the invocation of a (called) module, the called module cannot change the state transition. For many practical applications this does not create a problem. However, for some practical cases it is a problem, which must be resolved. If we remove the line FSM_stack(stack_ pointer)<=N_S; then after a hierarchical return the top register of the *FSM_stack* contains the code of the state where the terminated module was called. This enables us to provide correct transitions to the next state because all logic conditions that might have changed in the called module have already received the proper values. However, this gives rise to another problem; namely it is necessary to avoid both repeating invocations of the same module in the state where it has to be called and generating unnecessary outputs (see Fig. 5.20). The following code overcomes the problem:

```
-- see VHDL description for stacks
elsif pop = '1' then
  stack_pointer <= stack_pointer - 1;
  return_flag <= '1';
else FSM_stack(stack_pointer) <= N_S;
  return_flag <= '0';
end if;
```

The signal *return_flag* permits module invocation and output operations to be activated during a hierarchical call and to be avoided during a hierarchical return [28]. Indeed, the *return_flag* is equal to 1 only in a clock cycle when the signal stack_ pointer is decremented. As soon as the currently active (called) module is being terminated, the control flow will be returned to the point of the calling module from which this (called) module was called. Thus, the top of the *M_stack* will contain

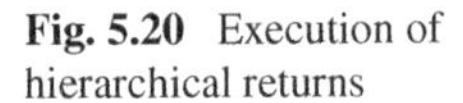

Fig. 5.20 Execution of hierarchical returns

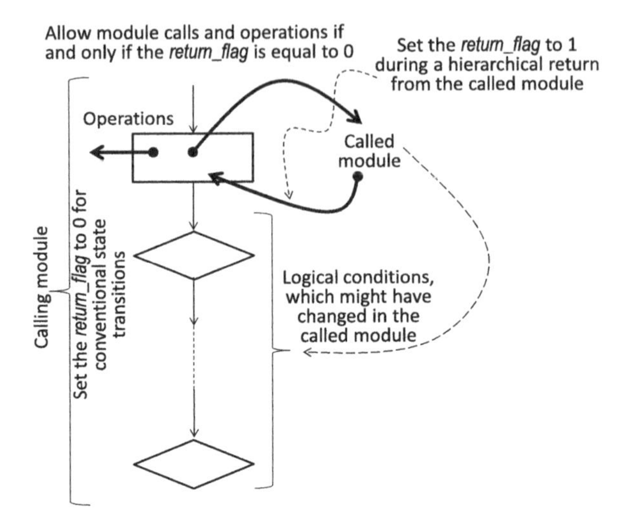

the code of the calling module (z →) and the top of the *FSM_stack* will store the code of the calling state (a→). The *return_flag* enables us to eliminate the second call of the same module (and the second activation of the relevant output signals). This is achieved with the aid of the following lines that have to be inserted in the process that describes transitions and operations (see also Fig. 5.20):

```
when state_with_module_call
  if return_flag='0' then push<='1'; -- specifying operations and calling the next module
  else push<='0';                    -- no operation and no module call is involved
  end if;
```

Finally, the proposed technique permits logic conditions to be tested after terminating the called module, which might alter these conditions.

5.7.2 Providing Multiple Entry Points to HGSs

Any considered above invocation of hierarchical modules activates a new HGS starting from the node *Begin* (somehow associated with the state a_0) and, as a rule, this node does not contain micro operations. Skipping the node a_0 removes one clock cycle from a hierarchical call. However, in this case, the relevant HGS might require multiple entry points and the particular entry point will be chosen by the group of rhomboidal/triangular nodes (enclosed in an ellipse in Fig. 5.21) tested in the calling module (z →). Here NM_FS is the first state of the next module. The description of the stacks has to be slightly modified:

```
FSM_stack(stack_pointer+1) <= NM_FS;
```

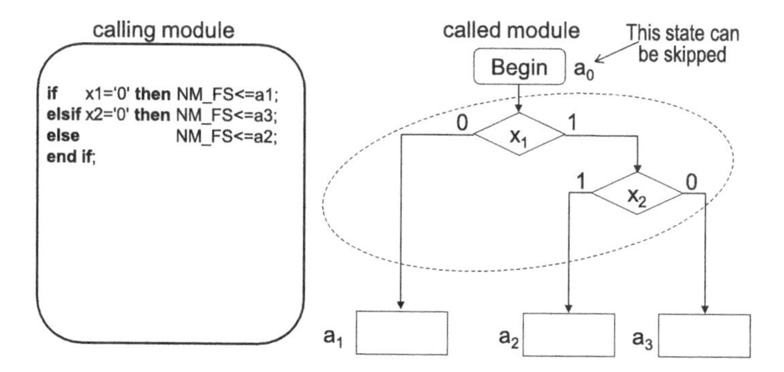

Fig. 5.21 Providing multiple entry points for HGSs

5.7.3 *Fast Stack Unwinding*

Some HGSs are called recursively just before the node *End* and as soon as the node *End* is reached a sequence of recursive calls has to be terminated. Such termination can be done during only one clock cycle through the use of a fast stack unwinding technique. Indeed, at the end of recursive calls, the line:

```
if pop='1' then stack_pointer <= stack_pointer – 1;
```

is executed repeatedly until the stack_pointer receives the value assigned at the beginning of the sequence with recursive calls. Repeated execution of the line stack_pointer <= stack_pointer–1; requires multiple clock cycles. To eliminate such redundant cycles the code above is changed as follows:

```
if pop='1' then stack_pointer <= stack_pointer – unwinding;
```

where the signal *unwinding* is calculated as

```
unwinding <= stack_pointer - saved_sp + 1;
```

and an assignment saved_sp<=stack_pointer has to be done during the first invocation of the module in a sequence of potential hierarchical (recursive) calls. Thus, redundant clock cycles for hierarchical returns can be avoided.

5.8 Practical Applications

This section presents some of practical examples in which HFSMs and PHFSMs can efficiently be used. Let us discuss at the beginning such applications that require traversing $\mathcal{N}$-ary trees (see Sect. 3.4.3 and [32]). Let us consider $\mathcal{N}$-ary tree ($\mathcal{N} = 4$) from Fig. 3.12. This tree can store a set of data that are linked in accordance with given relationships. For example, the tree in Fig. 5.22 holds the following set of integers: 60, 12, 31, 56, 0, 9, 63, 28, 6, 1, 58, 15, 2, 62, 48,

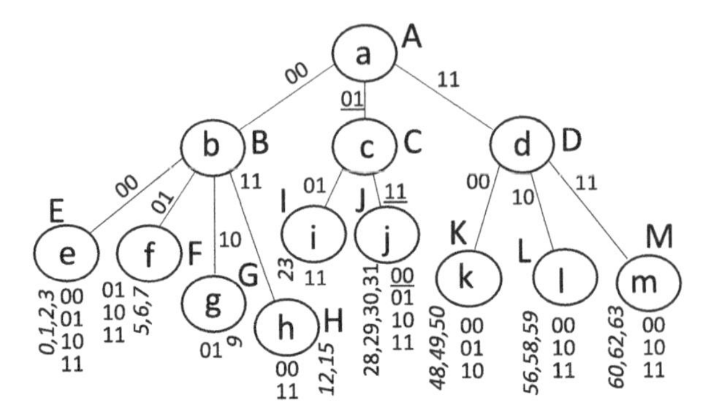

Fig. 5.22 An example of $\mathcal{N}$-ary tree ($\mathcal{N} = 4$) from Fig. 3.12 that can now be used for data sort

49, 7, 29, 50, 5, 3, 30, 59, 23. Let us consider the binary codes of the integers decomposed in G-bit groups ($G = 2$): *11*1100, *00*1100, *01*1111, *11*1000, *00*0000, *00*1001, *11*1111, *01*1100, *00*0110, *00*0001 *11*1010, *00*1111, *00*0010, *11*1110, *11*0000, *11*0001, *00*0111, *01*1101, *11*0010, *00*0101, *00*0011, *01*1110, *11*1011, *01*0111. The first group on the left-hand side is shown in *italic*. Let us use this group to allocate three children of the root for all the codes found: *00*, *01*, and *11* leading to the children b, c and d, accordingly. Now the nodes b, c and d can be considered as roots of sub-trees for which the same rules have to be applied. Items from the last group are not expanded for new tree nodes, but are just associated with the leaves at depth 2 (these are the leaves e, f, g, h, i, j, k, l, m). Such a tree can easily be built, and it can be traversed [32] by applying either an iterative or a recursive procedure. Data attached to the leaves are ordered (the leftmost leaf contains the smallest values and the rightmost leaf—the greatest values). Thus, the tree may be used for data sorting, or for searching for particular items. For example, to check if the data item 28 is in the set you can execute three tests: one for the tree root and others for the nodes c and j (see underlined codes in Fig. 5.22).

$\mathcal{N}$-ary trees are involved in numerous practical applications (see, for example, [32]) and we will use them for sorting data by applying *two types* of modules: (1) for traversing the tree enabling all leaves to be found; and (2) for fast sort of data associated with the leaves. The *first module* will have two alternative implementations: *iterative* and *recursive*. The *second module* executes sequential (non-recursive) operations enabling reusable sorting networks described above (see Sect. 3.5) to be involved.

Suppose an $\mathcal{N}$-ary tree for sorting data has been built and it is necessary to extract the sorted data from the tree. The following recursive C function does this:

```
void traverse_tree(treenode* root, int depth)
{       depth++;
        if (root == 0) { depth--; return; }
        if (depth == max_depth) {    sort_and_print_leaf_data(root); depth--; return; }
        for (int i = 0; i < N; i++)
                    traverse_tree(root->node[i], depth);     -- recursive call
        depth--;                                                    }
```

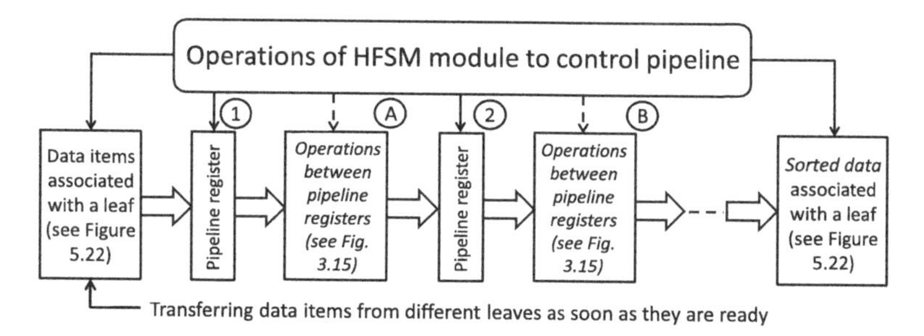

Fig. 5.23 A pipeline controlled by an HFSM module

where treenode is the following C structure (N is a constant $\mathcal{N}$):

```
struct treenode          {
  int* arrayTOsort;
  int count;
  treenode* node[N];     };
```

Similarly, an iterative function void iterative_traverse_tree(treenode* root, int depth) can be built for which the treenode structure has an additional field with a pointer to the parent node of the tree.

The functions traverse_tree and iterative_traverse_tree can be transformed to hardware circuits using the methods and tools described above. Different branches of the tree (such as with the local roots b, c, d) can be traversed concurrently and, thus, the PHFSM described in Sect. 5.4 can be applied directly allowing different modules to be executed in parallel. Eventual data dependency between the modules is avoided by storing sub-trees in different memory blocks. Also, any module allows a pipeline to be created. For example, the function sort_and_print_leaf_data(root); in the C code above sorts data associated with the tree leaves. Figure 5.23 demonstrates a pipeline implemented in an HFSM module.

As soon as the function traverse_tree finds the sub-set of data with the smallest values (e.g. node e in Fig. 5.22), all items are transferred to the input of the leftmost pipeline register in Fig. 5.23 (see the number 1 enclosed in a circle). At the next iteration, a subsequent sub-set (e.g. node f in Fig. 5.22) is transferred and the results of operations with the first sub-set are stored in the next pipeline register of Fig. 5.23 (see the number 2 enclosed in a circle). Subsequent iterations are executed similarly. Examples of operations between pipeline registers are given in Fig. 3.15.

Any HFSM module has a unified interface. However, the implementations of modules may be different. For example, the recursive function traverse_tree(treenode* root, int depth) can easily be replaced by the iterative function iterative_traverse_tree(treenode* root, int depth). Such a technique is indispensable for experiments and comparisons. Note that a complete synthesizable VHDL code of HFSM implementation for traversing binary trees can be found in [5].

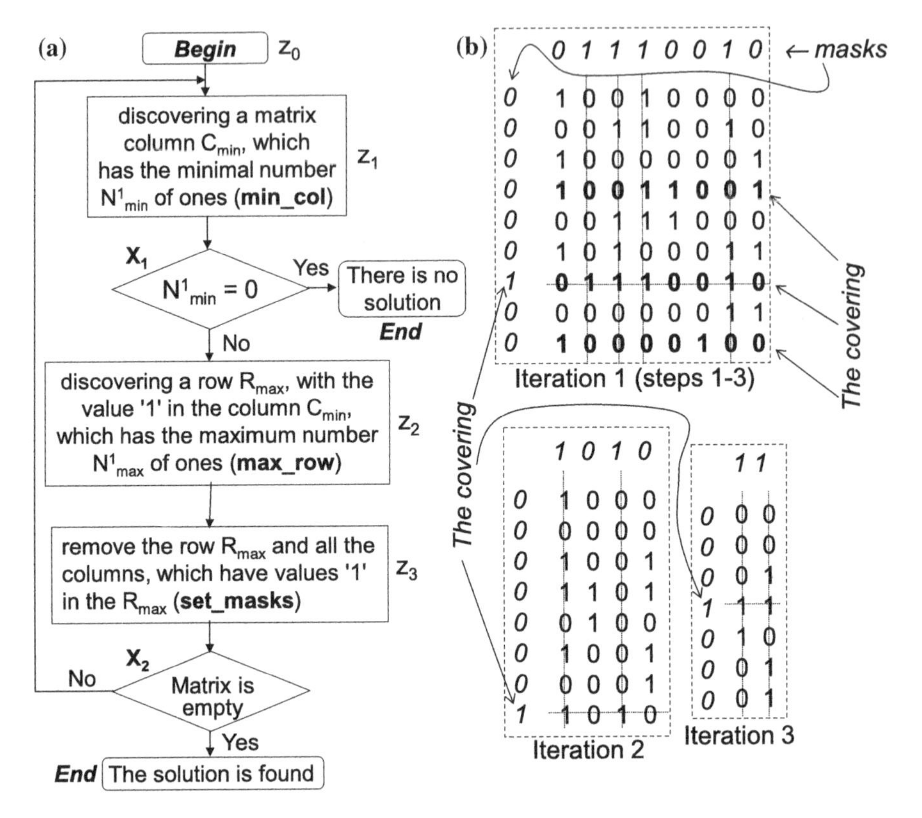

Fig. 5.24 Approximate matrix covering algorithm (**a**); iterations of the algorithm applied to the given binary matrix (**b**)

The second example is taken from the scope of combinatorial search. Suppose we want to find a minimal row cover of a given binary matrix, i.e. the minimum number of rows such that in conjunction they have at least one value '1' in each column. The approximate algorithm [33] that allows this problem to be solved requires the following sequence of steps (see Fig. 5.24a):

1. discovering a matrix column C_{min}, with the minimal Hamming weight N^1_{min} (if $N^1_{min} = 0$ then the covering does not exist);
2. discovering a row R_{max}, with the value '1' in the column C_{min}, with the maximum Hamming weight N^1_{max};
3. including the row R_{max} in the solution and removing this row and all the columns, which have values '1' in the R_{max};
4. repeating the steps 1–3 until the matrix is empty or there is a column with only zeros meaning that the solution does not exist.

Figure 5.24b gives an example from [1] of a particular matrix to which the steps in Fig. 5.24a have been applied. These steps can be realized in the respective HFSM modules that involve fast parallel computations from Sects. 3.7–3.9.

The main module is z_0 and it calls the modules z_1, z_2, and z_3.The module z_1 (see Fig. 5.24a) executes step 1 and outputs the values N^1_{min} and C_{min}. The module z_2 (see Fig. 5.24a) finds R_{max}. The module z_3 updates masks that are used to indicate the rows and columns that have been removed (see Fig. 5.24a) and, thus, the same masked (reduced) matrix is taken for subsequent steps. The steps above are repeated until the covering is found or until it is concluded that the solution does not exist.

Clearly all such steps can easily be implemented in general-purpose processors, which, as a rule, operate with significantly higher clock frequency than FPGAs. However, FPGA-based (and HFSM-based) implementations might have advantages that are listed below:

- Very fast parallel computations (such as that are needed to calculate Hamming weights) can easier be executed without communication overheads that usually take place when similar accelerators are involved in processor-based implementations.
- Concurrent operations are possible. For example, Hamming weights can be found for all the matrix columns (that are kept in FPGA registers accessed in parallel) and N^1_{min} may be found in a combinational process that describes functionality of the circuit shown in Fig. 3.16 (see Sect. 3.6). This permits, in particular, the steps 1 and 2 above to be completed within one clock cycle.
- Fast hardware components that provide support for combinatorial search can be created as a part of more complicated systems implemented in the same FPGA.

The last example with an HFSM demonstrates a pipeline with such operations between pipeline registers that are executed sequentially and require more than one clock cycle. Indeed, any operation in Fig. 5.23 (see letters A and B enclosed in circles) can be either combinational or sequential. Combinational operations do not use sequential control circuits but they might involve excessive propagation delays. For certain applications (such as that are discussed in Sects. 3.5 and 3.6) sequential operations executed with high clock frequency might be better because they enable resource consumption to be significantly reduced and the required performance to be achieved and adjusted with other characteristics of the developed system such as communication overheads. HFSMs can be used to control pipelines in which operations between pipeline registers are sequential and each particular operation is described by the relevant HGS. Let us look at Fig. 5.25 which gives some examples of operations (see letters A and B enclosed in circles). For instance, the operation A is an iterative sorting that may be controlled by HGSs in Fig. 5.8a and b. Some of possible types of the operation B are listed in Fig. 5.25.

Let us consider now a particular example depicted in Fig. 5.26 which enables two subsets (composed of N data items each) to be sorted in a pipeline requiring approximately N/2 clock cycles for each pipeline stage (step) and functioning at a high frequency. Besides, the final set (up to $2 \times N$ items) contains only non-repeated positive values.

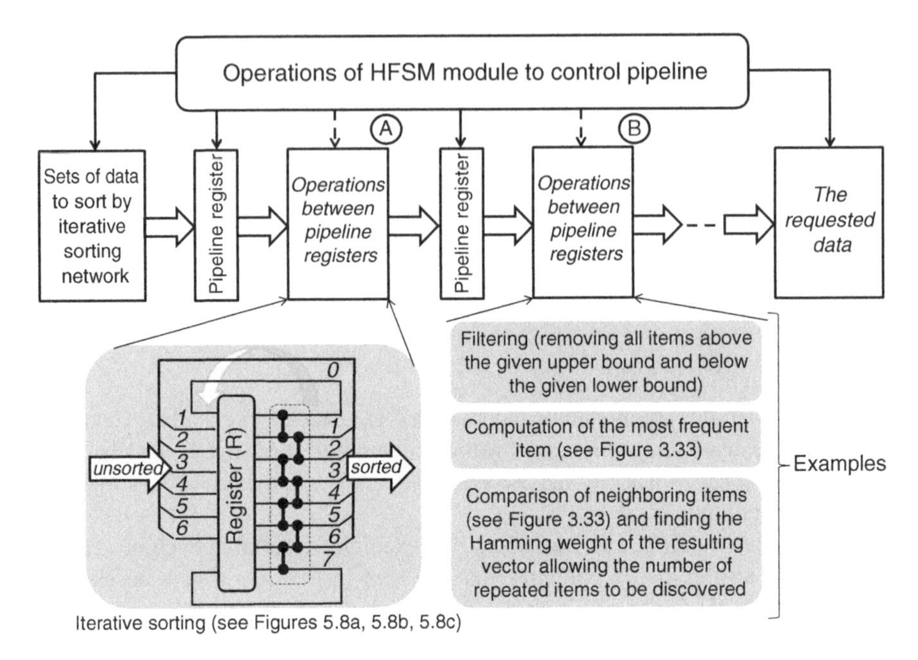

Fig. 5.25 Examples of sequential operations between pipeline registers

There are three modules between the pipeline registers in Fig. 5.26a. The first module examines availability of data on the outputs of the left block (designated as SOURCE) and executes the module z_1 activating in parallel two iterative sorters (for N items each) shown in Fig. 5.8c (see also Fig. 5.25). Such operation requires at most N/2 clock cycles with a combinational path delay equal to the delay of two comparators. For the sake of simplicity we assume in Fig. 5.26b that if there are no input data then the previously recorded data have to be sorted first and only after that a new signal indicating availability of new data sets may appear. This pre-condition can easily be avoided but the HGS in Fig. 5.26b becomes more complicated. The second module z_2 removes all the repeated items. It is done as follows: (a) comparison of neighboring items (see Fig. 3.33) and finding the Hamming weight of the resulting vector allowing the number of repeated items to be discovered; (b) keeping just one item from any set of the repeated items; (c) resorting with the aid of the circuits from Chap. 3. Since operations (a) and (b) can be done combinationally, the number of the required clock cycles is also N/2 at maximum. The last module z_3 exchanges items in the two sorted subsets as it is shown in Fig. 5.26c using the method [34] and then resorts two N-item sets giving the final sorted set with $2 \times N$ items. Note that the network [34] is also described in [35] with additional details that, in particular, prove that the considered here method is correct. It is easy to show that such operation also requires not more than N/2 clock cycles. Finally $2 \times N$ sorted items are transferred to the block designated in Fig. 5.26a as DESTINATION. Figure 5.26c illustrates the intended functionality on a simple example of two 4-item sets: 7,3,9,3 and 2,8,2,1. Note

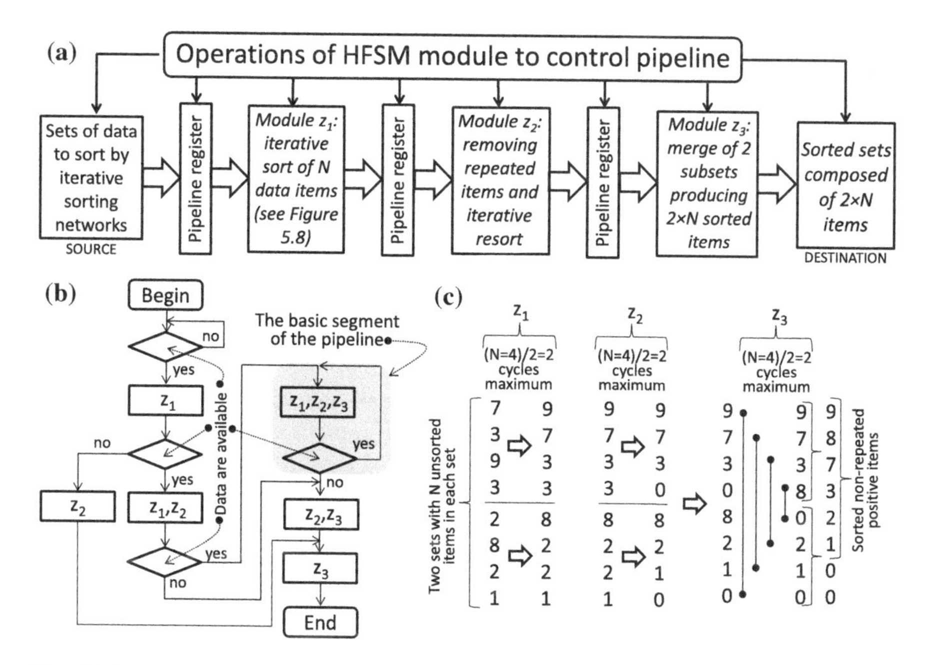

Fig. 5.26 Pipelined sort with sequential modules between pipeline registers (**a**); HGS that controls the pipeline (**b**); an example of sort (**c**)

that unnecessary zeros may easily be discarded. Alternatively, the module z_2 may explicitly indicate the number of items in the final set.

As you can see transferring data between the pipeline registers is executed after not more than N/2 clock cycles and this is faster than $2 \times N/2$ clock cycles needed for the circuit in Fig. 3.14 if this circuit is directly used for sorting $2 \times N$ data items. Hence, pipelined implementation in Fig. 5.26 is faster (by a factor of about 2) than using the circuit in Fig. 3.14 in a similar pipeline in such a way that instead of 3 modules z_1, z_2, z_3 just one module is used that controls the circuit in Fig. 3.14 for $2 \times N$ items.

It should be noted in conclusion that many additional examples can be found in [1, 5, 36].

It is known that HFSMs can be configured either statically or dynamically. In the last case the behavior of an HFSM may be changed during run-time. Methods [37] can be applied for such purposes and they permit HFSM circuits to be built from reloadable memories that determine the desired functionality. The memories (that are embedded or distributed FPGA blocks) can be reloaded during execution time and, thus, the operations of the HFSM can be changed in accordance with the requirements that might depend on some factors (e.g. weather conditions, surrounding temperature, faults in some units, etc.). Since HFSMs are composed of modules that can be replaced if required, different control algorithms specified by the modules can be selected during execution time in order to adjust parameters of the controlled devices. Let us look at Fig. 5.27 which shows one possible

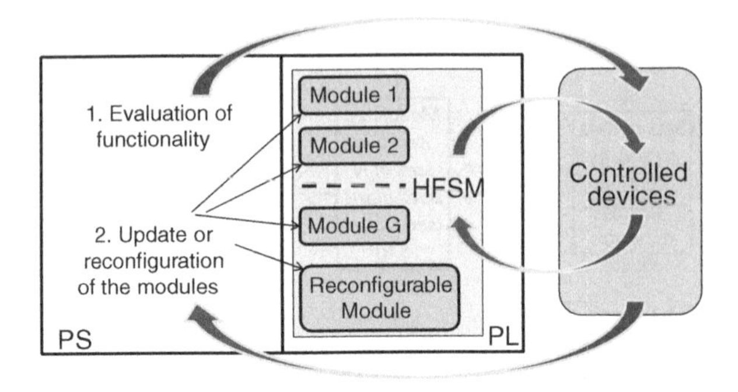

Fig. 5.27 Using an HFSM for an intelligent control

way that enables an intelligent control of external devices to be implemented in APSoCs discussed in Sect. 4.5. The HFSM is composed of different modules aimed at control of external devices (i.e. controlled devices) and some of the modules implement alternative or competitive algorithms. Thus, we can apply the strategy "try, test and replace if required". Besides, any module can be updated with an improved version without modifications in surrounding modules. Indeed, it is sufficient to change the relevant microoperation (z_i) which indicates an entry point to the module with a new microoperation (z_j) indicating an entry point to another (alternative) module (from the set 1, 2,…, G). Figure 5.27 demonstrates a potential intelligent control. The PS (see Sect. 4.5) evaluates the functionality of the controlled devices and verifies if different established requirements are satisfied. If from the result of evaluation the PS makes a conclusion that some modes or algorithms applied to the controlled devices may be improved then the set of active modules implemented in the PL can be updated and some of such modules may be reconfigured using the methods [37]. Update means a replacement of a currently active module with a currently idle module that is expected to be used more efficiently. Reconfiguration means changes in the selected module functionality through reloading its memories.

For some applications the described above intelligent system permits a better satisfaction of different requirements to be achieved. Besides, HFSMs are capable to realize a very high-speed control which also might be beneficial for some practical cases.

References

1. Sklyarov V, Skliarova I (2013) Hardware implementations of software programs based on HFSM models. Comput Electr Eng 39(7):2145–2160
2. Carrano FM, Henry T (2012) Data abstraction and problem solving with C++: walls and mirrors, 6th edn. Prentice Hall, New Jersey
3. Cormen TH, Leiserson CE, Rivest RL, Stain C (2009) Introduction to algorithms, 3rd edn. MIT Press, Cambridge
4. Skliarova I, Sklyarov V (2009) Recursion in reconfigurable computing: a survey of implementation approaches. In: Proceedings of the 19th international conference on field-programmable logic and applications, FPL 2009, Prague

5. Skliarova I, Sklyarov V, Sudnitson A (2012) Design of FPGA-based circuits using hierarchical finite state machines. TUT Press, Tallinn
6. Baranov S (2008) Logic and system design of digital systems. TUT Press, Tallinn
7. Sklyarov V (1999) Hierarchical finite-state machines and their use for digital control. IEEE Trans VLSI Syst 7(2):222–228
8. Baranov S (1994) Logic synthesis for control automata. Kluwer Academic Publishers, Norwell
9. Sklyarov V (1983) Finite state machines with stack memory and their automatic design. In: Proceedings of USSR conference on computer-aided design of computers and systems, 1983 (in Russian)
10. Sklyarov V (1984) Synthesis of finite state machines based on matrix lsi. Science and Techniques, Minsk (in Russian)
11. Sklyarov V (2010) Synthesis of circuits and systems from hierarchical and parallel specifications. In: Proceedings of the 12th biennial baltic electronics conference, Tallinn
12. Sklyarov V, Skliarova I (2008) Design and implementation of parallel hierarchical finite state machines. In: Proceedings of the 2nd international conference on communications and electronics, Hoi An
13. Lyshevski SE (2003) Hierarchical finite state machines and their use in hardware and software design. In: Goddard WA, Brenner DW, Lyshevski SE, Iafrate GJ (eds) Handbook of nanoscience engineering and technology. CRC Press, Boca Raton
14. Marcon CAM, Calazans NLV, Moraes FG (2002) Requirements, primitives and models for systems specification. In: Proceedings of the 15th symposium on integrated circuits and systems design, Porto Alegre
15. del Moral BA, Zafra JMJ, Gómez JFR, Mesa RS, Muñoz RM, Trinidad AR, Moreno JJL, and The International Medusa Team (2010) New control system for space instruments. Application for medusa experiment. In: Proceedings of the 7th international planetary probe workshop, Barcelona
16. Neishaburi MH, Zilic Z (2011) Hierarchical trigger generation for post-silicon debugging. In: Proceedings of the international symposium on VLSI design, automation and test, Taiwan
17. Perez-Rodriguez R, Caeiro-Rodriguez M, Anido-Rifon L, Llamas-Nistal M (2010) Execution model and authoring middleware enabling dynamic adaptation in educational scenarios scripted with PoEML. J Univ Comput Sci 16(19):2821–2840
18. Hu W, Zhang Q, Mao Y (2011) Component-based hierarchical state machine - a reusable and flexible game AI technology. In: Proceedings of the 6th IEEE joint international conference on information technology and artificial intelligence, Chongqing
19. Jenihhin M, Gorev M, Pesonen V, Mihhailov D, Ellervee P, Hinrikus H, Bachmann M, Lass J (2011) EEG analyzer prototype based on FPGA. In: Proceedings of the 7th international symposium on image and signal processing and analysis, Dubrovnik
20. Mihhailov D, Sklyarov V, Skliarova I, Sudnitson A (2011) Acceleration of recursive data sorting over tree-based structures. Electron Electr Eng 7(113):51–56
21. Ninos S, Dollas A (2008) Modeling recursion data structures for FPGA-based implementation. In: Proceedings of the 18th international conference on field-programmable logic and applications, Heidelberg
22. Malakonakis P, Dollas A (2011) Exploitation of parallel search space evaluation with fpgas in combinatorial problems: the eternity II case. In: Proceedings of the 21st international conference on field-programmable logic and applications, Crete
23. Muñoz DM, Llanos CH, Ayala-Rincón M, van Els RH (2008) Distributed approach to group control of elevator systems using fuzzy logic and FPGA implementation of dispatching algorithms. Eng Appl Artif Intell 21(8):1309–1320
24. Sklyarov V, Skliarova I, Neves A (2009) Modeling and implementation of automatic system for garage control. In: Proceedings of ICROS-SICE international joint conference, Fukuoka
25. Harel D (1987) Statecharts: a visual formalism for complex systems. Sci Comput Program 8(3):231–274

26. Gajski DD, Abdi S, Gerstlauer A, Schirner G (2009) Embedded system design. Springer, New York
27. Sklyarov V, Skliarova I, Pimentel B (2005) FPGA-based implementation and comparison of recursive and iterative algorithms. In: Proceedings of the 15th international conference on field-programmable logic and applications, Tampere
28. Sklyarov V (2004) FPGA-based implementation of recursive algorithms. Microprocess Microsyst 28(5–6):197–211 Special issue on FPGAs: applications and designs
29. Skliarova I, Sklyarov V (2010) Reconfiguration technique for adaptive embedded systems. In: Proceedings of the 3rd international conference on intelligent and advanced systems, Kuala Lumpur
30. Patterson DA, Hennessy JL (2009) Computer Organization and Design. Morgan Kaufmann Publishers, Burlington
31. Sklyarov V, Skliarova I (2006) Recursive and iterative algorithms for n-ary search problems. In: Debenham J (ed) Proceedings of the 19th IFIP world computer congress, Santiago de Chile
32. Rosen KH, Michaels JG, Gross JL, Grossman JW, Shier DR (eds) (2000) Handbook of discrete and combinatorial mathematics. CRC Press, Boca Raton
33. Zakrevskij A, Pottosin Y, Cheremisiniva L (2008) Combinatorial algorithms of discrete mathematics. TUT Press, Tallinn
34. Alekseev VE (1969) Sorting algorithms with minimum memory. Kibernetika 5(5):99–103
35. Knuth DE (2011) The art of computer programming, vol 3: sorting and searching. Addison-Wesley, New York
36. Sklyarov V, Skliarova I (2013) Parallel processing in FPGA-based digital circuits and systems. TUT Press, Tallinn
37. Sklyarov V (2002) Reconfigurable models of finite state machines and their implementation in FPGAs. J Syst Architect 47(14–15):1043–1064

Part II
Methods for Optimization of Finite State Machines for FPGA-Based Circuits and Systems

Chapter 6
Hardware Reduction in Logic Circuits of Moore FSM

Abstract The Chapter is devoted to the problems of optimization of Moore FSM logic circuits implemented with FPGAs. The general characteristic is given for methods of functional and structural decomposition. Distinctive features of FPGA are analyzed allowing the number of look-up table (LUT) elements in logic circuits of Moore FSMs to be decreased. The classification of optimization methods are given for Moore FSM including: (1) the transformation of state codes into codes of the classes of pseudoequivalent states (PES); (2) presentation of state codes as concatenations of codes of PES and collections of microoperations; (3) replacement of logical conditions (input variables of FSM) by additional variables. All discussed methods are illustrated by examples. The chapter is written together with PhD student Olena Hebda (University of Zielona Gora, Poland).

6.1 General Characteristic of Existing Methods

One of the main problems connected with implementing logic circuits of control units is the problem of hardware reduction [1, 2]. Solution of this problem allows decreasing the chip area occupied by the FSM logic circuit. The positive back effects of this problem's solving is increasing for performance and decreasing for power consuming of the logic circuit [3–6]. If the propagation time of logic elements in use is not diminished, then the increasing for performance is possible only by the decreasing for the number of layers in the combinational part of a control unit [7]. The structural diagram of Moore FSM includes two combinational blocks and a register RG (Fig. 6.1) [8].

A block of input memory functions (BIMF) implements the functions $D_r \in \Phi$, where the system of input memory is represented as

$$\Phi = \Phi(T, X). \quad (6.1)$$

V. Sklyarov et al., *Synthesis and Optimization of FPGA-Based Systems*, Lecture Notes in Electrical Engineering 294, DOI: 10.1007/978-3-319-04708-9_6,

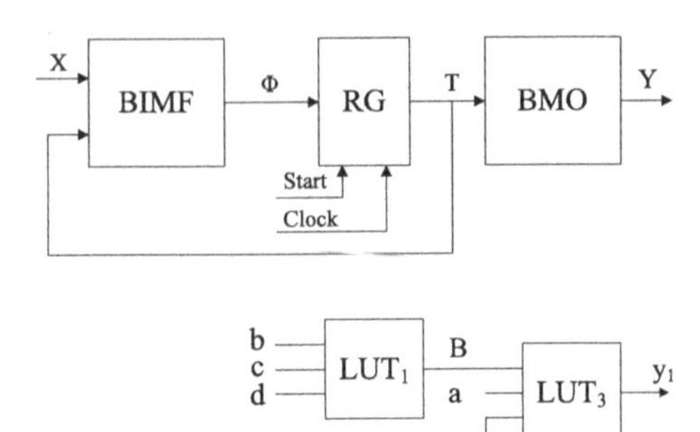

Fig. 6.1 Structural diagram of Moore FSM

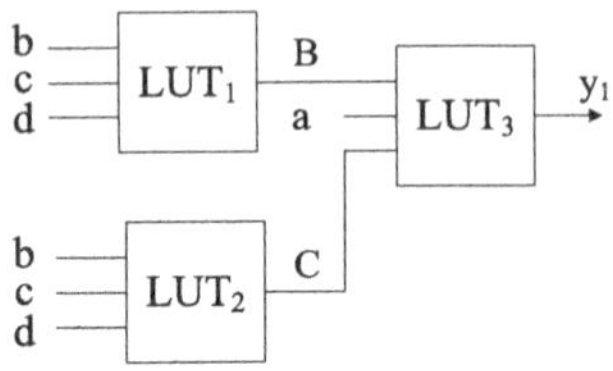

Fig. 6.2 Logic circuit for function y_1

As a rule, there is $\Phi = \{D_1, \ldots, D_R\}$ because the register RG is implemented using D flip-flops [9]. The minimum number of bits in the register is determined by the following equation: $R = \lceil \log_2 M \rceil$. A block of microoperations (BMO) generates functions $y_n \in Y$, where the system of microoperations is represented as

$$Y = Y(T). \tag{6.2}$$

For the sake of compactness, let us name the methods of hardware reduction as optimization methods. The existing optimization methods for Moore FSM can be divided by universal and specialized. The universal methods are applied for optimization of FSM implemented with arbitrary logic elements and for arbitrary GSA. This group includes such methods as functional and structural decomposition of FSM.

The functional decomposition is based on the Shannon's expansion [9, 10]. Let us consider the following example. Let it be necessary to construct the logic circuit for the following function $y_1 = abcd \vee ab\bar{c}\bar{d} \vee \bar{a}\bar{b}cd \vee \bar{a}\bar{b}\bar{c}\bar{d}$. Let only LUTs having three inputs ($S = 3$) can be used for implementing the circuit. Let us expand the function by the variable a. It gives the following result:

$$y_1 = a\left(bcd \vee b\bar{c}\bar{d}\right) \vee \bar{a}\left(\bar{b}cd \vee \bar{b}\bar{c}\bar{d}\right). \tag{6.3}$$

It is enough three LUTs for implementing the circuit corresponding to (6.3). The resulting circuit has two layers (Fig. 6.2).

In this circuit, the logic element LUT_1 implements the function $B = bcd \vee b\bar{c}\bar{d}$, whereas LUT_2 the function $C = \bar{b}cd \vee \bar{b}\bar{c}\bar{d}$. If the decomposition is not used, then each term of the (6.1) is implemented using two LUTs. To implement the disjunction of these terms, two LUTs are used with $S = 3$ inputs. Therefore, the resulting circuit without decomposition includes 10 LUTs and has four layers.

The functional decomposition was applied for implementing logic circuits with NAND gates [1]. Next, it found application for implementing FSM circuits with FPGA chips [11, 12]. In the last time, the functional decomposition is widely used for implementing FSM circuits with CPLD chips [13–17]. The number of logic elements can be decreased if the functional decomposition is used together

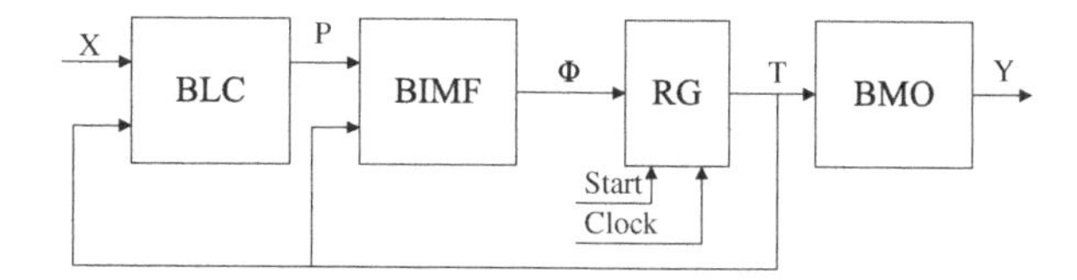

Fig. 6.3 Structural diagram of MPY Moore FSM

with factoring of Boolean functions representing FSM circuits [18]. The factoring assumes use of some conjunctions in different functions of the system (6.1).

The structural decomposition is based on increasing for the number of structure levels in the FSM logic circuit [9, 19]. The methods of structural decompositions are the following:

1. Replacement of the logical conditions.
2. Encoding of the collections of microoperations.
3. Encoding of the fields of compatible microoperations.
4. Transformations of objects.

Let us discuss main ideas of these methods.

Replacement of the logical conditions. This method targets decreasing of the numbers of arguments in the FSM input memory functions. Let $L_m = |X(a_m)|$, where $X(a_m) \subseteq X$ is a set of logical conditions determining transitions from the state $a_m \in A$. Let us replace the set X by the set of additional variables $P = \{p_1, \ldots, p_G\}$, where there is $G = \max(L_1, \ldots, L_M)$. Let us name the Moore FSM (Fig. 6.1) PY Moore FSM. Let us denote the existence of the block BIMF by the symbol P, whereas the existence of the block BMO by Y. The replacement of logical conditions leads to the MPY Moore FSM shown in Fig. 6.3.

In the MPY Moore FSM, the block of logical conditions (BLC) implements the following system of functions:

$$P = P(T, X). \tag{6.4}$$

The block BIMF implements functions

$$\Phi = \Phi(T, P). \tag{6.5}$$

Let us point out that the block BMO still implements functions (6.2). This method can be applied if the condition $G << L$ takes place. In this case, the number of arguments in the system (6.4) decreases significantly in the comparison with their number in the system (6.1). As the results of investigations [19] prove, the decreasing for the number of arguments leads to decreasing for the number of LUT elements in the corresponding combinational circuit.

Encoding of the collections of microoperations. This method targets the hardware reduction in the block BMO. Let Q collections of microoperations (CMO) $Y_q \subseteq Y$ be placed into the operator vertices of GSA Γ. Let us encode the CMO $Y_q \subseteq Y$ by the binary codes $K(Y_q)$ having the following amount of bits:

$$R_Q = \lceil \log_2 Q \rceil. \tag{6.6}$$

Let us use the variables $z_r \in Z$ for encoding the collections $Y_q \subseteq Y$, where there is $|Z| = R_Q$. Now, the system of microoperations Y is represented as

$$Y = Y(Z). \tag{6.7}$$

If the condition $R_Q < R$ takes place, the system (6.7) leads to the circuit with fewer amounts of logic elements than in the circuit corresponding to the system (6.2),

Encoding of the fields of compatible microoperations. This method also targets the optimization of the block BMO. Let us remind that the microoperations $y_i, y_j \in Y$ are compatible if they do not belong to the same collections of microoperations $Y_q \subseteq Y$ $(q = \overline{1, Q})$. Let us find a partition of the set Y by the classes of compatible microoperations $Y^1, \ldots, Y^K$. The following conditions should take places in this partition:

$$\begin{aligned} &Y_n \cap Y_m = \emptyset \quad (n \neq m,\ m, n \in \{1, \ldots, K\}); \\ &\bigcup\nolimits_{k=1}^{K} Y_k; \\ &Y_k \neq \emptyset \quad (k = \overline{1, K}). \end{aligned} \tag{6.8}$$

Let it be $|Y_k| = N_k$ $(k = \overline{1, K})$. Let us Hencode microoperations $y_n \in Y^k$ by binary codes $K(y_n)$ having $R_k = \lceil \log_2 (N_k + 1) \rceil$ bits. Now, it is necessary R_D variables $z_r \in Z$ for encoding the microoperations $y_n \in Y$, where the value of R_D is determined by the following equation:

$$R_D = \sum_{k=1}^{K} R_k. \tag{6.9}$$

Transformations of objects. Both previously discussed methods cannot be directly applied in the Moore FSM. They can be used only together with the transformations of objects [20–23]. There are two different objects in the Moore FSM, namely its states and collections of microoperations. The transformation of the state codes into collections of microoperations with the following encoding of collections leads to the P_AY Moore FSM. The transformation of the state codes into collections of microoperations with the following encoding of the fields of compatible microoperations leads to the P_AD Moore FSM. Both these models possess the same structural diagram (Fig. 6.4).

The structural diagram includes a block of state transformation (BST), generating the following functions

$$Z = Z(T). \tag{6.10}$$

The functions (6.10) are used as the arguments of the system (6.7) implemented by the block BMO. There are other methods of object's transformations discussed further in this Chapter.

The methods of structural decomposition are connected with the idea of heterogeneous implementation of logic circuits [8]. In this case, different logic elements are used for implementing different structure parts of the FSM circuits. For example, such elements as either NAND gates, or PAL macrocells, or LUT elements

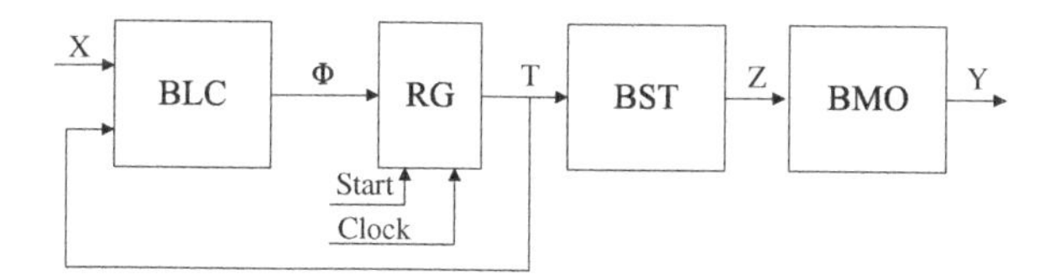

Fig. 6.4 Structural diagram of P_AY and P_AD Moore FSMs

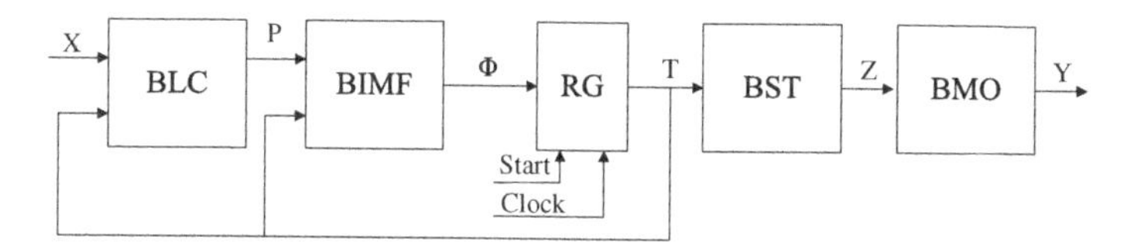

Fig. 6.5 Structural diagram of MP_AY Moore FSM

are used for implementing the circuit of BIMF. The multiplexers are used for implementing the block BLC. The decoders are used for implementing the circuit of BMO in P_AD Moore FSM. Obviously, the circuits of multiplexers and decoders are implemented using logic elements. But, multiplexers, as well as decoders, are library elements of any industrial CAD tools targeting CPLD or FPGA chips. Application of more complex library elements instead of either LUTs or PALs simplifies the design process.

Boolean functions of systems $Y(T)$, $Z(T)$, and $Y(Z)$ are determined for more than 50 % of possible input assignments. It is reasonable to use memory blocks (RAMs, PROMs) for implementing such functions. It is known that a single memory cell replaces at least one logic element. Because of it, application of memory blocks results in the significant reduction for hardware amount.

Let us point out that the discussed methods can be applied simultaneously. For example, mutual application of the replacement of logical conditions, objects' transformation and encoding of CMOs leads to the MP_AY Moore FSM (Fig. 6.5). To minimize the circuit, the functional decomposition of (6.5) can be used.

Obviously, the growth of the number of structure levels leads to increasing of the propagation time. But it is possible the positive side effect, namely, the decreasing for the number of layers of logic elements in the logic circuit of BIMF. It can compensate the previously mentioned negative effect.

The specialized optimization methods are based on the taking into account the peculiarities of: (a) logic elements in use; (b) a control algorithm used for implementing the resulting FSM; (c) an FSM model. Let us discuss the using of these peculiarities.

Using peculiarities of logic elements. The peculiarities of PAL macrocells are the significant number of inputs (up to 30) and very small number of product terms q (around 8). The first peculiarity allows using more than one source of the classes of pseudoequivalent states [15, 17, 18] leading to the hardware reduction in the block BIMF. The second peculiarity leads to the necessity of separate minimization for the input memory functions Φ [24, 25]. For minimization, it is enough to find such a variant when each sum-of-product (SOP) form includes not more than q terms for any function $D_r \in \Phi$ [26].

The main peculiarity of FPGA chips is existence of heterogeneous basis; it consists from lock-up table elements and embedded memory blocks. The modern LUTs have $S \leq 8$ inputs. A single LUT might implement a truth table of an arbitrary Boolean function depending on not more than S arguments. To optimize a combinational circuit implemented with LUTs, it is necessary to diminish the number of arguments, as well as the number of product terms in a Boolean function to be implemented [2].

Embedded memory blocks are used for implementing systems of Boolean functions specified for more than 50 % of possible input assignments. Therefore, it is reasonable to use EMBs for implementing the logic circuit of BMO [27, 28]. The peculiarity of EMB is its reconfigurability [5] assuming changing the numbers of address inputs S_A and cell outputs t_F under the constant size of a block. The number of outputs t_F cannot be arbitrary; it belongs to some fixed set $S(t_F)$. For the up-to-day FPGAs, there is $S(t_F) = \{1, 2, 4, 8, 18, 36, 72\}$ [5]. The size (the number of cells) of a EMB is determined as

$$V_o = 2^{S_A} \cdot t_F. \tag{6.11}$$

Because the value of V_o is constant for given FPGA chip, then the decreasing for the value of parameter S_A by 1 leads to doubling for the number of outputs of the EMB.

Nowadays, the FPGAs are used for which there is $V_o = 16k$ (bits) [7]. These EMBs have the following configurations: $16k \times 1$, $8k \times 2$, $4k \times 4$, $2k \times 8$, $1k \times 16$, 512×36, 256×72, bits. The following expression can be used for determining the number of outputs t_{FR} of EMBs implementing the circuit of BMO for PY Moore FSM:

$$t_{FR} = \lceil V_o/M \rceil. \tag{6.12}$$

It was always assumed that the replacement of the PY- model by the P_AY-model allowed decreasing for the number of memory blocks in the circuit of BMO [8]. In this case, the number of outputs t_{FQ} of EMB is determined as

$$t_{FQ} = \lceil V_o/Q \rceil. \tag{6.13}$$

But if the following condition

$$t_{FR} \geq N \tag{6.14}$$

takes place, then the numbers of EMBs in use are equal for PY and P_AY FSMs. In both cases, only one EMB is necessary. Therefore, the use of the encoding of the collections of microoperations leads to decreasing of performance without any hardware reduction. It means that such a peculiarity of EMB as existence of fixed outputs allows refusing from the approach which was always treated as reasonable.

The block BMO of PY Moore FSM can be presented as a table having $M \times N$ bits. On the other hand, the block EMB can be presented as a table having $2^R \times t_{FR} = V_o$ bits (Fig. 6.6). As follows from Fig. 6.6, it is quite possible the existence of free (unused) resources in a EMB; it could be either cells, or, outputs, or both. These free resources are determined as $\Delta M = 2^R - M$ (for cells)

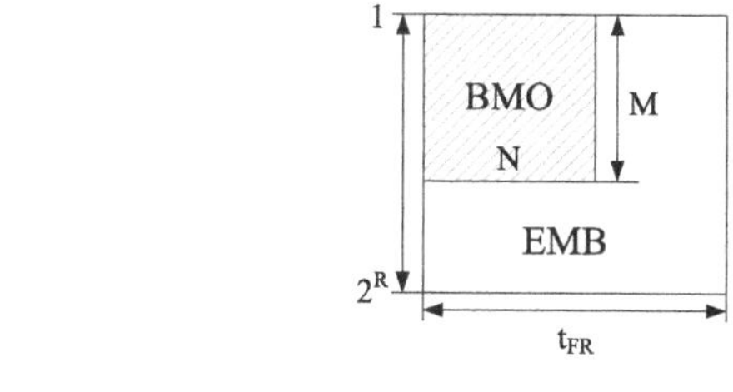

Fig. 6.6 Relation of characteristics for BMO and EMB

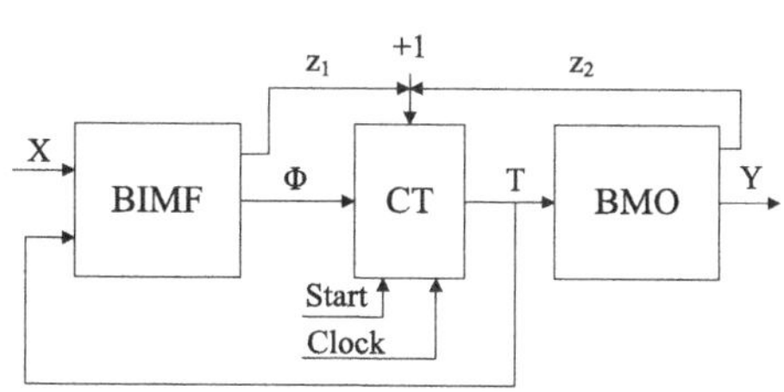

Fig. 6.7 Structural diagram of $P_{CT}Y$ Moore FSM

and $\Delta t = t_{FR} - N$ (for outputs). They can be used for decreasing in the number of LUTs in the logic circuit of BIMF.

Using peculiarities of control algorithms. As it is shown in [29], it is possible to replace the register RG by the counter CT. It has sense if an initial GSA includes not less than 75 % of operator vertices. So called compositional microprogram control units (CMCU) are discussed in [29]. The CMCU can be viewed as Moore FSMs because their outputs are represented by the system (6.2). The logic synthesis of CMCU is based on constructing operator linear chains (OLC) representing some sequences of operator vertices.

This idea can be developed by introducing the conditional vertices into OLCs [30]. Such an approach leads to the Moore $P_{CT}Y$ Moore FSM (Fig. 6.7). Let us discuss the rules used for state assignment of $P_{CT}Y$ Moore FSMs. Let it be an unconditional transition $< a_m, a_s >$ for the states $a_m \in A$ from the same OLC.

In this case the state codes are determined by the following expression:

$$K(a_s) = K(a_m) + 1. \tag{6.15}$$

To organize the transitions (6.15), a special variable z_1 is generated. This variable is used for incrementing the content of the counter CT.

If there is a conditional transition from a_m into a_s and the condition (6.15) takes place, then the variable z_2 is generated to increment the content of CT.

If for some transition $< a_m, a_s >$ the condition (6.15) is violated, then there is $z_1 = z_2 = 0$. In this case the next state code is determined by functions Φ. This approach allows decreasing for the number of structure table's rows in comparison with this value for an equivalent PY FSM. Let us point out that now there are no design methods of $P_{CT}Y$ FSMs targeting either CPLD or FPGA.

Usage peculiarities of FSM model. There are two specifics of Moore FSM which can be used for its circuit's optimization: (1) the dependence of output functions only from state variables $T_r \in T$ and (2) the existence of classes of pseudoequivalent states.

The first specific allows implementing the logic circuit of BMO using only EMBs. If all existed EMBs of a particular FPGA chip are used in a project, then

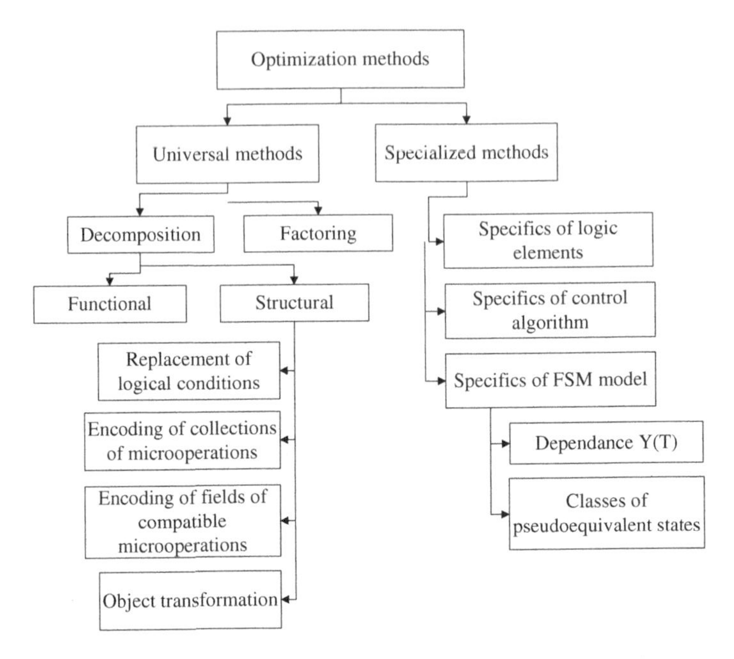

Fig. 6.8 Classification of the optimization methods of Moore FSM logic circuits

the circuit of BMO is implemented using LUTs. In this case, the states should be assigned in the manner leading to minimizing the number of LUTs in the circuit of BMO. In the ideal case, only N of LUTs are enough for implementing the circuit of BMO. The state assignment methods for this case are considered in [31].

The second specific allows decreasing for the number of Moore FSM structure table's rows up to this value of equivalent Mealy FSM [32]. Three main approaches can be used for reaching this goal: (a) the optimal state assignment; (b) the transformation of the states codes into the codes of classes of pseudoequivalent states; (c) the transformation of initial GSA. In this Chapter, we discuss the second approach, whereas two others can be found in [30].

The classification of the optimization methods of Moore FSM logic circuits is shown in Fig. 6.8. Let us point out that these methods are used simultaneously. Only the simultaneous approach can result in a circuit with minimum amounts of LUTs and EMBs. Some of optimization methods are discussed in this chapter.

6.2 Object Transformation in Moore FSM

As it is shown in [8], the optimal state encoding does not always leads to decreasing for the number of ST rows up to H_0, where H_0 is the number of ST rows for equivalent Mealy FSM. In this case, it is reasonable to replace the state codes into the codes of classes of pseudoequivalent states [19].

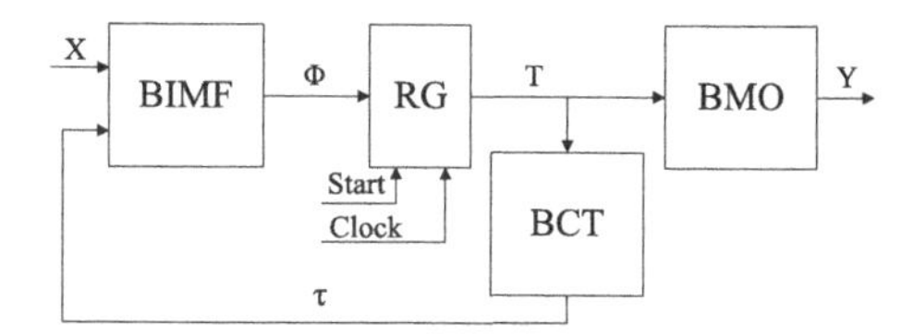

Fig. 6.9 Structural diagram of P_BY Moore FSM

Let the partition $\Pi_A = \{B_1, \ldots, B_I\}$ of the set of states A by the classes of pseudoequivalent states be found for some Moore FSM. Let us encode each class $B_i \in \Pi_A$ by the binary code $K(B_i)$ having the following amount of bits:

$$R_B = \lceil \log_2 I \rceil. \tag{6.16}$$

Let us use the variables $\tau_r \in \tau$ for encoding the classes $B_i \in \Pi_A$, where $|\tau| = R_B$.

Let us transform the codes of states $a_m \in B_i$ into the codes of corresponding classes $B_i \in \Pi_A$. To do such a transformation, it is necessary to include the special block of code transformer (BCT) into the Moore FSM. The proposed approach leads to P_BY Moore FSM shown in Fig. 6.9.

In the P_BY Moore FSM, the blocks BIMF and BCT implement the following systems of functions:

$$\Phi = \Phi(\tau, X), \tag{6.17}$$

$$\tau = \tau(T). \tag{6.18}$$

Let us compare the systems (6.2) and (6.18). The comparison shows that functions of these systems have the same nature. The functions τ and Y depend only on the state variables $T_r \in T$. Therefore, it is reasonable to use EMBs for implementing systems (6.2) and (6.18).

Let the following condition take place

$$t_{FR} \geq N + R_B. \tag{6.19}$$

In this case, only one block of EMB is enough for implementing both systems (6.2) and (6.18). If there is $t_{FR} < N$, then $N(Y)$ of memory blocks are necessary for implementing the BMO's logic circuit:

$$N(Y) = \lceil N/t_{FR} \rceil. \tag{6.20}$$

Let the following condition take place:

$$N(Y) \cdot t_{FR} - N \geq R_B. \tag{6.21}$$

In this case, the block BCT is implemented using the same EMBs as the block BMO. If the condition (6.20) is violated, then some part of the functions $\tau_r \in \tau$ is implemented with EMBs, whereas the other part with LUTs. Let us point out that our analysis of standard benchmarks [33] shows that the condition (6.19) takes place for the overwhelming majority of practical control algorithms.

Let us point out that the register RG is shown for the purposes of explanation. In reality, the flip-flops of RG are distributed among the LUTs of a logic circuit. If

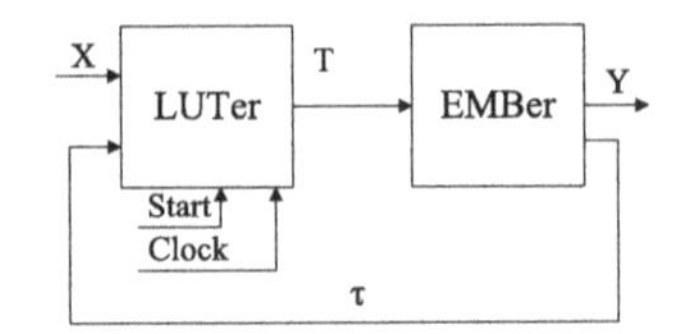

Fig. 6.10 Implementation of $\mathrm{P_BY}$ Moore FSM with FPGA

a LUT element implements the function $D_r \in \Phi$, then its output is connected with the flip-flop of the corresponding macrocell. Therefore, the practical structural diagram of the $\mathrm{P_BY}$ Moore FSM is shown in Fig. 6.10. This circuit has only two levels. Let us point out that the condition (6.21) should take place for the $\mathrm{P_BY}$ Moore FSM. In Fig. 6.10, the block LUTer denotes the collection of look-up table elements implementing the system (6.17), whereas the block EMBer denotes the collection of embedded memory blocks implementing the systems (6.2) and (6.18).

The synthesis method for $\mathrm{P_BY}$ Moore FSM includes the following steps:

1. Marking the initial GSA Γ by the states of Moore FSM and constructing the set of states A.
2. Finding the partition $\Pi_A = \{B_1, \ldots, B_I\}$.
3. Encoding of the states $a_m \in A$ and the classes $B_i \in \Pi_A$.
4. Constructing the reduced structure table of Moore FSM.
5. Constructing the system (6.17).
6. Constructing the table of block of microoperations.
7. Constructing the table of block of code transformer.
8. Implementing the FSM logic circuit for given FPGA chip.

Let us discuss an example of synthesis for the Moore FSM $\mathrm{P_BY}(\Gamma_1)$, where the part (Γ_i) means that the given model is synthesized using a GSA Γ_i. The graph-scheme of algorithm Γ_1 is shown in Fig. 6.11.

The states $a_m \in A$ are already shown in Fig. 6.11. Thus, the following information about sets and their parameters can be derived from Fig. 6.11: $A = \{a_1, \ldots, a_8\}$, $M = 8$; $X = \{x_1, \ldots, x_4\}$; $L = 4$; $Y = \{y_1, \ldots, y_5\}$, and $N = 5$. Obviously, there are $R = 3$, $T = \{T_1, T_2, T_3\}$, and $\Phi = \{D_1, D_2, D_3\}$. Analysis of GSA Γ_1 allows constructing the partition $\Pi_A = \{B_1, \ldots, B_4\}$. Therefore, there are the following values: $I = 4$, $R_B = 2$. It gives the set of variables $\tau = \{\tau_1, \tau_2\}$.

Let us encode the states $a_m \in A$ in an arbitrary manner. Let us use the following codes: $K(a_1) = 000$, $K(a_2) = 001$, ..., $K(a_8) = 111$. Let us encode the classes $B_i \in \Pi_A$ using the frequency principle [19]. In this case, the more states a class includes the more zeros its code contains. In the case of FSM $\mathrm{P_BY}(\Gamma_1)$, there are the classes $B_1 = \{a_1\}$, $B_2 = \{a_2, a_3, a_4\}$, $B_3 = \{a_5, a_6\}$, and $B_4 = \{a_7, a_8\}$. Using the frequency principle produces the following class codes: $K(B_1) = 11$, $K(B_2) = 00$, $K(B_3) = 01$, and $K(B_4) = 10$.

To construct the reduced structure table, it is necessary to find the system of generalized formulae of transitions (GFT) [19]. A generalized formula of transitions has the following form:

$$B_i \rightarrow \bigvee_{h=1}^{H_m} X_h a_s (i = \overline{1, I}). \tag{6.22}$$

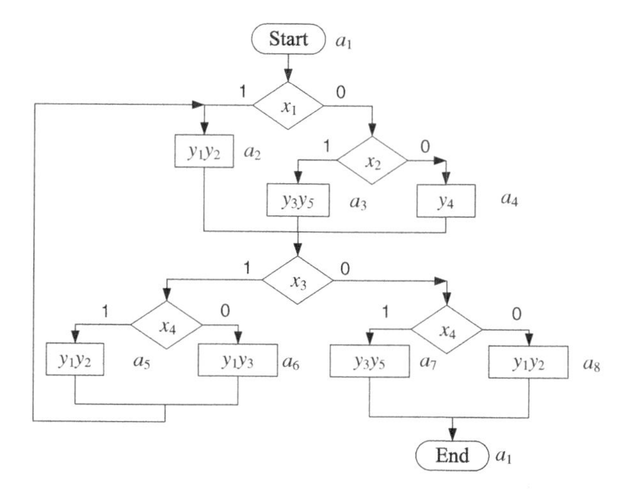

Fig. 6.11 The graph-scheme of algorithm Γ_1

Table 6.1 Reduced structure table of Moore FSM $P_BY(\Gamma_1)$

B_i	$K(B_i)$	a_s	$K(a_s)$	X_h	Φ_h	h
B_1	11	a_2	001	x_1	D_3	1
		a_3	010	$\bar{x}_1x_2$	D_2	2
		a_4	011	$\bar{x}_1\bar{x}_2$	D_2D_3	3
B_2	00	a_5	100	x_3x_4	D_1	4
		a_6	101	$x_3\bar{x}_4$	D_1D_3	5
		a_7	110	$\bar{x}_3x_4$	D_1D_2	6
		a_8	111	$\bar{x}_3\bar{x}_4$	$D_1D_2D_3$	7
B_3	01	a_2	001	1	D_3	8

In (6.22), the symbol H_m stands for the number of transitions from any state $a_m \in B_i$, the symbol X_h corresponds to a conjunction of input variables $x_l \in X$, determining the transition from the state $a_m \in B_i$ into the next state $a_s \in A$. In the case of $P_BY(\Gamma_1)$ FSM, there is the following GFT:

$$\begin{aligned} &B_1 \rightarrow x_1a_2 \vee \bar{x}_1x_2a_3 \vee \bar{x}_1\bar{x}_2a_4; \\ &B_2 \rightarrow x_3x_4a_5 \vee x_3\bar{x}_4a_6 \vee \bar{x}_3x_4a_7 \vee \bar{x}_3\bar{x}_4a_8; \\ &B_3 \rightarrow a_2; \quad B_4 \rightarrow a_1. \end{aligned} \tag{6.23}$$

The reduced structure table has the following columns: B_i, $K(B_i)$, a_s, $K(a_s)$, X_h, Φ_h, h. In the case of Moore FSM $P_BY(\Gamma_1)$, the reduced structure table includes $H_0 = 8$ rows (Table 6.1).

The connection between the system (6.23) and Table 6.1 is obvious. Let us point out that the transitions are not considered for the class B_4. It is connected with the fact that there is $K(a_1) = 00$, and, therefore, there is $D_1 = D_2 = 0$. Also, let us point out that there is $H = 20$ for the Moore FSM $PY(\Gamma_1)$.

The content of Table 6.1 is used for deriving the system (6.17). After minimizing, this system is the following:

$$\begin{aligned} &D_1 = \bar{\tau}_1\bar{\tau}_2; \quad D_2 = \tau_1\tau_2\bar{x}_1 \vee \bar{\tau}_1\bar{\tau}_2x_3; \\ &D_3 = \tau_1\tau_2x_1 \vee \tau_1\tau_2\bar{x}_2 \vee \bar{\tau}_1\bar{\tau}_2\bar{x}_4 \vee \bar{\tau}_1\tau_2. \end{aligned} \tag{6.24}$$

Table 6.2 Table of BMO for Moore FSM $P_BY(\Gamma_1)$

$K(a_m)$			Microoperations					m
T_1	T_2	T_3	y_1	y_2	y_3	y_4	y_5	
0	0	0	0	0	0	0	0	1
0	0	1	1	1	0	0	0	2
0	1	0	0	0	1	0	1	3
0	1	1	0	0	0	1	0	4
1	0	0	1	1	0	0	0	5
1	0	1	1	0	1	0	0	6
1	1	0	0	0	1	0	1	7
1	1	1	1	1	0	0	0	8

Table 6.3 Table of BCT for Moore FSM $P_BY(\Gamma_1)$

$K(a_m)$			$K(B_i)$		m	i
T_1	T_2	T_3	τ_1	τ_2		
0	0	0	1	1	1	1
0	0	1	0	0	2	2
0	1	0	0	0	3	2
0	1	1	0	0	4	2
1	0	0	0	1	5	3
1	0	1	0	1	6	3
1	1	0	1	0	7	4
1	1	1	1	0	8	4

Table of BMO (Table 6.2) is constructed in the trivial way using both codes $K(a_m)$ and collections of microoperations $Y_q \subseteq Y$ from GSA Γ_1.

The table of BCT (Table 6.3) includes the columns $K(a_m)$, $K(B_i)$, m, i. The column m includes the subscript of a state (as for Table 6.2), whereas the column i includes the subscript of a block B_i, where there is $a_m \in B_i$.

The implementing logic circuit of P_BY Moore FSM is reduced to the implementing the system (6.24) by the LUTer, whereas both Tables 6.2 and 6.3 are implemented by the EMBer. Let the FSM logic circuit be implemented using LUTs having $S = 3$ inputs. Let the possible configurations of EMB include the configuration 8×8, bits. Let us denote the number of literals in the function $D_r \in \Phi$ as $L(D_r)$. Let the following condition take place:

$$L(D_r) \leq S. \tag{6.25}$$

In this case, the part of the logic circuit corresponding to the function $D_r \in \Phi$ is implemented using one LUT. If the condition (6.25) is violated, then the method of functional decomposition should be used for function $D_r \in \Phi$.

The following values can be found for Moore FSM $P_BY(\Gamma_1)$: $L(D_1) = 2$, $L\ (D_2) = 4$, and $L(D_3) = 5$. Therefore, both the functions D_2 and D_3 should be decomposed. It leads to the following system of Boolean functions:

$$\begin{aligned} D_2 &= \tau_1(\tau_2\bar{x}_1) \vee \bar{\tau}_1(\bar{\tau}_2 x_3) = \tau_1\Phi_1 \vee \bar{\tau}_1\Phi_2; \\ D_3 &= \tau_1(\tau_2 x_1 \vee \tau_2 x_2) \vee \bar{\tau}_1(\bar{\tau}_2\bar{x}_4 \vee \tau_2) = \tau_1\Phi_3 \vee \bar{\tau}_1\Phi_4. \end{aligned} \tag{6.26}$$

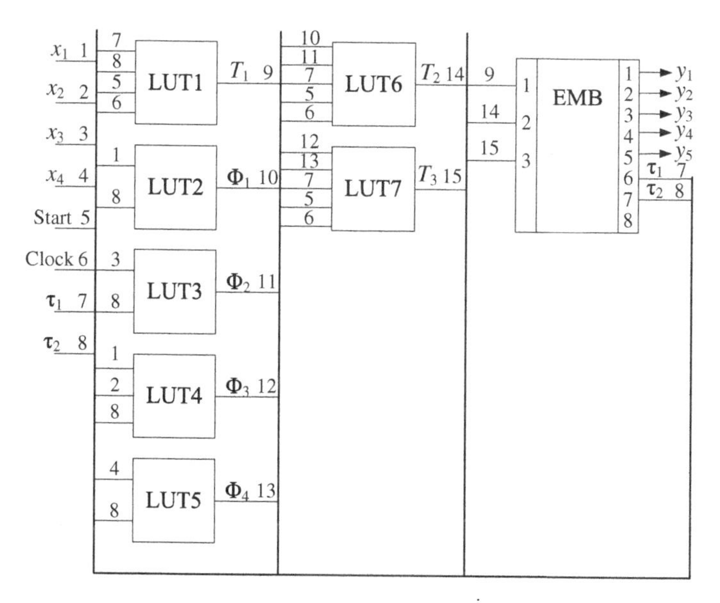

Fig. 6.12 Logic circuit of Moore FSM $P_BY(\Gamma_1)$

Because there is $t_{FR} = 8$, the condition (6.19) takes place. Therefore, only a single EMB is necessary for implementing the circuit of EMBer. The resulting logic circuit is shown in Fig. 6.12.

As follows from Fig. 6.12, seven LUTs are used for implementing the circuit of BIMF. The combinational outputs of elements LUT2–LUT5 are used, whereas the registered outputs of elements LUT1, LUT6 and LUT7 are used. The pulses *Start* and *Clock* are connected correspondingly with synchronization and clearing inputs of logic elements 1, 6 and 7. The block LUTer has two layers of LUTs, whereas the block EMBer uses only a single EMB.

It can be shown that the block LUTer of Moore FSM $PY(\Gamma_1)$ consists from 34 LUTs having $S = 3$ inputs; it has four layers of logic elements. Therefore, the application of the method of objects' transformation allows, in the discussed case, decreasing for both the number of LUTs (in 4,85 times) and the propagation time of the resulting FSM logic circuit (in 2 times). Let us point out that there are a lot of different methods of objects' transformation [8], but they are beyond the scope of this chapter.

6.3 Expansion of State Codes for Moore FSM

Let the parameters t_{FR} and t_{FY} be found for some PY Moore FSM (for the given FPGA chip). Let the following condition take place

$$\left\lceil \frac{N}{t_{FR}} \right\rceil > \left\lceil \frac{N}{t_{FY}} \right\rceil. \tag{6.27}$$

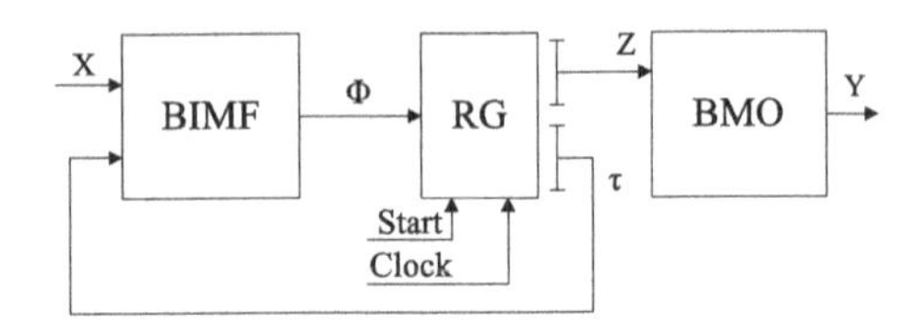

Fig. 6.13 Structural diagram of Moore FSM with extension of state codes

In this case, it is reasonable to encode the collections of microoperations $Y_q \subseteq Y$ and use the variables $z_r \in Z$ as address inputs for the block BMO. Let us discuss an approach which can be viewed as the development of ideas from [34, 35].

Let us find the partition $\Pi_A = \{B_1, \ldots, B_I\}$ and encode the classes $B_i \in \Pi_A$ by binary codes $K(B_i)$ having R_B bits. Let us encode the collections $Y_q \subseteq Y$ by binary codes $K(Y_q)$ having R_Q bits. The value of R_B is determined by (6.16), whereas the value of R_Q by (6.6). Let us use the variables $\tau_r \in \tau$ for encoding of the classes $B_i \in \Pi_A$, whereas the variables $z_r \in Z$ for encoding of the collections of microoperations.

Let a collection $Y_q \subseteq Y$ be generated for a state $a_m \in B_i$. Let us represent the state code $K(a_m)$ as the following expression:

$$K(a_m) = K(B_i) * K\left(Y_q\right). \tag{6.28}$$

In (6.28), the sign "*" denotes the concatenation of codes. The representation (6.28) is named the extension of state codes [35]. This representation allows obtaining the structural diagram of $P_{BY}Y$ Moore FSM (Fig. 6.13).

In the $P_{BY}Y$ Moore FSM, the block BIMF implements $R_B + R_Q$ functions forming the system (6.7). If the condition (6.27) takes place, the block BMO of $P_{BY}Y$ Moore FSM requires fewer embedded memory blocks in comparison with either PY or P_BY Moore FSMs. Let the following condition take place:

$$R < R_B + R_Q. \tag{6.29}$$

In this case, the block BIMF implements more functions than in the cases of both PY or P_BY Moore FSMs.

The proposed synthesis method of $P_{BY}Y$ FSM includes the following steps:

1. Marking the initial GSA and forming the set of states A.
2. Finding the partition $\Pi_A = \{B_1, \ldots, B_I\}$.
3. Encoding of classes $B_i \in \Pi_A$ and collections of microoperations $Y_q \subseteq Y$. Finding the extended state codes.
4. Constructing the reduced structure table.
5. Constructing the system of functions $D_r \in \Phi$.
6. Constructing the table of BMO.
7. Implementing FSM circuit for a given FPGA chip.

Let us discuss an example of synthesis for the Moore FSM $P_{BY}Y(\Gamma_2)$, where the initial GSA Γ_2 is shown in Fig. 6.14.

Let us analyse the characteristics of Moore FSM PY(Γ_2). The set of states A includes $M = 9$ elements; therefore, there is $R = 4$. The following collections of microoperations can be derived from the operator vertices of GSA Γ_2: $Y_2 = \{y_1, y_2\}$, $Y_3 = \{y_3, y_5\}$, $Y_4 = \{y_4\}$, $Y_5 = \{y_3, y_4\}$, $Y_6 = \{y_2, y_5\}$. Besides,

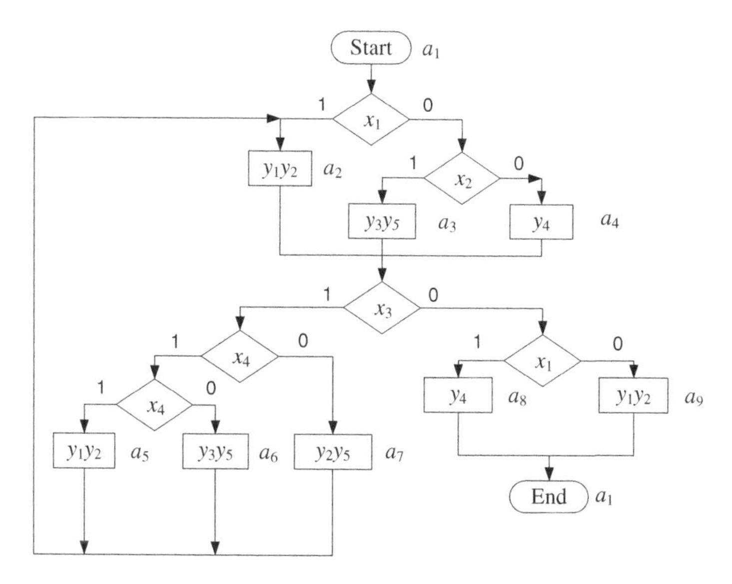

Fig. 6.14 Initial graph-scheme of algorithm Γ_2

the start vertex corresponds to the empty collection $Y_1 = \emptyset$. Therefore, there are $Q = 6$, $R_Q = 3$, and $Z = \{z_1, z_2, z_3\}$. The partition Π_A includes $I = 4$ blocks, namely: $B_1 = \{a_1\}$, $B_2 = \{a_2, a_3, a_4\}$, $B_3 = \{a_5, a_6, a_7\}$, and $B_4 = \{a_8, a_9\}$. Therefore, there are $R_B = 2$ and $\tau = \{\tau_1, \tau_2\}$.

Let an FPGA chip in use include embedded memory blocks having the following configurations: 16×4 and 8×8 (bits). Therefore, each EMB has $V_o = 64$ bits. Using expressions (6.12) and (6.13), respectively, the following values can be found: $t_{FR} = 4$ and $t_{FQ} = 8$. Because there is $N = 5$, then there are $\left\lceil \frac{N}{t_{FR}} \right\rceil = 2$ and $\left\lceil \frac{N}{t_{FQ}} \right\rceil = 1$. It means that the condition (6.26) takes place. Therefore, it is reasonable to use the approach of state expansion.

Let us point out one specific of the P_{BY}Y Moore FSM. Using the pulse *Start*, the zero code (all zeros) corresponding to the initial state a_1 should be loaded into the register RG. According with (6.28), there is $K(a_1) = K(B_1) * K(Y_1)$. Therefore, both the class $B_1 \in \Pi_A$ and the collection $Y_1 = \emptyset$ should be encoded by zero codes. Let us use the frequency principle for encoding of both classes of pseudoequivalent states and collections of microoperations. For collections of microoperations this principle can be formulated as the following one: the more operator vertices contain a collection $Y_q \subseteq Y$, the more zeros its code includes.

Let us encode the classes $B_i \in \Pi_A$ in the following manner: $K(B_1) = 00$, $K(B_2) = 01$, $K(B_3) = 10$, and $K(B_4) = 11$. Let us encode the collections of microoperations $Y_q \subseteq Y$ in the following manner: $K(Y_1) = 000$, $K(Y_2) = 001$, $K(Y_3) = 010$, $K(Y_4) = 100$, $K(Y_5) = 011$, and $K(Y_6) = 101$. Using codes of both classes of pseudoequivalent states and collections of microoperations, the extended state codes can be found for states $a_m \in A$ (Fig. 6.15).

Fig. 6.15 Extended state codes of Moore FSM $P_{BY}Y(\Gamma_2)$

$\tau_1\tau_2$ \ $z_1z_2z_3$	000	001	010	011	100	101	110	111	
00	a_1	*	*	*	*	*	*	*	B_1
01	*	a_2	a_3	*	a_4	*	*	*	B_2
10	*	a_5	*	a_6	*	a_7	*	*	B_3
11	*	a_9	*	*	a_8	*	*	*	B_4
	Y_1	Y_2	Y_3	Y_5	Y_4	Y_6			

Table 6.4 Reduced structure table of Moore FSM $P_{BY}Y(\Gamma_2)$

B_i	$K(B_i)$	a_s	$K(a_s)$	X_h	Φ_h	h
B_1	00	a_2	01001	x_1	D_2D_5	1
		a_3	01010	$\bar{x}_1x_2$	D_2D_4	2
		a_4	01100	$\bar{x}_1\bar{x}_2$	D_2D_3	3
B_2	01	a_5	10001	$x_3x_4x_5$	D_1D_5	4
		a_6	10011	$x_3x_4\bar{x}_5$	$D_1D_4D_5$	5
		a_7	10101	$x_3\bar{x}_4$	$D_1D_3D_5$	6
		a_8	11101	$\bar{x}_3x_1$	$D_1D_2D_3$	7
		a_9	11001	$\bar{x}_3\bar{x}_1$	$D_1D_2D_5$	8
B_3	10	a_2	01001	1	D_2D_5	9

In Fig. 6.15, the sign "*" marks the codes (6.28) which do not correspond to the states $a_m \in A$ for the Moore FSM $P_{BY}Y(\Gamma_2)$. The reduced structure table of $P_{BY}Y$ Moore FSM is constructed using the same approach as for P_BY Moore FSM. For the discussed example, the system of GFT includes the following formulae:

$$\begin{aligned}
&B_1 \rightarrow x_1a_2 \vee \bar{x}_1x_2a_3 \vee \bar{x}_4\bar{x}_2a_4;\\
&B_2 \rightarrow x_3x_4x_5a_5 \vee x_3x_4\bar{x}_5a_6 \vee x_3\bar{x}_4a_7 \vee \bar{x}_3x_1a_8 \vee \bar{x}_3\bar{x}_1a_9;\\
&B_3 \rightarrow a_2; \quad B_4 \rightarrow a_1.
\end{aligned} \tag{6.30}$$

The system (6.30) includes 10 terms, but the transitions from the states of the class B_4 are not listed in the table. Because of it, the reduced structure table of the Moore FSM $P_{BY}Y(\Gamma_2)$ includes only $H_0 = 9$ rows (Table 6.4).

The system of input memory functions $D_r \in \Phi$ is derived from this table. These functions can be minimized. After minimizing, the system $D_r \in \Phi$ is the following in the discussed case:

$$\begin{aligned}
D_1 &= \bar{\tau}_1\tau_2;\\
D_2 &= \bar{\tau}_2 \vee \bar{\tau}_1\tau_2\bar{x}_3;\\
D_3 &= \bar{\tau}_1\bar{\tau}_2\bar{x}_1\bar{x}_2 \vee \bar{\tau}_1\tau_2x_3\bar{x}_4 \vee \bar{\tau}_1\tau_2\bar{x}_3x_1;\\
D_4 &= \bar{\tau}_1\bar{\tau}_2\bar{x}_1x_2 \vee \bar{\tau}_1\tau_2x_3x_4x_5;\\
D_5 &= \bar{\tau}_1\bar{\tau}_2x_1 \vee \bar{\tau}_1\tau_2x_3 \vee \bar{\tau}_1\tau_2\bar{x}_1 \vee \tau_1\bar{\tau}_2.
\end{aligned} \tag{6.31}$$

Let LUTs having $S = 4$ inputs be used for implementing the logic circuit of BIMF. Analysis of the system (6.31) shows that there are $L(D_1) = 2$, $L(D_2) = 3$,

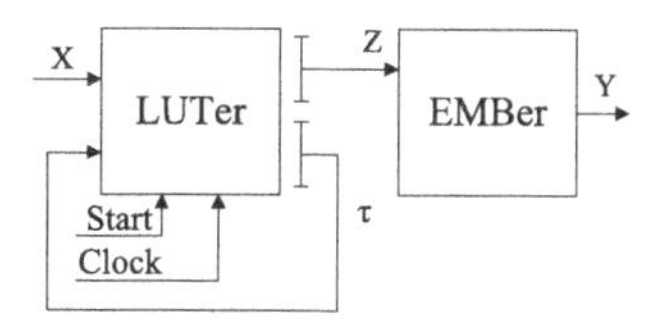

Fig. 6.16 Structural diagram of $P_{BY}Y$ Moore FSM implemented with FPGA

Table 6.5 Reduced structure table of Moore FSM $P_{BY}Y(\Gamma_2)$

$K(Y_q)$			Microoperations					q
z_1	z_2	z_3	y_1	y_2	y_3	y_4	y_5	
0	0	0	0	0	0	0	0	1
0	0	1	1	1	0	0	0	2
0	1	0	0	0	1	0	1	3
0	1	1	0	0	1	1	0	5
1	0	0	0	0	0	1	0	4
1	0	1	0	1	0	0	1	6
1	1	0	0	0	0	0	0	*
1	1	1	0	0	0	0	0	*

$L(D_3) = 6$, $L(D_4) = 7$, and $L(D_5) = 4$. Therefore, only three LUTs are used for implementing a subcircuit for functions D_1, D_2, and D_5. The functions D_3 and D_4 should be decomposed. Let us represent them in the following manner:

$$\begin{aligned} D_3 &= \bar{\tau}_1(\bar{\tau}_2\bar{x}_1\bar{x}_2 \vee \tau_2\bar{x}_3x_1) \vee \bar{\tau}_1\tau_2x_3\bar{x}_4 = \bar{\tau}_1\Phi_1 \vee \Phi_2; \\ D_4 &= \bar{\tau}_1(\bar{\tau}_2x_1x_2) \vee \bar{\tau}_1(\tau_2x_3x_4\bar{x}_5) = \bar{\tau}_1\Phi_3 \vee \bar{\tau}_1\Phi_4. \end{aligned} \tag{6.32}$$

As in the case of P_BY Moore FSM, the structure of $P_{BY}Y$ Moore FSM can be represented as a composition of LUTer and EMBer (Fig. 6.16). The table of BMO is the same as the table of EMBer (Table 6.5).

The logic circuit of Moore FSM $P_{BY}Y(\Gamma_2)$ is shown in Fig. 6.17. In this circuit, the block LUTer consists from 9 LUTs, whereas the block EMBer is implemented using only one EMB.

Let us point out that there is $H = 21$ for the Moore FSM $PY(\Gamma_2)$. If the FSM states are encoded in the natural order ($K(a_1) = 0000$, $K(a_2) = 0001$, ...), then it is necessary 15 LUTs for implementing the logic circuit of BIMF and two EMBs for implementing BMO. The EMB should have the configuration 16×4, because there is $R = 4$ for the Moore FSM $PY(\Gamma_2)$.

Let us use the principle of optimal state assignment [2] for the discussed example. Let us denote the model based on this principle as P_oY. For optimal state assignment, the codes of states belonging to a single class of pseudoequivalent states should be placed in the minimal possible amount of generalized intervals of coding space. The optimal state codes for the FSM $P_oY(\Gamma_2)$ are shown in the Karnaugh map (Fig. 6.18).

The following class codes can be found from Fig. 6.18: $K(B_1) = 00{*}{*}$, $K(B_2) = 01{*}{*}$, $K(B_3) = 11{*}{*}$, $K(B_4) = 10{*}{*}$. The reduced structure table of Moore FSM $P_oY(\Gamma_2)$ includes $H_0 = 9$ rows (Table 6.6).

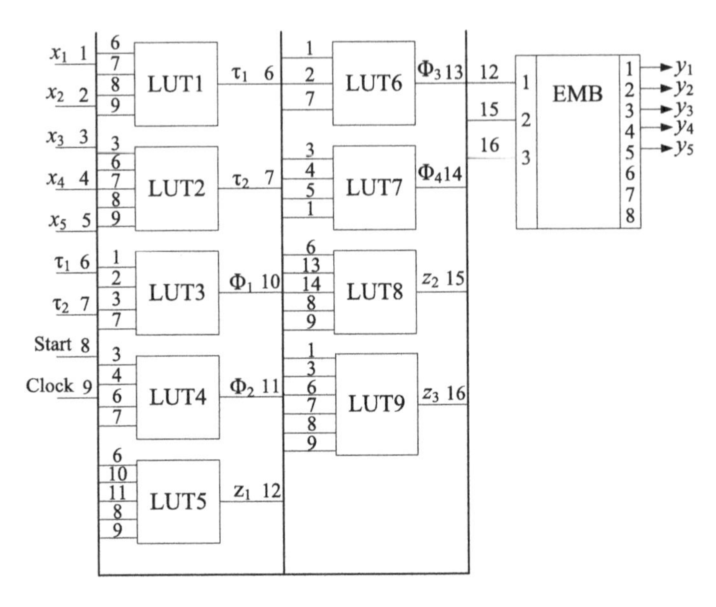

Fig. 6.17 Logic circuit of Moore FSM $P_{BY}Y(\Gamma_2)$

Fig. 6.18 Optimal state codes for Moore FSM $P_oY(\Gamma_2)$

T_3T_4 \ T_1T_2	00	01	11	10
00	a_1	a_2	a_5	a_8
01	*	a_3	a_6	a_9
11	*	*	*	*
10	*	a_4	a_7	*

Table 6.6 Reduced structure table of Moore FSM $P_oY(\Gamma_2)$

B_i	$K(B_i)$	a_s	$K(a_s)$	X_h	Φ_h	h
B_1	00**	a_2	0100	x_1	D_2	1
		a_3	0101	$\bar{x}_1x_2$	D_2D_4	2
		a_4	0110	$\bar{x}_1\bar{x}_2$	D_2D_3	3
B_2	01**	a_5	1100	$x_3x_4x_5$	D_1D_2	4
		a_6	1101	$\bar{x}_5x_3x_4$	$D_1D_2D_4$	5
		a_7	1110	$x_3\bar{x}_4$	$D_1D_2D_3$	6
		a_8	1000	$\bar{x}_3x_1$	D_1	7
		a_9	1001	$\bar{x}_3\bar{x}_1$	D_1D_4	8
B_3	11**	a_2	0100	1	D_2	9

This table is a base for constructing the system of functions $D_r \in \Phi$. After minimizing, this system is the following:

$$
\begin{aligned}
D_1 &= B_2 = \bar{T}_1T_2;\\
D_2 &= B_1 \vee B_2x_3 \vee B_3;\\
D_3 &= B_1\bar{x}_1\bar{x}_2 \vee B_2x_3\bar{x}_4;\\
D_4 &= B_1\bar{x}_1x_2 \vee B_2x_3x_4\bar{x}_5 \vee B_2\bar{x}_3\bar{x}_1.
\end{aligned}
$$

Table 6.7 Characteristics of FSMs

Type of FSM	Number of LUTs	Number of layers	Number of EMBs
$P_{BY}Y$	9	2	1
PY	15	3	2
P_oY	10	3	2

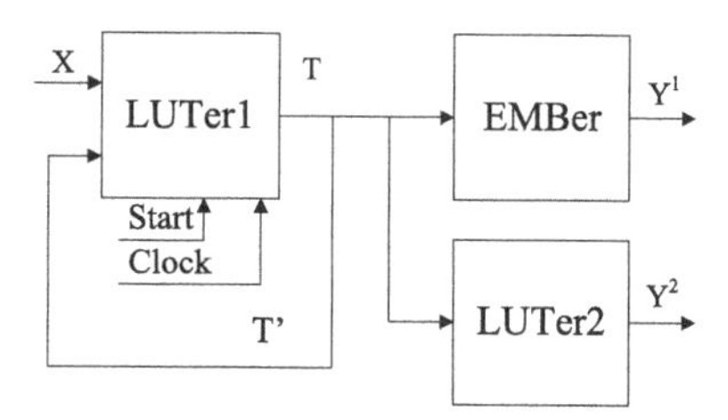

Fig. 6.19 Structural diagram of P_oY_M Moore FSM

It can be shown that it is necessary 10 LUTs having $S = 4$ inputs for implementing the system Φ. The circuits for each function D_1 and D_2 are implemented using only one LUT, the circuit for D_3 needs three LUTs (it has two layers), the circuit for D_4 requires 5 LUTs and includes three layers. The characteristics of logic circuits for P_oY, PY and $P_{BY}Y$ FSMs for GSA Γ_2 are listed in Table 6.7.

Therefore, for the case of GSA Γ_2, the logic circuit of Moore FSM $P_{BY}Y$ contains the least amount of LUTs and EMBs. Besides, this circuit has the least value of the propagation time. Of course, this conclusion cannot be done for the common case.

The number of EMBs in the circuit of BMO can be diminished if t_{FR} of microoperations $y_n \in Y$ are implemented with embedded memory blocks, whereas the other $N - t_{FR}i$ of microoperations with LUTs [36]. Let us explain this idea for the general case.

Let us represent the set Y in the form $Y = Y^1 \cup Y^2$, where $Y^1 \cap Y^2 = \emptyset$. Let the set Y^1 include N_1 elements, where

$$N_1 = t_{FR} \cdot (N(Y) - 1). \tag{6.33}$$

Let us point out that the parameter $N(Y)$ is determined by the expression (6.19). It is clear that the set Y^2 includes the remaining N_2 elements:

$$N_2 = N - t_{FR} \cdot (N(Y) - 1). \tag{6.34}$$

In this case, the P_oY Moore FSM, for example, is represented as a circuit with a mixed memory (Fig. 6.16).

Let us use the symbol P_oY_M for denoting the FSM shown in Fig. 6.19. In the P_oY_M Moore FSM, the block LUTer1 implements the system Φ:

$$\Phi = \Phi\left(T', X\right). \tag{6.35}$$

In the expression (6.35), the set $T' \subseteq T$ is a set of input variables sufficient for encoding of the classes $B_i \in \Pi_A$. For example, the set T' includes only two elements in the FSM $P_oY(\Gamma_2)$, namely, there is the set $T' = \{T_1, T_2\}$. At the same time, some LUTs are used for implementing the register RG.

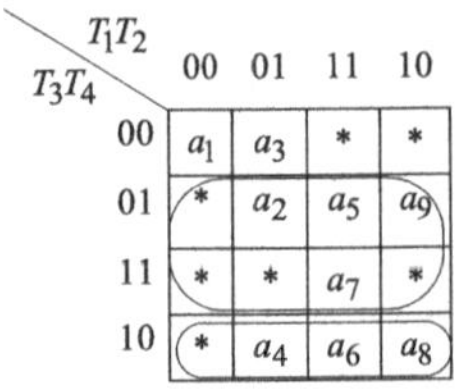

Fig. 6.20 Refined state codes for Moore FSM $P_oY_M(\Gamma_2)$

T_3T_4 \ T_1T_2	00	01	11	10
00	a_1	a_3	*	*
01	*	a_2	a_5	a_9
11	*	*	a_7	*
10	*	a_4	a_6	a_8

The block EMBer implements microoperations $y_n \in Y^1$, whereas the block LUTer2 implements microoperations $y_n \in Y^2$. The microoperations $y_n \in Y^2$ are chosen in such a way that the corresponding circuits will be implemented with the fewest amounts of LUTs.

For example, the following formulae can be obtained from analysis of GSA Γ_2 and Karnaugh map (Fig. 6.18):

$$\begin{aligned}
y_1 &= A_2 \vee A_5 \vee A_9 = T_2\bar{T}_3\bar{T}_4 \vee \bar{T}_2\bar{T}_3T_4;\\
y_2 &= A_2 \vee A_5 \vee A_7 \vee A_9;\\
y_3 &= A_3 \vee A_6;\\
y_4 &= A_4 \vee A_6 \vee A_8;\\
y_5 &= A_3 \vee A_7.
\end{aligned} \tag{6.36}$$

In the discussed example, there is $R = S = 4$. Therefore, the logic circuit for any function of the system (6.36) is implemented using a single LUT. If there is $R > S$, then the states should be rearranged inside the Karnaugh map. The rearrangement should be executed in such a manner that to diminish the value of $A(y_n)$, where $A(y_n)$ is the number of arguments in the SOP of a function $y_n \in Y$. Obviously, the places of states $a_m \in B_i$ might be changed only within the range of the map's columns occupied by these states after the execution of the optimal state assignment. Let us name this approach a refined state assignment. The result of refined state assignment for the FSM $P_oY(\Gamma_2)$ is shown in Fig. 6.20.

In the case of refined state assignment, the system (36) is represented as the following one:

$$\begin{aligned}
&y_1 = \bar{T}_3T_4; \quad y_2 = T_4; \quad y_3 = T_2\bar{T}_3\bar{T}_4 \vee T_1T_2\bar{T}_4;\\
&y_4 = T_3T_4; \quad y_5 = T_2\bar{T}_3T_4 \vee T_3T_4.
\end{aligned} \tag{6.37}$$

Analysis of the system (6.37) shows that it is enough to use the state variable T_4 for implementing the circuit for y_2. The circuits for microoperations y_1 and y_4 are implemented using only one LUT having $S \geq 2$. It is enough to have only one LUT with $S \geq 3$ for implementing the circuit for the function y_5. It is enough to have only one LUT having $S \geq 4$ for implementing the circuit for function y_3. Because there are $N_1 = 4$ and $N_2 = 1$, it is reasonable to choose the set $Y^2 = \{y_2\}$. Let us point out that the rearrangement of states in the Karnaugh map leads to changing the SOPs for functions $D_r \in \Phi$. In turn, it can lead to increasing for the number of LUTs in the circuit of the block LUTer1 in comparison with the initial block LUTer.

Table 6.8 Table of replacement of logical conditions of FSM MPY(Γ_1)

a_m	a_1	a_2	a_3	a_4	a_5	a_6	a_7	a_8
$K(a_m)$	000	001	010	011	100	101	110	111
P_1	x_1	x_3	x_3	x_3	–	–	–	–
P_2	x_2	x_4	x_4	x_4	–	–	–	–

6.4 Synthesis of Moore FSM with Replacement of Logical Conditions

Let us consider an example of synthesis for Moore FSM $MPY(\Gamma_1)$, where the GSA Γ_1 is shown in Fig. 6.11. The following sets of logical conditions can be derived from the GSA Γ_1: $X(a_1) = \{x_1, x_2\}$, $X(a_2) = X(a_3) = X(a_4) = \{x_3, x_4\}$, and $X(a_5) = \cdots = X(a_8) = \emptyset$. Therefore, there is $G = 2$. It means that the following set $P = \{p_1, p_2\}$ should be formed. Let the states have the following codes: $K(a_1) = 000$, $K(a_2) = 001$, ..., $K(a_8) = 111$. Let us construct the table of replacement of logical conditions for the FSM MPY(Γ_1) (Table 6.8).

Using Table 6.8, it is possible to find the following functions of the system (6.4):

$$\begin{aligned} P_1 &= A_1x_1 \vee A_2x_3 \vee A_3x_3 \vee A_4x_3; \\ P_2 &= A_1x_2 \vee A_2x_4 \vee A_3x_4 \vee A_4x_4. \end{aligned} \quad (6.38)$$

As it is shown in [1], the states codes of the states $a_m \in A$ having $X(a_m) = \emptyset$ might be considered as insignificant; they might be used for minimizing functions of the system (6.4). Using this possibility, the following system of functions can be obtained:

$$\begin{aligned} P_1 &= \bar{T}_1\bar{T}_2x_1 \vee T_2x_3 \vee T_1x_3; \\ P_2 &= \bar{T}_1\bar{T}_2x_2 \vee T_2x_4 \vee T_1x_4. \end{aligned} \quad (6.39)$$

To get the system (6.5), it is necessary to construct the transformed structure table of Moore FSM [27]. To do it, the column X_h of the initial structure table should be replaced by the column P_h. The replacement rule is obvious: if a variable $x_l \in X$ is situated on the intercrossing of the column a_m and the row p_g of the table of replacement of logical conditions, then the variable p_g replaces the logical condition x_l for the structure table's part with transitions from the state a_m. In the case of Moore FSM MPY(Γ_1), the transformed structure table includes 19 rows. The fragment of the table for states a_1 and a_2 is shown in Table 6.9.

The following fragment of the system (6.4) can be derived from Table 6.9: $D_1 = \bar{T}_1\bar{T}_2T_3$; $D_2 = \bar{T}_1\bar{T}_2\bar{p}_1$; $D_3 = \bar{T}_1\bar{T}_2\bar{T}_3(p_1 \vee \bar{p}_1\bar{p}_2) \vee \bar{T}_1\bar{T}_2T_3\bar{p}_2$. These functions are minimized.

Using the classes of pseudoequivalent states, it is possible to simplify the functions P and Φ. The structural diagram of MP$_B$Y Moore FSM is shown in Fig. 6.21. In the MP$_B$Y Moore FSM, the block LUTer1 implements the system

$$P = P(\tau, X). \quad (6.40)$$

Table 6.9 Fragment of transformed structure table for Moore FSM MPY(Γ_1)

a_m	$K(a_m)$	a_s	$K(a_s)$	P_h	Φ_h	h
a_1						
(–)	000	a_2	001	p_1	D_3	1
		a_3	010	$\bar{p}_1 p_2$	D_2	2
		a_4	011	$\bar{p}_1 \bar{p}_2$	$D_2 D_3$	3
a_2						
$(y_1 y_2)$	001	a_5	100	$p_1 p_2$	D_1	4
		a_6	101	$p_1 \bar{p}_2$	$D_1 D_3$	5
		a_7	110	$\bar{p}_1 p_2$	$D_1 D_2$	6
		a_8	111	$\bar{p}_1 \bar{p}_2$	$D_1 D_2 D_3$	7

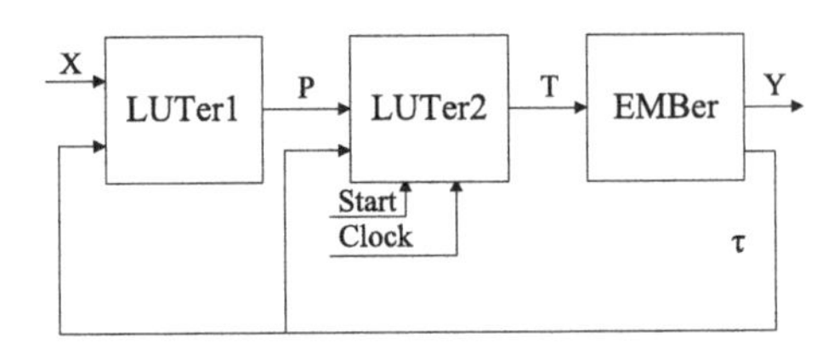

Fig. 6.21 Structural diagram of MP_BY Moore FSM

The block LUTer2 of the MP_BY Moore FSM implements the register RG and system of input memory functions represented in the following form:

$$\Phi = \Phi(\tau, P). \tag{6.41}$$

As in the case of P_BY Moore FSM, the block EMBer implements the systems $Y(T)$ and $\tau(T)$.

The proposed synthesis method for MP_BY Moore FSM includes the following steps:

1. Marking the states and constructing the set A.
2. Constructing the partition $\Pi_A = \{B_1, \ldots, B_I\}$ of the set of states by the classes of pseudoequivalent states.
3. Encoding of the states $a_m \in A$.
4. Constructing the reduced table of replacement of logical conditions.
5. Optimal encoding of the classes $B_i \in \Pi_A$
6. Constructing the reduced transformed structure table
7. Constructing the systems (6.40) and (6.41).
8. Constructing the table of EMBer.
9. Implementing the FSM logic circuit with given logic elements.

Let us consider an example of synthesis for Moore FSM MPY(Γ_1). Such elements as the set of internal states A, the partition Π_A and the classes $B_i \in \Pi_A$ are obtained before (see Sect. 6.2). Let us encode the states $a_m \in A$ in the trivial way: $K(a_1) = 000$, $K(a_2) = 001$, …, $K(a_8) = 111$.

Obviously, the transitions for all states $a_m \in B_i$ depend on the same logical conditions. This rule can be represented by the following expression:

$$a_m, a_s \in B_i \rightarrow X(a_m) = X(a_s). \tag{6.42}$$

Table 6.10 Reduced table of replacement of logical conditions for Moore FSM MPY(Γ_1)

B_i	B_1	B_2	B_3	B_4
P_1	x_1	x_3	–	–
P_2	x_2	x_4	–	–

Table 6.11 Reduced transformed structure table of Moore FSM MP$_B$Y(Γ_1)

B_i	$K(B_i)$	a_s	$K(a_s)$	P_h	Φ_h	h
B_1	11	a_2	001	p_1	D_3	1
		a_3	010	$\bar{p}_1 p_2$	D_2	2
		a_4	011	$\bar{p}_1 \bar{p}_2$	$D_2 D_3$	3
B_2	00	a_5	100	$p_1 p_2$	D_1	4
		a_6	101	$p_1 \bar{p}_2$	$D_1 D_3$	5
		a_7	110	$\bar{p}_1 p_2$	$D_1 D_2$	6
		a_8	111	$\bar{p}_1 \bar{p}_2$	$D_1 D_2 D_3$	7
B_3	01	a_2	001	1	D_3	8

This property allows replacing the states $a_m \in B_i$ in the table of replacement of logical conditions by the corresponding class $B_i \in \Pi_A$. Therefore, the resulting table of replacement includes fewer rows than the corresponding table for MPY Moore FSM. In the case of Moore FSM MPY(Γ_1), the reduced table of replacement of logical conditions is represented by Table 6.10.

The following system of equations can be derived from Table 6.10:

$$\begin{aligned} P_1 &= B_1 x_1 \vee B_2 x_3; \\ P_2 &= B_1 x_2 \vee B_2 x_4. \end{aligned} \tag{6.43}$$

Because of the equality $X(B_3) = X(B_4) = \emptyset$, both codes of classes $K(B_3)$ and $K(B_4)$ can be used for minimizing the system (6.43). Let us name the encoding of classes $B_i \in \Pi_A$ leading to minimizing the system of additional variables as optimal encoding. One of the variants of optimal encoding is a trivial one, namely: $K(B_1) = 00$, …, $K(B_4) = 11$. This variant of class encoding leads to the following system of equations:

$$\begin{aligned} P_1 &= \bar{\tau}_1 x_1 \vee \tau_1 x_3; \\ P_2 &= \bar{\tau}_1 x_2 \vee \tau_1 x_4. \end{aligned} \tag{6.44}$$

The logic circuit for any function of the system (6.44) is implemented with LUTs having $S = 3$ inputs.

Let us use Table 6.1 for constructing the reduced transformed structure table of Moore FSM MP$_B$Y(Γ_1) (Table 6.11).

Using Table 6.11, the system of input memory functions is constructed (after minimization):

$$\begin{aligned} D_1 &= \bar{\tau}_1 \tau_2; \\ D_2 &= \bar{\tau}_1 \bar{\tau}_2 \bar{p}_1 \vee \bar{\tau}_1 \tau_2 \bar{p}_1 = \bar{\tau}_1 \bar{p}_1; \\ D_3 &= \bar{\tau}_1 \bar{\tau}_2 p_1 \vee \bar{\tau}_1 \bar{p}_2 \vee p_1 \bar{p}_2. \end{aligned} \tag{6.45}$$

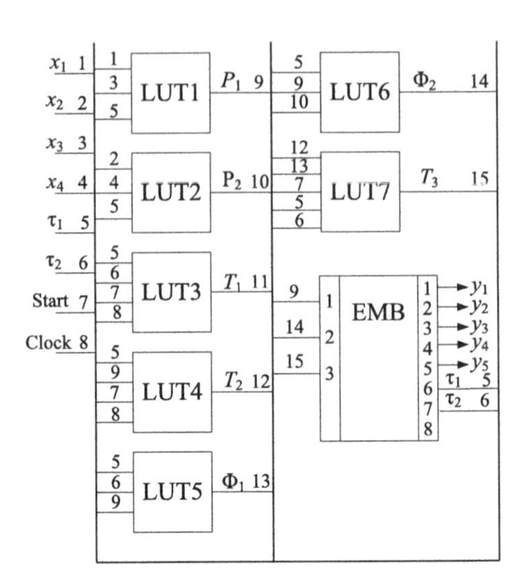

Fig. 6.22 Logic circuit of Moore FSM $MP_BY(\Gamma_1)$

Obviously, the tables for both blocks BCT and BMO are the same for FSMs $\mathrm{MP_BY}(\Gamma_1)$ and $\mathrm{P_BY}(\Gamma_1)$, as well as the circuit of the block EMBer. The logic circuit of FSM $\mathrm{MP_BY}(\Gamma_1)$ is shown in Fig. 6.22.

Two subfunctions ($\Phi_1 = \bar{\tau}_1\bar{\tau}_2 p_1$ and $\Phi_2 = \bar{\tau}_1\bar{p}_2 \vee \bar{p}_1\bar{p}_2$) are used for implementing the circuit for function D_3. Comparison of logic circuits showed in Figs. 6.12 and 6.22 shows that they include the same amount of LUTs. But the circuit of $\mathrm{MP_BY}(\Gamma_1)$ FSM has more layers. Therefore, there is no sense in the replacement of logical conditions for the case of GSA Γ_1. So, this example is shown only to demonstrate how the proposed method is applied.

Comparison of functions (6.26) and (6.45) shows that the replacement of logical conditions leads to simplifying the functions $D_r \in \Phi$. This conclusion is true for the general case [19]. In the general case, the replacement of logical conditions is reasonable for FSMs of average and large complexity having $M \geq 200$, $L \geq 50$, $G \approx 6$. But this problem should be investigated.

Some optimization methods targeting optimization of the block of replacement of logical conditions are discussed in the works [8, 37]. These methods are based on the encoding of logical conditions. It leads to introducing the block of encoding of the logical conditions (BELC). For example, the structural diagram of $\mathrm{MP_{BL}Y}$ Moore FSM is shown in Fig. 6.23. In the expression $\mathrm{MP_{BL}Y}$, the subscript L shows that the encoding of logical conditions is applied in the particular model of FSM.

In the $\mathrm{MP_{BL}Y}$ Moore FSM, the block BLC implements the functions

$$P = P(Z, X), \tag{6.46}$$

whereas the block BELC the functions

$$Z = Z(T). \tag{6.47}$$

The number of functions Z is determined by the parameter R_x

$$R_x = \lceil \log_2 (L + 1) \rceil. \tag{6.48}$$

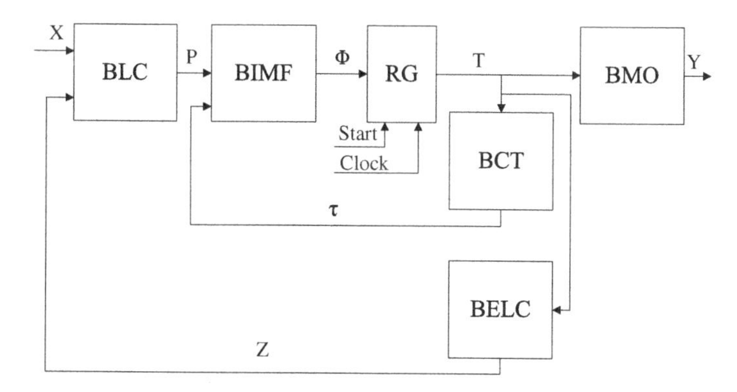

Fig. 6.23 Structural diagram of Moore $MP_{BL}Y$ Moore FSM

Let us point out that the synthesis methods for Moore FSM based on this idea are not published. We treat these methods as one of possible directions of further investigations.

References

1. Baranov SI (1994) Logic synthesis of control automata. Kluwer Academic Publishers, Boston
2. De Micheli G (1994) Synthesis and optimization of digital circuits. McGraw-Hill, New York
3. Grout I (2008) Digital systems design with FPGAs and CPLDs. Elsevier, Oxford University Press, Inc, Amsterdam
4. Jenkins J (1995) Design with FPGAs and CPLDs. Prentice Hall, New York
5. Maxfield C (2004) The design warrior's guide to FPGAs. Elsevier, Amsterdam
6. Zeidman B (2002) Designing with FPGAs and CPLDs. CMP Books, Lawrence
7. Maxfield C (2008) FPGAs: instant access. Elsevier, Oxford
8. Barkalov AA, Titarenko LA (2009) Synthesis of microprogrammed automata with customized and programmable VLSI. UNITEX, Donetsk (in Russian)
9. Baranov S, Sklyarov V (1986) Digital devices on programmable LSI with matrix structure. Radio i Swiaz, Moscow (in Russian)
10. Łuba T, Rawski M, Jachna Z (2002) Functional decomposition as a universal method for logic synthesis of digital circuits. In: Proceeding of IX international conference MIXDES'02, pp 285–290
11. Łuba T (1994) Multi-level logic synthesis based on decomposition. Microprocess Microsyst 18(8):429–437
12. Łuba T, Selvaraj H (1995) A general approach to boolean functions decomposition and its application in fpga-based synthesis. VLSI Des 3(3):289–300
13. Kania D (2004) The logic synthesis for the PAL-based complex programmable logic devices. Zeszyty naukowe Politechniki Ślaskiej, Gliwice (in Polish)
14. Kania D (2011) Efficient technology mapping method for pal-based devices. In: Adamski M, Barkalov A, Wegrzyn M (eds) Design of digital systems and devices. Springer, Berlin, pp 145–163
15. Kania D, Czerwinski R (2012) Area and speed oriented synthesis of FSMs for PAL-based CPLDs. Microprocess Microsyst 36(1):45–61

16. Kania D, Milik A (2010) Logic synthesis based on decomposition for CPLDs. Microprocess Microsyst 34(1):28–38
17. Opara A, Kania D (2010) Decomposition-based logic synthesis for PAL-based CPLDs. Int J Appl Math Comput Sci 20(2):367–384
18. Baranov S (2008) Logic and system design of digital systems. TUT Press, Tallinn
19. Barkalov A, Titarenko L (2009) Logic synthesis for FSM-based control units. Springer, Berlin
20. Barkalov A, Barkalov A (2001) Optimization of logic circuit of Moore FSM with programmable LSI. Control Syst Mach 6:38–41 (in Russian)
21. Barkalov A, Barkalov A (2002) Synthesis of control units with transformation of objects. Control Syst Mach 6:41–44 (in Russian)
22. Barkalov A, Barkalov A (2005) Design of Mealy FSMs with transformation of object codes. Int J Appl Math Comput Sci 15(1):151–158
23. Barkalov A, Titarenko L, Barkalov A (2012) Structural decomposition as a tool for the optimization of an FPGA-based implementation of a Mealy FSM. Cybern Syst Anal 48(2):313–323
24. Solovjov V, Klimowicz A (2008) Logic design of digital systems on the base of programmable logic devices. Hot line-Telecom, Moscow (in Russian)
25. Solovjov VV (2001) Design of digital systems using the programmable logic integrated circuits. Hot line-Telecom, Moscow (in Russian)
26. Palagin A, Barkalov A, Usifov S, Shvets A (1992) Synthesis of microprogrammed automata with FPLDs. IC NAC Ukraine, Preprint 92:18–26 (in Russian)
27. Borowik G (2007) Finite state machine synthesis for FPGA structure with embedded memory blocks. PhD thesis, WUT, Warszawa (in Polish)
28. Rawski H, Tomaszewicz P, Borowski G, Luba T (2011) Logic synthesis method of digital circuits designed for implementation with embedded memory blocks on FPGAs. In: Wegrzyn M, Adamski M, Barkalov A (eds) Design of digital systems and devices. Springer, Berlin, pp 121–144
29. Barkalov A, Titarenko L (2008) Logic synthesis for compositional microprogram control units. Springer, Berlin
30. Barkalov AA (2002) Synthesis of control units with PLDs. Donetsk National Technical University, Donetsk (in Russian)
31. Achasova SN (1987) Algorithms of synthesis of automata on programmable arrays. Radio i Swiaz, Moscow (in Russian)
32. Barkalov A (1998) Principles of logic optimization for a Moore microprogrammed automaton. Cybern Syst Anal 34(1):54–61
33. Yang S (1991) Logic synthesis and optimization benchmarks user guide. Technical report, Microelectronics center of North Carolina
34. Barkalov A, Titarenko L, Hebda O (2010) Matrix implementation of Moore FSM with expansion of coding space. Meas Autom Monit 56(7):694–696
35. Barkalov A, Titarenko L, Hebda O, Soldatov K (2009) Matrix implementation of Moore FSM with encoding of collections of microoperations. Radioelectron Inf 4:4–8
36. Barkalov A, Matvienko A, Tsololo S (2011) Optimization of logic circuit of Moore FSM with FPGAs. IC NAC Ukraine 10:22–29 (in Russian)
37. Barkalov A, Zelenjova I (2001) Optimization of logic circuit of control unit with replacement of variables. Control Syst Mach 1:75–78 (in Russian)

Chapter 7
Design of FSMs with Embedded Memory Blocks

Abstract Chapter deals with design of Moore FSMs based on using embedded memory blocks (EMB). The methods of trivial EMB-based implementation of logic circuits of both Moore and Mealy FSMs are discussed. In this case, only one EMB is enough for implementing the circuit. Next, the optimization methods are discussed based on replacement of logical conditions as well as encoding of the collections of microoperations. The considered methods are based on encoding the rows of FSM's structure table. All these methods lead to two-level models of Mealy FSMs and to three-level models of Moore FSMs. Next, these methods are combined together for further optimizing the hardware amount in FSM logic circuits. The last section considers applying PES-based methods in EMB-based Moore FSMs. All discussed methods are illustrated by examples. The chapter is written together with PhD Malgorzata Kolopienczyk (University of Zielona Gora, Poland).

7.1 Trivial Implementation of Mealy and Moore FSMs

The majority of FPGAs include three main blocks: look-up table (LUT) elements connected with programmable flip-flops, embedded memory blocks (EMB), and a matrix of programmable interconnections [7, 9]. One LUT together with a flip-flop forms a logic element (LE), two LEs form a slice, two slices form a configurable logic block (CLB). The fast interconnections are used inside a CLB [20], but it is a very rear situation when only one CLB is enough for implementing an FSM logic circuit. The flip-flop of LE can be bypassed, so the output of LUT can be either registered or combinational. As a rule, the number of LUT's inputs is rather small ($S \leq 6$) [1, 20]. If the number of arguments of a Boolean function exceeds the number of LUT's inputs, then more than one LUT is necessary to implement the corresponding combinational circuit. In this case, the methods of functional decomposition are used [8, 10, 13]. It leads to increasing for the number of layers of logic in a resulting circuit and to complication for interconnections. In turn, it results in increasing for the propagation time and power consumption [17, 19]. To

V. Sklyarov et al., *Synthesis and Optimization of FPGA-Based Systems*,
Lecture Notes in Electrical Engineering 294, DOI: 10.1007/978-3-319-04708-9_7,

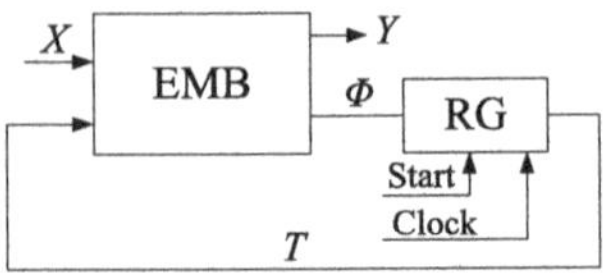

Fig. 7.1 Structural diagram of Mealy FSM U_1

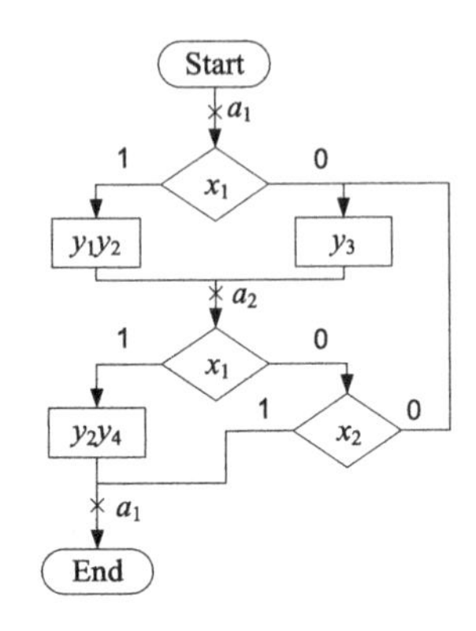

Fig. 7.2 Graph-scheme of algorithm Γ_3

improve the parameters of an FSM circuit, the embedded memory blocks should be used for implementing some its parts [5, 11, 12, 15, 18].

As it is mentioned before, the EMBs of up-to-day FPGAs have a property of configurability. It means that such parameters as the number of cells and their outputs can be changed [7, 9]. Typical configurations of modern EMBs are the following: 16 K $\times$ 1, 8 K $\times$ 2, 4 K $\times$ 4, 2 K $\times$ 8, 1 K $\times$ 18, 512 $\times$ 36 (bits) [7, 9].

So, the modern EMBs are very flexible and can be tuned to meet demands of a particular design project. Let an EMB contain V cells and t_F outputs. Let V_o be a number of cells for the number of outputs $t_F = 1$. The number V can be determined as

$$V = \lceil V_o/t_F \rceil. \tag{7.1}$$

Let us discuss a case when a single EMB is enough for implementing an FSM's logic circuit. Let the following condition take place:

$$2^{L+R}(R+N) \leq V_o. \tag{7.2}$$

In this case a Mealy FSM can be implemented in a trivial way [6] using only one EMB and R flip-flops forming the register (Fig. 7.1). Let us denote this circuit as a Mealy FSM U_1.

In FSM U_1, the EMB implements functions

$$Y = Y(X,T), \tag{7.3}$$

$$\Phi = \Phi(X,T). \tag{7.4}$$

The circuit of RG is implemented using R logic elements whose flip-flops are programmed as D flip-flops.

Let us consider an example of FSM design for a GSA Γ_3 (Fig. 7.2).

Table 7.1 Structure table of Mealy FSM $U_1(\Gamma_3)$

a_m	$K(a_m)$	a_s	$K(a_s)$	X_h	Y_h	Φ_h	h
a_1	0	a_2	1	x_1	y_1y_2	D_1	1
		a_2	1	$\bar{x}_1$	y_3	D_1	2
a_2	1	a_1	0	x_1	y_2y_4	–	3
		a_1	0	$\bar{x}_1x_2$	–	–	4
		a_2	1	$\bar{x}_1\bar{x}_2$	y_3	D_1	5

Table 7.2 Transformed structure table of Mealy FSM $U_1(\Gamma_3)$

$K(a_m)$	X	Y	Φ	v
T_1	x_1x_2	$y_1y_2y_3y_4$	D_1	
0	00	0010	1	1
0	01	0010	1	2
0	10	1100	1	3
0	11	1100	1	4
1	00	0010	1	5
1	01	0000	0	6
1	10	0101	0	7
1	11	0101	0	8

The GSA Γ_3 is marked by the states of Mealy FSM using the rules from [2]. The following sets and their parameters can be derived from the GSA Γ_3: $A = \{a_1, a_2\}$, $M = 2$, $X = \{x_1, x_2\}$, $L = 2$, $Y = \{y_1, \ldots, y_4\}$, $N = 4$, $R = 1$, $T = \{T_1\}$ and $\Phi = \{D_1\}$.

Let the symbol $U_i(\Gamma_j)$ mean that the model U_i of FSM is used for synthesis of a control unit represented by a GSA Γ_j. To apply the model $U_1(\Gamma_3)$, the following conditions should take places: $t_F \geq 5$, $S_A = 3$. Here the symbol S_A stands for the number of address inputs of EMB. The design method for FSM $U_1(\Gamma_j)$ includes all steps used for designing a Mealy FSM [2], as well as one additional step. This step is reduced to some transformation of the initial structure table.

In the case of FSM $U_1(\Gamma_3)$, the structure table includes $H_1(\Gamma_3) = 5$ rows (Table 7.1). In this table, the trivial state codes are used ($K(a_1) = 0$, $K(a_2) = 1$). To design the logic circuit, the initial structure table should be transformed.

The transformed structure table should include V_1 rows:

$$V_1 = 2^{R+L}. \tag{7.5}$$

This table includes the following columns: $K(a_m)$, X, Y, Φ, v, where v is a number of a row. In the discussed example, the transformed structure table has $V_1(\Gamma_3) = 8$ rows (Table 7.2).

In the transformed structure table, the columns $K(a_m)$ and X determine the address of a cell, whereas the columns Y and Φ determine its content. Each row h of the initial structure table corresponds to $n(h)$ cells of EMB:

$$n(h) = 2^{L-L_h}. \tag{7.6}$$

In the expression (7.6), the symbol L_h stands for the number of logical conditions from the row number h. The transitions from the state $a_m \in A$ are represented by $H(L)$ rows of the transformed structure table:

$$H(L) = 2^L. \tag{7.7}$$

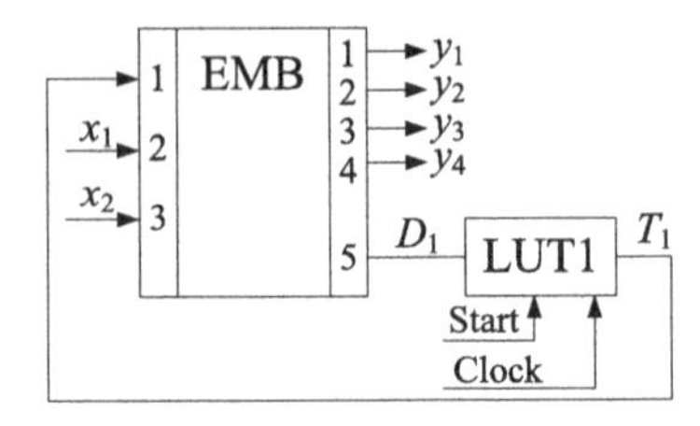

Fig. 7.3 Logic circuit of Mealy FSM $U_1(\Gamma_3)$

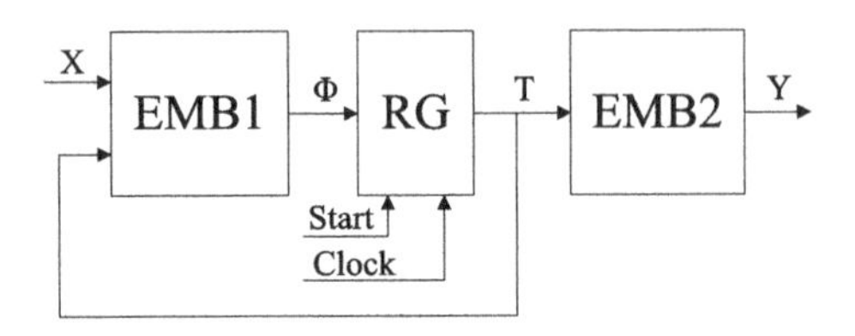

Fig. 7.4 Structural diagram of CityplaceMoore FSM U_2

Let $H(a_m)$ be the number of transitions from the state $a_m \in A$. If there is $H(a_m) < H(L)$, then the contents of some cells are identical. For example, the row 1 of Table 7.1 includes only x_1. Because of the equality $x_1 = x_1x_2 \vee x_1\bar{x}_2$, the rows 3 and 4 of Table 7.2 include the same data for columns Y and Φ. All other rows of initial structure table are transformed in this very way.

The functional circuit of FSM $U_1(\Gamma_3)$ in shown in Fig. 7.3. To stress the fact that flip-flops of particular logic elements are used, the corresponding LUTs are connected with pulses Start and Clock (Fig. 7.4).

Now, let us consider the trivial EMB-based implementation for a Moore FSM. If the condition (7.2) takes place, then the model U_1 can be used for Moore FSM. Let us do not discuss this trivial case. Let the following conditions take places:

$$2^{L+R}(R+N) > V_o; \tag{7.8}$$

$$R \cdot 2^{L+R} \leq V_o; \tag{7.9}$$

$$N \cdot 2^R \leq V_o. \tag{7.10}$$

The condition (7.8) shows that it is impossible to use the model U_1. The condition (7.9) shows that the circuit for system Φ can be implemented using a single EMB. The condition (7.10) shows that the circuit for system

$$Y = Y(T) \tag{7.11}$$

can be implemented using a single EMB. Therefore, the structural diagram for Moore FSM U_2 can be obtained (Fig. 7.5) using conditions (7.9)–(7.10).

In the model U_2, the block EMB1 implements the system of input memory functions (7.3), whereas the block EMB2 the system of microoperations (7.11). The design method for FSM U_2 includes the following steps:

1. Constructing the set of states A.
2. State assignment.

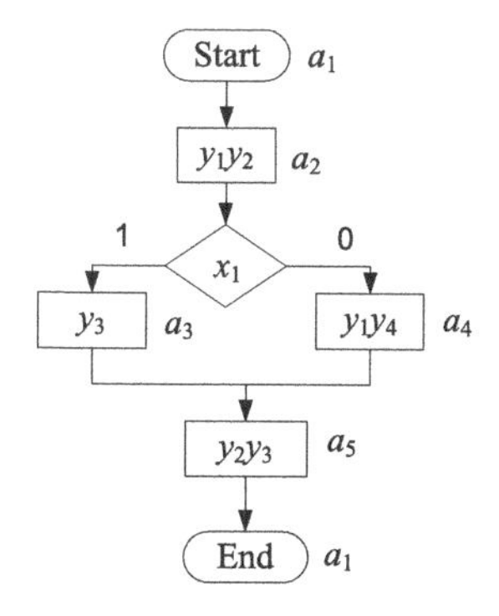

Fig. 7.5 Graph-scheme of algorithm Γ_4

Table 7.3 Structure table of Moore FSM $U_2(\Gamma_4)$

a_m	$K(a_m)$	a_s	$K(a_s)$	X_h	Φ_h	h
a_1	000	a_2	001	1	D_3	1
a_2 (y_1y_2)	001	a_3	010	x_1	D_2	2
		a_4	011	$\bar{x}_1$	D_2D_3	3
a_3 (y_3)	010	a_5	100	1	D_1	4
a_4 (y_1y_4)	011	a_5	100	1	D_1	5
a_5 (y_2y_3)	100	a_1	000	1	–	6

3. Constructing the structure table.
4. Transformation of the structure table.
5. Constructing the table of microoperations.
6. Implementing the FSM logic circuit with EMBs and LUTs of a given FPGA chip.

Let us discuss an example of design for Moore FSM $U_2(\Gamma_4)$, where the GSA Γ_4 is shown in Fig. 7.5.

The GSA Γ_4 is marked by the states of Moore FSM using the rules from [2]. The following sets and their parameters can be derived from the GSA Γ_4: $A = \{a_1, \ldots, a_5\}$, $M = 5$, $X = \{x_1\}$, $L = 1$, $Y = \{y_1, \ldots, y_4\}$, $N = 4$, $R = 3$, $T = \{T_1, T_2, T_3\}$, and $\Phi = \{D_1, D_2, D_3\}$. Let us encode the states $a_m \in A$ in the following manner: $K(a_1) = 000, \ldots, K(a_5) = 100$. Using these codes and GSA Γ_4, the structure table of FSM $U_2(\Gamma_4)$ can be constructed (Table 7.3).

This table includes $H_2(\Gamma_4) = 6$ rows. The column a_m of this table contains the current state $a_m \in A$, as well as the set of microoperations $Y(a_m) \subseteq Y$ generated in this state.

The transformed structure table of Moore FSM U_2 includes V_2 rows where $V_2 = V_1$. The transformed structure table includes the columns $K(a_m)$, X, Φ, v. In the case of Moore FSM $U_2(\Gamma_4)$ this table includes $V_2(\Gamma_4) = 16$ rows. Because the rows 11–16 contain only zeros, they are not shown in Table 7.4.

To make the connection between Tables 7.3 and 7.4 more obvious, the last includes the column h. This column shows the numbers of rows of structure table

Table 7.4 Transformed structure table of Moore FSM $U_2(\Gamma_4)$

$K(a_m)$	X	Φ	v	h
$T_1T_2T_3$	x_1	$D_1D_2D_3$		
000	0	001	1	1
000	1	001	2	1
001	0	011	3	3
001	1	010	4	2
010	0	100	5	4
010	1	100	6	4
011	0	100	7	5
011	1	100	8	5
100	0	000	9	6
100	1	000	10	6

Table 7.5 Table of microoperations of Moore FSM $U_2(\Gamma_4)$

$K(a_m)$	Y	m
$T_1T_2T_3$	$y_1y_2y_3y_4$	
000	0000	1
001	1100	1
010	0010	3
011	1001	2
100	0110	4

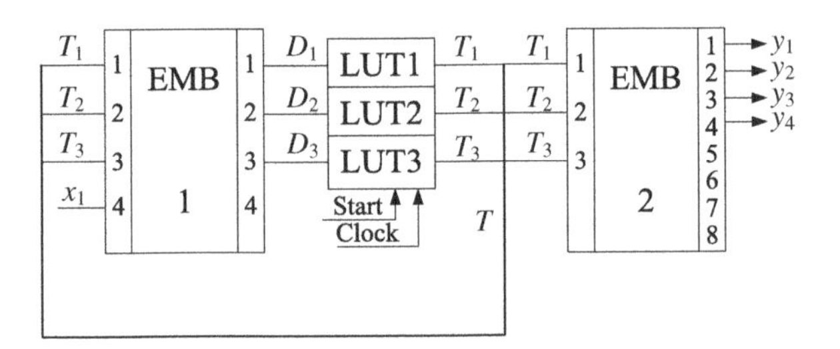

Fig. 7.6 Logic circuit of Moore FSM $U_2(\Gamma_4)$

corresponding to the rows of transformed structure table. For example, the rows 1 and 2 of Table 7.4 correspond to the row 1 of Table 7.3.

The table of microoperations contains the columns $K(a_m)$, Y, m. In the case of FSM $U_2(\Gamma_4)$ it should include 8 rows. Only five of them are shown in Table 7.5. To construct this table, data from the column a_m of the structure table are used.

Let an FPGA chip in use include EMB having configurations 16×4 and 8×8. The first of them is used for implementing the transformed structure table. Both configurations can be used for implementing the table of microoperations. Let us choose the configuration 8×8 for implementing the system Y. The logic circuit of Moore FSM $U_2(\Gamma_4)$ is shown in Fig. 7.6.

In this circuit, LUT1–LUT3 are used for implementing the register RG. Let us point out that both blocks EMB_1 and EMB_2 have unused resourses.

It is known, that conditions (7.2), (7.9) and (7.10) take places only for very simple FSMs [16]. If these conditions are violated, the different methods of structural decomposition [3, 4] should be applied for optimizing EMB-based circuits of FSMs.

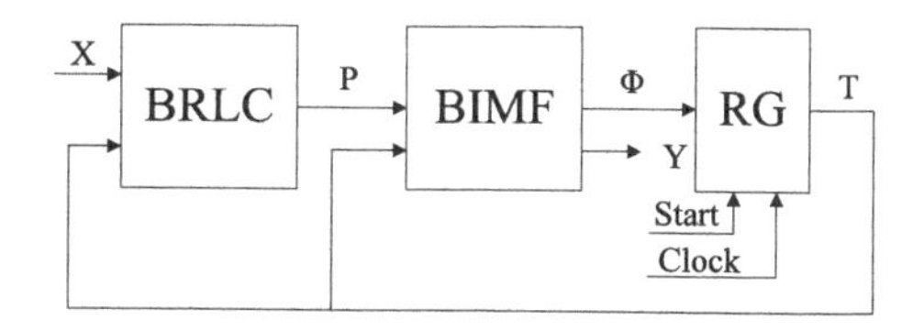

Fig. 7.7 Structural diagram of *MP* Mealy FSM

7.2 Structural Decomposition of FSMs

To diminish the numbers of LUTs in FSM logic circuits, the methods of structural decomposition can be used. The structural decomposition is resulted in increasing for the number of structural levels in an FSM circuit. There are the following methods of structural decomposition [2, 3, 14]: (1) replacement of logical conditions; (2) encoding of collections of microoperations; (3) encoding of the fields of compatible microoperations; (4) encoding of the rows of structure table. Let us discuss these methods.

Let $X(a_m)$ be a set of logical conditions determining transitions from the state $a_m \in A$, and let

$$G = \max(|X(a_1)|, \ldots, |X(a_M)|). \tag{7.12}$$

If the following condition takes place

$$G << L, \tag{7.13}$$

then the method of replacement of logical conditions [2] can be applied. Let $P = \{p_1, \ldots, p_G\}$ be a set of additional variables used for the replacement of logical conditions. To execute the replacement, a special table of replacement of logical conditions should be constructed. In this table, the columns are marked by variables $p_g \in P$, whereas the rows by states $a_m \in A$. So, the table includes G columns and M rows. If a variable $p_g \in P$ replaces a logical condition $x_l \in X$ in a state $a_m \in A$, then the symbol x_l should be written on the intersection of the row a_m and column p_g of the table. To minimize the hardware amount for a logic circuit used for the replacement, the distribution of logical conditions is executed in such a manner that each variable $x_l \in X$ is always placed in the same column of the table. Of course, such a distribution is not always possible. The following system can be derived from the table of replacement of logical conditions:

$$P = P(T, X). \tag{7.14}$$

The structural diagram of MP Mealy FSM is shown in Fig. 7.7. The symbol *M* in *MP* denotes existence of the block of replacement of logical conditions (BRLC) and the symbol *P* the block of input memory functions (BIMF).

In MP Mealy FSM, the block BRLC is implemented with LUTs. It generates the functions (7.14). The block BIMF can be implemented using either LUTs or EMBs. It implements the functions

$$\Phi = \Phi(T, P); \tag{7.15}$$

$$Y = Y(T, P). \tag{7.16}$$

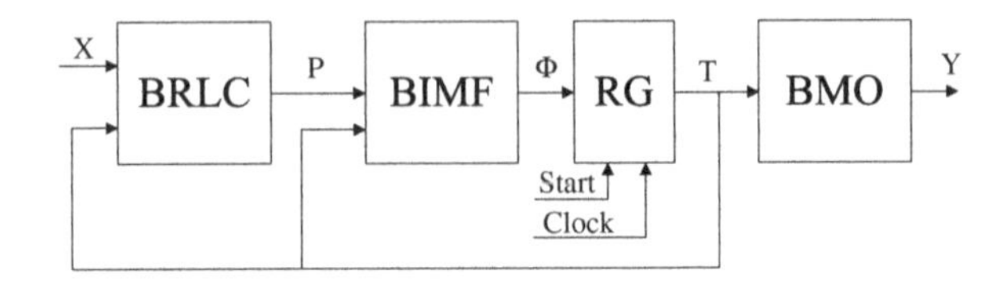

Fig. 7.8 Structural diagram of *MPY* Moore FSM

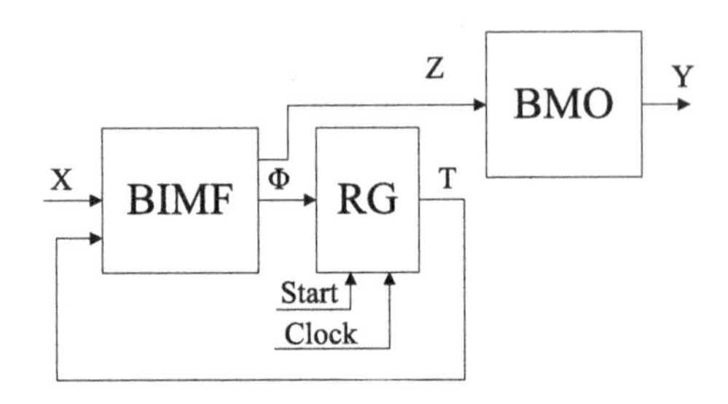

Fig. 7.9 Structural diagram of *PY* Mealy FSM

The structural diagram of *MPY* Moore FSM is shown in Fig. 7.8. The symbol Y denotes the existence of the block of microoperations (BMO).

In *MPY* Moore FSM, the block BRLC generates functions (7.14), the block BIMF implements functions (7.15) and the block BMO generates the functions (7.11). Both blocks BIMF and BMO can be implemented using either LUTs or EMBs.

The design methods of *MP* and *MPY* FSMs are discussed in the next chapter of this book.

Let T_o different collections of microoperations $Y_t \subseteq Y$ be written in operator vertices of a GSA Γ. Let us encode each collection Y_t by a binary code $K(Y_t)$ having R_Y bits, where

$$R_Y = \lceil \log_2 (T_o + 1) \rceil. \tag{7.17}$$

The value 1 is added to T_o to take into account the empty collection $y_o = \emptyset$. Let us use the variables $z_r \in Z$ for encoding of the collections $Y_t \subseteq Y$, where $|Z| = R_Y$.

Application of this approach leads to the model of *PY* Mealy FSM shown in Fig. 7.9. The symbol Y denotes the block of microoperations BMO.

In *PY* Mealy FSM, the block BIMF implements the system (7.3) and system of functions

$$Z = Z(T, X). \tag{7.18}$$

The block BMO implements functions

$$Y = Y(Z). \tag{7.19}$$

The circuit of BMO can be implemented using either LUTs or EMBs.

In the case of Moore FSM, this method is not used. It is connected with the fact that output functions of Moore FSM depend only on its inputs. This method can be applied together with the method of object transformation [3], but we do not discuss this approach in our book.

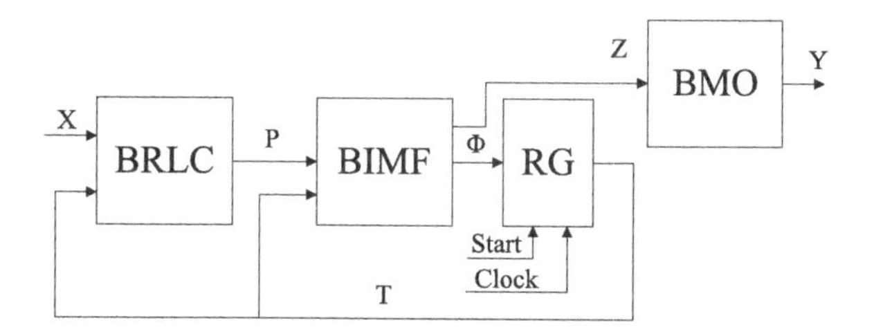

Fig. 7.10 Structural diagram of *MPY* Mealy FSM

The joint application of the replacement of logical conditions and encoding of collections of microoperations leads to the *MPY* Mealy FSM (Fig. 7.10).

In *MPY* Mealy FSM, the block BIMF generates functions (7.15) and functions

$$Z = Z(T, P). \tag{7.20}$$

Let us name microoperations y_i, $y_j \in Y$ compatible, if they do not written in the same operator vertices of the initial GSA [3]. Let the set of microoperations Y be divided by the classes of compatible microoperations and represented as

$$Y = Y^1 \cup Y^2 \cup \cdots \cup Y^K. \tag{7.21}$$

In (7.21), Y^k is a class number k of compatible microoperations ($k = \overline{1, K}$). Let it be $|Y^k| = N_k$. Let us encode the microoperations $y_n \in Y^k$ by binary codes $K(y_n)$ having R_k bits, where

$$R_k = \lceil \log_2 (N_k + 1) \rceil. \tag{7.22}$$

The value 1 is added to N_k to take into account the fact that no microoperation $y_n \in Y^k$ belongs to any collection of microoperations $Y_t \subseteq Y$. To encode all microoperations, it is enough R_D variables $z_r \in Z$, where

$$R_D = R_1 + R_2 + \cdots + R_K. \tag{7.23}$$

The set Z can be represented as $Z = Z^1 \cup Z^2 \cup \ldots \cup Z^k$; the variables $z_r \in Z^k$ are used for encoding compatible microoperations $y_n \in Y^k$. After encoding the microoperations, the system Y is represented as the following collection of subsystems:

$$\begin{aligned} Y^1 &= Y(Z^1); \\ &\vdots \qquad ; \\ Y^K &= Y(Z^K). \end{aligned} \tag{7.24}$$

Microoperations $y_n \in Y^k$ are generated by a decoder DC_k having R_k inputs and N_k outputs ($k = \overline{1, K}$).

The totality of decoders forms a block BD. Application of this method leads to *PD* Mealy FSM shown in Fig. 7.11.

In *PD* Mealy FSM, the block BIMF is implemented using either LUTs or EMBs. It generates functions (7.3) and (7.20). The block BD is implemented using LUTs; it implements the system (7.24).

Of course, this method can be applied together with the replacement of logical conditions. It leads to *MPD* Mealy FSM. If the block BMO (Fig. 7.10) is replaced

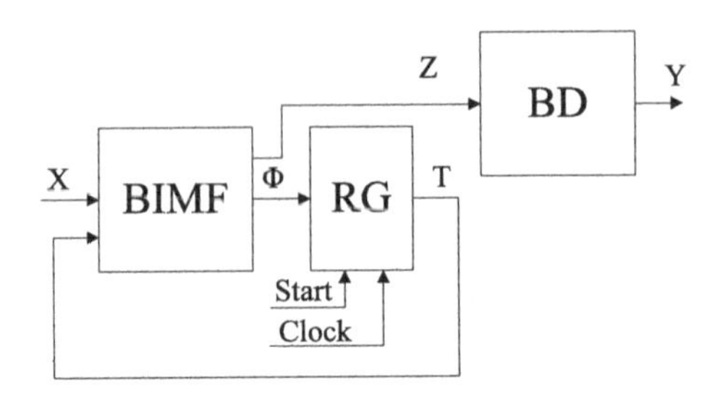

Fig. 7.11 Structural diagram of *PD* Mealy FSM

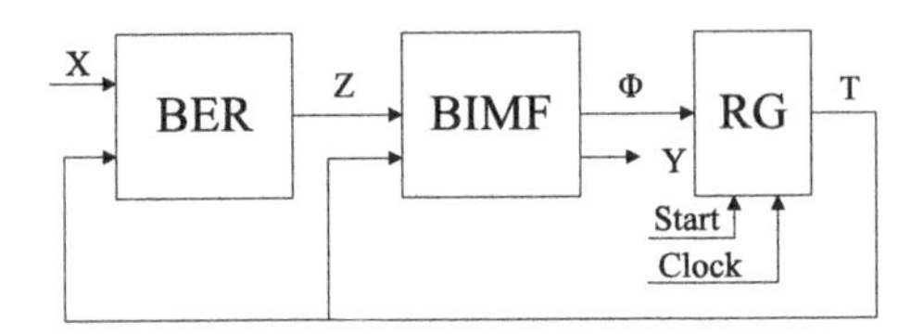

Fig. 7.12 Structural diagram of *PH* Mealy FSM

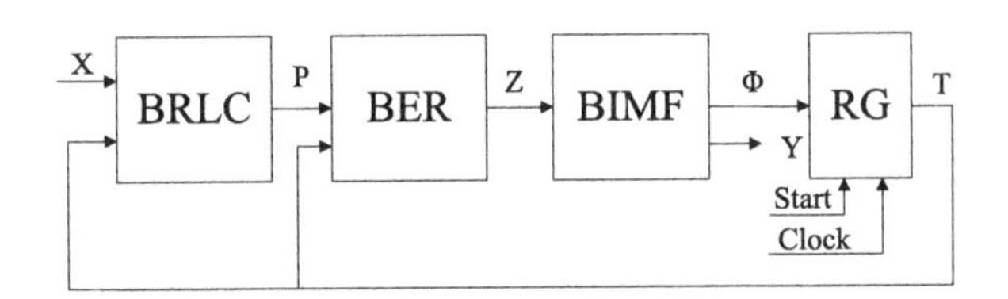

Fig. 7.13 Structural diagram of *MPH* Mealy FSM

by the block BD, then *MPY* Mealy FSM is transformed into *MPD* Mealy FSM. As in the previous case, this approach is not directly used in Moore FSM.

Let us encode each row F_h of a Mealy FSM structure table by binary codes $K(F_h)$ having R_H bits, where

$$R_H = \lceil \log_2 H \rceil. \tag{7.25}$$

Let us use the variables $z_r \in Z$ for this encoding, where there is $|Z| = R_H$. It leads to *PH* Mealy FSM (Fig. 7.12).

In *PH* Mealy FSM, the block of encoding of rows (BER) implements functions (7.18); the block BIMF implements the functions (7.19) and the input memory functions represented as the following system:

$$\Phi = \Phi(Z). \tag{7.26}$$

In *MPH* Mealy FSM, the block BRLC implements the system (7.14), the block BER the system (7.20), the block BIMF the systems (7.19) and (7.26) (Fig. 7.13). The same approach can be used in the case of Moore FSM. Let us discuss design methods and examples for some of models discussed in Sect. 7.2.

If EMBs are used for implementing FSM circuit, only models *PY* and *PH* can be used for Mealy FSM, whereas only the model *PH* can be used for Moore FSM.

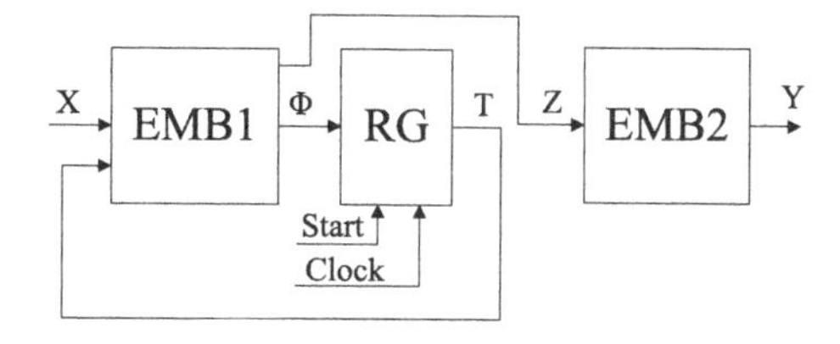

Fig. 7.14 Structural diagram of EMB-based *PY* Mealy FSM

7.3 Design of Mealy FSM with Encoding of the Collections of Microoperations

The structural diagram of EMB-based *PY* Mealy FSM is shown in Fig. 7.14.

To stress the fact that both blocks BIMF and BMO are implemented with EMBs, let us denote the FSM (Fig. 7.14) as PY_m Mealy FSM. In PY_m FSM, the block EMB1 implements functions (7.3) and (7.18); the block EMB2 implements functions (7.19). The design method of PY_m Mealy FSM includes the following steps:

1. Constructing the set of states A.
2. State assignment.
3. Constructing the structure table of Mealy FSM.
4. Encoding of collections of microoperations.
5. Constructing the transformed structure table.
6. Constructing the table of BIMF.
7. Constructing the table of BMO.
8. Implementing the FSM logic circuit with EMBs and LUTs.

The model PY_m can be applied if the following conditions take places:

$$2^{L+R}(R + R_Y) \leq V_o; \tag{7.27}$$

$$N \cdot 2^{R_Y} \leq V_o. \tag{7.28}$$

Let us discuss an example of design for $PY_m(\Gamma_5)$ where the GSA Γ_5 is shown in Fig. 7.15. This GSA is marked by states of Mealy FSM using the rules from [2].

The following sets and their parameters can be found for Mealy FSM $U_1(\Gamma_5)$: $A = \{a_1, \ldots, a_5\}$, $M = 5$, $X = \{x_1, x_2, x_3\}$, $L = 3$, $Y = \{y_1, \ldots, y_7\}$, $N = 7$, $R = 3$, $T = \{T_1, T_2, T_3\}$, $\Phi = \{D_1, D_2, D_3\}$. Let us encode the states $a_m \in A$ in the trivial way: $K(a_1) = 000, \ldots, K(a_5) = 100$.

Let the FPGA chip in use have $V_o = 384$(bits) and let the following configurations of EMB exist: 64×6 and 32×12 (bits). For the FSM $U_1(\Gamma_5)$, there is $2^{L+R}(R + N) = 640$ bits. So, the model $U_1(\Gamma_5)$ cannot be used.

There are $T_o = 5$ collections of microoperations in the operator vertices of GSA Γ_5: $Y_1 = \{y_1, y_2, y_3\}$, $Y_2 = \{y_1, y_4\}$, $Y_3 = \{y_2, y_5\}$, $Y_5 = \{y_3, y_6, y_7\}$. Also, there are no microoperations generated during the transition from a_4 into a_1. Therefore, there is the collection $y_0 = \emptyset$ in the discussed case. Its existence should be taken

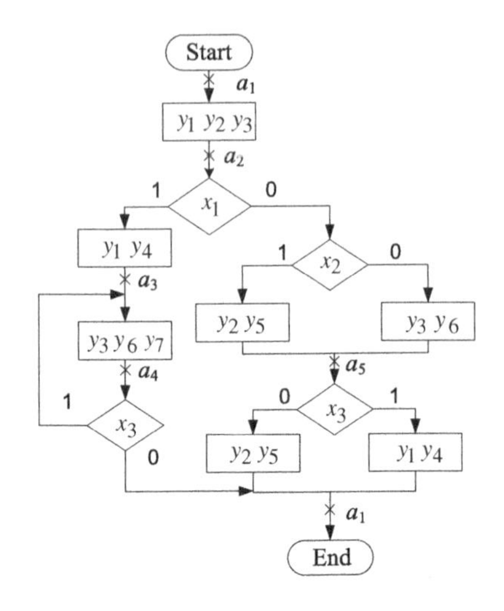

Fig. 7.15 Initial GSA Γ_5

Table 7.6 Structure table of Mealy FSM $U_1(\Gamma_5)$

a_m	$K(a_m)$	a_s	$K(a_s)$	X_h	Y_h	Φ_h	h
a_1	000	a_2	001	1	$y_1y_2y_3$	D_3	1
a_2	001	a_3	010	x_1	y_1y_4	D_2	2
		a_5	100	$\bar{x}_1x_2$	y_2y_5	D_1	3
		a_5	100	$\bar{x}_1\bar{x}_2$	y_3y_6	D_1	4
a_3	010	a_4	011	1	$y_3y_6y_7$	D_2D_3	5
a_4	011	a_4	011	x_3	$y_3y_6y_7$	D_2D_3	6
		a_1	000	$\bar{x}_3$	–	–	7
a_5	100	a_1	000	x_3	y_2y_5	–	8
		a_1	000	$\bar{x}_3$	y_1y_4	–	9

into account for finding the number of additional variables. Using (7.17), it can be found that there are $R_Y = 3$ and $Z = \{z_1, z_2, z_3\}$.

Let us check the conditions (7.27)–(7.28). In the case of GSA Γ_5, the expression (7.27) is the following: $384 = 384$. The expression (7.28) produces the inequality $56 < 384$. So, both conditions take places and the model $PY_m(\Gamma_5)$ can be used.

The structure table of Mealy FSM $U_1(\Gamma_5)$ includes $H_1(\Gamma_5) = 9$ rows (Table 7.6).

Because EMBs are used for implementing all combinational parts of the FSM logic circuit, the collections of microoperations can be encoded in the arbitrary manner. Let us encode them in the following manner: $K(Y_0) = 000$, $K(Y_1) = 001, \ldots, K(Y_5) = 101$.

To construct the transformed structure table, the column Y_h of initial table should be replaced by columns Y_t and $K(Y_t)$. Obviously, the number of rows in the transformed structure table is the same as for the initial structure table of Mealy FSM. The transformed structure table of Mealy FSM $PY_m(\Gamma_5)$ is represented by Table 7.7. The transformed structure table is used for constructing the table of BIMF. In the case of PY_m FSM, this table includes the following columns: $K(a_m)$, X, Z, Φ, v. The columns $K(a_m)$ and X form an address of the cell of EMB. The columns Z and Φ determine contents of the cells. In the case of Mealy FSM $PY_m(\Gamma_5)$, there is $L = 3$ and, therefore, $H(L) = 8$. So, the table of BIMF includes 40 rows

Table 7.7 Transformed structure table of Mealy FSM $PY_m(\Gamma_5)$

a_m	$K(a_m)$	a_s	$K(a_s)$	X_h	Y_t	$K(Y_t)$	Φ_h	h
a_1	000	a_2	001	1	Y_1	001	D_3	1
a_2	001	a_3	010	x_1	Y_2	010	D_2	2
		a_5	100	$\bar{x}_1 x_2$	Y_3	011	D_1	3
		a_5	100	$\bar{x}_1\bar{x}_2$	Y_4	100	D_1	4
a_3	010	a_4	011	1	Y_5	101	D_2D_3	5
a_4	011	a_4	011	x_3	Y_5	101	D_2D_3	6
		a_1	000	$\bar{x}_3$	Y_0	000	–	7
a_5	100	a_1	000	x_3	Y_3	011	–	8
		a_1	000	$\bar{x}_3$	Y_2	010	–	9

Table 7.8 The part of table of BIMF (for state a_2)

$K(a_m)$	X	Z	Φ	v	h
$T_1T_2T_3$	$x_1x_2x_3$	$z_1z_2z_3$	$D_1D_2D_3$		
001	000	100	100	9	4
001	001	100	100	10	4
001	010	011	100	11	3
001	011	011	100	12	3
001	100	010	010	13	2
001	101	010	010	14	2
001	110	010	010	15	2
001	111	010	010	16	2

Table 7.9 Table of BMO for Mealy FSM $PY_m(\Gamma_5)$

$K(Y_t)$	Y_t	v
$z_1z_2z_3$	$y_1y_2y_3y_4y_5y_6y_7$	
000	0 0 0 0 0 0 0	1
001	1 1 1 0 0 0 0	3
010	1 0 0 1 0 0 0	2
011	0 1 0 0 1 0 0	4
100	0 0 1 0 0 1 0	5
101	0 0 1 0 0 1 1	6
110	0 0 0 0 0 0 0	7
111	0 0 0 0 0 0 0	8

where there is some useful data. Let us point out, that there is $V_1(\Gamma_5) = 64$. So, 24 rows of the table include only zeros. The part of the table of BIMF is represented by Table 7.8.

Table 7.8 represents transitions from the state $a_2 \in A$. The column h is added to show the connection between the rows of Tables 7.7 and 7.8.

The table of BMO includes the following columns: $K(Y_t)$, Y_t, v. It is constructed in a trivial way. This table includes Z_o rows where

$$Z_o = 2^{R_Y}. \tag{7.29}$$

In the case of Mealy FSM $PY_m(\Gamma_5)$, this table contains $Z_o = 8$ rows (Table 7.9).

The logic circuit of Mealy FSM $PY_m(\Gamma_5)$ is shown in Fig. 7.16.

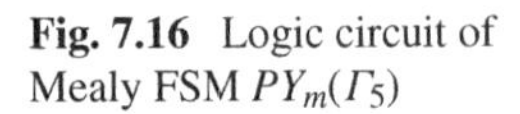

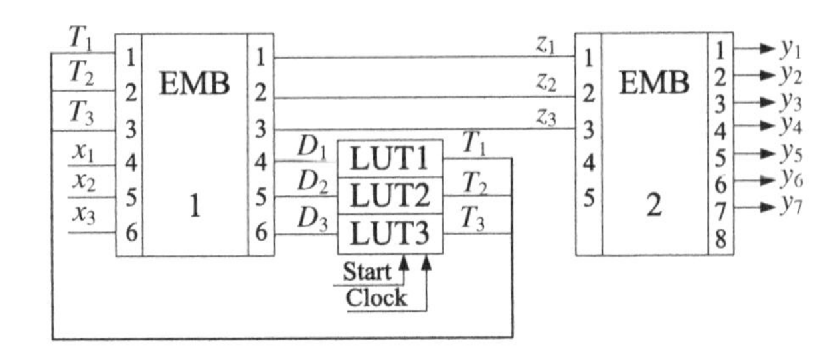

Fig. 7.16 Logic circuit of Mealy FSM $PY_m(\Gamma_5)$

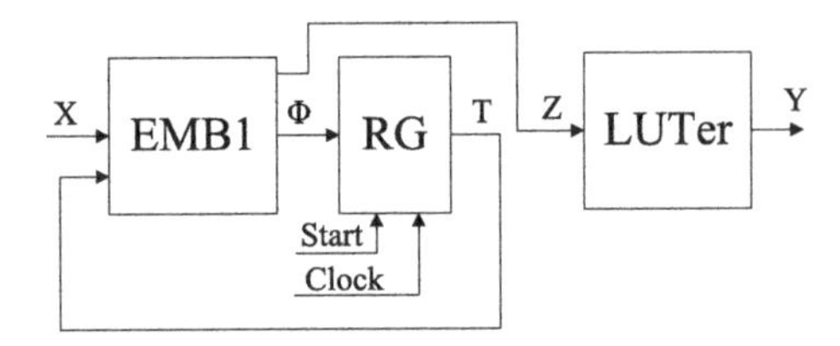

Fig. 7.17 Structural diagram of EMB-based *PD* Mealy FSM

As you can see from Fig. 7.16, the circuit of BIMF is implemented using EMB with the configuration 64×6 bits, whereas the circuit of BMO is based on EMB with configuration 32×8 bits.

7.4 Design of Mealy FSM with Encoding of the Fields of Compatible Microoperations

The structural diagram of EMB-based *PD* Mealy FSM is shown in Fig. 7.17.

In this model, the block EMB implements functions (7.3) and (7.20). The block LUTer implements functions (7.24). The circuits of both register RG and LUTer are implemented using LUTs. The design method of *PD* Mealy FSM includes the same steps as the method for PY_m Mealy FSM. The only difference is for the step 4.

In the case of *PD* Mealy FSM, the step 4 is connected with finding and encoding of the fields of compatible microoperations. The model of *PD* Mealy FSM can be applied if the following condition takes place:

$$2^{L+R}(R+R_D) \leq V_o. \tag{7.30}$$

Let us discuss an example of design for Mealy FSM $PD(\Gamma_5)$. The GSA Γ_5 is shown in Fig. 7.15. Let us point out that all sets and their parameters are already found is Sect. 7.3. Let it be $K(a_1) = 000, \ldots, K(a_5) = 100$.

Let us find the partition $\Pi_Y = \{Y^1, \ldots, Y^K\}$ for the set Y. In the case of FSM $PD(\Gamma_5)$, the following classes of compatible microoperations can be formed: $Y^1 = \{y_1, y_5, y_6\}$, $Y^2 = \{y_2, y_4, y_7\}$ and $Y^3 = \{y_3\}$. Therefore, $K = 3$, $N_1 = N_2 = 3$ and $N_3 = 1$. Using (7.22) and (7.23), the following values can be found:

Table 7.10 Transformed structure table of Mealy FSM $PD(\Gamma_5)$

a_m	$K(a_m)$	a_s	$K(a_s)$	X_h	Y_t	$K(Y_t)$	Φ_h	h
a_1	000	a_2	001	1	Y_1	01011	D_3	1
a_2	001	a_3	010	x_1	Y_2	01100	D_2	2
		a_5	100	$\bar{x}_1x_2$	Y_3	10010	D_1	3
		a_5	100	$\bar{x}_1\bar{x}_2$	Y_4	11001	D_1	4
a_3	010	a_4	011	1	Y_5	11111	D_2D_3	5
a_4	011	a_4	011	x_3	Y_5	11111	D_2D_3	6
		a_1	000	$\bar{x}_3$	Y_0	00000	–	7
a_5	100	a_1	000	x_3	Y_3	10010	–	8
		a_1	000	$\bar{x}_3$	Y_2	01100	–	9

$R_1 = R_2 = 2$, $R_3 = 1$ and $R_D = 5$. It means that $Z^1 = \{z_1, z_2\}$, $Z^2 = \{z_3, z_4\}$ and $Z^3 = \{z_5\}$.

Let the FPGA chip in use have $V_o = 512$ (bits) and let the configuration 64×8 exist. It was found that $V_o \geq 640$ is required for implementing the logic circuit of Mealy FSM $U_1(\Gamma_5)$. So, the model $U_1(\Gamma_5)$ cannot be used. The expression (7.30) produces the following equality: $512 = 512$. Therefore, the model $PD(\Gamma_5)$ can be used.

Let us encode the microoperations $y_n \in Y^k$ in the following way: $K(y_1) = K(y_2) = 01$, $K(y_4) = K(y_5) = 10$, $K(y_6) = K(y_7) = 11$ and $K(y_3) = 1$.

The structure table of Mealy FSM $U_1(\Gamma_5)$ is represented by Table 7.6. The transformed structure table of Mealy FSM $PD(\Gamma_5)$ has the same columns as Table 7.7. For the discussed example, it is represented by Table 7.10.

Let us explain how the column $K(Y_t)$ is filled. The code $K(Y_t)$ of a collection of microoperations $Y_t \subseteq Y$ can be represented as a concatenation of codes $K(y_n)^k$ $(k = \overline{1, K})$, where $y_n \in Y_t$:

$$Y_t = K(y_n)^1 * K(y_n)^2 * \cdots * K(y_n)^K. \tag{7.31}$$

In (7.31), the sign * denotes the concatenation.

For example, there is $Y_1 = \{y_1, y_2, y_3\}$. As we know, there are $K(y_1) = 01$, $K(y_2) = 01$ and $K(y_3) = 1$. Therefore, the first row of Table 7.10 should include the code 01011 in the column $K(Y_t)$. The collection $Y_4 = \{y_3, y_6\}$ and $y_3 \in Y^3$, $y_6 \in Y^1$. So, this collection does not include microoperations $y_n \in Y^2$. It means that the code $K(\emptyset)^2 = 00$ should be used. It gives the code $K(Y_4) = 11001$. All other codes $K(Y_t)$ are formed in this very manner.

The table of BIMF for *PD* FSM is constructed on the base of transformed structure table. In the case of FSM $PD(\Gamma_5)$, this table includes $V_1(\Gamma_5) = 64$ rows. The part of this table is represented by Table 7.11. This table describes transitions from the state $a_2 \in A$.

In the case of *PD* FSM, there is no need in table of BMO. The system (7.24) can be derived from the table with codes of microoperations. In the discussed example, the following system can be derived from the codes of microoperations: $y_1 = \bar{z}_1z_2$, $y_2 = \bar{z}_3z_4$, $y_3 = z_5$, $y_4 = z_1\bar{z}_2$, $y_6 = z_1z_2$, $y_7 = z_3z_4$.

The logic circuit of FSM $PD(\Gamma_5)$ is shown in Fig. 7.18.

Table 7.11 The part of table of BIMF for FSM $PD(\Gamma_5)$

$K(a_m)$	X	Z	Φ	v	h
$T_1T_2T_3$	$x_1x_2x_3$	$z_1z_2z_3z_4z_5$	$D_1D_2D_3$		
001	000	11001	100	9	4
001	001	11001	100	10	4
001	010	10010	100	11	3
001	011	10010	100	12	3
001	100	01100	010	13	2
001	101	01100	010	14	2
001	110	01100	010	15	2
001	111	01100	010	16	2

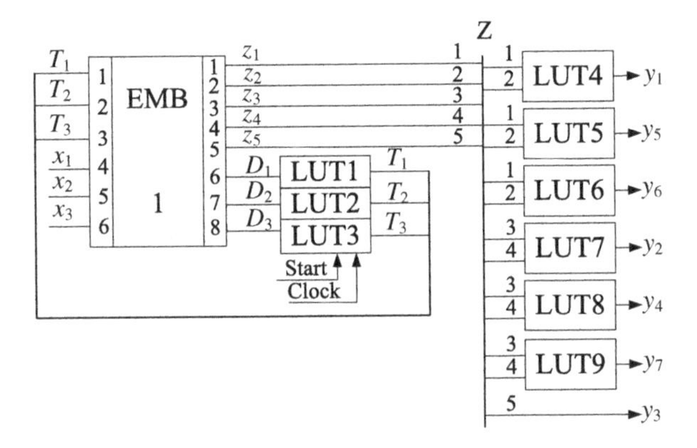

Fig. 7.18 Logical circuit of Mealy FSM $PD(\Gamma_5)$

The circuit of BIMF is implemented using a single EMB having 64×8 bits. Three logic elements are used for implementing the circuit of RG. At last, six LUTs are used to implement the circuit of BMO. Obviously, the equation $y_3 = z_5$ is implemented without LUTs.

7.5 Design of Mealy FSM with Encoding of the Rows of Structure Table

The structural diagram of EMB-based *PH* Mealy FSM is shown in Fig. 7.19.

In this model, the block EMB1 implements functions (7.18), the block EMB2 functions (7.19) and (7.26). The design method of *PH* FSM includes the following steps:

1. Constructing the set of states A.
2. State assignment.
3. Constructing the structure table of Mealy FSM.
4. Encoding of the rows of structure table.
5. Constructing the transformed structure table.

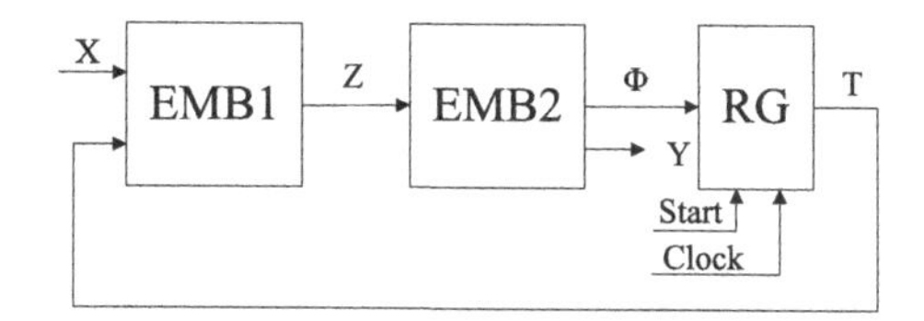

Fig. 7.19 Structural diagram of EMB-based *PH* Mealy FSM

Table 7.12 Transformed structure table of Mealy FSM $PH(\Gamma_5)$

a_m	$K(a_m)$	a_s	$K(a_s)$	X_h	$K(F_h)$	h
a_1	000	a_2	001	1	0000	1
a_2	001	a_3	010	x_1	0001	2
		a_5	100	$\bar{x}_1 x_2$	0010	3
		a_5	100	$\bar{x}_1 \bar{x}_2$	0011	4
a_3	010	a_4	011	1	0100	5
a_4	011	a_4	011	x_3	0101	6
		a_1	000	$\bar{x}_3$	0110	7
a_5	100	a_1	000	x_3	0111	8
		a_1	000	$\bar{x}_3$	1000	9

6. Constructing the table of BER.
7. Constructing the table of BIMF.
8. Implementing the logic circuit of FSM using EMBs and LUTs.

The model of *PH* FSM is applied if the following conditions take places:

$$2^{L+R} \cdot R_H \leq V_o; \tag{7.32}$$

$$(N + R) \cdot 2^{R_H} \leq V_o. \tag{7.33}$$

Let us discuss an example of design for Mealy FSM $PH(\Gamma_5)$. As in the previous examples, there are the following sets and their parameters: $A = \{a_1, \ldots, a_5\}$, $M = 5$, $X = \{x_1, x_2, x_3\}$, $L = 3$, $Y = \{y_1, \ldots, y_7\}$, $N = 7$, $R = 3$, $T = \{T_1, T_2, T_3\}$ and $\Phi = \{D_1, D_2, D_3\}$. Let us encode the states $a_m \in A$ in the trivial way: $K(a_1) = 000, \ldots, K(a_5) = 100$.

Let the FPGA chip in use have $V_o = 256$ bits with configurations 256×1, 128×2, 64×4, 32×8 and 16×16 (bits). Because 640 bits are necessary for implementing the logic circuit of $U_1(\Gamma_5)$, this model cannot be used.

Because there is $H_1(\Gamma_5) = 9$, then $R_H = 4$ and $Z = \{z_1, \ldots, z_4\}$. For the FSM $PH(\Gamma_5)$, the relations (7.32)–(7.33) are the following: $64 \times 4 = 256$ and $10 \times 16 < 256$. Therefore, the model $PH(\Gamma_5)$ can be used.

The structure table of the FSM $U_1(\Gamma_5)$ is represented by Table 7.6. As it was pointed out, the set of rows $F = \{F_1, \ldots, F_9\}$. Let us encode the rows $F_h \in F$ in the trivial way: $K(F_1) = 0000$, $K(F_2) = 0001, \ldots, K(F_9) = 1000$.

To construct the transformed structure table, it is enough to replace the columns Y_h and Φ_h of the initial structure table by the column $K(F_h)$. Obviously, this column contains a code for corresponding row. The transformed structure table of the Mealy FSM $PH(\Gamma_5)$ is represented by Table 7.12.

Table 7.13 The part of table of BER for FSM $PH(\Gamma_5)$

$K(a_m)$	X	$K(F_h)$	v	h
$T_1T_2T_3$	$x_1x_2x_3$	$z_1z_2z_3z_4$		
001	000	0011	9	4
001	001	0011	10	4
001	010	0010	11	3
001	011	0010	12	3
001	100	0001	13	2
001	101	0001	14	2
001	110	0001	15	2
001	111	0001	16	2

Table 7.14 Table of BIMF of Mealy FSM $PH(\Gamma_5)$

$K(F_h)$	Φ	Y	h
$z_1z_2z_3z_4$	$D_1D_2D_3$	$y_1y_2y_3y_4y_5y_6y_7$	
0000	001	1 1 1 0 0 0 0	1
0001	010	1 0 0 1 0 0 0	2
0010	100	0 1 0 0 1 0 0	3
0011	100	0 0 1 0 0 1 0	4
0100	011	0 0 1 0 0 1 1	5
0101	011	0 0 1 0 0 1 1	6
0110	000	0 0 0 0 0 0 0	7
0111	000	0 1 0 0 1 0 0	8
1000	000	1 0 0 1 0 0 0	9

The transformed structure table is used for constructing the table of BER. In the case of *PH* FSM, this table includes the following columns: $K(a_m)$, X, $K(F_h)$, v. For the Mealy FSM $PH(\Gamma_5)$ this table has $V_1(\Gamma_5) = 64$ rows. The part of this table (for the state $a_2 \in A$) is represented by Table 7.13.

The table of BIMF includes the following columns: $K(F_h)$, Φ, Y, h. The first column contains the address of the cell of EMB2. The contents of cell are determined by columns Φ and Y. This table is filled in the trivial way. The contents of columns Φ and Y are taken from the structure table. In the case of FSM $PH(\Gamma_5)$, this table includes $H_1(\Gamma_5) = 9$ rows (Table 7.14).

In this circuit, the block BER is implemented by EMB1 and the block BIMF is implemented by EMB2. The content of EMB1(EMB2) is taken from Table 7.13 (Table 7.14). The logical circuit of Mealy FSM $PH(\Gamma_5)$ is shown in Fig. 7.20.

Let the condition (7.23) be violated for some GSA Γ_j and FPGA chip in use. Let the following conditions are satisfied:

$$(R + R_Y) \cdot 2^{R_H} \leq V_o; \tag{7.34}$$

$$N \cdot 2^{R_Y} \leq V_o. \tag{7.35}$$

In this case, the model of *PHY* Mealy FSM (Fig. 7.21) can be used. All blocks of *PHY* Mealy FSM are implemented using EMBs. The block BER implements the system $Z(T, X)$, the block BIMF implements the system $\Phi(Z)$ and

$$Z^1 = Z^1(Z). \tag{7.36}$$

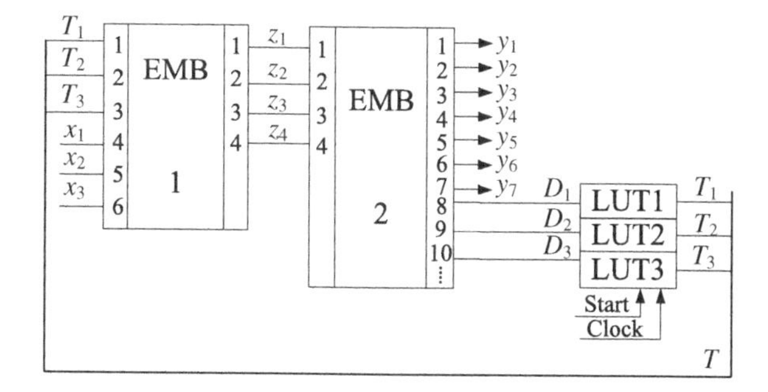

Fig. 7.20 Logic circuit of Mealy FSM $PH(\Gamma_5)$

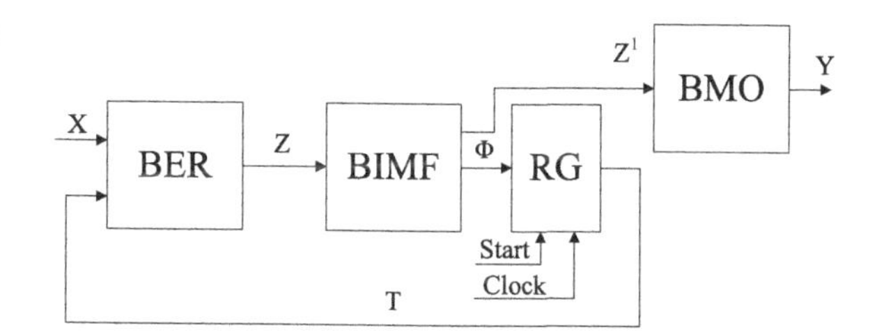

Fig. 7.21 Structural diagram of *PHY* Mealy FSM

The variables $z_r \in Z^1$ are used for encoding of collections of microoperations $Y_t \subseteq Y$.The number of these variables (R_Y) is determined by (7.17). The block BMO implements the following system

$$Y = Y\left(Z^1\right). \tag{7.37}$$

Let us point out that the system (7.37) is the same as the system (7.19).

The design methods of *PHY* Mealy FSM includes the following steps:

1. Constructing the set of states A.
2. State assignment.
3. Constructing the structure table of Mealy FSM.
4. Encoding of the rows of structure table.
5. Constructing the transformed structure table.
6. Constructing the table of BER.
7. Encoding of the collection of microoperations.
8. Constructing the table of BIMF.
9. Constructing the table of BMO.
10. Implementing the logic circuit of FSM with a particular FPGA chip.

Let us discuss an example of design for Mealy FSM $PHY(\Gamma_5)$. Let us point out that sets A, X, Y, T and Φ are already found, as well as their parameters. Let us encode the states $a_m \in A$ in the trivial way: $K(a_1) = 000, \ldots, K(a_5) = 100$.

The structure table of the FSM $U_1(\Gamma_5)$ is represented by Table 7.6. Let us use the same codes $K(F_h)$ as for the case of $PH(\Gamma_5)$. It allows constructing the transformed structure table of Mealy FSM $PHY(\Gamma_5)$ which is the same as Table 7.12.

Table 7.15 Table of BIMF of Mealy FSM $PH(\Gamma_5)$

$K(F_h)$	Φ	$K(Y_t)$	h
$z_1z_2z_3z_4$	$D_1D_2D_3$	$z_5z_6z_7$	
0000	001	001	1
0001	010	010	2
0010	100	011	3
0011	100	100	4
0100	011	101	5
0101	011	101	6
0110	000	000	7
0111	000	011	8
1000	000	010	9

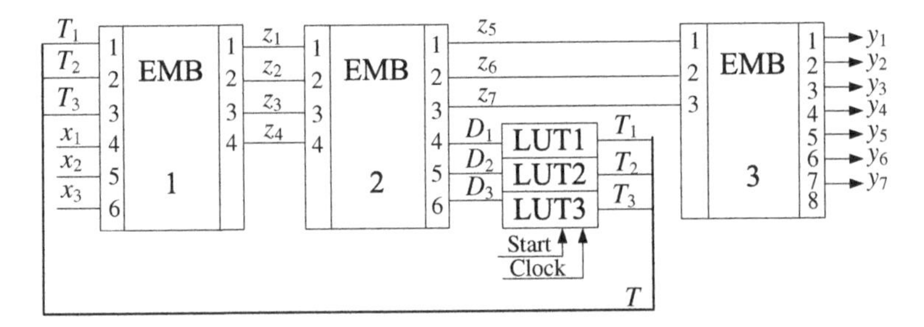

Fig. 7.22 Logic circuit of Mealy FSM $PHY(\Gamma_5)$

As it was found before, there are $T_0 = 5$ collections of microoperations for the case of FSM $U_1(\Gamma_5)$. There are the following collections of microoperations: $Y_1 = \{y_1, y_2, y_3\}$, $Y_2 = \{y_1, y_4\}$, $Y_3 = \{y_2, y_5\}$, $Y_4 = \{y_3, y_6\}$, and $Y_5 = \{y_3, y_6, y_7\}$. Because of it, there is $R_Y = 3$. Let us encode the collections $Y_t \subseteq Y$ in the trivial way: $K(Y_0) = 000$, $K(Y_1) = 001$, ..., $K(Y_5) = 100$. Obviously, $Y_0 = \emptyset$.

The table of BER for FSM $PHY(\Gamma_j)$ is the same as for Mealy FSM $PH(\Gamma_j)$. The table of BIMF includes the following columns: $K(F_h)$, Φ, $K(Y_t)$, h. This table is constructed in the trivial way. It is similar to the table of BIMF for $PHY(\Gamma_j)$ FSM. In the case of FSM $PHY(\Gamma_5)$, this table includes $H_1(\Gamma_5) = 9$ rows (Table 7.15).

As follows from Table 7.15, the set Z^1 includes the variables z_5–z_7. The table of BMO is constructed in the same way as it is done for PY FSM. The logic circuit of Mealy FSM $PHY(\Gamma_5)$ is shown in Fig. 7.22.

This circuit includes three levels of EMBs. It is the slowest implementation of FSM equivalent to $U_1(\Gamma_5)$. Let us point out that this FSM can be implemented using only one level of EMBs having configuration 64×4 (Fig. 7.23).

Let us denote a Mealy FSM with the single-level structure as P Mealy FSM. It has the structural diagram shown in Fig. 7.24. The number of EMBs in the P Mealy FSM is determined as

$$I = \left\lceil \frac{R+N}{t_F} \right\rceil. \tag{7.38}$$

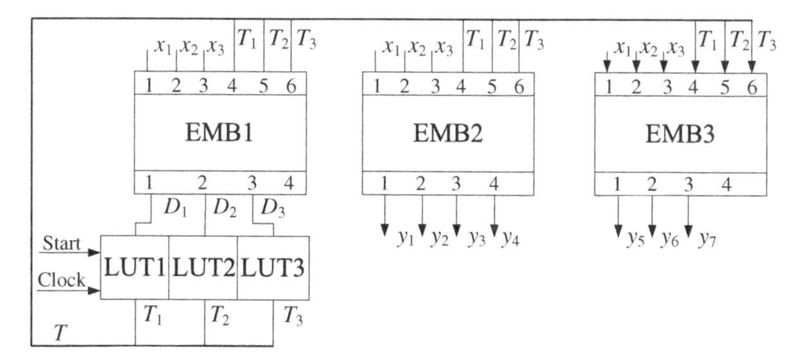

Fig. 7.23 Single-level circuit of Mealy FSM $U_1(\Gamma_5)$

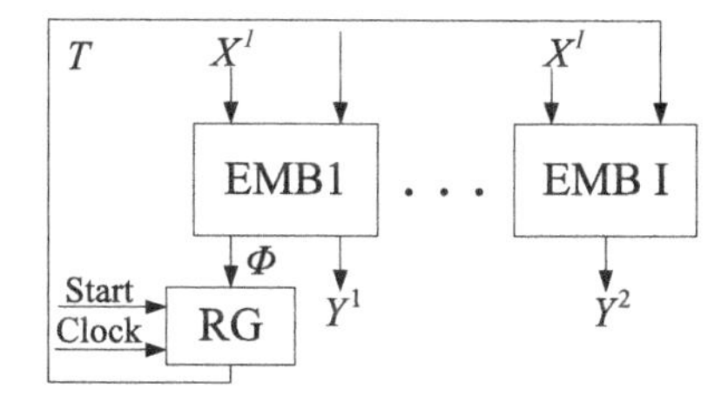

Fig. 7.24 Structural diagram of *P* Mealy FSM

In *P* Mealy FSM, a block EMBi implements microoperations $y_n \in Y^i$, where $Y^i \subseteq Y$. As follows from (7.38), the following condition should take place:

$$Y^i \cap Y^j = \emptyset (i \neq j; i,j \in \{1, \ldots, I\}). \tag{7.39}$$

Obviously, it is possible the situation when only a set $X^i \subseteq X$ is required for implementing functions $y_n \in Y^i$ $(i = \overline{1, I})$. It means that different EMBs of *P* FSM will require different amount of cells. Therefore, they will have different values of t_F. If this fact is taken into account, the number *I* can be decreased in comparison with this parameter for the FSM shown in Fig. 7.24. Let us point out that *P* Mealy FSM can be used as an alternative for all FSMs discussed in this chapter.

7.6 Optimization of BIMF Based on Pseudoequivalent States of Moore FSM

One of the specific features of Moore FSM is the existence of classes of pseudoequivalent states [3]. The states a_m, $a_s \in A$ are pseudoequivalent states if outputs of corresponding operator vertices are connected with the input of the same vertex of GSA Γ. Lets us find the partition Π_A of the set *A* by classes of pseudoequivalent states $B_1, \ldots, B_I$.

Two approaches can be used for optimizing the BIMF of Moore FSM. The first of them is the optimal state assignment. In this case, the states are encoded in such

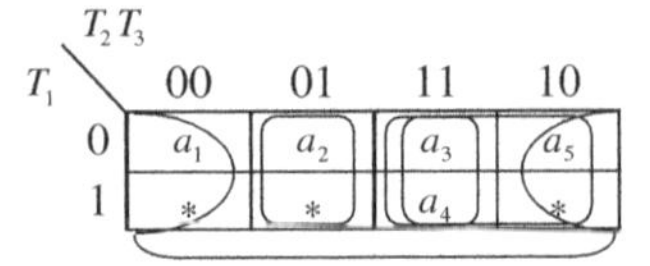

Fig. 7.25 Optimal state codes for Moore FSM $U_2(\Gamma_4)$

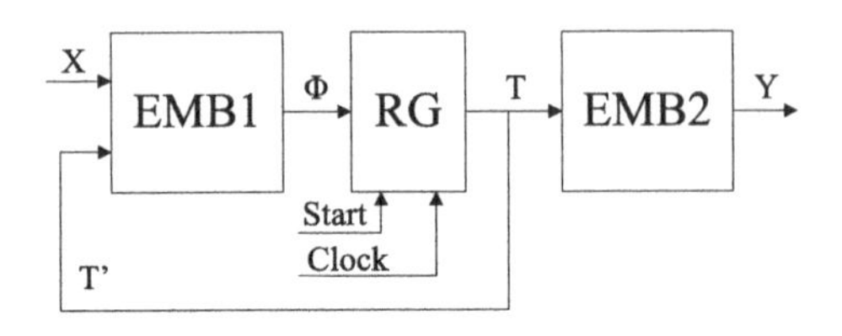

Fig. 7.26 Structural diagram of P_EY Moore FSM

a way that each class $B_i \in \Pi_A$ is represented by minimum possible amount of generalized intervals of R-dimensional Boolean space. The second approach is connected with encoding of the classes $B_i \in \Pi_A$. Let us discuss these approaches and corresponding models of EMB-based Moore FSMs.

The following classes $B_i \in \Pi_A$ can be found from GSA Γ_4: $B_1 = \{a_1\}$, $B_2 = \{a_2\}$, $B_3 = \{a_3, a_4\}$ and $B_4 = \{a_5\}$. So there is the partition $\Pi_A = \{B_1 \ldots B_4\}$ with $I = 4$. Let us encode the states $a_m \in A$ in the optimal way (Fig. 7.25).

The transitions from the state a_5 do not present in the structure table because they are executed automatically (using only pulse Clock). Because of it, the code of state a_5 can be treated as "don't care" and can be included into the cubes for other classes $B_i \in \Pi_A$.

Taking it into account, the following codes can be obtained for classes $B_i \in \Pi_A$ in the discussed case: $K(B_1) = **0$, $K(B_2) = *01$, $K(B_3) = *11$. So, the value of T_1 is not significant to determine the classes $B_i \in \Pi_A$. In the general case this approach leads to P_EY Moore FSM (Fig. 7.26).

In the P_EY FSM, the BIMF is represented by the block EMB1. It implements the system

$$\Phi = \Phi(T', X). \tag{7.40}$$

The block EMB2 implements the system (7.11). The following condition should take place for implementing this model:

$$R \cdot 2^{L+R_E} \leq V_0. \tag{7.41}$$

The value of R_E is determined by the cardinality value of the set $T' \subseteq T$. The proposed design method for P_EY Moore FSM is the following one:

1. Constructing the set of states A.
2. Optimal state assignment.
3. Constructing the transformed structure table
4. Constructing the table of BIMF.
5. Constructing the table of microoperations.
6. Implementation of the FSM logic circuit.

Table 7.16 Transformed structure table of Moore FSM $P_EY(\Gamma_4)$

B_i	$K(B_i)$	a_s	$K(a_s)$	X_h	Φ_h	h
B_1	**0	a_2	001	1	D_3	1
B_2	*01	a_3	011	x_1	D_2D_3	2
		a_4	111	$\bar{x}_1$	$D_1D_2D_3$	3
B_3	*11	a_5	010	1	D_2	4

Table 7.17 Table of BIMF of Moore FSM $P_EY(\Gamma_4)$

$K(B_i)$	X	Φ	v	h
T_2T_3	x_1	$D_1D_2D_3$		
00	0	001	1	1
00	1	001	2	1
01	0	111	3	3
01	1	011	4	2
10	0	001	5	1
10	1	001	6	1
11	0	010	7	4
11	1	010	8	4

Let us discuss an example of $P_EY(\Gamma_4)$ FSM's design. The GSA Γ_4 is shown in Fig. 7.5. Let the EMB in use have the following configurations: 32×1, 16×2, 8×4 (bits). Because there is $R = 3$, the configuration 8×4 should be chosen. But because of $R + L = 4$, the number of cells should be equal to 16 for $t_F = 4$. It is $t_F = 2$ for $V = 16$. So, the model $PY(\Gamma_4)$ cannot be used in the discussed case.

Two first steps have been already executed. The applying optimal state assignment gives the value $R_E = 2$. Now there is $2^{1+2} \times 3 = 24 < 32$. It means that the condition (7.40) is satisfied and the model $P_EY(\Gamma_4)$ can be applied.

To construct the transformed structure table of P_EY Moore FSM, it is necessary to construct the system of generalized formulae of transitions for classes $B_i \in \Pi_A$. This system does not include the class $B_4 \in \Pi_A$ because the state $a_5 \in B_4$ is connected only with state $a_1 \in A$. The following system can be derived from GSA Γ_4:

$$\begin{aligned} &B_1 \rightarrow a_2;\ B_3 \rightarrow a_5; \\ &B_2 \rightarrow x_1 a_3 \vee \overline{x_1} a_4. \end{aligned} \tag{7.42}$$

The transformed structure table of Moore FSM $P_EY(\Gamma_4)$ is represented by Table 7.16.

The table of BIMF contains the columns $K(B_i)$, X, Φ, v. In the case of FSM $P_EY(\Gamma_4)$ this table includes 8 rows (Table 7.17).

Four lines of Table 7.17 correspond to the row number 1 of the transformed table. Two lines of Table 7.17 correspond to the row number 4 of the transformed table. The connection between Tables 7.16 and 7.17 is transparent. The table of microoperations is constructed in the same manner as for the Moore FSM U_2. In the discussed case, this table includes 8 rows (Table 7.18).

The logic circuit of FSM $P_EY(\Gamma_4)$ is shown in Fig. 7.27. The circuits of BIMF and BMO are implemented using EMBs having configurations 8×4.

Table 7.18 Table of microoperations of Moore FSM $P_EY(\Gamma_4)$

$K(a_m)$	Y	m
$T_1T_2T_3$	$y_1y_2y_3y_4$	
000	0000	1
001	1100	2
010	0110	3
011	0010	4
100	0000	5
101	0000	6
110	0000	7
111	1001	8

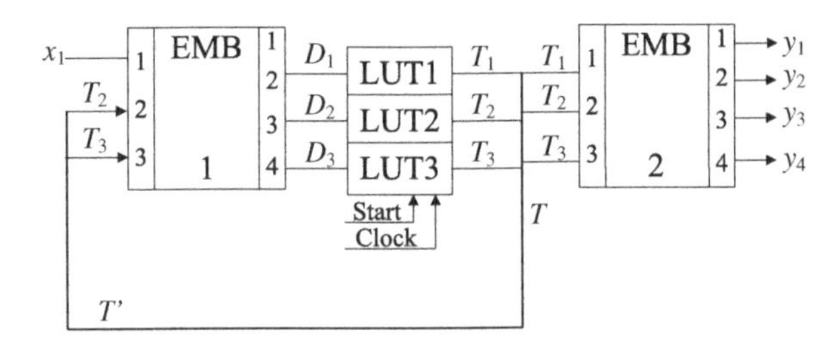

Fig. 7.27 Logic circuit of Moore FSM $P_EY(\Gamma_4)$

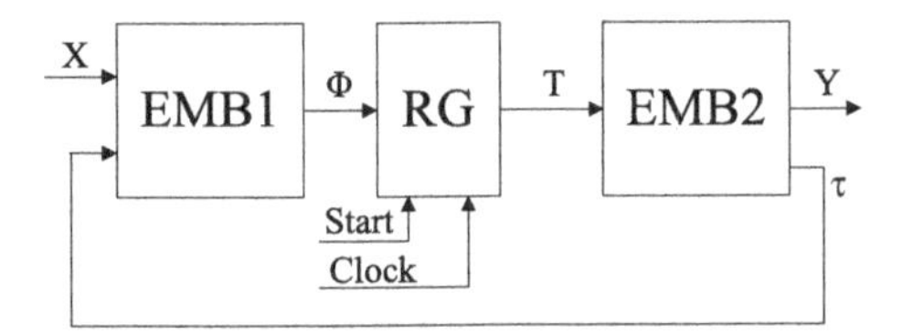

Fig. 7.28 Structural diagram of P_CY Moore FSM

It is quite possible the situation when $T' = T$ after the execution of optimal state assignment [3]. In this case the following approach can be used.

Let us encode each class $B_i \in \Pi_A$ by a binary code $K(B_i)$ having R_B bits:

$$R_B = \lceil \log_2 \mathrm{I} \rceil. \tag{7.43}$$

Let us use the variables $\tau_r \in \tau$ for the encoding, where $|\tau| = R_B$. Let the following condition take place:

$$\begin{aligned} &2^{R_B+L} \cdot R \leq V_0; \\ &2^R \cdot (N + R_B) \leq V_0. \end{aligned} \tag{7.44}$$

In this case, we propose to use the P_CY Moore FSM. Its structural diagram is shown in Fig. 7.28.

In P_CY Moore FSM, the block EMB1 corresponds to BIMF. It implements the system of input memory functions

$$\Phi = \Phi(\tau, X). \tag{7.45}$$

Table 7.19 Transformed structure table of Moore FSM $P_CY(\Gamma_4)$

B_i	$K(B_i)$	a_s	$K(a_s)$	X_h	Φ_h	h
B_1	00	a_2	001	1	D_3	1
B_2	01	a_3	010	x_1	D_2	2
		a_4	011	$\bar{x}_1$	D_2D_3	3
B_3	10	a_5	100	1	D_1	4

Table 7.20 Table of BIMF of Moore FSM $P_CY(\Gamma_4)$

$K(B_i)$	X	Φ	v	h
$\tau_1\tau_2$	x_1	$D_1D_2D_3$		
00	0	001	1	1
00	1	001	2	1
01	0	011	3	3
01	1	010	4	2
10	0	100	5	4
10	1	100	6	4
11	0	000	7	0
11	1	000	8	0

The block EMB2 implements the circuit of BMO. It generates the functions (7.11) and the system of additional variables

$$\tau = \tau(X). \tag{7.46}$$

The proposed design method for P_CY Moore FSM includes the following steps:

1. Constructing the set of states A.
2. State assignment.
3. Finding the partition $\Pi_A = \{B_1 \ldots B_4\}$.
4. Encoding of the classes $B_i \in \Pi_A$.
5. Constructing the transformed structure table
6. Constructing the table of BIMF.
7. Constructing the table of BMO.
8. Implementation of the FSM logic circuit.

Let us discuss an example of design for $P_CY(\Gamma_4)$ Moore FSM. The set A includes $M = 5$ elements and there is $R = 3$. Let us encode the states $a_m \in A$ in the trivial way: $K(a_1) = 000$, …, $K(a_5) = 100$.

There is the partition $\Pi_A = \{B_1, \ldots, B_4\}$ with $\mathrm{I} = 4$. It gives $R_B = 2$. Let us encode the classes $B_i \in \Pi_A$ in the trivial way: $K(B_1) = 00$, …, $K(B_4) = 11$.

To construct the transformed structure table, the system of generalized formulae of transitions should be derived from a GSA Γ_j. In the discussed case, this system is represented by (7.40). The transformed structure table of P_CY Moore FSM includes the same columns as its counterpart for P_EY Moore FSM (Table 7.19).

The table of BIMF is constructed on the base of the transformed structure table. In the discussed case, it is represented by Table 7.20. The table of BMO includes the additional column τ (Table 7.21). If $a_m \in B_i$, then the row corresponding to the state a_m includes the code $K(B_i)$.

Table 7.21 Table of BMO of Moore FSM $P_C Y(\Gamma_4)$

$K(a_m)$	Y	τ	m
$T_1T_2T_3$	$y_1y_2y_3y_4$	$\tau_1\tau_2$	
000	0000	00	1
001	1100	01	2
010	0010	10	3
011	1001	10	4
100	0110	11	5
101	0000	00	6
110	0000	00	7
111	1001	00	8

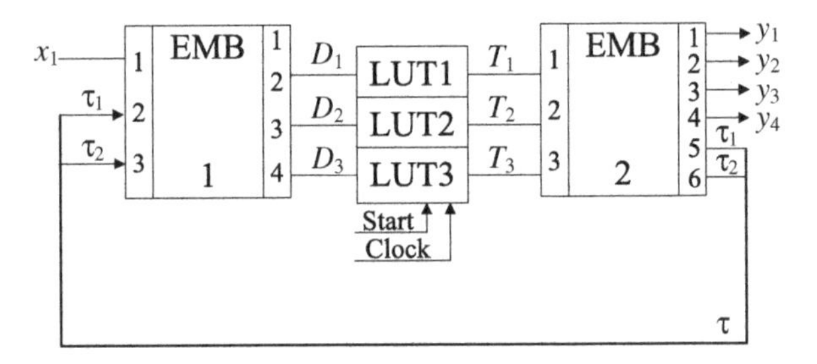

Fig. 7.29 Logic circuit of FSM $P_C Y(\Gamma_4)$

The logic circuit of Moore FSM $P_C Y(\Gamma_4)$ is shown in Fig. 7.29. In this circuit, the configuration 8×6 is required for EMB implementing the BMO. The same configuration is required for implementing the BIMF. But only three outputs are used in this case.

Of course, the examples discussed in this Chapter are very simple. They merely illustrate the main ideas which can be used for optimizing logic circuits of EMB-based finite state machines.

References

1. ALTERA (2013) Website of the Altera Corporation
2. Baranov S (1994) Logic synthesis of control automata. Kluwer, Boston
3. Barkalov A, Titarenko L (2009) Logic synthesis for FSM-based control units. Springer, Berlin
4. Barkalov A, Titarenko L, Barkalov A (2012) Structural decomposition as a tool for the optimization of an FPGA-based implementation of a mealy FSM. Cybern Syst Anal 48(2):313–323
5. Cong J, Yan K (2000) Synthesis for FPGAs with embedded memory blocks. In: Proceedings of the 2000 ACM, SIGDA 8th international symposwium on FPGAs, pp 75–82
6. Garcia-Vargas I, Senhadji-Navarro R, Civit-Balcells A, Guerra-Gutierrezz P (2007) ROM-based finite state machine implementation in low cost FPGAs. In: IEEE international simposium on industrial electronics, Vigo, pp 2342–2347
7. Grout I (2004) Digital system design with FPGAs and CPLDs. Elsevier, Amsterdam
8. Kim T, Vella T, Brayton R, Sangiovanni-Vincentalli A (1997) Synthesis of finite state machines: functional optimization. Kluwer, Boston
9. Maxfield C (2004) The design Warrior's guide to FPGAs, Academic Press Inc, Orlando

10. Nowicka M, Łuba T, Rawski M (1999) FPGA-based decomposition of boolean functions: algorithms and implementation. In: Proceedings of the 6th international conference on ACS, pp 502–509
11. Rawski M, Tomaszewicz P, Borowik G, Luba T (2011) Logic synthesis method of digital circuits designed with Embedded Memory Blocks of FPGAs. In: Adamski M et al. (eds) Design of digital systems and devices, Springer, Berlin, pp 121–144
12. Rawski M, Selvaraj H, Luba T (2005) An application of functional decomposition in ROM-based FSM implementation in FPGA devices. J Syst Archit 51(6–7):423–434
13. Scholl C (2001) Functional decomposition with application to FPGA synthesis. Kluwer, Norwell
14. Sklyarov V (1984) Synthesis of FSMs based on matrix LSI. Science and Technique, Minsk
15. Sklyarov V (2000) Synthesis and implementation of RAM-based finite state machines in FPGAs. In: Proceedings of conference on field programmable logic, Villach, pp 718–728
16. Sklyarova I, Sklyarov V, Sudnitson A (2008) Design of FPGA-based circuits using hierarchical finite state machines. TUT Press, Tallinn
17. Sutteer G, Todorowich E, Lopez-Buedo S, Boemo E (2002) Lower-power FSMs in FPGA: encoding alternatives. Lecture notes in computer science 2451, Springer, Berlin
18. Tiwari A, Tomko K (2004) Saving power by mapping finite state machines into embedded memory blocks in FPGAs. In: Proceedings of design automation and test in Europe, vol 2. pp 916–921
19. Wu X, Pedram M, Wang L, Multi-code state assignment for low-power design. In: IEEE proceedings on circuits, devices and systems, vol 147. pp 271–275
20. XILINX (2013) Website of the Xilinx Corporation

Chapter 8
Optimization of FSMs with Embedded Memory Blocks

Abstract Chapter is devoted to optimization of logic circuits of EMB-based FSMs. First of all, the design methods based on the replacement of logical conditions are discussed for both Moore and Mealy FSMs. Next, the proposed optimization methods are presented. These methods are based on splitting the set of logical conditions. This approach allows decreasing the number of LUTs in the circuit of the block of replacement of logical conditions. In the case of Moore FSM, the optimization methods are based on optimal state assignment, as well as the transformation of state codes into codes of the classes of PES. All discussed methods are illustrated by examples. The chapter is written together with PhD Malgorzata Kolopienczyk (University of Zielona Gora, Poland).

8.1 Trivial Implementation of MP Mealy FSMs

Let us find the value of G determined by (7.12) and let us form the set $P = \{P_1; \ldots, P_G\}$ for some FSM $U_1(\Gamma_j)$. In this case, it is possible to use the model $MP(\Gamma_j)$. Its structural diagram is shown in Fig. 8.1.

In the MP Mealy FSM, a block LUTer represents the block of replacement of logical conditions (BRLC) from the structural diagram shown in Fig. 7.7. It is implemented using LUT elements of FPGA chip. The LUTer implements the system (7.14) which can be represented as the following:

$$\begin{aligned} P_1 &= P_1\left(\mathrm{T}, X^1\right) \\ &\vdots \\ P_G &= P_G\left(\mathrm{T}, X^G\right). \end{aligned} \tag{8.1}$$

In (8.1), the set X^g includes logical conditions $x_l \in X$ replaced by the variable $P_g \in P$. It is quite possible that the following relation is true:

$$X^i \cap X^j \neq \emptyset (i \neq j; i,j \in \{1, \ldots, G\}). \tag{8.2}$$

V. Sklyarov et al., *Synthesis and Optimization of FPGA-Based Systems*,
Lecture Notes in Electrical Engineering 294, DOI: 10.1007/978-3-319-04708-9_8,

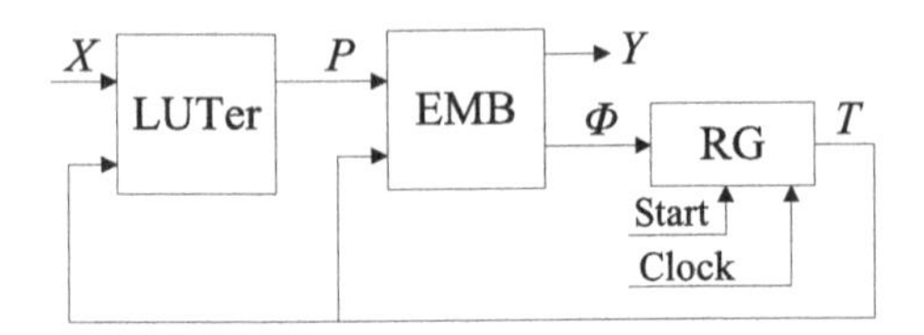

Fig. 8.1 Structural diagram of MP Mealy FSM

A block EMB represents the block of input memory functions (BIMF) from Fig. 7.7. It implements the systems (7.15 and 7.16).

The design method of Mealy FSM *MP*(Γ_j) includes the following steps:

1. Constructing the set of states A.
2. State assignment.
3. Constructing the structure table of FSM $U_1(\Gamma_j)$.
4. Replacement of logical conditions.
5. Constructing the system (8.1).
6. Constructing the transformed structure table.
7. Constructing the table of BIMF.
8. Implementing the FSM logic circuit with EMB and LUTs of a particular FPGA chip.

The, model of Mealy FSM *MP*(Γ_j) can be applied if the following condition takes place:

$$2^{G+R}(R+N) \leq V_o. \tag{8.3}$$

Let us discuss an example of design for Mealy FSM *MP*(Γ_6). The GSA Γ_6 is shown in Fig. 8.2.

This GSA is marked by states of Mealy FSM using the rules from [1]. The following sets and their characteristics can be found for Mealy FSM $U_1(\Gamma_6)$: $A = \{a_1, \ldots, a_4\}$, $M = 4$, $X = \{x_1, \ldots, x_6\}$, $L = 6$, $Y = \{y_1, \ldots, y_6\}$, $N = 6$, $R = 2$, $T = \{T_1, T_2\}$ and $\Phi = \{D_1, D_2\}$. Let us encode the states $a_m \in A$ in the trivial way: $K(a_1) = 00, \ldots, K(a_4) = 11$.

Let the FPGA chip in use have $V_0 = 128$ bits and let the following configurations of EMB exist: 128×1 64×2, 32×4, 16×8 (bits). For the FSM $U_1(\Gamma_6)$, the following relation takes place: $2^{L+R}(R+N) = 2^8 \cdot 8 = 2048 > 128$. It means that the model $U_1(\Gamma_6)$ cannot be used.

The structure table of Mealy FSM $U_1(\Gamma_6)$ includes $H_1(\Gamma_6) = 10$ rows (Table 8.1). As follows from Table 8.1, there are four sets of logical conditions $X(a_m) : X(a_1) = \{x_1, x_2\}, X(a_2) = \{x_3, x_4\}, X(a_3) = \{x_5, x_6\}, X(a_4) = \emptyset$. Obviously, it defines the value $G = 2$ and, therefore, there is the set $P = \{P_1, P_2\}$. Let us form the table of replacement of logical conditions for Mealy FSM *MP*(Γ_6) (Table 8.2). Let us point out that the condition (8.3) takes place for the given example.

There are no equal logical conditions in the sets $X(a_m) \subseteq X$ for the discussed case. Because of it, the distribution of the logical conditions among the variables $P_g \in P$ is executed in the trivial way. If there is $X(a_i) \cap X(a_j) \neq \emptyset$, then the distribution should be executed in such a way that the intersection for any pair of sets $X^g \left(g = \overline{1, G}\right)$ has the minimum capacity [1].

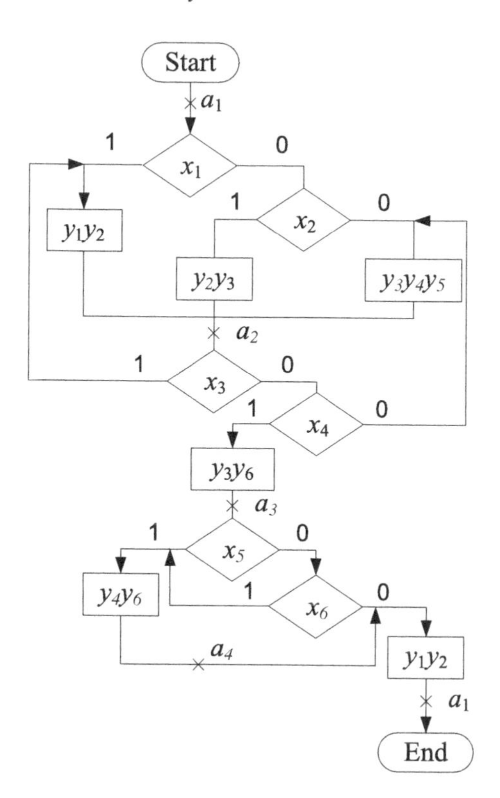

Fig. 8.2 Initial graph-scheme of algorithm Γ_6

Table 8.1 Structure table of Mealy FSM $U_1(\Gamma_6)$

a_m	$K(a_m)$	a_s	$K(a_s)$	X_h	Y_h	Φ_h	h
a_1	00	a_2	01	x_1	y_1y_2	D_2	1
		a_2	01	$\bar{x}_1x_2$	y_2y_3	D_2	2
		a_2	01	$\bar{x}_1\bar{x}_2$	$y_3y_4y_5$	D_2	3
a_2	01	a_2	01	x_3	y_1y_2	D_2	4
		a_3	10	$\bar{x}_3x_4$	y_3y_6	D_1	5
		a_2	01	$\bar{x}_3\bar{x}_4$	$y_3y_4y_5$	D_2	6
a_3	10	a_4	11	x_5	y_4y_6	D_1D_2	7
		a_4	11	$\bar{x}_5x_6$	y_4y_6	D_1D_2	8
		a_1	00	$\bar{x}_5\bar{x}_6$	y_1y_2	–	9
a_4	11	a_1	00	1	y_1y_2	–	10

Table 8.2 Table of replacement of logical conditions for Mealy FSM $MP(\Gamma_6)$

a_m	a_1	a_2	a_3	a_4
P_1	x_1	x_3	x_5	–
P_2	x_2	x_4	x_6	–

Table 8.3 Transformed structure table of Mealy FSM MP (Γ_6)

a_m	$K(a_m)$	a_s	$K(a_s)$	P_h	Y_h	Φ_h	h
a_1	00	a_2	01	P_1	y_1y_2	D_2	1
		a_2	01	$\bar{P}_1P_2$	y_2y_3	D_2	2
		a_2	01	$\bar{P}_1\bar{P}_2$	$y_3y_4y_5$	D_2	3
a_2	01	a_2	01	P_1	y_1y_2	D_2	4
		a_3	10	$\bar{P}_1P_2$	y_3y_6	D_1	5
		a_2	01	$\bar{P}_1\bar{P}_2$	$y_3y_4y_5$	D_2	6
a_3	10	a_4	11	P_1	y_4y_6	D_1D_2	7
		a_4	11	$\bar{P}_1P_2$	y_4y_6	D_1D_2	8
		a_1	00	$\bar{P}_1\bar{P}_2$	y_1y_2	–	9
a_4	11	a_1	00	1	y_1y_2	–	10

The following system of equations can be derived from Table 8.2:

$$\begin{aligned} P_1 &= A_1x_1 \vee A_2x_3 \vee A_3x_5; \\ P_2 &= A_1x_2 \vee A_2x_4 \vee A_3x_6. \end{aligned} \tag{8.4}$$

If the variables $A_m(m = \overline{1,M})$ are replaced by corresponding state codes, the system (8.4) is transformed into the following one:

$$\begin{aligned} P_1 &= \bar{T}_1\bar{T}_2x_1 \vee \bar{T}_1T_2x_3 \vee T_1\bar{T}_2x_5; \\ P_2 &= \bar{T}_1\bar{T}_2x_2 \vee \bar{T}_1T_2x_4 \vee T_1\bar{T}x_6. \end{aligned} \tag{8.5}$$

Obviously, the system (8.5) represents the system (8.1) for the case of Mealy FSM $PM(\Gamma_6)$. The logic circuit corresponding to system (8.5) should be implemented using look-up table elements.

The transformed structure table of $PM(\Gamma_j)$ includes all columns that its counterpart for the Mealy $U_1(\Gamma_j)$. But the column X_h is replaced by the column P_h. The transformation is executed in the obvious way. For the discussed example, the transformation leads to Table 8.3.

This table is the base for constructing the table of BIMF containing the following columns: $K(a_m)$, P, Y, Φ, v. The columns $K(a_m)$ and P form the addresses of cells. The table of BIMF includes $V(\Gamma_6) = 16$ rows for the discussed case. The number of rows $H(P)$ for representing transitions from a state $a_m \in A$ is determined as:

$$H(P) = 2^G. \tag{8.6}$$

In the discussed case, there is $H(P) = 4$. The table of BIMF is represented by Table 8.4. The logic circuit of FSM $PM(\Gamma_6)$ is shown in Fig. 8.3. In the discussed example, they use LUTs having $S = 5$ inputs. In this case, each function $P_g \in P$ is implemented using only a single LUT. In the common case, each function $P_g \in P$ is implemented using only a single LUT if the following condition takes place:

$$R + |X^g| \leq S(g = \overline{1,G}). \tag{8.7}$$

So, the logic circuit of $MP(\Gamma_6)$ consists from 4 LUTs and one block EMB. Let us point out that two LUTs are used for implementing the circuit of RG.

Table 8.4 Table of BIMF of Mealy FSM $MP(\Gamma_6)$

$K(a_m)$	P	Y	Φ	v	h
T_1T_2	P_1P_2	$y_1y_2y_3y_4y_5y_6$	D_1D_2		
00	00	001110	01	1	3
00	01	011000	01	2	2
00	10	110000	01	3	1
00	11	110000	01	4	1
01	00	001110	01	5	6
01	01	001001	10	6	5
01	10	110000	01	7	4
01	11	110000	01	8	4
10	00	110000	00	9	9
10	01	000101	11	10	8
10	10	000101	11	11	7
10	11	000101	11	12	7
11	00	110000	00	13	10
11	01	110000	00	14	10
11	10	110000	00	15	10
11	10	110000	00	16	10

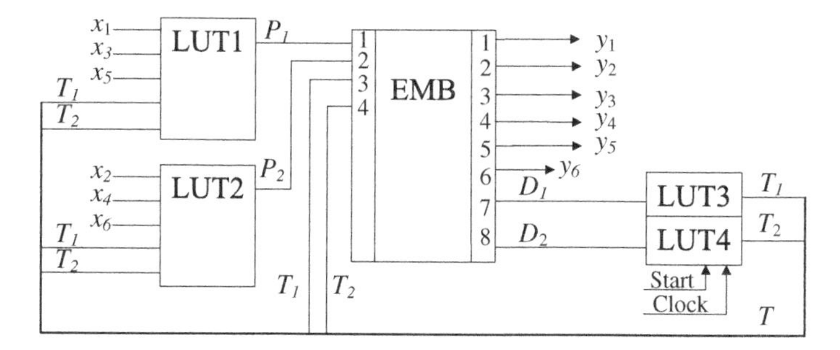

Fig. 8.3 Logic circuit of Mealy FSM MP (Γ_6)

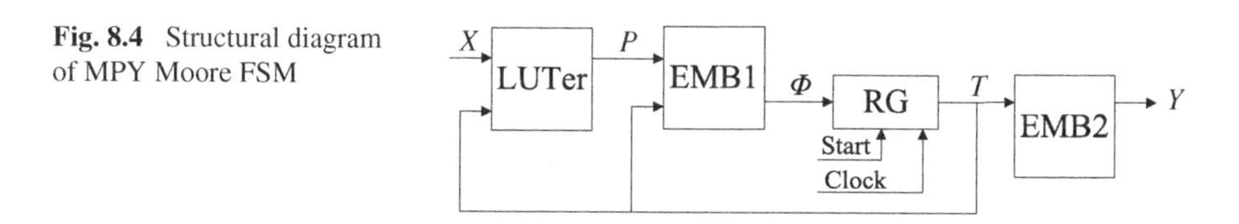

Fig. 8.4 Structural diagram of MPY Moore FSM

The structural diagram of MPY Moore FSM is shown in Fig. 8.4. A block LUTer represents BRLC, a block EMB1 represents BIMF, and a block EMB2 represents BMO.

As in the case of MP Mealy FSM, the LUTer implements the system (8.1). The EMB1 implements functions (7.15), whereas the EMB2 the system (7.11). This model can be applied if the following conditions take places:

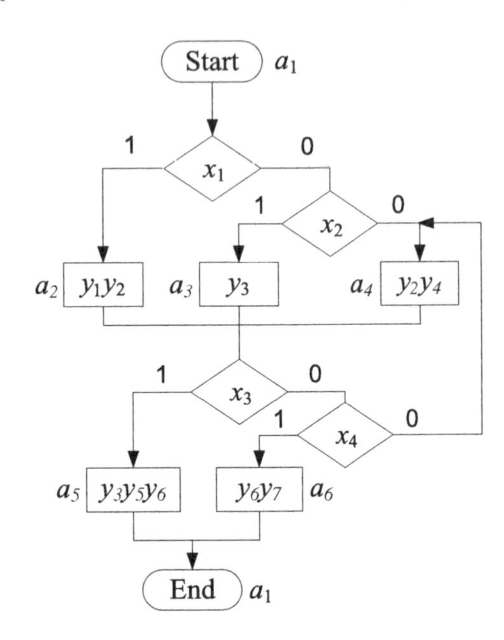

Fig. 8.5 Initial GSA Γ_7

$$2^{G+R} \cdot R \leq V_0; \tag{8.8}$$

$$2^R \cdot N \leq V_0. \tag{8.9}$$

The design method of Moore FSM $MPY(\Gamma_j)$ includes the following steps:

1. Constructing the set of states A.
2. State assignment.
3. Constructing the structure table of FSM $U_2(\Gamma_j)$.
4. Replacement of logical conditions.
5. Constructing the system (8.1).
6. Constructing the transformed structure table.
7. Constructing the table of BIMF.
8. Constructing the table of BMO.
9. Implementing the FSM logic circuit with EMBs and LUTs of a particular FPGA chip.

Let us discuss an example of design for Moore FSM $MPY(\Gamma_7)$. The GSA Γ_7 is shown in Fig. 8.5.

This GSA is marked by states of Moore FSM using the rules from [1]. The following sets and their characteristics can be found for Moore FSM $U_2(\Gamma_7): A = \{a_1, \ldots, a_6\}$, $M = 6$, $X = \{x_1, \ldots, x_4\}$, $L = 4$, $Y = \{y_1, \ldots, y_7\}$, $N = 7$, $R = 3$, $T = \{T_1, T_2, T_3\}$ and $\Phi = \{D_1, D_2, D_3\}$.

Let the FPGA chip in use have $V_0 = 128$ bits and let the following configurations of EMBs exist: 128×1 64×2, 32×4 and 16×8 (bits). For the FSM

Table 8.5 Structure table of FSM $U_2(\Gamma_7)$

a_m	$K(a_m)$	a_s	$K(a_s)$	X_h	Φ_h	h
a_1-	000	a_2	001	x_1	D_3	1
		a_3	010	$\bar{x}_1x_2$	D_2	2
		a_4	011	$\bar{x}_1\bar{x}_2$	D_2D_3	3
$a_2\ y_1y_2$	001	a_5	100	x_3	D_1	4
		a_6	101	$\bar{x}_3x_4$	D_1D_3	5
		a_4	011	$\bar{x}_3\bar{x}_4$	D_2D_3	6
$a_3\ y_3$	010	a_5	100	x_3	D_1	7
		a_6	101	$\bar{x}_3x_4$	D_1D_3	8
		a_4	011	$\bar{x}_3\bar{x}_4$	D_2D_3	9
$a_4\ y_2y_4$	011	a_5	100	x_3	D_1	10
		a_6	101	$\bar{x}_3x_4$	D_1D_3	11
		a_4	011	$\bar{x}_3\bar{x}_4$	D_2D_3	12
$a_5\ y_3y_5y_6$	100	a_1	000	1	–	13
$a_6\ y_6y_7$	101	a_1	000	1	–	14

Table 8.6 Table of replacement of logical conditions for Moore FSM $MPY(\Gamma_7)$

a_m	a_1	a_2	a_3	a_4	a_5	a_6
P_1	x_1	x_3	x_3	x_3	–	–
P_2	x_2	x_4	x_4	x_4	–	–

$U_2(\Gamma_7)$ the following relation takes place: $2^{L+R} \cdot R = 128 \times 3 > 128$. Therefore, this model cannot be used in the discussed case.

Let us encode the $a_m \in A$ in the trivial way, namely: $K(a_1) = 000, \ldots, K(a_6) = 101$. The structure table of FSM $U_2(\Gamma_7)$ includes $H_2(\Gamma_7) = 14$ rows (Table 8.5).

As follows from this table, there are the following sets $X(a_m) \subseteq X : X(a_1) = \{x_1, x_2\}, X(a_2) = X(a_3) = X(a_4) = \{x_3, x_4\}, X(a_5) = X(a_6) = \emptyset$. Obviously, there is $G = 2$ and $P = \{P_1, P_2\}$.

Let us check the conditions (8.8) and (8.9). For the discussed case there are:

$$2^5 \cdot 3 = 96 < 128;$$

$$2^3 \cdot 7 = 56 < 128.$$

It means that the model $MPY(\Gamma_7)$ can be used.

The table of replacement of logical conditions is represented by Table 8.6.

The following system of equations can be derived from Table 8.6:

$$\begin{aligned} P_1 &= A_1x_1 \vee (A_2 \vee A_3 \vee A_4)x_3; \\ P_2 &= A_1x_2 \vee (A_2 \vee A_3 \vee A_4)x_4. \end{aligned} \tag{8.10}$$

If variables $A_m \in A$ are replaced by corresponding conjunctions, the system (8.10) represents the system (8.1) for the given example.

The transformed structure table of Moore FSM $MPY(\Gamma_j)$ is constructed in the same way as its counterpart for Mealy FSM $MP(\Gamma_j)$. In the discussed example, it is represented by Table 8.7. This table is the base for constructing the table of BIMF. The table of BIMF contains the following columns: $K(a_m)$, P, Φ, v. In the discussed example,

Table 8.7 Transformed structure table of Moore FSM $MPY(\Gamma_7)$

a_m	$K(a_m)$	a_s	$K(a_s)$	P_h	Φ_h	h
a_1 –	000	a_2	001	P_1	D_3	1
		a_3	010	$\bar{P}_1P_2$	D_2	2
		a_4	011	$\bar{P}_1\bar{P}_2$	D_2D_3	3
$a_2\ y_1y_2$	001	a_5	100	P_1	D_1	4
		a_6	101	$\bar{P}_1P_2$	D_1D_3	5
		a_4	011	$\bar{P}_1\bar{P}_2$	D_2D_3	6
$a_3\ y_3$	010	a_5	100	P_1	D_1	7
		a_6	101	$\bar{P}_1P_2$	D_1D_3	8
		a_4	011	$\bar{P}_1\bar{P}_2$	D_2D_3	9
$a_4y_2y_4$	011	a_5	100	P_1	D_1	10
		a_6	101	$\bar{P}_1P_2$	D_1D_3	11
		a_4	011	$\bar{P}_1\bar{P}_2$	D_2D_3	12
$a_5\ y_3y_5y_6$	100	a_1	000	1	–	13
$a_6\ y_6y_7$	101	a_1	000	1	–	14

Table 8.8 Part of table of BIMF for FSM $MPY(\Gamma_7)$

$K(a_m)$	P	Φ	v	h
$T_1T_2T_3$	P_1P_2	$D_1D_2D_3$		
010	00	011	9	9
010	01	101	10	8
010	10	100	11	7
010	11	100	12	7

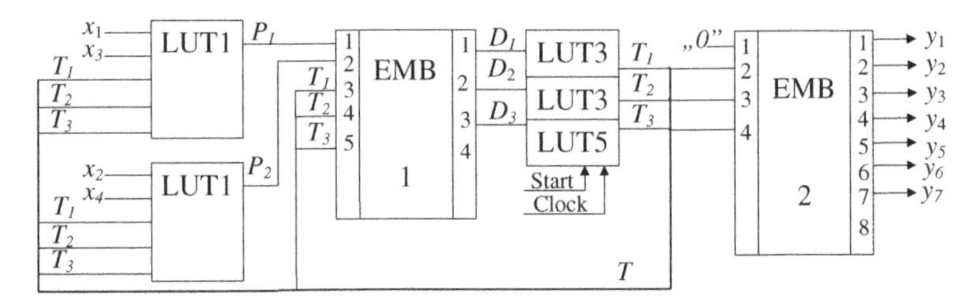

Fig. 8.6 Logic circuit of Moore FSM $MPY(\Gamma_7)$

this table includes $V_2(\Gamma_7) = 32$ rows. Using (8.6), it can be found that $H(\mathrm{P}) = 4$. The part of table of BIMF for the state $a_3 \in A$ is represented by Table 8.8. Other parts of the table are not shown but they are constructed in the same manner as Table 8.8.

The logic circuit of FSM $MPY(\Gamma_7)$ is shown in Fig. 8.6. As in the previous case, the circuit of BRLC is implemented using LUTs with $S = 5$. In this circuit, the EMB having configuration 32×4 (bits) is used for implementing the logic circuit of BIMF. The EMB having configuration 16×8 is used for implementing the circuit of BMO. Because there is $R = 3$, the most significant digit of address is equal to zero. The condition (8.7) is true for both variables P_1 and P_2. Due to it, only two LUTs are used in the circuit of BRLC.

8.2 Optimization of LUTer

Let us discus the case when the following condition takes place:

$$S_A > G + R. \tag{8.11}$$

In (8.11), the value S_A determines the number of address bits of EMB for given t_F. If (8.11) is true, then the following condition takes place:

$$S_0 = S_A - (G + R) > 0. \tag{8.12}$$

The value S_0 is equal to the number of "free" address inputs of EMB. These inputs are not connected with neither variables $T_r \in T$ or $P_g \in P$. We propose to use them for optimization of the circuit of LUTer.

Let us start from *MP* Mealy FSM. Let us represent the set X as $X^1 \cup X^2$ where the following condition takes place:

$$\begin{aligned} &X^1 \cap X^2 = \emptyset; X^1 \cup X^2 = X; \\ &\left|X^1\right| = S_0; \left|X^2\right| = L - S_0. \end{aligned} \tag{8.13}$$

Let us replace the logical conditions $x_l \in X^2$ by additional variables $P_g \in P$. It leads to M_0P Mealy FSM shown in Fig. 8.7.

In M_0P FSM, the LUTer executes the replacement of logical conditions $x_l \in X^2$. It implements the system

$$P = P\left(T, X^2\right). \tag{8.14}$$

The EMB implements functions

$$Y = Y\left(T, P, X^1\right); \tag{8.15}$$

$$\Phi = \Phi\left(T, P, X^1\right). \tag{8.16}$$

The following design method is proposed for M_0P Mealy FSM.

1. Constructing the set of states A.
2. State assignment.
3. Constructing the structure table of Mealy FSM.
4. Partitioning the set X and finding the sets X^1 and X^2.
5. Replacement of logical conditions $x_l \in X^2$.

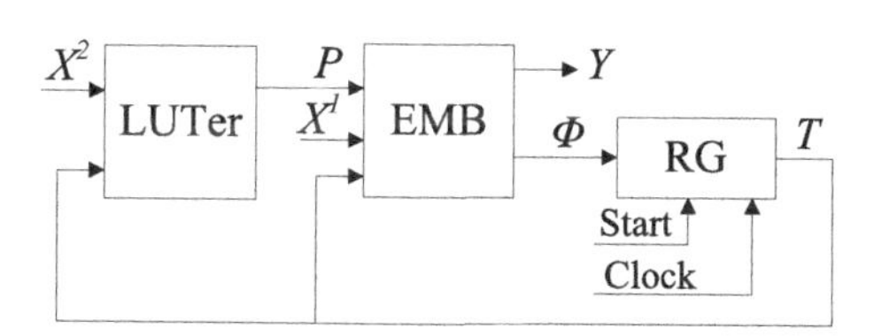

Fig. 8.7 Structural diagram of M_0P Mealy FSM

Table 8.9 Transformed structure table of Mealy FSM $M_0P(\Gamma_6)$

a_m	$K(a_m)$	a_s	$K(a_s)$	X_h^1	P_h	Y_h	Φ_h	h
a_1	00	a_2	01	x_1	1	y_1y_2	D_2	1
		a_2	01	$\bar{x}_1$	P_1	y_2y_3	D_2	2
		a_2	01	$\bar{x}_1$	$\bar{P}_1$	$y_3y_4y_5$	D_2	3
a_2	01	a_2	01	x_2	1	y_1y_2	D_2	4
		a_3	10	$\bar{x}_2$	P_1	y_3y_6	D_1	5
		a_2	11	$\bar{x}_2$	$\bar{P}_1$	$y_3y_4y_5$	D_2	6
a_3	10	a_4	11	x_3	1	y_4y_6	D_1D_2	7
		a_4	11	$\bar{x}_3$	P_1	y_4y_6	D_1D_2	8
		a_1	00	$\bar{x}_3$	$\bar{P}_1$	y_1y_2	–	9
a_4	11	a_1	00	1	1	y_1y_2	–	10

6. Constructing the transformed structure table of Mealy FSM $M_0P(\Gamma_j)$.
7. Constructing the table of BIMF.
8. Implementing the logic circuit of FSM for a given FPGA chip.

Let us discus an example of design for Mealy FSM $M_0P(\Gamma_6)$ where the GSA Γ_6 is shown in Fig. 8.2. Let FPGA chip in use include EMBs having the following configurations: 512×1, 256×2, 128×4, 64×8 (bits). For Mealy FSM $U_1(\Gamma_6)$, there is $N + R = 8$. Therefore, the configuration 64×8 can be used with $S_A = 6$. There is $R = 2$, then it is possible to have $G + S_0 = 4$.

The structure table of Mealy FSM $U_1(\Gamma_6)$ is represented by Table 8.1. The following sets can be derived from this table: $X(a_m) \subseteq X : X(a_1) = \{x_1, x_2\}, X(a_2) = \{x_3, x_4\}$, $X(a_3) = \{x_5, x_6\}$. Let us represent each set $X(a_m) \subseteq X$ as $X(a_m)^1 \cup X(a_m)^2$ where $X(a_m)^1 \cap X(a_m)^2 = \emptyset$ $\left(m = \overline{1, M}\right)$. Let us find the sets X^1 and X^2 using the rules:

$$X^1 = \bigcup_{m=1}^{M} X(a_m)^1; \tag{8.17}$$

$$X^2 = \bigcup_{m=1}^{M} X(a_m)^2. \tag{8.18}$$

Let us construct the following sets of logical conditions: $X(a_1)^1 = \{x_1\}, X(a_1)^2 = \{x_2\}$, $X(a_2)^1 = \{x_3\}$, $X(a_2)^2 = \{x_4\}$, $X(a_3)^1 = \{x_5\}$ and $X(a_3)^2 = \{x_6\}$. It leads to the sets $X^1 = \{x_1, x_3, x_5\}$ and $X^2 = \{x_2, x_4, x_6\}$. Obviously, there is $G = 1$. The additional variable is determined by the following equation:

$$P_1 = A_1x_2 \vee A_2x_4 \vee A_3x_6.$$

The transformed structure table of Mealy FSM $M_0P(\Gamma_j)$ includes the following columns: $a_m, K(a_m), a_s, K(a_s), X_h^1, P_h, Y_h, \Phi_h, h$. In the discussed case, it is represented by Table 8.9.

The table of BIMF includes the following columns: $K(a_m), X^1, P, Y, \Phi, v, h$. The first three columns create the address of some cell of EMB. This table includes $V_3(\Gamma_j)$ rows where:

Table 8.10 Part of table of BIMF of Mealy FSM $M_0P(\Gamma_6)$

$K(a_m)$	X^1	P	Y	Φ	v	h
T_1T_2	$x_1x_2x_3$	P_1	$y_1y_2y_3y_4y_5y_6$	D_1D_2		
00	000	0	001110	01	1	3
00	000	1	011000	01	2	2
00	001	0	001110	01	3	3
00	001	1	011000	01	4	4
00	010	0	001110	01	5	3
00	010	1	011000	01	6	2
00	011	0	001110	01	7	3
00	011	1	011000	01	8	2
00	100	0	110000	01	9	1
00	100	1	110000	01	10	1
00	101	0	110000	01	11	1
00	101	1	110000	01	12	1
00	110	0	110000	01	13	1
00	110	1	110000	01	14	1
00	111	0	110000	01	15	1
00	111	1	110000	01	16	1

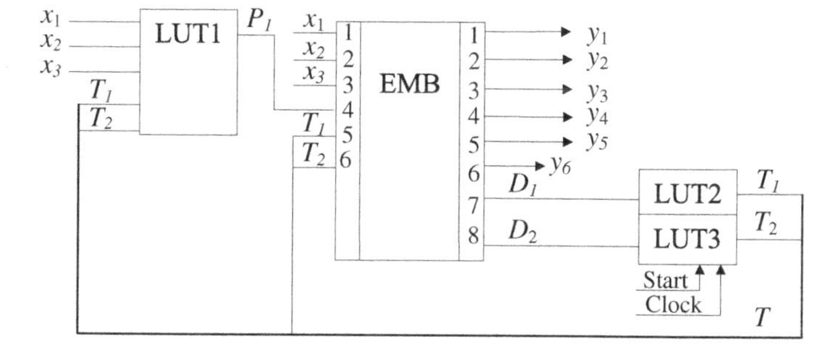

Fig. 8.8 Logic circuit of Mealy FSM $M_0P(\Gamma_6)$

$$V_3(\Gamma_j) = 2^{S_0+G+R}. \tag{8.19}$$

It can be found that $V_3(\Gamma_8) = 64$. Transitions from each state $a_m \in A$ of $M_0P(\Gamma_j)$ are represented by $H_3(\Gamma_j)$ rows, where

$$H_3(\Gamma_j) = 2^{S_0+G}. \tag{8.20}$$

In the case of FSM $M_0P(\Gamma_6)$, there is $H_3(\Gamma_6) = 16$. The part of the table of BIMF for FSM $M_0P(\Gamma_6)$ is represented by Table 8.10. This table contains the transitions from state $a_1 \in A$.

The logic circuit of FSM $M_0P(\Gamma_6)$ is shown in Fig. 8.8. The LUTs having $S = 5$ are used for implementing the logic circuit of LUTer. In the discussed example, it is enough one LUT for LUTer. Two more LUTs are used for implementing the register RG.

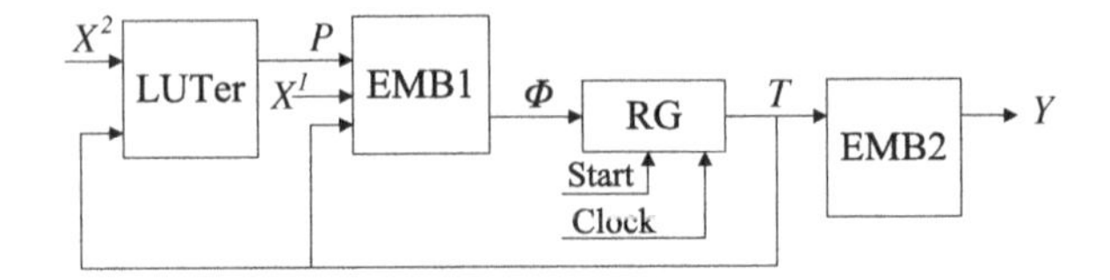

Fig. 8.9 Structural diagram of M_0PY Moore FSM

The same approach can be used for optimizing the LUTer of MPY Moore FSM. Its structural diagram is shown in Fig. 8.9. As in the previous case, the LUTer corresponds to the letter M_0 is the record "M_0PY". In the M_0PY Moore FSM, the LUTer implements system (8.14), whereas the EMB1 implements system (8.16). The EMB2 implements the system of microoperations $Y(T)$.

Let us point out the LUTer corresponds to BRLC, the EMB1 to BIMF, and the EMB2 to BMO. This model can be used if the following condition takes place:

$$2^{G+S_0+R;} \cdot R \leq V_0. \tag{8.21}$$

The design method for Moore FSM $M_0PY(\Gamma_j)$ includes the following steps:

1. Constructing the set of states A.
2. State assignment.
3. Constructing the structure table of FSM $PY(\Gamma_j)$.
4. Partitioning the set X by subsets X^1 and X^2.
5. Replacement of logical conditions $x_l \in X^2$.
6. Constructing the transformed structure table.
7. Constructing the table of BIMF.
8. Constructing the table of BMO.
9. Implementing the FSM logic circuit.

Let us discus an example of design for Moore FSM $M_0PY(\Gamma_7)$. All sets and their parameters have been found before. Let us use the trivial codes of internal states $K(a_1) = 000, \ldots, K(a_6) = 101$. The structure table of FSM $U_2(\Gamma_7)$ (Table 8.5) is the same for $PY(\Gamma_7)$ Moore FSM.

Let an FPGA chip in use include EMBs having the following configurations: 256×1, 128×2, 64×4, 32×8 (bits). In this case the model $U_2(\Gamma_7)$ cannot be applied. For Moore FSM $MPY(\Gamma_7)$, there is $G = 2$. So, there is $G = 2$. So, the model $MPY(\Gamma_7)$ can be applied but 160 bits of EMB are not used. Let us try to use the model $M_0PY(\Gamma_7)$.

There is $X(a_5) = X(a_6) = \emptyset$, then the configuration $X(a_m) \subseteq X$ can be chosen. In this case $|X^1| = 1$ and $S_A - R = 3$. So, two possibilities can be used for the replacement of logical conditions: (1) $G = 2$ and $|X'| = 1$ and (2) $G = 1$ and $|X'| = 2$. There are the following sets $X(a_m) \subseteq X : X(a_1) = \{x_1, x_2\}$, $X(a_2) = X(a_3) = X(a_4) = \{x_3, x_4\}$ and $X(a_5) = X(a_6) = \emptyset$. Let us divide the set X by following subsets: $X^1 = \{x_2, x_4\}$ and $X^2 = \{x_1, x_3\}$. It gives $P = \{P_1\}$ and the following equation can be found:

$$P_1 = A_1x_1 \vee A_2x_3 \vee A_3x_3 \vee A_4x_3. \tag{8.22}$$

Table 8.11 Transformed structure table of Moore FSM $M_0PY(\Gamma_7)$

a_m	$K(a_m)$	a_s	$K(a_s)$	X_h^1	P_h	Φ_h	h
a_1 (-)	000	a_2	001	1	P_1	D_3	1
		a_3	010	x_2	$\bar{P}_1$	D_2	2
		a_4	011	$\bar{x}_2$	$\bar{P}_1$	D_2D_3	3
a_2 (y_1y_2)	001	a_5	100	1	P_1	D_1	4
		a_6	101	x_4	$\bar{P}_1$	D_1D_3	5
		a_4	011	$\bar{x}_4$	$\bar{P}_1$	D_2D_3	6
a_3 (y_3)	010	a_5	100	1	P_1	D_1	7
		a_6	101	x_4	$\bar{P}_1$	D_1D_3	8
		a_4	011	$\bar{x}_4$	$\bar{P}_1$	D_2D_3	9
a_4 (y_2y_4)	011	a_5	100	1	P_1	D_1	10
		a_6	101	x_4	$\bar{P}_1$	D_1D_3	11
		a_4	011	$\bar{x}_4$	$\bar{P}_1$	D_2D_3	12
a_5 $(y_3y_5y_6)$	100	a_1	000	1	1	–	13
a_6 (y_6y_7)	101	a_1	000	1	1	–	14

Table 8.12 Part of table of BIMF of Moore FSM $M_0PY(\Gamma_7)$

$K(a_m)$	X^1	P	Φ	v	h
$T_1T_2T_3$	x_2x_4	P_1	$D_1D_2D_3$		
000	00	0	011	1	3
000	00	1	001	2	1
000	01	0	011	3	3
000	01	1	001	4	1
000	10	0	010	5	2
000	10	1	001	6	1
000	11	0	010	7	2
000	11	1	001	8	1

The structure table of Moore FSM $U_2(\Gamma_7)$ is represented by Table 8.5. Let us transform it to get Table 8.11. The table of BIMF includes 64 rows. The transitions from each state $a_m \in A$ are represented by 8 rows. The part of table of BIMF is represented by Table 8.12. It shows transitions from the state $a_1 \in A$.

8.3 Optimization of LUTer Based on Pseudoequivalent States

Let us discuss a situation when LUTs used for implementing the circuit of LUTer have $S = 4$. In the case of Moore FSM $M_0PY(\Gamma_7)$, the LUTer is represented by Eq. (8.22). It can be represented as the following one:

$$P_1 = \bar{T}_1\bar{T}_2\bar{T}_3x_1 \vee \bar{T}_1\bar{T}_2T_3x_3 \vee \bar{T}_1T_2\bar{T}_3x_3 \vee T_1\bar{T}_2\bar{T}_3x_3; \tag{8.23}$$

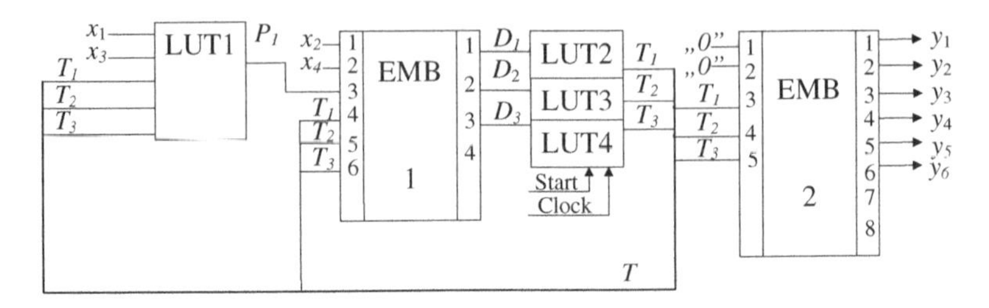

Fig. 8.10 Logic circuit of Moore FSM $M_0PY(\Gamma_7)$

Fig. 8.11 Logic circuit of LUTer for $S = 4$

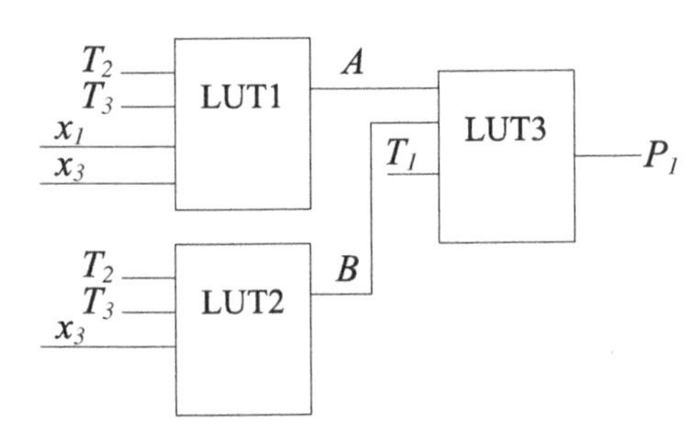

Fig. 8.12 Optimal state codes for Moore FSM $U_2(\Gamma_7)$

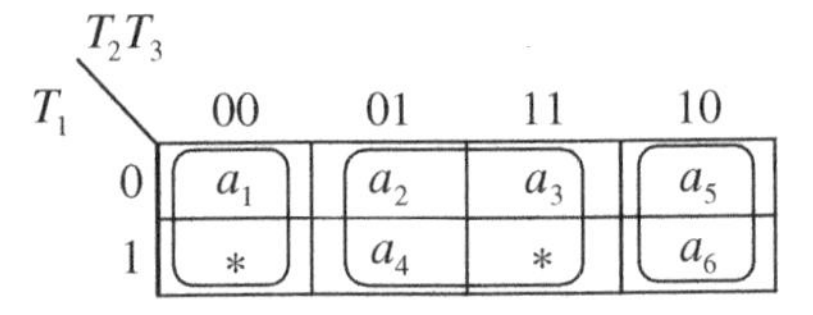

If there is $S = 4$, then the expression (8.23) should be transformed using the rules of functional decomposition [5, 6]. The transformed equation is the following:

$$P_1 = \bar{T}_1(\bar{T}_2\bar{T}_3x_1 \vee \bar{T}_2T_3x_3 \vee T_2\bar{T}_3x_3) \vee T_1(\bar{T}_2\bar{T}_3x_3); \tag{8.24}$$

The Eq. (8.25) corresponds to the logic circuit of LUTer shown in Fig. 8.11.

In Fig. 8.11, there are $A = \bar{T}_2\bar{T}_3x_1 \vee \bar{T}_2T_3x_3 \vee T_2\bar{T}_3x_3$ and $B = \bar{T}_2\bar{T}_3x_3$. This circuit has 2 levels and uses 3 LUTs with $S = 4$. Let us try to improve this circuit using pseudoequivalent states of Moore FSM [2, 3].

In the case of Moore FSM $U_2(\Gamma_7)$, there is the partition $\Pi_A = \{B_1, B_2, B_3\}$ where $B_1 = \{a_1\}$, $B_2 = \{a_1, a_2, a_3\}$ and $B_3 = \{a_5, a_6\}$. Let us encode the states $a_m \in A$ as it is shown in Fig. 8.12.

The logic circuit of Moore FSM $M_0PY(\Gamma_7)$ is shown in Fig. 8.10.

Now, the Eq. (8.22) can be represented as

$$P_1 = \bar{T}_2\bar{T}_3x_1 \vee T_3x_3. \tag{8.25}$$

This equation corresponds to the single-level logic circuit shown in Fig. 8.13a.

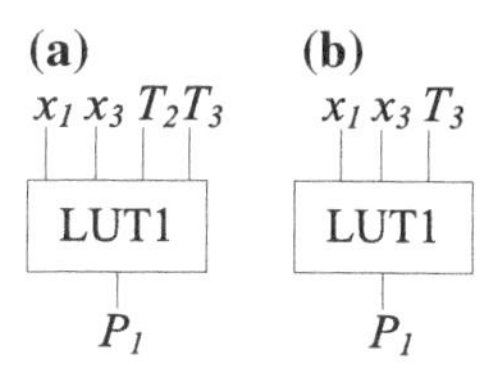

Fig. 8.13 Optimal circuit of LUTer

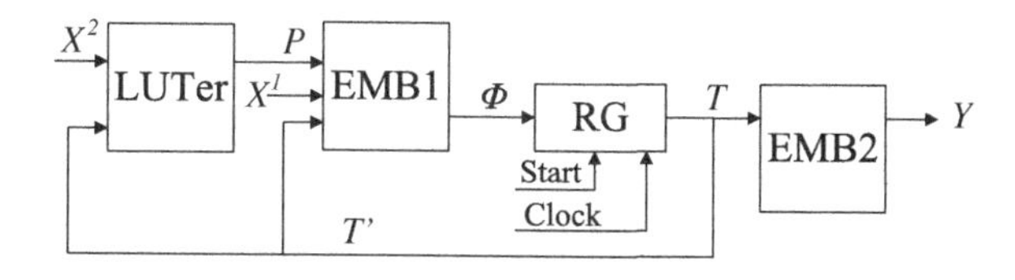

Fig. 8.14 Structural diagram of Moore FSM $M_EP_EY(\Gamma_j)$

Because of the equality $X(a_5) = X(a_6) = \emptyset$, their codes can be treated as "don't care" for function P_1. It allows obtaining the following equation:

$$P_1 = \bar{T}_3 x_1 \vee T_3 x_3. \tag{8.26}$$

The circuit for Eq. (8.26) can be implemented as a single-level circuit even for case of $S = 3$ (Fig. 8.13b).

The following codes $K(B_i)$ of classes of pseudoequivalent states $B_i \in \Pi_A$ can be derived from the Karnaugh map (Fig. 8.12): $K(B_1) = {*}00$, $K(B_2) = {*}{*}1$ and $K(B_3) = {*}10$. It means that input memory functions $D_r \in \Phi$ depend on variables $T_r \in T'$ where $T' \subset T$. In the discussed case, there is $T' = \{T_2, T_3\}$.

The discussed approach is based on the optimal state assignment for Moore FSM [2]. It leads to $M_EP_EY(\Gamma_j)$ Moore FSM (Fig. 8.14), where subscript "E" shows the usage of optimal state assignment.

In this model, the LUTer implements the system

$$P = P\left(T', X^2\right). \tag{8.27}$$

The EMB1 implements the system

$$\Phi = \Phi\left(T', P, X^2\right). \tag{8.28}$$

Let it be $R_E = |T'|$. In this case the proposed model can be used if the following condition takes place:

$$2^{G+S_0+R_E} \cdot R \leq V_0. \tag{8.29}$$

The proposed method for design of $M_EP_EY(\Gamma_j)$ Moore FSM includes the following steps:

1. Constructing the set of states A.
2. Constructing the partition $\Pi_A = \{B_1, \ldots, B_I\}$.
3. Optimal state assignment.
4. Constructing the structure table of $P_EY(\Gamma_j)$ Moore FSM.

Table 8.13 Structure table of Moore FSM $P_EY(\Gamma_7)$

B_i	$K(B_i)$	a_s	$K(a_s)$	X_h	Φ_h	h
B_1	*00	a_2	001	x_1	D_3	1
		a_3	011	$\bar{x}_1x_2$	D_2D_3	2
		a_4	101	$\bar{x}_1\bar{x}_2$	D_1D_3	3
B_2	**1	a_5	010	x_3	D_2	4
		a_6	110	$\bar{x}_3x_4$	D_1D_2	5
		a_4	101	$\bar{x}_3\bar{x}_4$	D_1D_3	6
B_3	*10	a_1	000	1	–	7

5. Partitioning the set X by subsets X^1 and X^2.
6. Replacement of logical conditions $x_l \in X^2$.
7. Constructing the transformed structure table.
8. Constructing the table of BIMF.
9. Constructing the table of BMO.
10. Implementing the FSM logic circuit.

Let us discuss an example of design of Moore FSM $M_EP_EY(\Gamma_7)$. The set of states A is constructed before, as well as the partition $\Pi_A = \{B_1, B_2, B_3\}$. Let us use the state codes from Fig. 8.12.

To construct the structure table of $P_EY(\Gamma_j)$ Moore FSM, let us form the system of generalized formulae of transitions [3]. In the discussed case, it is the following system:

$$\begin{aligned} B_1 &\to x_1a_2 \vee \bar{x}_1x_2a_3 \vee \bar{x}_1\bar{x}_2a_4; \\ B_2 &\to x_3a_5 \vee \bar{x}_3x_4a_6 \vee \bar{x}_3\bar{x}_4a_4; \\ B_3 &\to a_1. \end{aligned} \quad (8.30)$$

The structure table of Moore FSM $P_EY(\Gamma_7)$includes $H_E(\Gamma_7) = 7$ rows (Table 8.13). The table includes the following columns: B_i, $K(B_i)$, a_s, $K(a_s)$, X_h, Φ_h, h. The codes $K(B_i)$ of classes $B_i \in \Pi_A$ are taken from Fig. 8.12, as well as the codes of states $a_m \in A$.

Let the FPGA chip in use have EMBs with the following configurations: 128×1, 64×2, 32×4, 16×8 (bits). Because there is $R = 3$, we should choose the configuration 32×4 for implementing the circuit of BIMF. There is $T' = \{T_2, T_3\}$ and, therefore, $R_E = 2$. For given configuration, there is $S_A = 5$. It means that three inputs can be used for logical conditions $x_l \in X^1$ and additional variables $P_g \in P$. Let us make the following partition of the set of logical conditions X: $X^1 = \{x_2, x_4\}$ and $X^2 = \{x_1, x_3\}$. It gives the set $P = \{P_1\}$. Using Table 8.13, the following equation can be found:

$$P_1 = B_1x_1 \vee B_2x_3 = \bar{T}_2\bar{T}_3x_1 \vee T_3x_3. \quad (8.31)$$

Let $X(B_i)$ be a set of logical conditions determining transitions from states $a_m \in B_i$ $\left(i = \overline{1, I}\right)$. Because $X(B_3) = \emptyset$, the codes of states a_5, $a_6 \in B_3$ can be treated as "don't cares". It gives the final form of the system (8.27) for the given example:

$$P_1 = \bar{T}_3x_1 \vee T_3x_3. \quad (8.32)$$

Table 8.14 Transformed structure table of Moore FSM $M_EP_EY(\Gamma_7)$

B_i	$K(B_i)$	a_s	$K(a_s)$	P_h	X_h^1	Φ_h	h
B_1	*00	a_2	001	P_1	1	D_3	1
		a_3	011	$\bar{P}_1$	x_2	D_2D_3	2
		a_4	101	$\bar{P}_1$	$\bar{x}_2$	D_1D_3	3
B_2	**1	a_5	010	P_1	1	D_2	4
		a_6	110	$\bar{P}_1$	x_4	D_1D_2	5
		a_4	101	$\bar{P}_1$	$\bar{x}_4$	D_1D_3	6
B_3	*10	a_1	000	1	1	–	7

Table 8.15 Part of the table of BIMF for Moore FSM $M_EP_EY(\Gamma_7)$

$K(B_i)$	P	X^1	Φ	v	h
T_2T_3	P_1	x_2x_4	$D_1D_2D_3$		
00	0	00	101	1	3
00	0	01	101	2	3
00	0	10	011	3	2
00	0	11	011	4	2
00	1	00	001	5	1
00	1	01	001	6	1
00	1	10	001	7	1
00	1	11	001	8	1

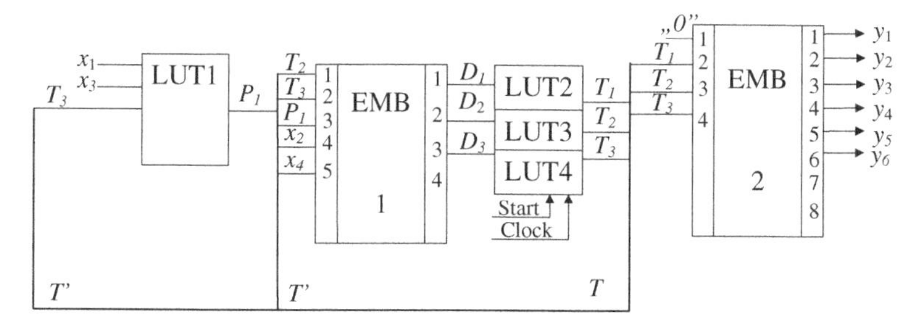

Fig. 8.15 Logic circuit of FSM $M_EP_EY(\Gamma_7)$

The transformed structure table of M_EP_EY Moore FSM includes the following columns: B_i, $K(B_i)$, a_s, $K(a_s)$, P_h, X_h^1, Φ_h, h. In the discussed case, it is represented by Table 8.14.

The table of BIMF of M_EP_EY Moore FSM includes the following columns: $K(B_i)$, P, X^1, Φ, h. The column $K(B_i)$, P, X^1 create the address of a cell inside the EMB. In the discussed case, the transitions from each class $B_i \in \Pi_A$ are represented by 8 rows of this table. The transitions from the class $B_1 \in \Pi_A$ are represented by Table 8.15.

The table of BMO is always the same for given GSA. It includes the columns $K(a_m)$, Y, m. The address of a cell is determined by the state code $K(a_m)$. Let the FPGA chip in use include LUTs having $S = 3$. In this case the Eq. (8.32) needs only a single LUT for implementation. The logic circuit of FSM $M_EP_EY(\Gamma_7)$ is shown in Fig. 8.15.

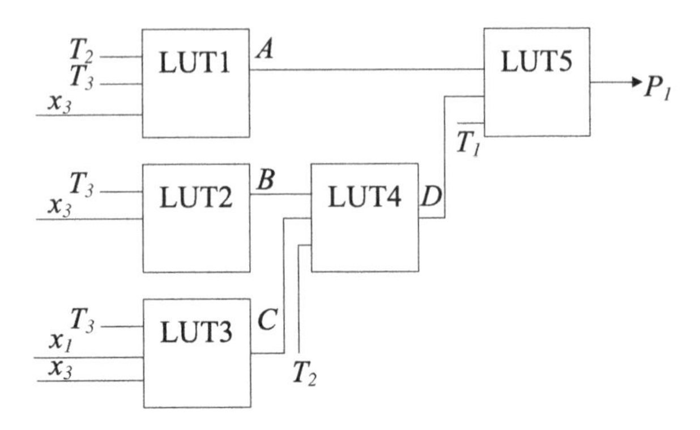

Fig. 8.16 Logic circuit of BRLC based on trivial state codes

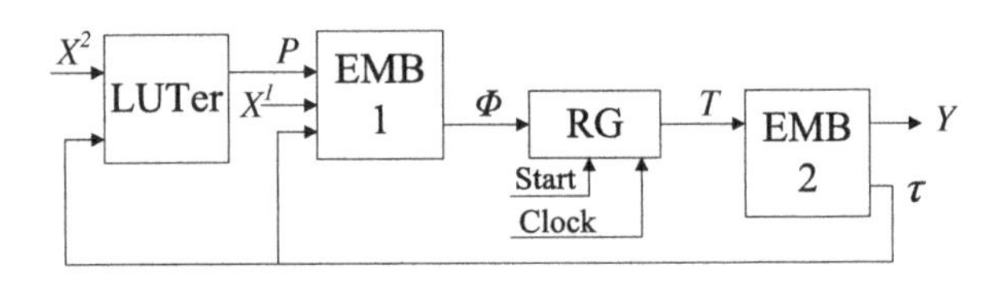

Fig. 8.17 Structural diagram of $M_C P_C Y$ Moore FSM

Due to the optimal state encoding, the circuit of BRLC is implemented using only one LUT with $S = 3$. To implement the Eq. (8.25) with LUTs having three inputs, it should be decomposed:

$$\begin{aligned} P_1 &= T_1\left(T_2\bar{T}_3x_3\right) \vee \bar{T}_1\left(T_2\left(\bar{T}_3x_3\right) \vee \bar{T}_2\left(\bar{T}_3x_1 \vee T_3x_3\right)\right) \\ &= T_1A \vee \bar{T}_1\left(T_2B \vee \bar{T}_2C\right) = T_1A \vee \bar{T}_1D. \end{aligned}$$

This equation corresponds to the circuit having three layers formed by 5 LUTs (Fig. 8.16).

So, the proposed approach allows 5 times reduction for the hardware of BRLC, as well as 3 times acceleration for the propagation time. Of course, it is true only for the given example.

It is known that sometimes the optimal state encoding is not possible [3]. In the case of replacement of logical conditions, it can lead to increasing both hardware amount and propagation time of BRLC. In this case the following approach is proposed.

Let us encode each class $B_i \in \Pi_A$ by a binary code $K(B_i)$ having R_B bits:

$$R_B = \lceil \log_2 I \rceil. \tag{8.33}$$

Let us use the variables $\tau_r \in \tau$ for the encoding, where $|\tau| = R_B$

Let the following condition takes place:

$$2_R(N + R_B) \leq V_0. \tag{8.34}$$

In this case, we propose to use the $M_C P_C Y$ Moore FSM. Its structural diagram is shown in Fig. 8.17. In this FSM, the LUTer implements the system of additional variables

$$P = P\left(\tau, X^2\right). \tag{8.35}$$

The block EMB1 implements input memory functions

$$\Phi = \Phi\left(\tau, X^1, P\right). \tag{8.36}$$

The block EMB2 implements microoperations $y_n \in Y$ and the system

$$\tau = \tau(T). \tag{8.37}$$

The proposed model can be used if the condition (8.22) takes place together with the following condition

$$2^{G+S_0+R_B} \cdot R \leq V_0. \tag{8.38}$$

The proposed design method for $M_C P_C Y$ Moore FSM includes the following steps:

1. Constructing the set of states A.
2. Constructing the partition $\Pi_A = \{B_1, \ldots, B_I\}$.
3. State assignment.
4. Encoding of the classes $B_i \in \Pi_A$.
5. Constructing the structure table of $P_C Y$ Moore FSM.
6. Partitioning the set X by subsets X^1 and X^2.
7. Replacement of logical conditions $x_l \in X^2$.
8. Constructing the transformed structure table
9. Constructing the table of BIMF.
10. Constructing the table of BMO.
11. Implementing the FSM logic circuit.

Let us discuss an example of design for FSM $M_C P_C Y(\Gamma_8)$. The GSA Γ_8 is shown in Fig. 8.18.

For the FSM $U_2(\Gamma_8)$, there are the following sets and their parameters: $A = \{a_1, \ldots, a_8\}$, $M = 8$, $X = \{x_1, \ldots, x_5\}$, $L = 5$, $Y = \{y_1, \ldots, y_6\}$, $N = 6$, $R = 3$, $T = \{T_1, T_2, T_3\}, \Phi = \{D_1, D_2, D_3\}$.

The following partition Π_A can be found for Moore FSM $U_2(\Gamma_8)$: $\Pi_A = \{B_1, \ldots, B_4\}$, where $B_1 = \{a_1\}$, $B_2 = \{a_2, a_3, a_4\}$, $B_3 = \{a_5, a_6, a_7\}$, $B_4 = \{a_8\}$. So there is $I = 4$ and $R_B = 2$. A state assignment is treated as optimal if any class $B_i \in \Pi_A$ is represented by a single generalized interval of R-dimensional Boolean space [2]. It is impossible to find such an outcome for Moore FSM $U_2(\Gamma_8)$. So, let us encode the states $a_m \in A$ in the trivial way: $K(a_1) = 000, \ldots, K(a_8) = 111$.

Because of $R_B = 2$, there is $\tau = \{\tau_1, \tau_2\}$. Let us encode the classes $B_i \in \Pi_A$ in the trivial way: $K(B_1) = 00, \ldots, K(B_4) = 11$. Let us point out that there is no need in the representing the transitions from states $a_8 \in B_4$ by the structure table. In the case of D flip-flops such transitions are executed automatically (using only pulse Clock).

Therefore, the code 11 can be treated as "don't care" input assignment. It can be used for minimizing FSM logic circuit.

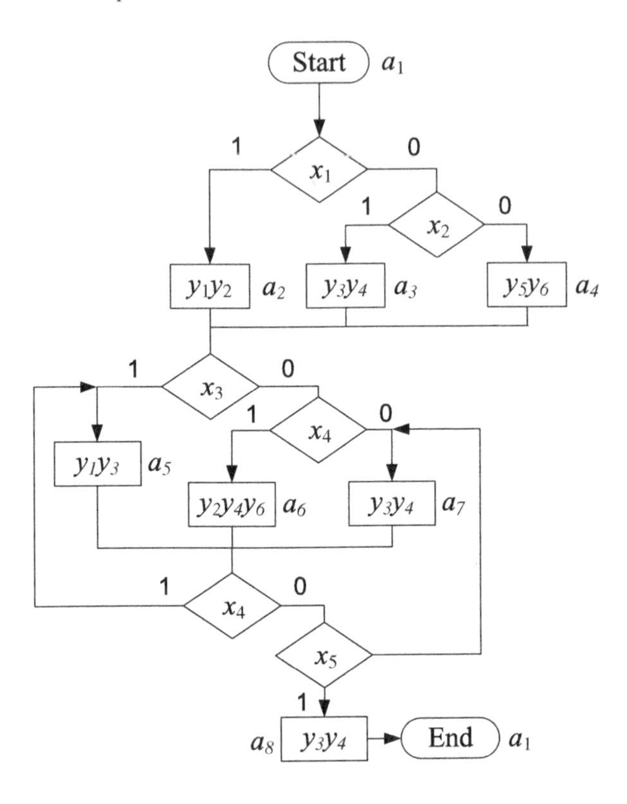

Fig. 8.18 Initial GSA Γ_8

Table 8.16 Structure table of Moore FSM $P_CY(\Gamma_8)$

B_i	$K(B_i)$	a_s	$K(a_s)$	X_h	Φ_h	h
B_1	00	a_2	001	x_1	D_3	1
		a_3	010	$\bar{x}_1x_2$	D_2	2
		a_4	011	$\bar{x}_1\bar{x}_2$	D_2D_3	3
B_2	01	a_5	100	x_3	D_1	4
		a_6	101	$\bar{x}_3x_4$	D_1D_3	5
		a_7	110	$\bar{x}_3\bar{x}_4$	D_1D_2	6
B_3	10	a_5	100	x_4	D_1	7
		a_8	111	$\bar{x}_4x_5$	$D_1D_2D_3$	8
		a_7	110	$\bar{x}_4\bar{x}_5$	D_1D_2	9

To construct the structure table of P_CY Moore FSM, it is necessary to construct the system of generalized formulae of transitions. In the discussed case, this system is the following one:

$$\begin{aligned} B_1 &\to x_1a_2 \vee \bar{x}_1x_2a_3 \vee \bar{x}_1\bar{x}_2a_4; \\ B_2 &\to x_3a_5 \vee \bar{x}_3x_4a_6 \vee \bar{x}_3\bar{x}_4a_7; \\ B_3 &\to x_4a_5 \vee \bar{x}_4x_5a_8 \vee \bar{x}_4\bar{x}_5a_7. \end{aligned} \tag{8.39}$$

Let us point out that there is no formula for the class $B_4 \in \Pi_A$ in the system (8.39).

Table 8.17 Table of replacement of logical conditions for Moore FSM $M_CP_CY(\Gamma_8)$

B_i	B_1	B_2	B_3	B_4
P_1	x_1	x_4	x_4	–

Table 8.18 Transformed structure table of Moore FSM $M_CP_CY(\Gamma_8)$

B_i	$K(B_i)$	a_s	$K(a_s)$	X_h^1	P_h	Φ_h	h
B_1	00	a_2	001	1	P_1	D_3	1
		a_3	010	x_2	$\bar{P}_1$	D_2	2
		a_4	011	$\bar{x}_2$	$\bar{P}_1$	D_2D_3	3
B_2	01	a_5	100	x_3	1	D_1	4
		a_6	101	$\bar{x}_3$	P_1	D_1D_3	5
		a_7	110	$\bar{x}_3$	$\bar{P}_1$	D_1D_2	6
B_3	10	a_5	100	1	P_1	D_1	7
		a_8	111	x_5	P_1	$D_1D_2D_3$	8
		a_7	110	$\bar{x}_5$	P_1	D_1D_2	9

The system (8.39) includes $H_C(\Gamma_8) = 9$ terms. Obviously, the structure table of Moore FSM $P_CY(\Gamma_8)$ includes 9 rows (Table 8.16).

Let us use the FPGA chip including EMBs with the following configurations: 256×1, 128×2, 64×4, 32×8 (bits). It is necessary to use EMBs with $V_0 = 2^{3+8} \cdot 3 = 6144$ bits on the case of $U_2(\Gamma_8)$. Obviously, the replacement of logical conditions should be used in the discussed case.

In the discussed case, there is $R = 3$. It means that the configuration 64×4 must be used. Because there is $R_B = 2$, the EMB has $S_A - R_B = 6-2 = 4$ free inputs. It gives $|X^1| = 3$, $|X^2| = 2$ and $G = 1$. Let us represent the set X as $X^1 \cup X^2$ where $X^1 = \{x_2, x_3, x_5\}$ and $X^2 = \{x_1, x_4\}$.

Let us form the table of replacement of logical conditions for Moore FSM $M_CP_CY(\Gamma_8)$ (Table 8.17).

The following equation can be derived from Table 8.17 $P_1 = B_1x_1 \vee B_2x_4 \vee B_3x_4 = \bar{\tau}_1\bar{\tau}_2x_1 \vee \bar{\tau}_1\tau_2x_4 \vee \tau_1\bar{\tau}_2x_4$. This formula can be implemented as a single-level circuit using LUTs with $S \geq 4$. Let us point out that this equation can be simplified due to appropriate encoding of the classes $B_i \in \Pi_A$. For example, if there is $K(B_1) = 00$, $K(B_2) = 10$, $K(B_3) = 11$, $K(B_4) = 01$, then there is the following equation $P_1 = \bar{\tau}_1x_1 \vee \tau_1x_4$. It can be implemented using only one LUT having $S = 3$ inputs.

The transformed structure table of M_CP_CY FSM includes the following columns: B_i, $K(B_i)$, a_s, $K(a_s)$, X_h^1, P_h, Φ_h, h. It is Table 8.18 for Moore FSM $M_CP_CY(\Gamma_8)$. The table of BIMF includes the same columns as it is for the case of M_EP_EY FSM. In the discussed case, the transitions from each class $B_i \in \Pi_A$ are represented by 16 rows of the table. The table BMO includes the following columns: $K(a_m)$, Y, $K(B_i)$, m. The column $K(B_i)$ includes the code of class $B_i \in \Pi_A$ such that $a_m \in B_i$ (for the row number m of the table). In the case of Moore FSM $M_CP_CY(\Gamma_8)$ This table has $M = 8$ rows (Table 8.19).

The logic circuit of FSM $M_CP_CY(\Gamma_8)$ is shown in Fig. 8.19.

Table 8.19 Table of BMO of Moore FSM $M_C P_C Y(\Gamma_8)$

$K(a_m)$	Y	$K(B_i)$	m
$T_1T_2T_3$	$y_1y_2y_3y_4y_5y_6$	$\tau_1\tau_2$	
000	000000	00	1
001	110000	01	2
010	001100	01	3
011	000011	01	4
100	101000	10	5
101	010101	10	6
110	001100	10	7
111	101000	11	8

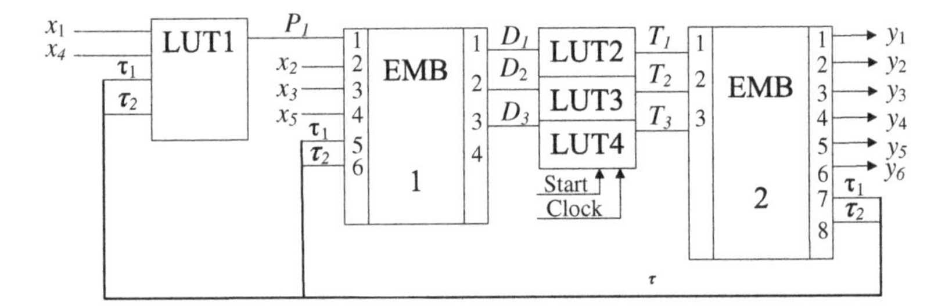

Fig. 8.19 Logic circuit of Moore FSM $M_C P_C Y(\Gamma_8)$

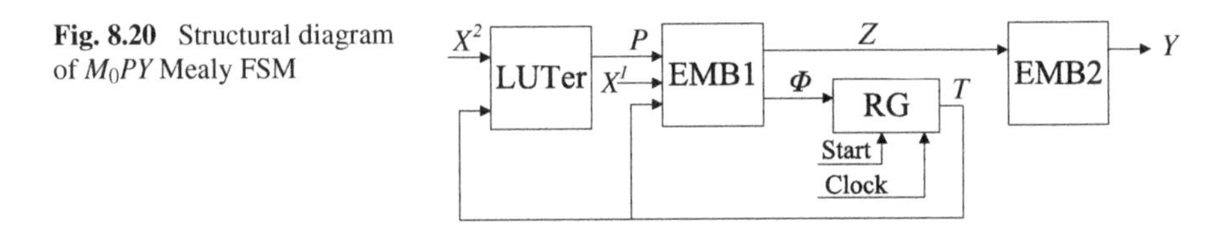

Fig. 8.20 Structural diagram of M_0PY Mealy FSM

8.4 Optimization of LUTer Based on Encoding of Collections of Microoperations

Three approaches can be used for the optimization of the circuit of BRLC of Mealy FSM. The first approach is based on the splitting the set X. The second is connected with the special state assignment. The third approach is based on the encoding of the logical conditions [4].

The splitting the set X by subsets X^1, X^2 is executed in the same manner as for M_0P Mealy FSM. It leads to M_0PY Mealy FSM shown in Fig. 8.20.

In this FSM, the LUTer implements the system (8.14). The block EMB1 implements the system (8.16) and the system of additional variables

$$Z = Z\left(T, P, X^1\right). \tag{8.40}$$

The block EMB2 implements the microoperations $y_n \in Y$ represented by the system

$$Y = Y(Z). \tag{8.41}$$

Fig. 8.21 Initial GSA Γ_9

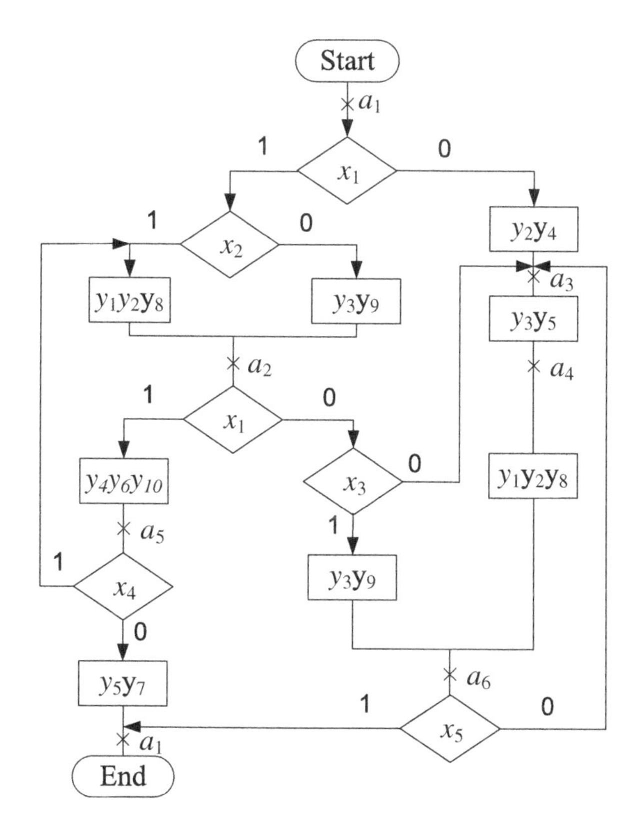

The following conditions should take places for the case of M_0PY Mealy FSM:

$$\begin{aligned} &2^{G+S_0+R}(R+R_Y) \leq V_0; \\ &2^{R_Y} \cdot N \leq V_0. \end{aligned} \tag{8.42}$$

The proposed design method for Mealy FSM $M_0PY(\Gamma_j)$ includes the following steps:

1. Constructing the set of states A.
2. State assignment.
3. Constructing the structure table of FSM $U_1(\Gamma_j)$.
4. Partitioning the set X by classes X^1 and X^2.
5. Replacement of logical conditions $x_l \in X^2$.
6. Encoding of the collections of microoperations.
7. Constructing the transformed structure table.
8. Constructing the table of BIMF.
9. Constructing the table of BMO.
10. Implementing the FSM logic circuit.

Let us discuss an example of design for Mealy FSM $M_0PY(\Gamma_9)$. The GSA Γ_9 is shown in Fig. 8.21. The following sets and their parameters can be found for Mealy FSM $U_1(\Gamma_9)$: $X=\{x_1,\ldots,x_5\}$, $L=5$, $Y=\{y_1,\ldots,y_{10}\}, N=10$, $A=\{a_1,\cdots,a_6\}$, $M=6$, $R=3$, $T=\{T_1,T_2,T_3\}$, $\Phi=\{D_1,D_2,D_3\}$. Let

Table 8.20 Structure table of Mealy FSM $U_1(\Gamma_9)$

a_m	$K(a_m)$	a_s	$K(a_s)$	X_h	Y_h	Φ_h	h
a_1	000	a_2	001	x_1x_2	$y_1y_2y_8$	D_3	1
		a_2	001	$x_1\bar{x}_2$	y_3y_9	D_3	2
		a_3	010	$\bar{x}_1$	y_2y_4	D_2	3
a_2	001	a_5	100	x_1	$y_4y_6y_{10}$	D_1	4
		a_6	101	$\bar{x}_1x_3$	y_3y_9	D_1D_3	5
		a_4	011	$\bar{x}_1\bar{x}_3$	y_3y_5	D_2D_3	6
a_3	010	a_4	011	1	y_3y_5	D_2D_3	7
a_4	011	a_6	101	1	$y_1y_2y_8$	D_1D_3	8
a_5	100	a_2	001	x_4	$y_1y_2y_8$	D_3	9
		a_1	000	$\bar{x}_4$	y_5y_7	–	10
a_6	101	a_1	000	x_5	–	–	11
		a_4	011	$\bar{x}_5$	y_3y_5	D_2D_3	12

Table 8.21 Replacement of logical conditions for Mealy FSM $M_0PY(\Gamma_9)$

a_m	a_1	a_2	a_3	a_4	a_5	a_6
P_1	x_1	x_1	–	–	x_4	x_5

us execute the state assignment in the trivial way: $K(a_1) = 000, \ldots, K(a_6) = 101$. The structure table of FSM $U_1(\Gamma_9)$ includes $H_1(\Gamma_9) = 12$ lines (Table 8.20).

Let the FPGA chip in use include EMBs with the following configurations: 512×1, 256×2, 128×4, 64×8, 32×16 (bits). To implement the circuit of FSM $U_1(\Gamma_9)$, it is necessary $V_0 = 2^8 \cdot 13 = 3328$ (bits). But the EMB in use have only 512 bits. It is necessary at least 7 blocks for implementing the FSM logic circuit. So, the replacement of logical conditions should be used.

There are $T_0 = 7$ different collections of microoperations in the vertices of GSA Γ_9. They are the following: $Y_1 = \emptyset, Y_2 = \{y_1, y_2, y_8\}, Y_3 = \{y_3, y_9\}$, $Y_4 = \{y_2, y_4\}, Y_5 = \{y_3, y_5\}, Y_6 = \{y_4, y_6, y_{10}\}, Y_7 = \{y_5, y_7\}$. It is enough $R_Y = 3$ variables $z_r \in Z$ for encoding of these collections.

Because of $R + R_Y = 6$, the configuration 64×8 should be chosen with $S_A = 6$. It gives $G = 1, S_0 = 2$; therefore, there is $|X^1| = 2, |X^2| = 3$.

Let us represent the set X as $X = X^1 \cup X^2$ where $X^1 = \{x_2, x_3\}$ and $X^2 = \{x_1, x_4, x_5\}$. The table of replacement of logical conditions for FSM $M_0PY(\Gamma_9)$ is represented by Table 8.21.

The following equation can be derived from Table 8.21:

$$P_1 = \bar{T}_1\bar{T}_2x_1 \vee T_1\bar{T}_2\bar{T}_3x_4 \vee T_1\bar{T}_2T_3x_5. \tag{8.43}$$

The circuit of LUTer requires LUTs with $S \geq 6$.

Let use encode the collections of microoperations $Y_t \subseteq Y$ in the trivial way: $K(Y_1) = 000, \ldots, K(Y_7) = 110$. Now, the transformed structure table of Mealy FSM $M_0PY(\Gamma_j)$ can by constructed. The table includes the columns $a_m, K(a_m), a_s, K(a_s), X_h^1, P_h, Z_h, \Phi_h, h$. The column Z_h contains additional variables $z_r \in Z$ which are equal to 1 in the code $K(Y_t)$ written in the h-th row of the table. The transformed table for FSM $M_0PY(\Gamma_9)$ is represented by Table 8.22. This table is a base for constructing the table of BIMF.

Table 8.22 Transformed table of Mealy FSM $M_0PY(\Gamma_9)$

a_m	$K(a_m)$	a_s	$K(a_s)$	X_h^1	P_h	Z_h	Φ_h	h
a_1	000	a_2	001	x_2	P_1	z_3	D_3	1
		a_2	001	$\bar{x}_2$	P_1	z_2	D_3	2
		a_3	010	1	$\bar{P}_1$	z_2z_3	D_2	3
a_2	001	a_5	100	1	P_1	z_1z_3	D_1	4
		a_6	101	x_3	$\bar{P}_1$	z_2	D_1D_3	5
		a_4	011	$\bar{x}_3$	$\bar{P}_1$	z_1	D_2D_3	6
a_3	010	a_4	011	1	1	z_1	D_2D_3	7
a_4	011	a_6	101	1	1	z_3	D_1D_3	8
a_5	100	a_2	001	1	P_1	z_3	D_3	9
		a_1	000	1	$\bar{P}_1$	z_1z_2	–	10
a_6	101	a_1	000	1	P_1	–	–	11
		a_4	011	1	$\bar{P}_1$	z_1	D_2D_3	12

Table 8.23 Part of table of BIMF for Mealy FSM $M_0PY(\Gamma_9)$

$K(a_m)$	P	X^1	Z	Φ	v	h
$T_1T_2T_3$	P_1	x_2x_3	$x_1x_2x_3$	$D_1D_2D_3$		
000	0	00	011	101	1	3
000	0	01	011	101	2	3
000	0	10	011	011	3	3
000	0	11	011	011	4	3
000	1	00	010	001	5	2
000	1	01	010	001	6	2
000	1	10	001	001	7	1
000	1	11	001	001	8	1

Table 8.24 Table of BMO for Mealy FSM $M_0PY(\Gamma_9)$

$K(Y_t)$	Y										t
$z_1z_2z_3$	y_1	y_2	y_3	y_4	y_5	y_6	y_7	y_8	y_9	y_{10}	
000	0	0	0	0	0	0	0	0	0	0	1
001	1	1	0	0	0	0	0	1	0	0	2
010	0	0	1	0	0	0	0	0	1	0	3
011	0	1	0	1	0	0	0	0	0	0	4
100	0	0	1	0	1	0	0	0	0	0	5
101	0	0	0	1	0	1	0	0	0	1	6
110	0	0	0	0	1	0	1	0	0	0	7
111	0	0	0	0	0	0	0	0	0	0	8

The table of BIMF includes the following columns: $K(a_m), P, X^1, Z, \Phi, v$. In the discussed case, transitions from each state $a_m \in A$ are represented by 8 rows of the table of BIMF. The part of this table is represented by Table 8.23. It describes the transitions from state $a_1 \in A$.

The table of BMO includes the columns $K(Y_t)$, Y, t. In the discussed case, this table is represented by Table 8.24.

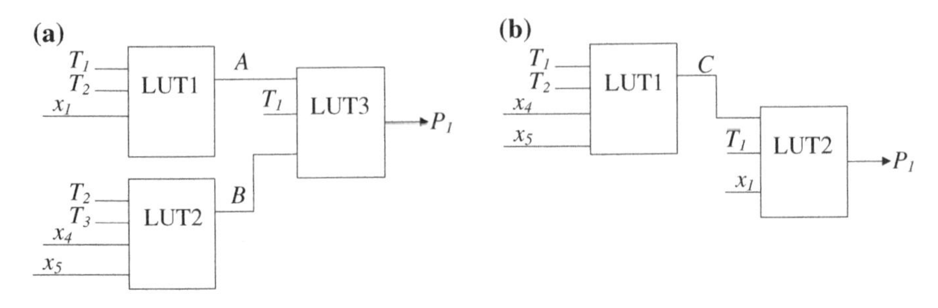

Fig. 8.22 Logical circuit of BRLC for Mealy FSM $M_0PY(\Gamma_7)$ (a) and $M_EPY(\Gamma_9)$ (b)

Fig. 8.23 Outcome of special state assignment for Mealy FSM $U_1(\Gamma_9)$

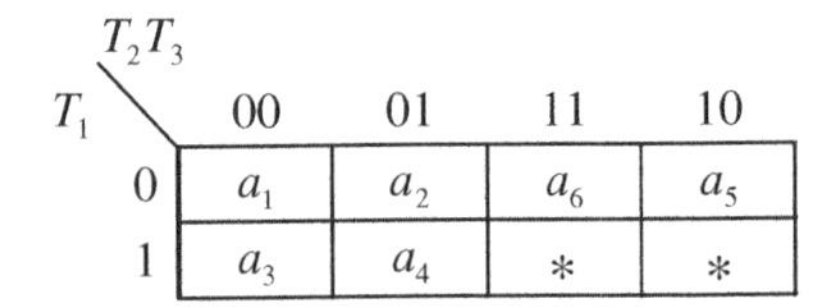

T_1 \ T_2T_3	00	01	11	10
0	a_1	a_2	a_6	a_5
1	a_3	a_4	*	*

Fig. 8.24 Structural diagram of M_EPY Mealy FSM

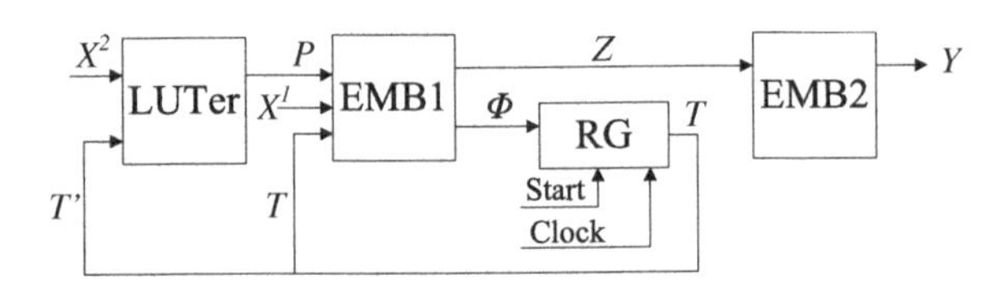

Let the FPGA chip in use include LUTs having $S = 4$. It means that the expression (8.43) should be transformed in the following way:

$$P_1 = \bar{T}_1\bar{T}_2x_1 \vee T_1\left(\bar{T}_2\bar{T}_3x_4 \vee \bar{T}_2T_3x_5\right) = A \vee T_1B. \tag{8.44}$$

The expression (8.44) corresponds to the circuit of BRLC shown in Fig. 8.22, a. The number of LUTs in the circuit of BRLC can be decreased due to the special state assignment [4]. In this case set A is represented as $A^1 \cup A^2$. The set A^1 includes states $a_m \in A$ with conditional transitions, as well as the initial state $a_1 \in A$. The set A^2 includes states $a_m \in A$ with the unconditional transitions. The state assignment starts from the states $a_m \in A^1$. The codes $K(a_m)$ for states $a_m \in A^1$ correspond to decimal numbers from 0 to $M_1 - 1$, where $|A^1| = M_1$.

It is enough R_E variables $T_r \in T$ for encoding of the states $a_m \in A^1$, where

$$R_E = \lceil \log_2 M_1 \rceil. \tag{8.45}$$

In the case of FSM $U_1(\Gamma_9)$, there are the following sets: $A^1 = \{a_1, a_2, a_5, a_6\}$ and $A^2 = \{a_3, a_4\}$. There is $R_E = 2$, so, the states $a_m \in A^1$ can be determined using only state variables T_2 and T_3 (Fig. 8.23).

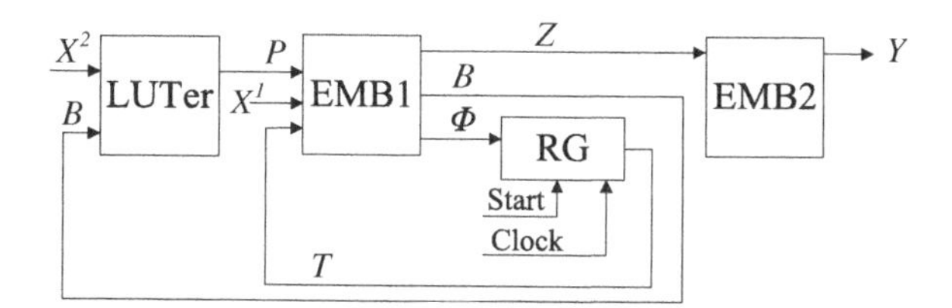

Fig. 8.25 Structural diagram of M_CPY Mealy FSM

Such an approach leads to M_EPY Mealy FSM. Its structural diagram is shown in Fig. 8.24. The set $T' \subseteq T$ includes R_E variables.

The only difference between the M_0PY and M_EPY FSMs is reduced to the system P. In the later case the additional variables are represented by the following system:

$$P = P\left(T', X_2\right). \tag{8.46}$$

In the discussed case, the following equation can be found for the function $P_1 \in P$:

$$P_1 = \bar{T}_2 x_1 \vee T_2 \bar{T}_3 x_4 \vee T_2 T_3 x_5 = \bar{T}_2 x_1 \vee C. \tag{8.47}$$

The logic circuit of BRLC for Mealy FSM $M_EPY(\Gamma_9)$ is show in Fig. 8.22b. It requires 1, 5 times less amount of LUTs than its counterpart for $M_0PY(\Gamma_9)$.

The only difference between design methods for M_0PY and M_EPY FSMs is reduced to the different state assignments. For M_EPY Mealy FSM, the special state assignment should be executed.

If $G = 1$, then the approach of encoding of logical conditions can be applied. Let the symbol $X(P)$ stand for the set of logical conditions replaced by the variable $P_1 \in P$. It is enough R_L variables for encoding of the logical conditions $x_l \in X(P)$:

$$R_L = \left\lceil \log_2 |X(P)| \right\rceil. \tag{8.48}$$

Let us use the variables $b_r \in B$ for encoding of logical conditions. This approach leads to M_CPY Mealy FSM (Fig. 8.25).

In M_CPY Mealy FSM, the LUTer implements system

$$P = P\left(B, X^2\right). \tag{8.49}$$

The block EMB1 implements system (8.16), (8.40) and

$$B = B\left(T, P, X^1\right). \tag{8.50}$$

The block EMB2 implements system (8.41).

The design method for Mealy FSM $M_CPY(\Gamma_j)$ includes all steps presented in the design method for M_0PY FSM. But the encoding of logical conditions is executed before the step of their replacement. Let us discuss an example of design for Mealy FSM $M_CPY(\Gamma_9)$.

The steps from 1 to 4 are executed before. There is the set $X(P) = X^2 = \{x_1, x_4, x_5\}$ with $L_P = 3$. These logical conditions can be encoded using $R_L = 2$ variables. It

Table 8.25 Transformed structure table of Mealy FSM $M_CPY(\Gamma_9)$

a_m	$K(a_m)$	a_s	$K(a_s)$	X_h^1	P_h	Z_h	B_h	Φ_h	h
a_1	000	a_2	001	x_2	P_1	z_3	–	D_3	1
		a_2	001	$\bar{x}_2$	P_1	z_2	–	D_3	2
		a_3	010	1	$\bar{P}_1$	z_2z_3	–	D_2	3
a_2	001	a_5	100	1	P_1	z_1z_3	–	D_1	4
		a_6	101	x_3	$\bar{P}_1$	z_2	–	D_1D_3	5
		a_4	011	$\bar{x}_3$	$\bar{P}_1$	z_1	–	D_2D_3	6
a_3	010	a_4	011	1	1	z_1	–	D_2D_3	7
a_4	011	a_6	101	1	1	z_3	–	D_1D_3	8
a_5	100	a_2	001	1	P_1	z_3	b_2	D_3	9
		a_1	000	1	$\bar{P}_1$	z_1z_2	b_2	–	10
a_6	101	a_1	000	1	P_1	–	b_1	–	11
		a_4	011	1	$\bar{P}_1$	z_1	b_1	D_2D_3	12

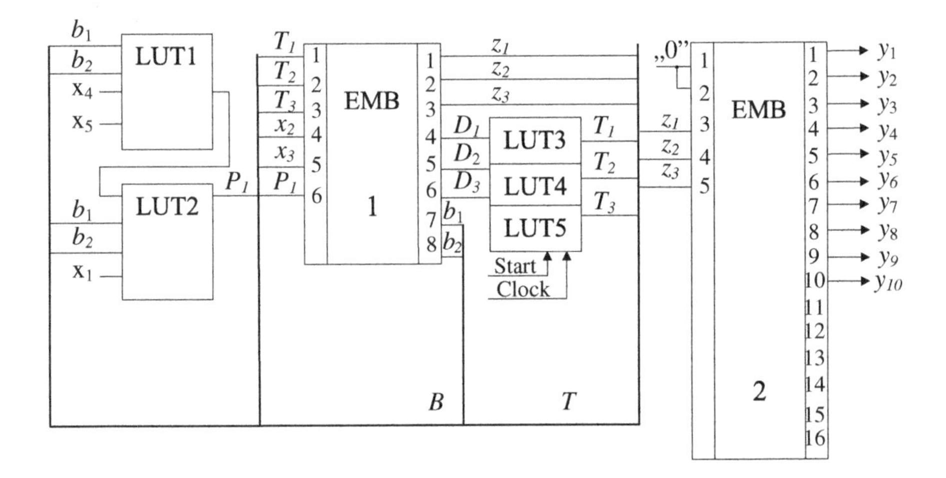

Fig. 8.26 Logic circuit of Mealy FSM $M_CPY(\Gamma_9)$

gives the set $B = \{b_1, b_2\}$. Let us encode the logical conditions in the following manner: $K(x_1) = 00$, $K(x_4) = 01$ and $K(x_5) = 10$. It gives the following equation for the block BRLC:

$$P_1 = \bar{b}_1\bar{b}_2x_1 \vee b_2x_4 \vee b_1x_5 = A \vee C. \tag{8.51}$$

Let us encode the collections of microoperations as for the $M_0PY(\Gamma_9)$.

The transformed structure table of M_CPY Mealy FSM includes all columns of its counterpart for M_0PY Mealy FSM. Also, it includes the column B_h with the variables $b_r \in B$ equal to 1 in the code $K(x_l)$ from the h-th row of the table.

The transformed structure table of Mealy FSM $M_CPY(\Gamma_9)$ is represented by Table 8.25. The table of BIMF includes an additional column B with $K(x_l)$. The

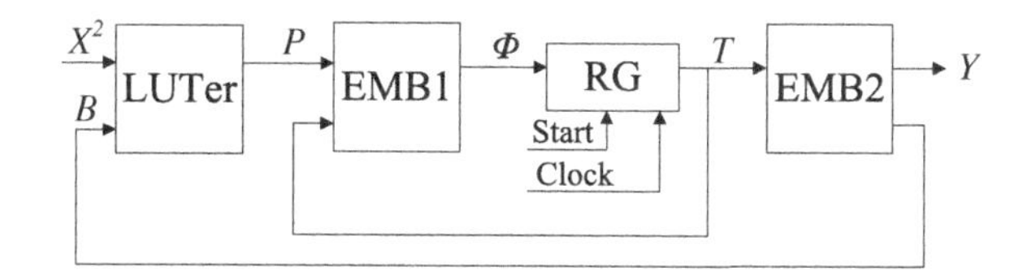

Fig. 8.27 Structural diagram of Moore FSM

table of BMO is the same as for Mealy FSM $M_0PY(\Gamma_9)$. The logic circuit of Mealy FSM $M_CPY(\Gamma_9)$ is shown in Fig. 8.26.

This approach can be used if the following condition takes place:

$$2^{R+S_0+G}(R + R_Y + R_L) \leq V_0. \tag{8.52}$$

It there is $G > 1$, then logical conditions from different sets $X(P_g)$ should be encoded using different variables $b_r \in B$. Let $X(P_g) \subseteq X$ is a set of logical conditions, it is enough $R_L{}^g$ variables where

$$R_L^g = \lceil \log_2 |X(Pg)| \rceil. \tag{8.53}$$

It gives the value of $R_L = R_L^1 + R_L^2 + \cdots + R_L^G$. This value should be used in (8.52).

The encoding of logical conditions can be used for optimizing the BRLC of Moore FSM. In this case, the codes $K(x_l)$ should be added into collections of microoperations. It leads to M_CPY Moore FSM shown in Fig. 8.27.

This approach can be used together with the method of optimal state assignment, as well as the encoding of the classes of pseudoequivalent states. We do not discuss these approaches in this chapter.

References

1. Baranov SI (1994) Logic synthesis of control automata. Kluwer, Boston
2. Barkalov A (1998) Principles of logic optimization for a Moore microprogrammed automaton. Cybern Syst Anal 34(1):54–61
3. Barkalov A, Titarenko L (2009) Logic synthesis for FSM-based control units. In: Number 53 in Lecture notes in electrical engineering. Springer, Heidelberg
4. Barkalov A, Zelenjova I (2000) Optimization of replacement of logical conditions for an automaton with bidirectional transitions. Autom Control Comput Sci 34(5):48–53
5. Łuba T (1994) Multi-level logic synthesis based on decomposition. Microprocess Microsyst 18(8):429–437
6. Rawski M, Selvaraj H, Łuba T (2005) An application of functional decomposition in ROM-based FSM implementation in FPGA devices. J Syst Archit 51(6–7):423–434

Chapter 9
Finite State Machines with Operational Implementation of Transitions

Abstract Chapter is devoted to the using the data-path for decreasing the number of LUTs in logic circuits of FPGA-based Moore FSMs. Firstly, the principle of operational implementation of interstate transitions is proposed. It is based on the usage of operational elements (adders, counters, shifters and so on) for calculating codes of the states of transitions. Next, the organization of FSM with operational implementation of interstate transitions is discussed. An example is given for application of the proposed method. Next, the base structure of synthesis process is proposed for Moore FSM with operational implementation of interstate transitions. The structure of the synthesis process depends on initial conditions such as set of operations or codes of FSM states. The typical structures are discussed for the operational automaton executing the transitions. Next, the method is shown based on mixture of traditional and proposed approaches for calculation of the codes of states of transitions. The last part of the chapter discusses the efficiency of proposed solution. The chapter is written together with PhD Roman Babakov (Donetsk National Technical University, Ukraine).

9.1 Conception of Operational Implementation of Transitions

The base for classical methods of FSM design is a proposed by Viktor Glushkov canonical method of structural synthesis. According to this principle, the logic circuit of BIMF is represented by a system of Boolean functions (SBF). In this system, both state variables and logical conditions are connected by Boolean operations such as negation, conjunction or disjunction.

The usage of SBF for representing a logic circuit is rather convenient because the corresponding synthesis methods, as well as different optimization techniques, are profoundly examined. There are very efficient methods targeting different logic elements (gates, PAL, PLA, CPLD, FPGA, and so on) [8]. Up-to-day CAD tools support synthesis based on Boolean functions. Moreover, a lot of industrial CAD packages include embedded tools targeting minimizing logic circuits [10, 12].

V. Sklyarov et al., *Synthesis and Optimization of FPGA-Based Systems*,
Lecture Notes in Electrical Engineering 294, DOI: 10.1007/978-3-319-04708-9_9,

The following peculiarities should be taken into account under implementing the BIMF circuit using SBF:

1. The complexity of an SBF is increased as far as the number of interstate transitions is increased. The growth for the number of interstate transitions results in increase for the number of product terms in the SBF to be implemented, as well as for the number of literals in these terms. It is connected with increasing number of state variables. Because the state codes are unique for each state, the growth of hardware amount with increasing the parameters of GSA is a natural process. This event can be partially compensated due to applying different optimization methods targeting FSMs [1, 7, 9]. In some cases, it can be obtained analytical dependences among the parameters of GSA and the number of logic elements required for implementing the corresponding FSM circuit. For example, it can be done for PALs or PLAs.
2. The exact minimization of SBF can be executed only by the complete enumeration for all possible solutions. As it is known, this problem is a *NC*-complete and finding the exact solution is a very time-consuming task [13]. There are a lot of heuristic methods decreasing the number of enumerations, but they do not guarantee finding the optimal solution [11].

Let us consider the proposed conceptual approach for constructing the BIMF. If some conditions take places, this approach allows restriction the complexity of BIMF logic circuit with the growth for the number of FSM transitions [2, 3].

The main goal of a control unit is the generation of a proper sequence of collections of microoperations entering a data-path (operational automaton) of some digital system. This sequence is determined by a particular GSA representing the control algorithm for executing some task. The operational automaton (OA) executes the required data processing. To do it, some operational blocks are used such as adders, multipliers, shifters, and so on. As a rule, the FSM processes logical conditions using either Boolean functions or truth tables. For example, the addresses of transitions are represented by a table corresponding to the control memory of the microprogram control unit with compulsory addressing of microinstructions. The only exception is a compositional microprogram control units, where some addresses of transitions are generated by incrementing a counter [5, 6].

As a rule, a state code is considered as a binary vector. But it is quite possible to consider it as some arithmetical value in some positional number system (mostly, in binary system). For example, the state code $K(a_i) = 11101011_2$ can be viewed either as the integer number 235, or the signed number -107, or as the two-complement number -21, or as the real number $-10.11_2 = -2.75_{10}$ and so on. Obviously, if state codes are treated as binary numbers, the different arithmetical, as well as logical, operations can be executed under these numbers.

Each transition of FSM can be viewed as a transformation of the code $K(a_i)$ of a current state into the code $K(a_j)$ of a state of transition. Let it be necessary, for example, to transform the state code $K(a_i) = 01010110_2 = 86_{10} = +86_{SM} = +86_{2C}$ into the code $K(a_j) = 10101001_2 = 169_{10} = -41_{SM} = -87_{2C}$. The subscript SM

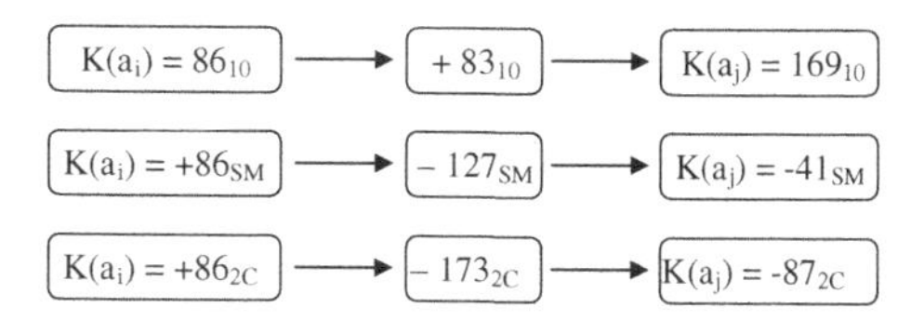

Fig. 9.1 Transformation of binary vectors as different numbers

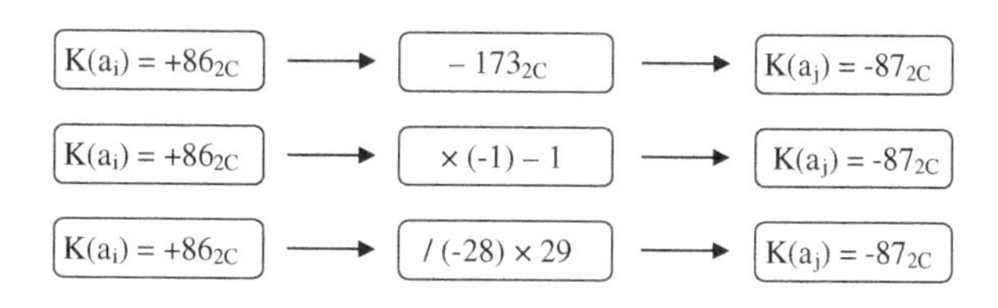

Fig. 9.2 Equivalent transformations of binary vectors as two's complements

means that the number is represented in the sign-and-magnitude form, whereas the subscript 2C in the two's complement form [8].

Interpreting the binary vectors of the state codes as numbers with different representation, the required transformation $K(a_i) \rightarrow K(a_j)$ can be executed, for example, in the ways shown in Fig. 9.1. Here the central blocks includes arithmetical operations necessary for transformation of $K(a_i)$ into $K(a_j)$. Each of these approaches gives the resulting binary vector $K(a_j)$.

Obviously, each of these transformations can be executed using different operations, as well as sequences of some operations. For example, three variants are shown in Fig. 9.2 for the transformation $K(a_i) \rightarrow K(a_j)$ if these codes are treated as two's complements. In the last case, the division of $K(a_i)$ by −28 is executed as the exact division (without saving the residue of division).

In some cases, the transformation can be executed using Boolean operations under the binary vector of a state code, as well as using some combination of arithmetical and logic operations. For example, the discussed transformation $K(a_i) \rightarrow K(a_j)$ can be executed by bit-wise inversion of the vector $K(a_i)$. From the design point of view, this operation leads to the circuit having maximum performance and minimum hardware amount in comparison with other discussed operations. But it is a particular case, which takes place only for discussed codes and only if they are represented as double-byte numbers.

The following two statements can be made on the base of previous discussion:

Statement 9.1 The transformation of FSM codes is possible with usage of arithmetical and logic operations, whose choice depends on the mathematical interpretation of corresponding binary vectors of state codes.

Statement 9.2 In the common case, there are a lot of transformation variants with arithmetical and logic operations. It can be chosen at least one variant leading to the logic circuit having either minimum hardware amount or maximum performance in comparison with other possible variants for given logic elements.

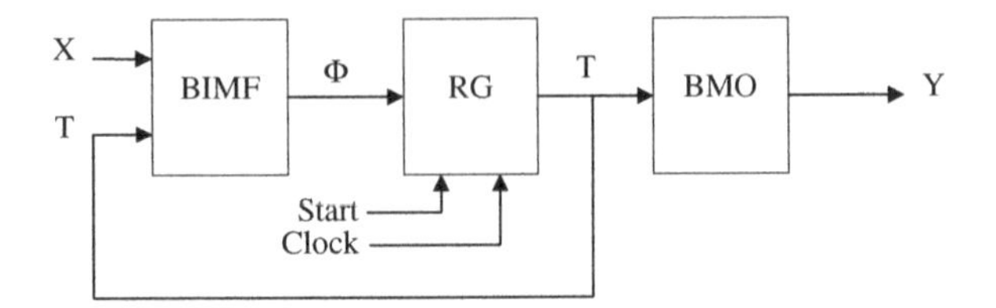

Fig. 9.3 Canonical structure diagram of Moore FSM

Let us name as *operational generation of transitions* this new approach of transformation of FSM codes using arithmetical and logic operations. This approach leads to the new structure (model) of FSM where the block BIMF is represented as a composition of combinational circuits executing different arithmetic and logic operations.

9.2 Organisation of FSM with Operational Generation of Transitions

The canonical structure diagram of Moore FSM [9] is shown in Fig. 9.3. In this model, the block BIMF implements the system of input memory functions Φ and generates a code of the next state entering the register RG. The code of the current state represented by state variables from the set T enters both blocks BIMF and BMO. The BMO is implemented using ROM; it keeps microoperations from the set Y. The microoperations enter the data-path and initiate execution of some primitive operations. The pulse *Start* is used for loading the code of initial state into RG. The pulse *Clock* causes changing the content of RG.

Let a Moore FSM corresponding to some GSA have the set of states $A = \{a_1,\ldots, a_M\}$. Let the GSA include branches corresponding to interstate transitions (conditional and unconditional) and forming the set $B = \{B_1,\ldots, B_V\}$. Let us name them as interstate branches. Obviously, each interstate branch corresponds to one unique row of the FSM structure table [9]. If all transitions are unconditional, then there is $V = M$, otherwise there is $V > M$.

Each interstate branch corresponds up to R product terms in the system of input memory functions, where R is the number of bits in state codes. It is clear, that the growth of the number of branches leads to the increasing for the hardware amount in the FSM logic circuit. This dependence is approximately linear.

Let us transform the model (Fig. 9.3) in the following manner.

1. Let us represent BIMF as a composition of combinational circuits $CC(O_i)$ $(i = \overline{1, Q})$.
   Each of them implements some unique arithmetical or logic operation (OP) $O_i \in O$ using both state variables T and logical conditions X as operands (Fig. 9.4). The outputs of circuits $CC(O_i)$ enter the multiplexer MX. The MX is controlled by the code of operation Ψ; it generates the input memory functions Φ to load the code of the next state into RG. Let us denote the operations implemented by circuits $CC(O_i)$ as operations of transitions. Let us name the

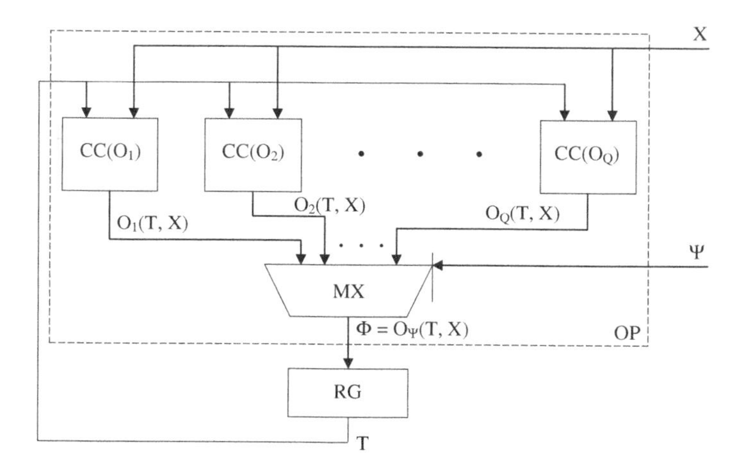

Fig. 9.4 Structure of operational part of Moore FSM

collection of the blocks {CC(O_1),..., CC(O_Q), MX) as operational part (OP) of an Moore FSM. The function of OP is generation of input memory functions Φ (code of the next state) on the base of state variables T (code of the current state), logical conditions X and code of operation Ψ:

$$\Phi = \Phi(T, X, \Psi). \tag{9.1}$$

It should be point out that the structure of OP (Fig. 9.4) is similar to the combinational part of operational automaton [9]. The register RG receives data from the output of multiplexer MX, whereas the output of RG enters the inputs of OP. Therefore, the register can be treated as a memory of OA. Because the block OP implements de facto the interstate transitions, let us name the pair <OP, RG> as *operational automaton of transitions* (OAT).

2. Let us introduce an additional block of operation of transition (BOT) in the structure of FSM. It is implemented using ROM (EMBs of FPGAs). The main function of the BOT is the generation of the code of operation of transition Ψ on the base of state variables T. Next, this code enters the input of multiplexer MX. Let us name the resulting FSM structure as the Moore FSM with operational automaton of transitions (FSM with OAT). The structure diagram of the Moore FSM with OAT is shown in Fig. 9.5. This FSM is based on the principle of operational generation of interstate transitions.
   The FSM with OAT operates in the following manner. In each cycle of operation, the register RG receives a code of the next state, represented by functions Φ. Now this code is represented by the state variables T and is treated as a code of the current state. Using this code, the block BMO generates microoperations Y. At the same time, the block BOT generates a code of operation Ψ entering OAT. Using this code, the OP generates new values of input memory functions executing one of Q possible operations of transitions.

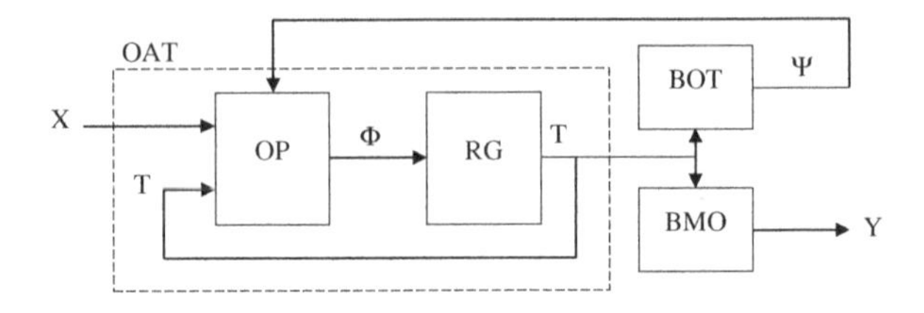

Fig. 9.5 Structure of FSM with OAT

Therefore, the FSM with OAT possesses the following peculiarities distinguishing it from the traditional Moore FSM:

1. A state code is interpreted not as a collection of bits, but as some arithmetical value represented in the binary number system.
2. The transformation of the current state code $K(a^t)$ into the next state code $K(a^{t+1})$, where $t = 0, 1$, is an automaton time, is executed using a set of arithmetical and logic operations.
3. The choice of operation of transition is a function of the current state code.

If there is an unconditional transition from the state a^t into the state a^{t+1}, then the function of transition depends only on the code of current state:

$$K(a^{t+1}) = O^t(K(a^t)). \tag{9.2}$$

In (9.2), the symbol O^t stands for operation of transition used for execution a transition from the state a^t.

If there are conditional transitions from the state a^t, then they can be made into some of the FSM states depending on the values of both the current state code and logical conditions sufficient for these transitions. The function of transformation is the following one:

$$K(a^{t+1}) = O^t(K(a^t), X^t). \tag{9.3}$$

In (9.3), the symbol $K(a^{t+1})$ stands for the code of the next state; X^t is a subset of the set of logical conditions checked for executing transitions from the current state; O^t is an operation of transition implementing a conditional transition from the current state.

Thus, the following sets can be found in the structural representation of the FSM with OAT:

1. The set of the state codes $K = \{K(a_1), K(a_2),\ldots,K(a_M)\}$, where M is the cardinal number of the set A. Only one transition corresponds to each state (it can be either unconditional or conditional).
2. The set of operations of transitions $O = \{O_1,\ldots,O_Q\}$, where $Q \leq M$. In the general case, each element of the set O allows different implementation (it can be implemented using different combinational circuits). The appropriate choice of a way for implementing each operation influences significantly the hardware characteristics of the final FSM circuit.

The relation $Q \leq M$ is based on the fact that the same operation can be used for executing more than one transition. For example, the transition from the state a_i

having the code 20 into the state a_j with the code 40 can be implemented using the operation "+20". The same is true for the transition from the state a_i having the code 34 into the state a_j with the code 54. In this case, the same operation of transition can be used for executing these transitions. It means that both transitions are executed using the same combinational circuit. Obviously, the same combinational circuit can be used for executing from 1 to M transitions. It is clear, that operations of transitions chosen for given GSA should provide executing all interstate transitions for given values of state codes.

The proposed FSM model can be represented by the following vector:

$$S = < K, O, Y >. \tag{9.4}$$

In accordance with [9], the synthesis of the logic circuit of FSM with OAT is reduced to constructing and physical implementation of all sets from (9.4). The microoperations are implemented using operational vertices of a particular GSA, whereas both the sets K and O are constructed in accordance with given optimization criteria.

9.3 Example of FSM Design

Let us discuss a design example for Moore FSM with OAT. The main goal of this example is only outlining the proposed principle. We do not consider such issues as minimizing the hardware amount and optimizing the FSM performance. It is connected with the fact that the example is very simple. Let a control algorithm be represented by GSA Γ_{10} (Fig. 9.6). Let us construct the set of operations of transitions including three elements. The operation O_1 is an operation of unconditional "sequential" transition corresponding to the "down" transition along the GSA to the next state. Let the operation O_1 correspond to the following expression:

$$O_1(a^t) = a^t - k_1, \tag{9.5}$$

where k_1 is some constant. Let us point out that the symbol a^t is treated as the code $K(a^t)$. It means that all similar expressions use the state codes.

The operation O_2 is an operation of conditional transition. This operation produces one of two possible results depending on the value of a logical condition to be checked:

$$O_2(a^t, x^t) = \begin{cases} 2a^t + k_2, & x^t = 0; \\ 2a^t - k_2, & x^t = 1, \end{cases} \tag{9.6}$$

where k_2 is some constant. This operation can be divided by two parts and represented as:

$$O_{2-0}(a^t) = 2a^t + k_2; \tag{9.7}$$

$$O_{2-1}(a^t) = 2a^t - k_2. \tag{9.8}$$

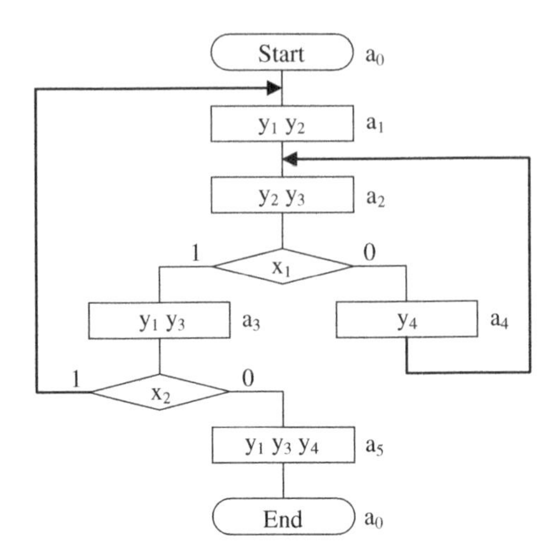

Fig. 9.6 Graph-scheme of algorithm Γ_{10}

The operation O_3 is an operation of unconditional "reverse" transition; it is a transition in the higher point of a GSA. In the discussed example, such a transition is executed from the state a_4 into the state a_2. Let us define this operation as the following one:

$$O_3(a^t) = a^t - k_3, \tag{9.9}$$

where k_3 is some constant.

Let us point out that we do not discuss how to choose the expressions (9.5) –(9.9). The special method should be applied for the choice. Obviously, quite different operations can be chosen for another GSA.

Let us construct a system of equations having number values of state codes as their roots. Each equation of the system corresponds to a unique transition, whereas the total number of equations is equal to the number of interstate transitions for a given GSA. Each transition is executed using one of the functions O_1–O_3.

In the general case, any from the appointed operations can be used for executing a transition. The only condition is the correct implementation of the transition. But let us follow the above chosen appointments for the operations of transitions. So, the transitions $a_0 \rightarrow a_1$, $a_1 \rightarrow a_2$, and $a_5 \rightarrow a_0$ should be executed using the operation O_1. The transitions $a_2 \rightarrow a_3$ and $a_3 \rightarrow a_1$ should be executed using the operation $O_{2\text{-}1}$, the transitions $a_2 \rightarrow a_4$ ï£¡ $a_3 \rightarrow a_5$ using the operation $O_{2\text{-}0}$. At last, the transition $a_4 \rightarrow a_2$ should be executed using the operation O_3. As a result, the following system of equations can be constructed:

$$\begin{cases} a_0 = O_1(a_5); & a_2 = O_3(a_4); \\ a_1 = O_1(a_0); & a_3 = O_{2-1}(a_2); \\ a_1 = O_{2-1}(a_3); & a_4 = O_{2-0}(a_2); \\ a_2 = O_1(a_1); & a_5 = O_{2-0}(a_3). \end{cases} \tag{9.10}$$

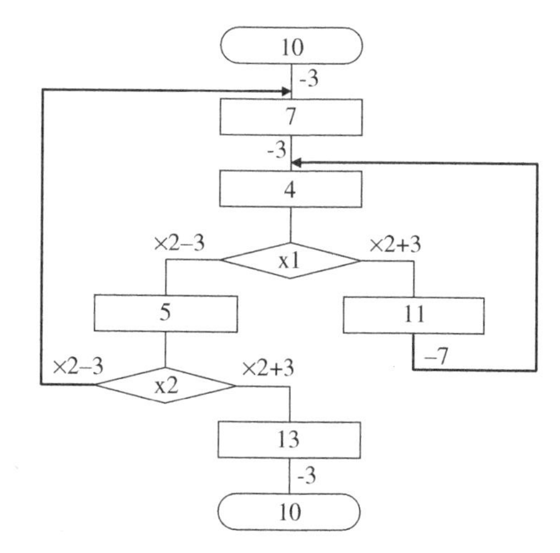

Fig. 9.7 Graph corresponding GSA $\Gamma_1 0$ with codes of states and operations of transitions

Obviously, the system (9.10) has a lot of possible solutions (roots). Let us choose the constants $k_1 = k_2 = 3$, $k_3 = 7$ for operations O_1–O_3, respectively. In this case, the following decimal values can be obtained corresponding to the state codes satisfying the system (9.10): $K(a_0) = 10$, $K(a_1) = 7$, $K(a_2) = 4$, $K(a_3) = 5$, $K(a_4) = 11$, $K(a_5) = 13$.

This solution can be shown as a following graph (Fig. 9.7). The rectangle vertices of this graph contain state codes, whereas it edges correspond to operations of transitions.

Despite the fact that there are only $M = 6$ states for Moore FSM corresponding to GSA Γ_{10}, the maximum decimal number used as a state code is equal to 13. Obviously, it is necessary 4 bits for state assignment. This value determines parameters of combinational blocks, RG and PROM of OAT.

One of two logical conditions (x_1 or x_2) is analyzed by the operation O_2. To choose one of these conditions, the well-known method of replacement of logical conditions can be used [1, 7]. To do it, let us add the field Z into the PROM of the block BOT. This field contains a code of logical condition to be chosen. In the discussed example, this field contains only one bit. To transfer only one of logical conditions on the input of OP, it is necessary to place a special multiplexer of logical conditions before the combinational block executing operation O_2. The multiplexer is controlled by the code from the field Z.

Let us encode operations O_1–O_3 in the following manner: $K(O_1) = 00$, $K(O_2) = 01$, and $K(O_3) = 10$. Let us encode the logical conditions using the following codes: $K(x_1) = 0$ and $K(x_2) = 1$. Let us construct a table reflecting the content of PROM of BOT (Table 9.1). In this table, the symbol "*" stands for the "don't care" value of a particular bit of PROM.

Let us point out that the content of PROM of the block BMO is constructed in a manner similar to constructing this content for the model shown in Fig. 9.1. To do it, both contents of operational vertices and values of state codes should be used. In the discussed example, the BMO should have four address inputs, whereas only

Table 9.1 Content of PROM of BOT (for GSA Γ_{10})

a_i	$K(a_i)$	$T_1T_2T_3T_4$	Ψ	Z
		0 0 0 0	**	*
		0 0 0 1	**	*
		0 0 1 0	**	*
		0 0 1 1	**	*
a2	4	0 1 0 0	01	0
a3	5	0 1 0 1	01	1
		0 1 1 0	**	*
a1	7	0 1 1 1	00	*
		1 0 0 0	**	*
		1 0 0 1	**	*
a0	10	1 0 1 0	00	*
a4	11	1 0 1 1	10	*
		1 1 0 0	**	*
a5	13	1 1 0 1	00	*
		1 1 1 0	**	*
		1 1 1 1	**	*

Table 9.2 Content of PROM of BMO (for GSA Γ_{10})

a_i	$K(a_i)$	$T_1T_2T_3T_4$	y_1 y_2 y_3 y_4
		0 0 0 0	* * * *
		0 0 0 1	* * * *
		0 0 1 0	* * * *
		0 0 1 1	* * * *
a_2	4	0 1 0 0	0 1 1 0
a_3	5	0 1 0 1	1 0 1 0
		0 1 1 0	* * * *
a_1	7	0 1 1 1	1 1 0 0
		1 0 0 0	* * * *
		1 0 0 1	* * * *
a_0	10	1 0 1 0	0 0 0 0
a_4	11	1 0 1 1	0 0 0 1
		1 1 0 0	* * * *
a_5	13	1 1 0 1	1 0 1 1
		1 1 1 0	* * * *
		1 1 1 1	* * * *

three inputs are necessary for the Moore FSM shown in Fig. 9.1. The content of BMO based on the one-hot encoding of microoperations is shown in Table 9.2.

The structure diagram of Moore FSM with operational automaton of transitions is shown in Fig. 9.8. The blocks O_1–O_3 implement the corresponding operations represented by expressions (9.5)–(9.9). The multiplexer of logical conditions MX_1 generates values of logical conditions in accordance with values of variables Z. The multiplexer of result MX_2 generates the values of input memory functions Φ in accordance with values of variables Ψ. Of course, this set of operational blocks is unique for each FSM with operational automaton of transitions. But the design method is general for any initial graph-scheme of algorithm.

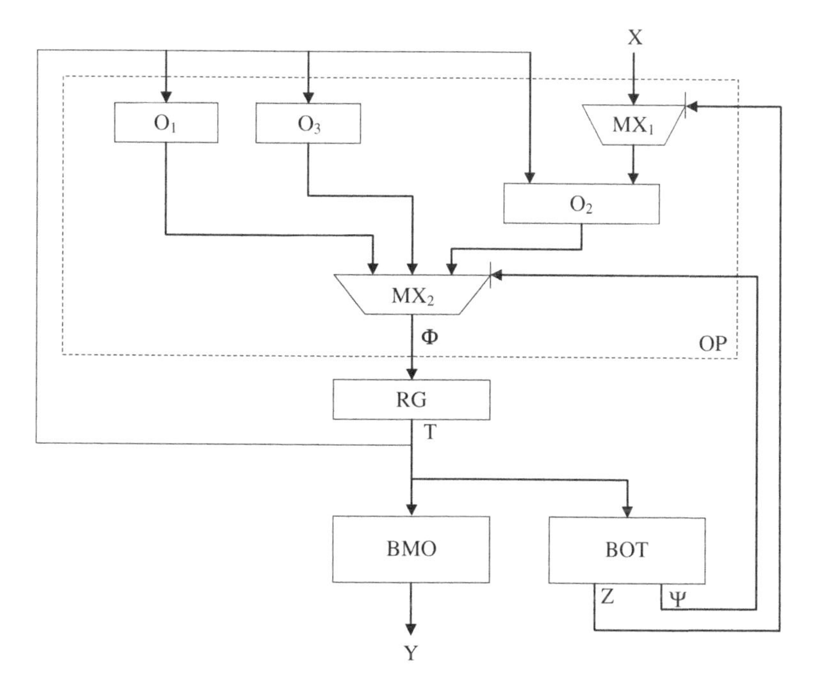

Fig. 9.8 Structure of Moore FSM with OAT (for GSA Γ_{10})

9.4 Structural Representation of Synthesis Process for FSM with OAT

There are some problems connected with the development of general synthesis method for FSM with OAT. First of all, it is necessary to determine the parameters of the FSM used as starting conditions for the process of synthesis. The following issues should be found:

- a fixed set of state codes used for constructing the set of operations of transitions;
- a fixed set of operations of transitions determined by available operational blocks (these blocks are viewed as library elements of a particular CAD);
- an optimization criterion used for choosing either state codes or operations of transitions.

Of course, different parameters can be chosen as starting, as well as more than one parameter can be chosen. This choice depends on a lot of particular factors. The final choice determines the peculiarities of synthesis process.

Apart from the proper choice of starting parameters, it should be taken into account that different stages of synthesis can be carried out by different tools. For example, different sets of operations of transitions can be constructed for given values of state codes. These sets differ in both hardware amount and performance of the final FSM circuit. On the other hand, the state codes can be chosen in different ways for the same set of operations. In this case, the choice influences the number of bits in state codes and, therefore, on hardware amount for blocks BOT and BMO.

Undoubtedly, the choice of this or that approach for implementing some stage of synthesis influences the final result, but it is very difficult to estimate this influence before getting the FSM circuit. Therefore, a designer of FSM with OAT should deal with a lot of variants for starting conditions' choice, as well as a lot of possibilities for executing the synthesis stages. At the same time, there are no precise preliminary knowledge about such important issues as the synthesis stages, their number and the order of their execution. Such a problem definition complicates the development of general approach for synthesis of FSM with OAT.

9.4.1 Base Structure of Synthesis Process for FSM with OAT

Let us consider the following approach for development of the general design method for FSM with OAT [4].

1. Let us form the set of main stages for design process. The following issues can be elements of this set:

   a. the choice of values for state codes;
   b. the constructing the set of operations of transitions;
   c. the constructing the set of interstate transitions for given GSA (if it is necessary, additional vertices can be introduced into initial GSA corresponding to idle FSM states);
   d. the data types used for interpreting state codes (for example, the state codes can be treated as either sign-and-magnitude integers or one's-complements, or two's-complements, or floating point numbers, or binary-encoded-decimal numbers and so on);
   e. the set of library elements used for implementing operations of transitions (the same OT can be implemented using blocks targeting either hardware minimization or performance maximization; this choice influences the area of a chip occupied by a particular circuit);
   f. the basic optimization criterion used for FSM circuit (hardware amount, performance, consumed power, reliability, and so on).

   Obviously, other synthesis steps can be added to meet some specifics of a particular project.
2. Let us represent the synthesis process as some directed graph. The graph nodes correspond to synthesis stages. The graph edges correspond to possible connections of synthesis process. The above mentioned stage of synthesis allows constructing the graph for FSM with OAT (Fig. 9.9). This graph is characterized by the comprehensive whole of interlacement and stable connections among its components. Due to it, the graph can be named as the *basic*

structure of synthesis process for FSM with OAT. The term "basic" means that this structure is not rigid; there is a possibility for either adding or deleting some nodes or edges.

Let us consider the connections between the elements of the basic structure.

The block 6 "Main optimization criterion" may influence practically all other blocks. The approach used for implementing all other blocks depends significantly on the optimization criterion in use. No another block might change the optimization criterion.

The block 7 "Level of optimization" allows restricting complexness of methods used in blocks 1–5.

The blocks 3 and 4 affect the generation process for state codes (block 1). Using state codes generated by the block 1, the set of operations of transitions can be developed (block 2).

Using the set of operations of transitions from block 2, it is possible to choose the format for code states (block 4) as well as the set of interstate transitions can be found (block 3) determining the values of state codes (block 1). Using blocks 2, 6 and 7, the choice of library elements is made for implementing the operations of transitions (block 5).

The block 8 "Functional circuit of FSM with OAT" has no outgoing edges. It can be viewed as a final node of the graph. It contains an outcome of the design process. The content of this block is determined by both the block 5 (the set of combinational circuits of operational part) and the block 1 (number of bits in both state codes and addresses of data in PROMs of BMO and BOT).

9.4.2 Refinement of Basic Structure of Synthesis Process

The block diagram (Fig. 9.9) represents only a collection of possible synthesis steps and their interconnections for FSM with OAT. To obtain a synthesis method which can be used in practice, it is necessary to refine the basic structure. The refinement is reduced to the following issues:

1. Some blocks (one or more) are chosen as the starting conditions for the synthesis. Either one of the blocks 1, 2, 6, 7 or the block pairs <1, 4> , <2, 4> can be chosen as the starting conditions. In the general case, the choice should provide implementing the FSM functional circuit. Next, the edges which enter into initial blocks are deleted from the graph. As a result, these synthesis steps do not depend on other steps.
2. The route is chosen leading from the initial nodes to the node 8. In the general case, the route can be either consecutive, or parallel, or iterative (having cycles).
3. The way of implementation is determined for each synthesis step. In the common case, it depends on outcomes of both previous points.

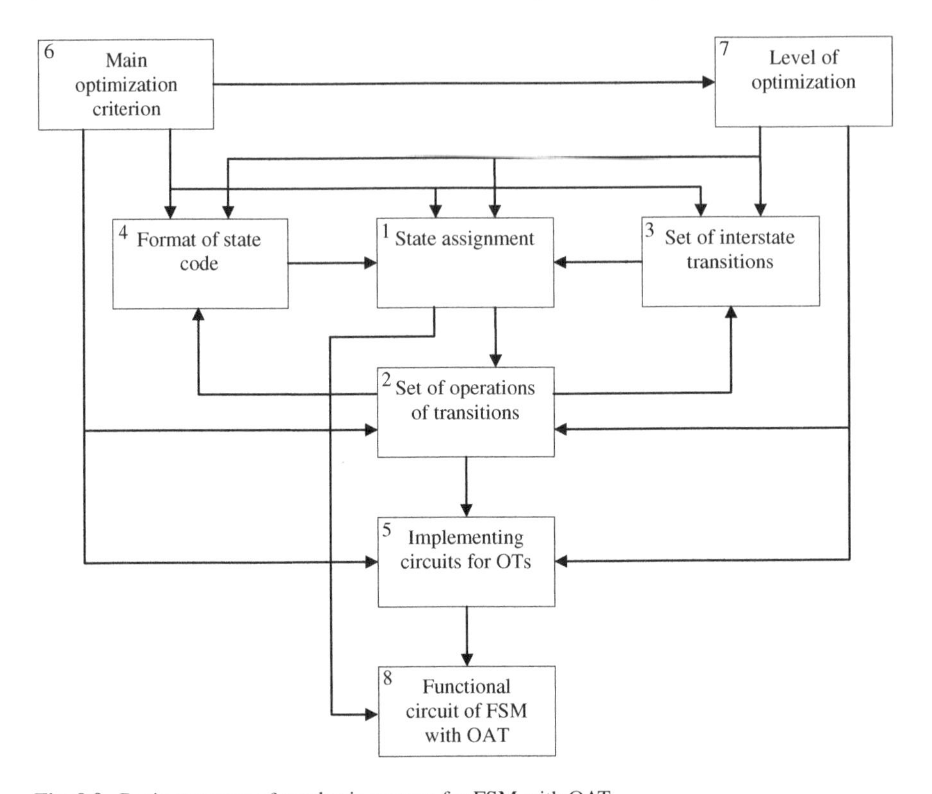

Fig. 9.9 Basic structure of synthesis process for FSM with OAT

Let us discuss some examples.

Example 9.1 Let the steps 1 and 6 be chosen as the starting conditions. It leads to the synthesis process shown in Fig. 9.10.

1. Both blocks 1 and 6 be starting; they have no ingoing edges. The nodes are deleted having outgoing edges connected with 1 and 6 (in this case, the nodes 3 and 4 are deleted).
2. The nodes 6 and 7 have no influence on the node 1. Of course, they can influence other synthesis steps. The outcome of this influence depends on specifics of each step's implementation.

Considering the interrelations of blocks 1 and 2 (Fig. 9.1), we can state that the *preliminary state assignment* takes place in this structure of synthesis process. The step of constructing the set of transitions is the following one.

The step-by-step implementation of each block produces an algorithm for synthesis of Moore FSM with OAT. Different algorithmic implementations are possible for most of blocks. Therefore, the structure (Fig. 9.10) might produce the variety of synthesis algorithms. Of course, it is true for any refinement of the basic structure of synthesis process.

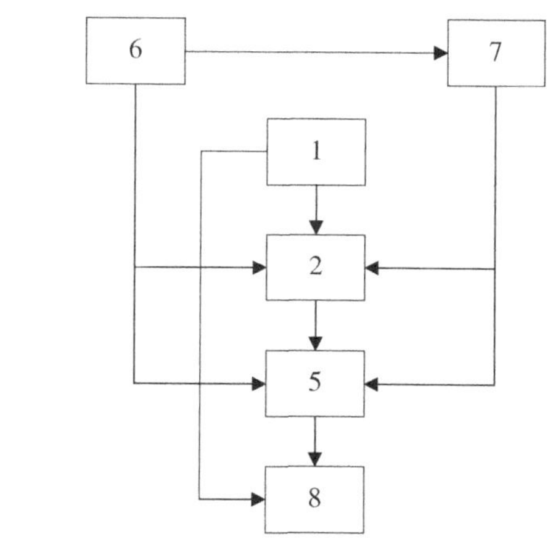

Fig. 9.10 Structure of synthesis process for initial blocks 1 and 6

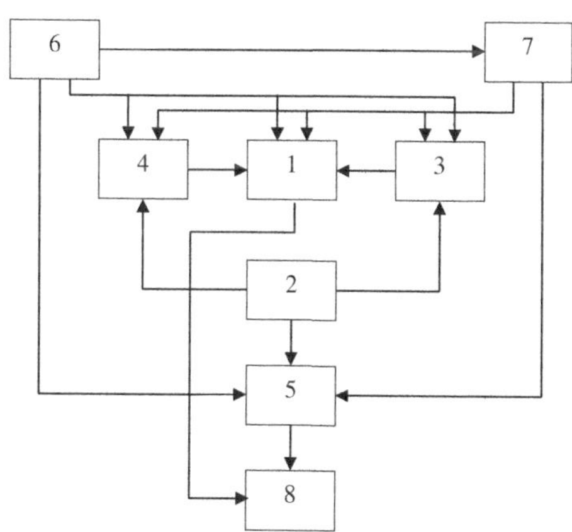

Fig. 9.11 Structure of synthesis process with starting blocks 2 and 6

The approach shown in Fig. 9.10 can be applied in the case when the state codes (block 1) are defined beforehand. It is possible, for example, when the content of BMO is obtained using these codes. The second possibility gives the preliminary design some blocks used in cases of structural decomposition of FSM [6].

Example 9.2 Let the steps 2 and 6 be chosen as starting. In this case, the basic structure takes on form shown in Fig. 9.11. It has the following specifics:

1. The block 6 does not affect the block 2 (both blocks are starting). It means that the appointed set of operations of transitions cannot be changed. If the set of OT is not sufficient for implementing all interstate transitions for all possible state codes, then it not possible to implement the synthesis.
2. Using the set of OT, the presentation of state codes can be chosen (block 4), as well as the set of interstate transitions (block 3). The choice can be done taking into account the optimization criterion (the influence of block 6 on blocks 3 and 4). Different level of optimization can be chosen (the influence of the block 7 on blocks 3 and 4).

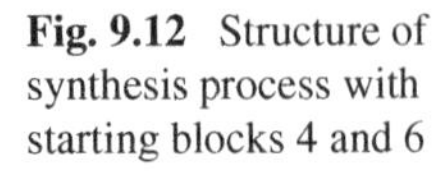

Fig. 9.12 Structure of synthesis process with starting blocks 4 and 6

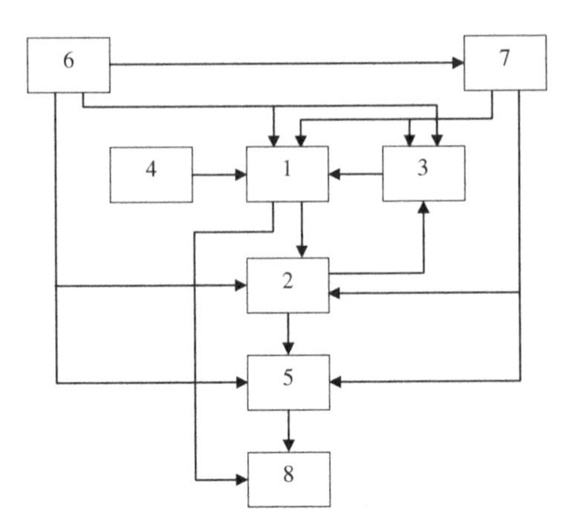

Considering the interrelation among the blocks 1 and 2, we can state that a *preliminary construction of the set of operations of transitions* can place in this model. This step is followed by the step of state assignment.

The structure shown in Fig. 9.11 corresponds to the situation when the set of operations of transitions is defined beforehand and cannot be changed. It is possible, for example, when the set of operations is implemented using either some standard ALU, or other standard equipment.

Example 9.3 Let the blocks 4 and 6 be taken as starting. In this case, the basic structure turns into the structure shown in Fig. 9.12.

As it can be seen, the blocks 1, 2 and 3 form a ring. It means that stages 1–3 can be multiple repeated during the synthesis process. Such a structure presumes different iterative implementations for given stages. It could improve the qualitative characteristics of synthesis outcomes. On the other hand, it can result in complication for these steps' implementations.

The structure shown in Fig. 9.12 corresponds to the situation when the stages of state assignment and constructing the set of interstate transitions are executed at the same time. This process can be named as a *concurrent state assignment and constructing the set of transitions*.

Thus, three following specified structures of synthesis process have been pointed out:

1. the structure with preliminary state assignment (Fig. 9.10);
2. the structure with preliminary constructing the set of operations of transitions (Fig. 9.11);
3. the structure with concurrent state assignment and constructing the set of interstate transitions (Fig. 9.12).

There are a lot of possible implementations for each block of each structure. It means that there are a lot of different possible synthesis processes. Besides, each structure can be modified due to introducing some new stages connected with applying either different constraints or optimization methods used for FSMs. As we know, each change in logic elements used for implementing the FSM logic circuits leads to development of new optimization methods based on peculiarities of these new elements.

9.5 Organization of Operational Automaton of Transitions

9.5.1 Typical Structure Models of Operational Automata

In the theory of structural synthesis, it is accepted to evaluate the structures of operational automata by such characteristics as hardware amount, performance, regularity, and universality. Different combinations of these characteristics found their embodiment in such structural models as the canonical automaton (C-OA), the automaton with individual microoperations (I-OA), the automaton with mutual microoperations (M-OA) and the automaton with either sequential or parallel combinational part (IM-OA) [9].

In C-OA, each microoperation of an algorithm is executed by a unique combinational circuit. Such an OA possesses the maximum values of both hardware amount and performance (an average number of microoperations executed during one cycle of operation), as well as the minimum value of propagation time among the equivalent OAs. Let us point out that operational automata are equivalent if they implement the same set of operations.

In I-OA, each word of information (operand) is processed by a unique combinational circuit. This individual circuit cannot include the same operational elements. For example, only one adder or shifter can be included into each CC. The additional multiplexers are introduced into I-OA in comparison with C-OA. It results in diminishing for the hardware amount and increasing the propagation time in comparison with C-OA.

There is only single CC in the M-OA, which is mutual for all registers keeping the possible operands. Both single-operand and double-operand microoperations can be executed by this CC for one cycle. The number of operational elements can be optimized for a given set of operations. It results in minimum values for both the hardware amount and performance. Only one microoperation can be executed during one cycle of operation. It means that the performance of M-OA does not exceed 1.

IM-automata allow execution up to three microoperations for one cycle. Either one single-operand and one double-operand microoperations can be executed for one cycle for two receiving registers (IM_P-OA, where the subscript p stands for parallel combinational part), or two single-operand and one double-operand microoperations can be executed for one cycle for the same receiving register (IM_S- OA, where the subscript s stands for sequential combinational part). The growth for the number of combinational circuits converges this model to I-OA, whereas the decreasing to M-OA. As a rule, IM-OAs possess the average values of characteristics in comparison with other models.

9.5.2 Organizational Specifics of OAT

The operational automaton of transitions as a part of Moore FSM possesses the following peculiarities:

1. It should be able to execute all operations for transforming the state codes under executing all transitions for a given GSA. Therefore, the set of OTs implemented by its circuitry is determined by the set of interstate transitions. In other words, it is determined by the initial GSA.
2. Only one register exists in OAT used for keeping the FSM state codes. This register is the only receiver for any OT. The initial data for executing an operation of transition include both the content of RG and values of logical conditions. The logical conditions are external; in general case, they are asynchronous in respect to the OAT.
3. In OA, the values of operands are random; they can be different for the same operation repeated once more. It can lead to some errors (for example, the overflow for adding or division by zero). These error situations are shown by flags (logical conditions), which are generated by a special block of OA and enter the circuit of FSM. In the case of OAT, the fixed state codes are used as the operands. Therefore, the OAT should be designed in the way excluding any error situation during the processing state codes. Each possible interstate transition should be processed without errors. So, there is no need in flags informing about errors. Therefore, the structure of OAT does not include a block for generation of logical conditions.

9.5.3 Organization of Combinational Part of OAT

The initial data for designing the combinational part of OA is the set of microoperations existed for an algorithm to be implemented. In general, it may be stated that the discussed above structures of OAs can be viewed as different approaches for the projection sets of microoperations on the set of combinational circuits. By analogy with the traditional OA of a digital system [7], it can be stated that the OAT implements some mapping of the set of interstate transitions into the set of combinational circuits. The following variants are possible for implementing such a mapping.

Individual implementation. In this case, each from H transitions corresponds to an individual OT. Some unique combinational circuit corresponds to each OT executing the required function for transformation of a current state code into a code of the next state. All blocks $CC(O_1)$–$CC(O_H)$ are connected with the multiplexer of result MX (Fig. 9.13). Let us name the OAT from Fig. 9.13 as OAT with combinational part of the type I (OATI). In OATI, the internal organization of combinational circuits can be different. If a CC corresponds to an unconditional transition,

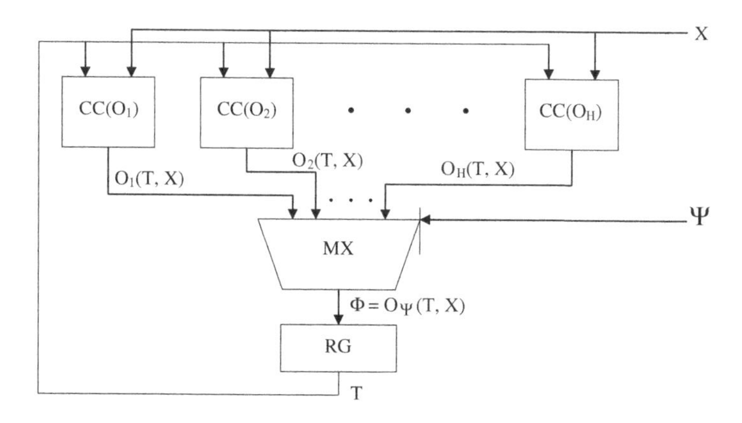

Fig. 9.13 Structure diagram of OAT with I-OA

then it can be implemented as a generator of a constant code of the next state. Its function does not depend on both the code of a current state and the values of logical conditions. If a CC corresponds to a conditional transition, then it can be implemented as a multiplexer having constants as its inputs and controlled by the corresponding logical conditions. The constants correspond to codes of next states.

The most hardware-consumed element of this OA is the multiplexer of result. If there are M states in a given FSM and M operations of transitions, that there are $R = \lceil \log_2 M \rceil$ bits in both state codes and codes of operations. In this case, the MX is an R-bit MX having M informational inputs and R control inputs. For FSM of average complexity, it is possible to have M = 200 [5]. In this case, there is a MX for 200 8-bits directions. Such a complex MX can be implemented only as a multilevel (cascaded) circuit. Obviously, it needs too much hardware and it is very slow.

Therefore, the main drawbacks of OATI are tremendous hardware amount and large propagation time increasing with the growth of the number of FSM states. It results in the high complexity for the circuit of BOT due to maximum possible number of bits in codes of operations. The main positive feature of this approach is the universality of a design process. It means that the design is the same for any GSA; it is reduced to the consecutive implementation of all CCs for all OTs finalized by design of a resulting MX.

Generalized implementation of operations of transitions. Let us name two or more transitions as pseudoequivalent transitions, if the same OT can be used for their implementation. For example, the transition from the state with the code 5 into the state with code 20 is executed by 2-bit left shifting the initial code. Using this very operation, a transition can be executed from the state with the code 8 into the state with code 32. These transitions form a class $B_i \in B$ of pseudoequivalent transitions, where there is $i = \overline{1, Q}$. Therefore, it is possible to form Q classes of pseudoequivalent transitions in a GSA.

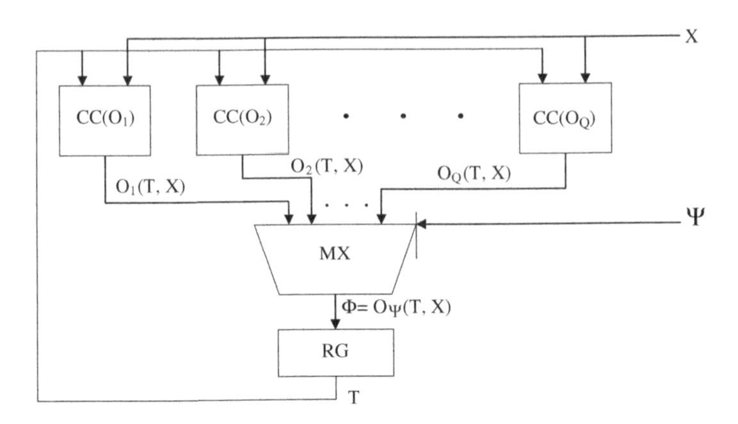

Fig. 9.14 Structure diagram of OATIM

Such a generalization of operations of transitions allows reducing the amount of CCs into the circuit of OAT. It leads to OAT having the operational part with IM-type (OATIM) (Fig. 9.14).

In comparison with OATI, this structure has the following peculiarities:

1. The number of CCs is decreased up to $Q \leq M$ due to existence of classes of pseudoequivalent transitions.
2. The internal structures of CCs are more complex. They implement some arithmetical and logic operations, rather than generation some constants. So, the propagation time for the operational part is increased.
3. Due to decreasing for the number of operations, the number of bits in codes of operations is decreased from $R_\Psi = \lceil \log_2 M \rceil$ to $R_\Psi = \lceil \log_2 Q \rceil$. It results in simplification for both BOT and MX of result.
4. The amount of classes of pseudoequivalent transitions depends on the values of state codes; it affects both the amount and complexity of combinational circuits. Therefore, there is a possibility for such a choice of state codes that the OATIM will include the minimum amount of CCs (the minimum hardware amount). But the problem of an appropriate choice of state codes is rather complex; it requires developing special algorithms.

General implementation of transitions. In some cases (for some GSAs), it is possible to reduce the set of OTs to two operations implemented by two combinational circuits. The first CC implements the unconditional transitions, whereas the second CC implements the conditional transitions. By analogy with IM-OA, two organizations are possible. One of them is OAT with sequential OP (OATS) and the second is OAT with parallel OP (OATP).

The OATS allows using the operation of unconditional transition independently, as well as a part of a complex operation of conditional transition. In the second case, the multiplexer of result is absent.

In the model with parallel OP, the operations of transitions are implemented by two CCs operating in parallel. The propagation time of FSM is smaller than in the previous case. It is determined by the maximum propagation time for one of the CCs. The MX of result is controlled by a single-bit code of operation Ψ generated by BOT.

Obviously, the necessary condition for applying this kind of OP is a possibility for implementing all interstate transitions by two combinational circuits. It is quite possible that such condition cannot take place for an arbitrary GSA. So, this model can be viewed as some "ideal model".

Generalizing discussed structures of OAT, the following conclusion can be done. The OATI possesses maximum hardware amount. Finding classes of pseudoequivalent transitions allows using models with generalized implementing operations of transitions. The synthesis outcome depends on optimization methods in use as well as on characteristics of a GSA to be implemented.

Canonical implementation of OAT. The traditional C-OA assumes existence of several registers receiving of data. Each register corresponds to its own collection of CCs. Because only one register is possible in OAT, then the structure of OATI can be treated as an OAT with canonical OP.

9.6 Synthesis Method for FSM with Supplemented Set of Operations of Transitions

In the case of preliminary constructing the set of operations of transitions, a designer should solve the problem connected with the choice of state codes on the base of these operations. The unique state codes should be chosen in such a way that any interstate transition can be executed using these operations. It is quite possible a situation when it is impossible to implement all transitions using only existing operations of transitions. In this case, the following actions can be done:

1. A part of states forming the set $A_1 \subseteq A$ is assigned using the acceptable range of codes.
2. The rest of state forming the set $A_2 = A \backslash A_1$ may be encoded using unused state codes from the acceptable range of codes.
3. The operations from the set O correspond to states from the set A_1.
4. There are no operations corresponding to states from the set A_2.

In this case, to design the circuit of OT, it is necessary to redefine the state codes from the set A_2. Next, the transitions from these states should be implemented. The following can be done to solve this problem.

1. Let us delete those rows of a structure table for whom both current and next states belong to A_1 (their codes are determined). It can be done if some operations of transitions correspond to these transitions. Let us name the resulting table as synthesizable table of transitions (STT).
2. Let us encode states from the set A_2 using arbitrary unique codes from the acceptable range.

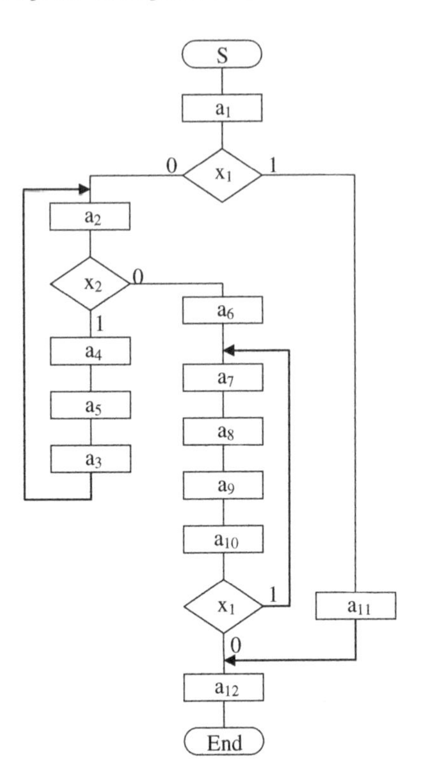

Fig. 9.15 The graph-scheme of algorithm Γ_{11}

3. Considering the state codes as binary vectors and using the STT, let us construct a system of Boolean functions implementing transitions from the states from A_2.
4. Let us treat the resulting SBF as a single operation of transition, O_{Q+1}. This operation supplements the existed set of OTs in such a way that all interstate transitions turn to be implemented. Let us design a combinational circuit corresponded to this system. Let us encode the operation O_{Q+1} by a unique code $\Psi(U_{Q+1})$.

Let us name the OAT with supplemented OT implementing transitions for the states from A_2 as *operational automaton of transitions with supplemented set of operations of transitions* (OATS). The organization of its operational part is similar to the one shown in Fig. 9.4.

The positive feature of OATS is a possibility for synthesis of FSM for any arbitrary GSA using any appointed set of OTs. The drawback of this approach is increasing for hardware amount with the growth for the number of transitions implemented by the supplemented OT.

Let us discuss an example of OATS synthesis for Moore FSM based on GSA Γ_{11} (Fig. 9.15). The approach proposed in [3] is used for the synthesis. The GSA Γ_{11} is characterized by the following parameters: it includes $M = 12$ collections of microoperations, a_1–a_{12} and $L = 2$ logical conditions, x_1–x_2. The distribution of

Table 9.3 Table of transitions (GSA Γ_{11})

a_i	$K(a_i)$	OT	a_j	$K(a_j)$	X
a_1	0	O_1	a_{11}	5	x_1
			a_2	3	$\bar{x}_1$
a_2	3	O_2	a_4	8	x_2
			a_6	6	$\bar{x}_2$
a_3	14	*	a_2	3	1
a_4	8	O_3	a_5	11	1
a_5	11	O_3	a_3	14	1
a_6	6	O_3	a_9	9	1
a_7	9	O_3	a_8	12	1
a_8	12	O_3	a_9	15	1
a_9	15	*	a_{10}	4	1
a_{10}	4	O_1	a_7	9	x_1
			a_{12}	7	$\bar{x}_1$
a_{11}	5	*	a_{12}	7	1

microoperations in operational vertices does not affect the design process. Because of it, there is no specification for contents of operational vertices. It means that the GSA Γ_{11} is rather abstract.

Let the following set of OTs be set up: $O = \{O_1, O_2, O_3\}$. Let its elements be determined as the following:

$$O_1: \quad A^{t+1} = \begin{cases} A^t + 3, \text{ if } x_1 = 0; \\ A^t + 5, \text{ if } x_1 = 1. \end{cases}$$

$$O_2: \quad A^{t+1} = \begin{cases} A^t + 3, \text{ if } x_2 = 0; \\ A^t + 5, \text{ if } x_2 = 1. \end{cases}$$

$$O_3: \quad A^{t+1} = A^t + 3.$$

Using our methodology, let us construct the structure table of Moore FSM for GSA Γ_{11} (Table 9.3). The table includes the following rows: a_i is the current state; $K(a_i)$ is the code of current state; OT is a code of operation of transition from the state a_i; a_j is the state of transition; $K(a_j)$ is the code of the state of transition; X are the logical conditions checked during the transition from a_i into a_j. In this table, codes correspond to all states, as well as to a majority of operations of transitions.

Let us point out that there are no operations of transitions for the following unconditional transitions: from a_3 into a_2, from a_9 into a_{10}, and from a_{11} into a_{12}. It is connected with the fact that there are no operation among the appointed ones capable to make the required transformation from $K(a^t)$ into $K(a^{t+1})$. Let us supplement the set of OT by the operation O_4, used for implemented the above mentioned uncoded transitions. To do it, let us represent the state codes as four-bit binary numbers using the variables T_1–T_4 for the encoding. For example, let it be $K(a_5) = 11_{10} = 1011_2$.

Let us construct the synthesized structure table where state codes are represented by corresponding binary values (Table 9.4). In this table, the column D includes the input memory functions used for loading the register RG.

Table 9.4 Synthesized structure table (for GSA Γ_{11})

a_i	$K(a_i)$	OT	a_j	$K(a_j)$	D	X
a_3	1110	*	a_2	0011	$D_3 D_4$	1
a_9	1111	*	a_{10}	0100	D_2	1
a_{11}	0101	*	a_{12}	0111	$D_2 D_3 D_4$	1

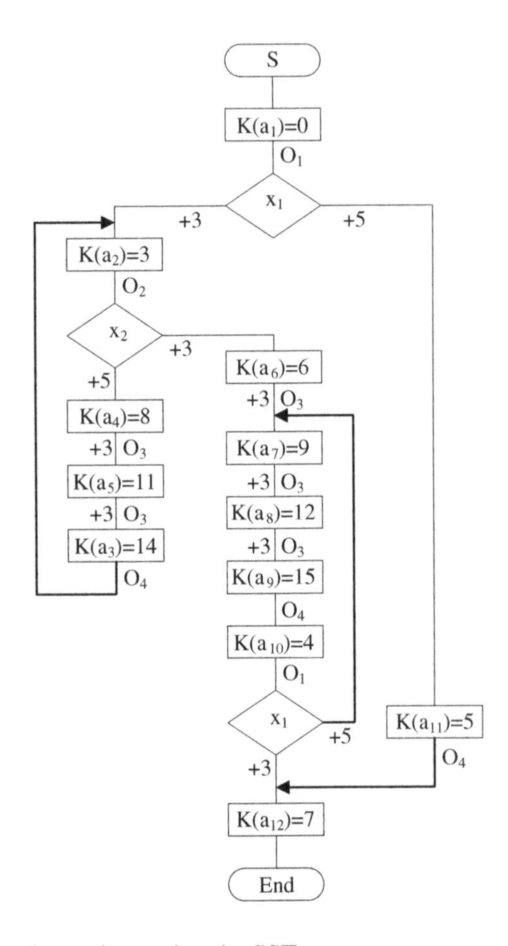

Fig. 9.16 State and operation codes for GSA Γ_{11}

Let us derive the following system of equations using the SST:

$$D_1 = 0;$$
$$D_2 = T_1T_2T_3T_4 \vee \bar{T}_1T_2\bar{T}_3T_4;$$
$$D_3 = T_1T_2T_3\bar{T}_4 \vee \bar{T}_1T_2\bar{T}_3T_4;$$
$$D_4 = D_3.$$

Let us design a combinational circuit corresponding to this system. It gives us the CC implementing the OT O_4. Now, the OT O_4 should be pointed in Table 9.3 instead of the sign "*". Let us point out that the CCs for operations O_1–O_3 are synthesized in a trivial way. The outcome of this approach is shown in Fig. 9.16.

Table 9.5 Content of PROM for the block BOT

Address $(K(a_i))$	Content (Ψ_1, Ψ_2)	Address $(K(a_i))$	Content (Ψ_1, Ψ_2)
0000	00	1000	11
0001	**	1001	11
0010	**	1010	**
0011	01	1011	10
0100	01	1100	10
0101	11	1101	**
0110	10	1110	11
0111	**	1111	11

To provide the correct operation of FSM with OAT, it is necessary to find the content of PROM for the block BOT. Using the variables Ψ_1, Ψ_2, let us encode the operations O_1–O_4 by the following binary codes: $K(O_1) = 00$, $K(O_2) = 01$, $K(O_3) = 10$, and $K(O_4) = 11$. The content of PROM is constructed on the base of state codes corresponding to addresses of cells, as well as codes of OTs corresponding to contents of the cells. For the GSA Γ_{11} the content of PROM for the block BOT is represented by Table 9.5.

Let us explain this table. The code 10 is written in the cell having the address 1011. It is done because the transition from the state a_5 having the code 1011 is executed using the OT O_3 having the code 10. There are arbitrary values in the cell with address 1101. This is done because there are no states having the code $1101_2 = 13_{10}$. There are arbitrary values in the cell with address 0111. This is done because the code 0111 corresponds to the final state a_{12} having no outgoing transitions.

The further synthesis of FSM is reduced to the synthesis of logic circuit of operational part. The content of PROM for BMO can be obtained if the collections of microoperations are known.

9.7 Investigation of Efficiency of FSM with OAT

The increase for hardware amount is proportional to the growth in the number of interstate transitions for the traditional Moore FSM. It is necessary to find a minimum set of operations of transitions which can be used for executing all transitions. If the ratio between numbers of transitions and operations is measured by tens (or even, hundreds), then the proposed approach allows tremendous saving of hardware in comparison with the traditional approaches. The block BIMF is the most complex block of an FSM. It is true for both the traditional Moore FSM and the FSM with OAT (where this block is represented by OP). Because of it, let us use the hardware amount in BIMF and OP for comparison of these two models. Let us choice the minimum hardware amount as an optimization criterion.

Let us compare the efficiency of Moore FSMs with BIMF and OP. Let us use the equivalent gates (EG) as standard units for the comparison. As it is adopted, one EG corresponds to a double-operand Boolean operation, such as NAND, for example.

Let a SBF having R of Boolean equations be used for implementing the circuit of BIMF. Without minimizing, each product term of this system is represented by a conjunction of $R_T = R_{LC} + R$ literals. Here the symbol R_{LC} stands for average number of logical conditions determining transitions for all states of a GSA. In this case, it is necessary to have H_1 equivalent gates for implementing each term, where

$$H_1 = R_T - 1 = R_{LC} + R - 1. \tag{9.11}$$

Let the number f terms in each equation is equal to V, where this number is equal to the number of interstate transitions for the given GSA. To connect these terms, it is necessary H_2 equivalent gates:

$$H_2 = V - 1. \tag{9.12}$$

For the system with R Boolean functions, the value $R \cdot H_1$ corresponds to hardware amount necessary for implementing one transition, whereas the value $R \cdot H_2$ for implementing all disjunctions in the system. Taking into account that there are V interstate transitions for a given FSM, the number of EGs in the circuit of Moore FSM is determined as:

$$H_K = R \cdot (V \cdot H_1 + H_2) = R \cdot (V \cdot (R_{LC} + R - 1) + V - 1). \tag{9.13}$$

As a rule, using some optimization methods leads to decreasing the value of H_K. To take it into account, let us introduce a coefficient k_1 into (9.13). This coefficient reflects the level of minimization. Now, the expression (9.13) is transformed into the following one:

$$H_K = k_1 \cdot R \cdot (V \cdot (R_{LC} + R - 1) + T - 1). \tag{9.14}$$

Let us estimate the hardware amount for FSM with OAT. Because the interstate transitions are executed by OAT, let us estimate necessary hardware amount in this block. Let us represent the hardware amount in OAT as result of summation for the following two components. First of them (H_3) is the hardware needed for implementing circuits for operations of transitions, the second component (H_4) is the hardware needed for implementing the multiplexer of result:

$$H_{OAT} = H_3 + H_4. \tag{9.15}$$

The value H_3 depends on the number Q of different OTs, as well as on their complexness. The rough estimate of the hardware amount in OAT can be made for some average amount of hardware (H_{OT}) for one operation of transition. In this case, the value H_3 is determined as:

$$H_3 = Q \cdot H_{OT}. \tag{9.16}$$

Adding one transition leads to increasing the value of H_K by

$$H_V = R \cdot H_1. \tag{9.17}$$

The value (9.17) is constant for given values of both R and R_{LC}. The added transition can be either unconditional or conditional. In both cases it is necessary to define the corresponding OT for executing this new transition. If the transition cannot be implemented using already existing combinational circuits, then some new OT should be introduced with corresponding CC. It results in increasing the value of H_{OAT} by the value H_{OT}. Neglecting the increasing for hardware in MX, the value of H_{OT} is constant for given values of both R and R_{LC}. Let us introduce a coefficient k_2 into the expression (9.16). The coefficient is used for expressing the value of H_{OT} through the value of H_V:

$$H_{OT} = k_2 \cdot H_V = k_2 \cdot R \cdot H_1. \tag{9.18}$$

The amount of operations of transitions Q is the most difficult for forecasting due to its dependence on both the structure of a GSA and types of operations of transitions for each particular case. It can be assumed that the value of Q is a function depending on V. But it is impossible to find the universal dependence $Q(V)$ for an arbitrary GSA. It is connected with existence of a lot of possibilities for the choice of types and quantity of OTs for each specific GSA.

After executing some investigations for GSA from LGSynth93, the following dependence $Q(V)$ has been found. If there is $V \in [0, 10]$, then each new OT is added for approximately 1–2 new transitions. It means that there is the dependence $Q \approx V/2$. If there is $V \in [10, 30]$, then each new OT is added for approximately 7–10 new transitions. For $V \in [30, 100]$, each new OT is added for approximately 10–20 new transitions, and so on. The obtained discrete dependence can be represented as some logarithmic function corresponded to the following expression:

$$Q = k_3 \cdot \ln(V/2) + 2. \tag{9.19}$$

In (9.19), the coefficient k_3 is obtained after carrying out a lot of experiments for different GSAs. As a rule, this coefficient belongs to the range from 2.5 to 5.

Taking into account expressions (9.18)–(9.19), the expression (9.16) can be represented as the following one:

$$H_3 = (k_3 \cdot \ln(V/2) + 2) \cdot k_2 \cdot R \cdot H_1. \tag{9.20}$$

The multiplexer of result generates the code of the next state having R bits. It is controlled by the code of operation having $R = \lceil \log_2 Q \rceil$. Each output of MX can be viewed as a SOP having Q terms. Taking into account the interterm disjunctions, the number of EGs necessary for implementing the MX is determined as the following:

$$H_4 = R \cdot \left(Q \cdot \left(\lceil \log_2 Q \rceil\right) + Q - 1\right). \tag{9.21}$$

Taking into account expressions (9.19)–(9.21), the expression (9.15) is transformed to the following one:

$$H_{OAT} = (k_3 \cdot ln(V/2) + 2) \cdot k_2 \cdot R \cdot H_1 + R \cdot \left(Q \cdot \left(\lceil \log_2 Q \rceil\right) + Q - 1\right). \tag{9.22}$$

Let us determine the efficiency of the circuit of OAT in comparison with the circuit of BIMF in the following manner:

$$E_{OAT} = H_K / H_{OAT}. \tag{9.23}$$

Table 9.6 Dependence $E_{OAT}(V)$

V	200	400	600	800	1000	1200	1400	1600	1800	2000
E_{OAT}	0.32	0.56	0.78	1.0	1.20	1.41	1.60	1.80	2.00	2.18

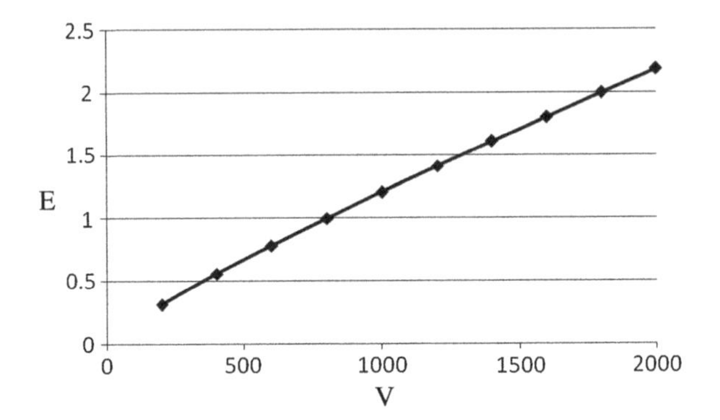

Fig. 9.17 Dependence $E_{OAT}(V)$

Table 9.7 Dependence $E_{OAT}(R)$

R	3	4	5	6	7	10	15	20	25	30
E_{OAT}	2.43	2.35	2.30	2.26	2.24	2.18	2.14	2.11	2.10	2.09

If there is $E_{OAT} > 1$, then the FSM with OAT is more efficient than the traditional Moore FSM (from the hardware point of view). Let us investigate the function (9.23) to find its dependence on different its components. Let the following parameters are constants having the following values: R = 10; V = 2000; $R_{LC} = 2$; $k_1 = 0.8$; $k_2 = 30$; $k_3 = 3.5$. The following diagrams are obtained as results of conducted investigations.

1. The dependence $E_{OAT}(V)$ is shown in both Table 9.6 and Fig. 9.17. Obviously, the function is a linear one for the given range of argument's values. For given constants, the FSM with OAT becomes more efficient starting from V > 800. The growth of V leads to the growth of the efficiency.
2. The dependence $E_{OAT}(R)$ is shown in both Table 9.7 and Fig. 9.18.
   As follows from the diagram, the investigated function is exponentially decreasing. For given range of its argument, it tends to a limit 2.05. For the whole diapason of R, the FSM with OAT is more efficient than the Moore FSM with BIMF.
3. The dependence $E_{OAT}(k_1)$ is shown in both Table 9.8 and Fig. 9.19.
   The diagram shows that the increasing of coefficient of minimization for the system of input memory functions leads to the linear increasing for the efficiency of FSM with OAT. If there is no minimization, then the maximum efficiency reaches the value 2.29. Only if the minimization simplifies the system of functions till 60 %, the FSM with BIMF turns to be more efficient than FSM with OAT.

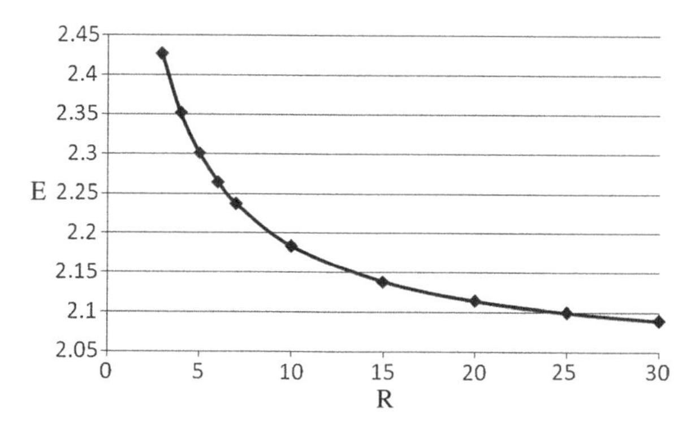

Fig. 9.18 Dependence $E_{OAT}(R)$

Table 9.8 Dependence $E_{OAT}(k_1)$

k_1	0.3	0.35	0.4	0.45	0.5	0.6	0.7	0.8	0.9	1.0
E_{OAT}	0.82	0.96	1.09	1.23	1.36	1.64	1.91	2.18	2.46	2.73

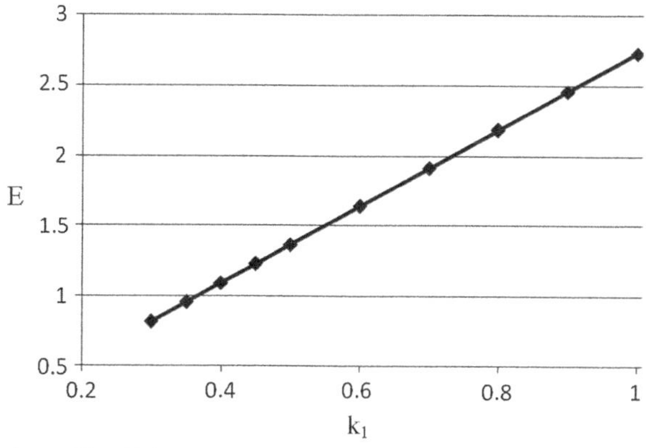

Fig. 9.19 Dependence $E_{OAT}(k_1)$

4. The dependence E_{OAT} (k_2) is shown in both Table 9.9 and Fig. 9.20. Analysis of this diagram shows that the efficiency of FSM with OAT is decreasing with the increasing of the average complexness of combinational circuits used for implementing the operations of transitions. For given values of arguments, the efficiency is getting lost starting from $k_2 > 65$.
5. The dependence $E_{OAT}(k_3)$ is shown in both Table 9.10 and Fig. 9.21. The exponentially decreasing nature of the diagram is explained by the influence of the value of coefficient k_3 on the logarithmic function (9.19). With the growth of k_3, the increasing of the amount of OTs is higher with the growth for the number of transitions. If there is $k_3 > 8$, then the amount of hardware in FSM with OAT exceeds this value for FSM with BIMF. To decrease the value of k_3, it is necessary to make a correct choice of the operations of transitions.

Table 9.9 Dependence $E_{OAT}(k_2)$

k_2	10	20	30	40	50	60	70	80	90	100
E_{OAT}	6.32	3.25	2.18	1.64	1.32	1.10	0.95	0.83	0.74	0.66

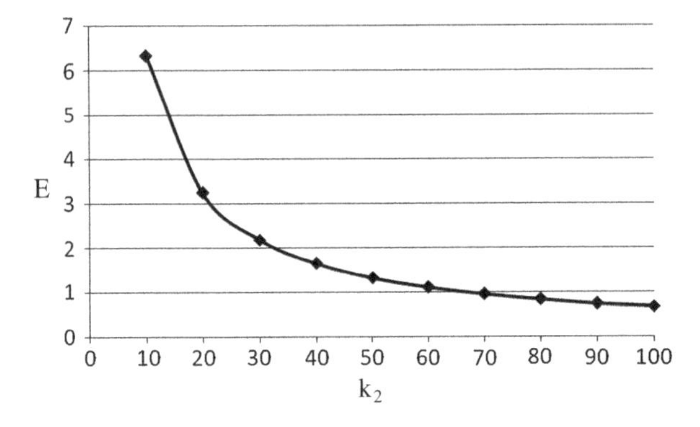

Fig. 9.20 Dependence $E_{OAT}(k_2)$

Table 9.10 Dependence $E_{OAT}(k_3)$

k_3	1	2	3	4	5	6	7	8	9	10
E_{OAT}	6.44	3.62	2.51	1.93	1.56	1.31	1.13	0.99	0.89	0.80

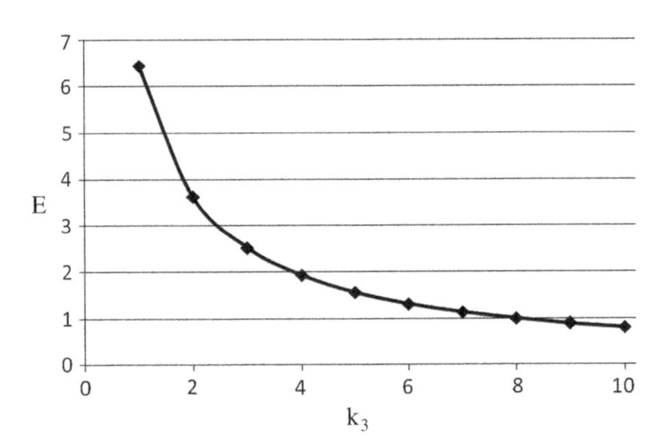

Fig. 9.21 Dependence $E_{OAT}(k_3)$

Integrally, the following conclusion can be made from analysis of diagrams shown in Figs. 9.17, 9.18, 9.19, 9.20, 9.21. There are the following factors leading to increasing the efficiency of FSM with OAT in comparison with FSM with BIMF:

- increasing for the number of interstate transitions;
- decreasing for the numbers of bits in state codes (in the ideal case this number should be minimum possible for a given GSA);

- implementing GSAs for which the system of input memory functions cannot be deeply minimized;
- decreasing the average complexness of combinational circuits used for implementing the operations of transitions (so, the operations leading to simple combinational circuits should be chosen);
- usage of special methods for constructing the set of operations of transitions allowing maximum decreasing for the growth of the quantity of operations of transitions for an arbitrary GSA with increasing for the number of interstate transitions.

References

1. Barkalov A (1998) Principles of logic optimization for a Moore microprogrammed automaton. Cybern Syst Anal 34(1):54–61
2. Barkalov A, Babakov R (2008) Organization of control units with operational addressing. Control Syst Mach 6:34–39 (in Russian)
3. Barkalov A, Babakov R (2011) Operational formation of state codes in microprogram automata. Cybern Syst Anal 2:193–199
4. Barkalov A, Babakov R (2011) Structural representation of syntheses process for control automata with operational automaton of transitions. Control Syst Mach 3:47–53
5. Barkalov A, Titarenko L (2008) Logic synthesis for compositional microprogram control units. Springer, Berlin
6. Barkalov A, Titarenko L (2009) Logic synthesis for FSM-based control units. Springer, Berlin
7. Barkalov A, Wegrzyn M (2006) Design of control units with programmable logic. UZ Press, Zielona Góra
8. De Micheli G (1994) Synthesis and optimization of digital circuits. McGraw–Hill, New York
9. Glushkov V (1962) Synthesis of digital automata. Fizmatgiz, Moscow (in Russian)
10. Grout I (2008) Digital systems design with FPGAs and CPLDs. Elsevier, Oxford
11. Kim T, Villa T, Brayton R, Sangiovanni-Vincentelli A (1997) Synthesis of finite state machines: functional optimization. Kluwer Academic Publishers, Boston
12. Maxfield C (2004) The design warrior's guide to FPGAs. Elsevier, Amsterdam
13. Zakrevskij A (1981) Logic synthesis for cascaded circuits. Nauka, Moscow

Appendix A
VHDL Constructions Used in the Book and Additional Support Materials

Abstract We present here concise information about used in the book synthesizable constructions and keywords (reserved words) of VHDL that are listed alphabetically and complemented with a brief informal description. We provide also a few useful tables (e.g. ASCII) with supplementary data needed for different chapters. Words written in *ITALIC SMALL CAPS* need to be replaced with user code.

Absolute Value—abs

This is a unary operator that is predefined for any numeric type and returns an absolute value of the operand. See examples in Sect. 2.2.

Aggregate

is a grouping of values to form an array or record expression. In *positional association*, the values are associated with elements from left to right. *Named association* indicates explicitly each value. Note that positional association cannot follow named association. Example:

```
-- Aggregates below are used in order to assign a record (positional association):
record_data <= ('0', '1', "01");            -- record_data is a signal of type my_packet (see below)
type my_packet is record                    -- see also records below
        first_bit, second_bit   : std_logic;
        data                    : std_logic_vector (1 downto 0);
end record;
```

V. Sklyarov et al., *Synthesis and Optimization of FPGA-Based Systems*,
Lecture Notes in Electrical Engineering 294, DOI: 10.1007/978-3-319-04708-9,

Elements can be grouped by named association, where the keyword **others** indicates the remaining elements:

```
(0=>bit3, 2=>bit2, 1=>bit1, 3=>bit0)                      -- an example of named association
A(7 downto 0) <=(7=>'0', 5 downto 4 => '0', others => '1'); -- an example of named association
```

Aggregates and Arrays

are shown on examples below. Two-dimensional arrays may be declared as follows (arrays and indices can be signals or variables):

```
type my_array is array (3 downto 0) of std_logic;-- type for a one-dimensional array
type my_packet is array (0 to 9) of my_array;    -- type for a two-dimensional array
signal my_data : my_packet;                      -- my_data is a two-dimensional array
```

Suppose the following declarations are done in the architecture.

```
type array2vect is array (0 to 1) of std_logic_vector(1 downto 0);
type array4vect is array (0 to 3) of array2vect;
signal table      : array4vect;                  -- table is a three-dimensional array
signal table_line : array2vect;                  -- table_line is a two-dimensional array
signal table_data: std_logic_vector(1 downto 0); -- table_data is a one-dimensional array
```

Different results can be tested in the architecture body using onboard LEDs:

```
led <= table(0)(0) & table(0)(1) & table(1)(0) & table(1)(1) &     -- there are 16 individual
       table(2)(0) & table(2)(1) & table(3)(0) & table(3)(1);      -- LEDs: led(15 downto 0)
```

The following statement assigns all the LEDs the value '1' (the LEDs are ON):

```
table <= (others=>(others=>(others=>'1')));
```

The next statements give results shown in the comments:

```
-- the result below is: led(15 downto 0) = 00 01 10 11 11 00 10 01
table <= (("00","01"),("10","11"),("11","00"),("10","01"));

-- sw(15 downto 0) are connected to led(15 downto 0) with the same indices
table <= ((sw(15 downto 14), sw(13 downto 12)), (sw(11 downto 10), sw(9 downto 8)),  -- #
          (sw(7 downto 6), sw(5 downto 4)), (sw(3 downto 2), sw(1 downto 0)));       -- #

-- the result below is: led(15 downto 0) = 00 00 01 11 01 11 00 00
table <= (1 to 2 =>(1=>(others=>'1'), 0=>"01"), others=>(others=>(others=>'0')));

-- the result below is: led(15 downto 0) =  01 10 11 00 00 01 10 11
table <=          (0 =>(0=>"01", 1=>"10"), 1 =>(0=>"11", 1=>"00"),
                   2 =>(0=>"00", 1=>"01"), 3 =>(0=>"10", 1=>"11"));

-- below sw(1 downto 0) control led(15 downto 14),
-- sw(3 downto 2) control led(13 downto 12), etc.
table <=          (0 =>(0=>sw(1 downto 0), 1=>sw(3 downto 2)),
                   1 =>(0=>sw(5 downto 4), 1=>sw(7 downto 6)),
                   2 =>(0=>sw(9 downto 8), 1=>sw(11 downto 10)),
                   3 =>(0=>sw(13 downto 12), 1=>sw(15 downto 14)));

-- sw(15 downto 0) in the process below are connected to the leds(15 downto 0) with the same indices
```

```
process(table)
begin
  for i in array4vect'range loop
    for j in array2vect'range loop
       led(i*4+j*2+1 downto i*4+j*2) <= table(i)(j);
    end loop;
  end loop;
end process;

-- sw(7 downto 4) in the code below control led(3 downto 0), for the assignment above marked with #
table_line <= table(2);
led <= (15 downto 4 => '0') & table_line(0) & table_line(1);

-- the table_line is linked with different groups of 4 switches depending on the values of two buttons
table_line <=      table(3) when buttons = "11" else          -- we assume the use of 2 buttons
                   table(2) when buttons = "10" else -- we assume the assignment above marked with #
                   table(1) when buttons = "01" else
                   table(0); -- the values of the switches are indicated below on the LEDs in the reverse order
led(3 downto 0) <= table_line(1)(0) & table_line(1)(1) & table_line(0)(0) & table_line(0)(1);
```

Alias

declaration permits an alternative name to be defined for an object. Alias declarations may be done in declarative parts. Alias declaration is done like the following: **alias** *<NEW NAME>* **is** *<EXISTING IDENTIFIER>;*

All

identifies all declarations within the package or library, for example: **use** ieee.std_logic_1164.**all**.

Architecture

is demonstrated on a general template below:

```
architecture <NAME OF ARCHITECTURE> of <NAME OF ENTITY> is
  -- declarative part
  -- declarations (of signals, components, functions, procedures)
  -- definitions (of types)
begin
        -- architecture body
end <NAME OF ARCHITECTURE>;
```

Array

is declared in the following general form:

```
type <NAME OF TYPE> is array <RANGE OF ARRAY> of <TYPE OF ELEMENTS>;
```

One-dimensional, two-dimensional, and three-dimensional arrays are shown above in ***aggregates and arrays***.

ASCII Table

provides an encoding for 128 characters. It is given in Table A.1 for 33 special characters (codes $0, \ldots, 31, 127$) and in Table A.2 for the remaining 95 printable characters (codes $32, \ldots, 126$).

Table A.1 ASCII codes for control characters

Code	Name
0	Nul
1	soh (start of heading)
2	stx (start of text)
3	etx (end of text)
4	eot (end of transmission)
5	enq (enquiry)
6	ack (acknowledge)
7	bel (bell)
8	bs (back space)
9	ht (horizontal tab)
10	lf (line feed)
11	vt (vert. tab)
12	ff (form feed)
13	cr (carriage return)
14	so (shift out)
15	si (shift in)
16	dle (data link escape)
17	dc1 (device control 1)
18	dc2 (device control 2)
19	dc3 (device control 3)
20	dc4 (device control 4)
21	nak (negative acknowledge)
22	syn (synchronous idle)
23	etb (end of transmitted block)
24	can (cancel)
25	em (end of medium)
26	sub (substitute)
27	esc (escape)
28	fsp (file separator)
29	gsp (group separator)
30	rsp (record separator)
31	usp (unit separator)
127	del (delete)

Table A.2 ASCII codes for 95 printable characters

<table>
<tr><th>code</th><th>+0</th><th>+1</th><th>+2</th><th>+3</th><th>+4</th><th>+5</th><th>+6</th><th>+7</th></tr>
<tr><td>32</td><td><space></td><td>!</td><td>"</td><td>#</td><td>$</td><td>%</td><td>&</td><td>'</td></tr>
<tr><td>40</td><td>(</td><td>)</td><td>*</td><td>+</td><td>,</td><td>-</td><td>.</td><td>/</td></tr>
<tr><td>48</td><td>0</td><td>1</td><td>2</td><td>3</td><td>4</td><td>5</td><td>6</td><td>7</td></tr>
<tr><td>56</td><td>8</td><td>9</td><td>:</td><td>;</td><td><</td><td>=</td><td>></td><td>?</td></tr>
<tr><td>64</td><td>@</td><td>A</td><td>B</td><td>C</td><td>D</td><td>E</td><td>F</td><td>G</td></tr>
<tr><td>72</td><td>H</td><td>I</td><td>J</td><td>K</td><td>L</td><td>M</td><td>N</td><td>O</td></tr>
<tr><td>80</td><td>P</td><td>Q</td><td>R</td><td>S</td><td>T</td><td>U</td><td>V</td><td>W</td></tr>
<tr><td>88</td><td>X</td><td>Y</td><td>Z</td><td>[</td><td>\</td><td>]</td><td>^</td><td>_</td></tr>
<tr><td>96</td><td>`</td><td>a</td><td>b</td><td>c</td><td>d</td><td>e</td><td>f</td><td>g</td></tr>
<tr><td>104</td><td>h</td><td>i</td><td>j</td><td>k</td><td>l</td><td>m</td><td>n</td><td>o</td></tr>
<tr><td>112</td><td>p</td><td>q</td><td>r</td><td>s</td><td>t</td><td>u</td><td>v</td><td>w</td></tr>
<tr><td>120</td><td>x</td><td>y</td><td>z</td><td>{</td><td>|</td><td>}</td><td>~</td><td></td></tr>
</table>

Assert

describes a condition that has to be evaluated and it is normally used to report warning and error messages (see Sect. 2.5 for details).

Attribute

is a named characteristic of certain objects and it permits constraints to be described directly in the code. In this book only predefined attributes (existing for types, arrays, and signals) have been used. They are defined in the form: type/array/signal'<*NAME OF ATTRIBUTE*>. An example of attribute event for signal clk is: **if** clk'event **and** clk = '1' **then**… (clock is changing from 0 to 1, i.e. the same as in **if** rising_edge(clk) **then** …) or **if** clk'event **and** clk = '0' **then**… (clock is changing from 1 to 0, i.e. the same as in **if** falling_edge(clk) **then** …). Examples of other useful attributes are:

- Test for an ascending range, for example: led_flash <= divided_clk **when not** led'ascending **else** '1'; If the LED is a descending range then the led_flash gets the divided clock value.
- The highest value of a type: integer'high.
- Test for the last value lv just before the last event on lv. For example: last_led <= lv(3)'last_value;
- Left-bound index signal'left. Let us consider the following example: internal_clock(internal_clock'left) where the type of internal_clock is std_logic_vector.
- Length of a dimension array'length. For example: my_RAM(i)'length.
- Position within a type: type'pos(…). For example: character'pos('A') returns the position of 'A' in the ASCII table, which is 65.
- Left downto/to right in an array array'**range**. For example: **for** i **in** input'**range loop**.

- Right downto/to left in an array array'reverse_range. For example: **for** i **in** input'reverse_range **loop**.
- Right-bound index signal'right For example: internal_clock(internal_clock'right) where the type of internal_clock is std_logic_vector.

The following lines give examples of user-defined attributes:

```
attribute LOC : string;                  -- specifying location constraints
attribute LOC of led: signal is "P2";    -- the led signal is assigned to the pin P2
attribute IOSTANDARD : string;           -- specifying input/output standard
attribute IOSTANDARD of led: signal is "LVCMOS33";    -- see the user constraints file for Nexys-4
```

Begin

marks the beginning of process/function/procedure statements or architecture body and the end of the respective declarative part in the process/function/procedure/architecture. It is also used in some other constructions such as in blocks and to describe multiple instances in generate statements.

Block

is a concurrent statement simplifying partition of a design and declared using the following basic format (examples are given in Sect. 2.4):

```
<OPTIONAL LABEL>: block (<OPTIONAL BOOLEAN GUARD EXPRESSION>) is
begin
        -- concurrent statements
end block <OPTIONAL LABEL>;
```

Body

is used with the **package** reserved word (see ***package***).

Buffer

is used for a port and enables the relevant signal to be read and written. Such port may have not more than one source and can be connected only to another buffer or linked to a signal with no more than one source. As distinct from **inout** ports, **buffer** ports cannot be connected to tri-state buses (see ***inout***) and they allow output signals declared as ports in a module to be also read in the module.

Case

statement has the following general form:

```
case < EXPRESSION > is
    when <VALUE OF THE EXPRESSION> => <STATEMENTS>;
    -- continue for other values: when <value of the expression> => <statements>;
    when others => <STATEMENTS>;
end case;
```

Case statements are used in processes, functions, and procedures and cannot be used directly in architectures (if required **when…else** can be applied instead in the architecture body). The following simple example demonstrates the use of the **case** statement:

```
process(A, B, C)            -- A,B,C are integers: signal A,B,C: integer range 0 to 7;
begin
  case (A+B+C) is                   -- A+B+C is the evaluated expression;
     when 1 to 3 | 5 | 10 => led <= '1'; -- when A+B+C = 1 or 2 or 3 or 5 or 10
     when others => led <= '0';
  end case;
  case (A+B+C)>12 is
     when true => led1 <= '1'; -- when A+B+C is greater than 12
     when others => led1 <= '0';
  end case;
end process;
```

Component

is a declaration in a higher-level entity enabling a lower-level entity to be instantiated. The following two templates can be used for component declaration and instantiation.

```
component <NAME OF COMPONENT>
generic (
  <NAME OF GENERIC> : <type> := <DEFAULT VALUE OF "NAME OF GENERIC">;
  <other generics...> );
port (
  <NAME OF PORT> : <mode of port such as in, out, inout, buffer> <type>;
  <other ports...> );
end component;

<NAME OF INSTANCE> : <NAME OF COMPONENT>
generic map ( < POSITIONAL OR NAMED ASSOCIATIONS > )
port map ( < POSITIONAL OR NAMED ASSOCIATIONS > );
```

Note that positional associations cannot follow named associations.

Component entities can be included in a library. The library named work is available by default. The following line demonstrates the use of this library without explicit component declarations:

```
u1: entity work.half_adder  port map(A, B, s2, s1);      -- using the default library work
```

A library with a different name can also be created (see Sect 2.6). The majority of examples in the book assume the use of the default library work without explicit component declarations.

Constant

can be declared in any declarative region. Constant values cannot be changed.

```
constant <NAME OF CONSTANT> : <type> := <USER VALUE>;
```

Examples:

```
constant line_with_equal_sign : string(1 to 3) := " = "; -- the symbol = is placed in between two spaces
constant ternary_vector     : std_logic_vector(5 downto 0) := "01-1-0";
constant my_integer         : integer := 7;
constant line1              : string(7 downto 1):="Index:" & CR; -- CR is a non-printing character
constant binary_constant    : std_logic_vector(5 downto 0) := "011100";
```

The following VHDL entity (test_const) gives an additional demonstration.

```
entity test_const is
  port ( sw     : in  std_logic_vector (2 downto 0);
         led    : out std_logic_vector (6 downto 0));
end test_const;

architecture Behavioral of test_const is
  constant binary        : std_logic_vector(6 downto 0)  := "0101010";
  constant octal         : std_logic_vector(6 downto 0)  := o"12" & '1';
  constant hexadecimal   : std_logic_vector(6 downto 0)  := x"a" & o"5";
  constant decimal       : integer                       := 63;
  type rom is array (0 to 3) of std_logic_vector (6 downto 0);
  constant ex            : rom :=(x"6" & o"3", x"8" & "101", '1' & o"45", o"3" & '0' & o"2");
begin
  led <= binary        when   sw = "001" else      -- the result: 0101010
         octal         when   sw = "010" else      -- the result: 0010101
         hexadecimal   when   sw = "011" else      -- the result: 1010101
         ex(0)         when   sw = "100" else      -- the result: 0110011
         ex(1)         when   sw = "101" else      -- the result: 1000101
         ex(2)         when   sw = "110" else      -- the result: 1100101
         ex(3)         when   sw = "111" else      -- the result: 0110010
         conv_std_logic_vector(decimal,7);          -- the result: 0111111
end Behavioral;
```

Conversion Functions

convert types (see ***type conversions*** and Sect. 2.2).

Downto

declares a direction in a range, for example, A(7 **downto** 0).

End

concludes descriptions (statements) in a process/function/procedure/architecture. It is also used in some other constructions such as **block** and **generate** statements.

Entity

describes inputs and outputs of the design module. It is demonstrated on a general template below:

```
entity <NAME OF ENTITY> is
generic (  <NAME OF GENERIC> : <type> := <USER VALUE>;
             <other generics if required...> );
port (  <NAME OF PORT> : <mode of port such as in, out, inout, buffer > <type>;
   <other ports if required...>
);
end <NAME OF ENTITY>;
```

Enumerated Type

may be user-defined, such as that is frequently needed for listing names of states in FSMs as it is shown in the example below:

```
type state_type is (init, run_state);      -- there are two states in the state_type: init and run_state
```

Exit

forces exiting from the innermost loop or from the loop with the indicated label. In the following example the variable count (declared as: **variable** count : integer **range** 0 **to** 4:= 0;) is always equal to 0 if the statement **exit** a is used and always equal to 4 when the statement without the label a (e.g. **exit**) is used:

```
a: for i in 0 to 3 loop                -- a is an optional label
  for j in 0 to 3 loop                 -- begin of the innermost loop
    if i = j then exit a;              -- count is always equal to 0 with the label a
        -- ..................................    -- count is always equal to 4 without the label a
    end if;
  end loop;                            -- end of the innermost loop
  count := count+1;
end loop a;                            -- a is an optional label
```

File

is a type that provides for interaction of the design with storage devices. An example with files is given in Sect. 2.6. In another example below the function read_array from Sect. 2.6 uses a while loop to read data from the file data.txt:

```
impure function read_array (input_data : in string) return my_array is
  file         my_file     : text is in input_data;
  variable     line_name: line;
  variable     a_name     : my_array;
  variable     index      : natural;
begin
  index := 0;
  while not endfile(my_file) loop        -- using the function endfile(file) [2]
    readline (my_file, line_name);
    read (line_name, a_name(index));
    index := index+1;                    -- index is incremented until the end of file is reached
  end loop;
  return a_name;
end function;
```

The next example demonstrates writing to a file:

```
library IEEE;
use IEEE.STD_LOGIC_1164.all;
use IEEE.STD_LOGIC_UNSIGNED.all;
use IEEE.STD_LOGIC_arith.all;
use IEEE.STD_LOGIC_TEXTIO.all;
use STD.TEXTIO.all;

entity WriteToFile is
generic (M        : integer := 32;
         N        : integer := 1024);
port ( const_bit   : out std_logic;
       bit_number : in integer range 0 to 14);
end WriteToFile;

architecture Behavioral of WriteToFile is          -- opening the file MyFile.txt for writing
  file generic_and_constants : text open write_mode is "MyFile.txt";
  constant oct_const: std_logic_vector(14 downto 0) :=o"37145";
begin

process(bit_number)         -- combinational process
  variable file_line : LINE;
begin
  write(file_line, string'("----------"));              -- preparing the string "----------"
  writeline(generic_and_constants, file_line);          -- recording the string "----------"
  write(file_line, string'("M = "));                    -- preparing the string "M = "
  write(file_line, M);                                  -- preparing the generic value M
  writeline(generic_and_constants, file_line);          -- recording the string and generic value M
  write(file_line, string'("N = "));                    -- preparing the string "N = "
  write(file_line, N);                                  -- preparing the generic value N

  writeline(generic_and_constants, file_line);          -- recording the string and generic value N
  write(file_line, string'("The maximum value of an integer: ")); -- preparing the string ...
```

```
    write(file_line, integer'high);                     -- preparing the value of the largest integer
    writeline(generic_and_constants, file_line);        -- recording the prepared string and value
    write(file_line, string'("Decimal value of octal constant: "));  -- preparing the string ...
    write(file_line, conv_integer(oct_const));          -- converting the octal constant to integer
    writeline(generic_and_constants, file_line);        -- recording the converted value
    for i in 3877 to 3879 loop              -- this loop records binary values of the index i in the file
        write(file_line, conv_std_logic_vector(i,12));-- converting integers to std_logic_vector
        writeline(generic_and_constants, file_line); -- recording the std_logic_vector
    end loop;
    const_bit <= hex_const(bit_number); -- other statements
     -- the remaining part of the code
  end process;

  -- the remaining part of the code

  end Behavioral;
```

Suppose the module WriteToFile is a component of another module in which it is declared as follows:

```
entity Top is
  generic (M: integer := 16; N: integer := 512);
  port -- descriptions of ports
end Top;

architecture Behavioral of Top is
begin
test: entity work.WriteToFile
        generic map (M, N)
        port map (const_bit, bit_number);

end Behavioral;
```

The file *MyFile.txt* is created during synthesis and is composed of the following lines:

```
--------------------
M = 16
N = 512
The maximum value of an integer: 2147483647
Decimal value of octal constant: 15973
111100100101
111100100110
111100100111
```

It contains those generic values that have been provided by the entity Top.

Files may be useful for such tasks as initializing arrays (ROMs), reading stimulus for simulations, etc. The predefined package textio in the library std collects useful functions, types and operations that permit to read/write files from the design. Additional details can be found in [1, 2].

For

statement permits to replicate logic in generate and loop constructions and may also serve some other purposes [1, 2]. Iterations **for-loop** can be used in processes, functions, and procedures. Suppose the following signal is declared:

```
signal vector : std_logic_vector(N-1 downto 0);
```

The following fragments demonstrate examples of **for** statements:

```
<OPTIONAL LABEL>: for i in vector'range loop  -- 1) elements of the vector from N-1 to 0 will be used
  -- statements that specify logic that has to be replicated
end loop <OPTIONAL LABEL>;
------------------------------------------------------------------------------------------------
for i in vector'reverse_range loop              -- 2) elements of the vector from 0 to N-1 will be used
  -- statements that specify logic that has to be replicated
end loop;
------------------------------------------------------------------------------------------------
for i in N-1 downto 0 loop                      -- 3) elements of the vector from N-1 to 0 will be used
  -- statements that specify logic that has to be replicated
end loop;
------------------------------------------------------------------------------------------------
for i in a downto b loop              -- 4) elements from a to b will be used (a≥b, a<N, b≥0)
  -- statements that specify logic that has to be replicated
end loop;
------------------------------------------------------------------------------------------------
for i in vector'left downto vector'right loop  -- 5) elements from N-1 to 0 will be used
  -- statements that specify logic that has to be replicated
end loop;
```

Function

computes a single value and always terminates with a **return** statement. It is declared according to the following general format:

```
function <NAME OF THE FUNCTION> (<LIST OF INPUT PARAMETERS>) return <TYPE> is
<DECLARATIVE PART>
begin
  -- sequential statements (the function body)
end <NAME OF THE FUNCTION>;
```

Input parameters can be unconstrained, i.e. they may have no bounds. The body of a function is similar to the body of a combinational process.

Section 2.4 is dedicated to functions with many simple examples.

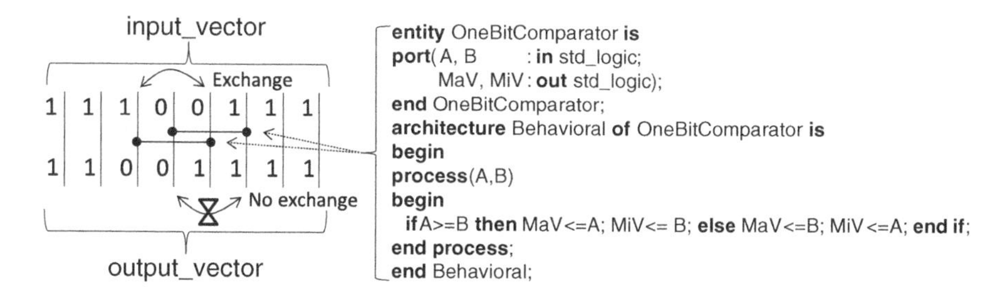

Fig. A.1 An example of input/output vectors and VHDL code for a 1-bit comparator

Generate

construction is used to instantiate an array of components allowing concurrent statements to be replicated. The modes **for** and **if** can be applied which is illustrated below on an example (Fig. A.1 shows the output_vector for the given input_vector):

```
entity Test_generic is
  generic( N              : integer := 8);
  port   ( input_vector   : in  std_logic_vector (N-1 downto 0);
           output_vector  : out std_logic_vector (N-1 downto 0));
end Test_generic;

architecture Behavioral of Test_generic is
begin

  example: for i in N-1 downto 0 generate
    exchange: if (i >= 2 and i <= 3) generate
       for_example: entity work.OneBitComparator
              port map( input_vector(i), input_vector(i+2), output_vector(i), output_vector(i+2));
    end generate exchange;

    copy: if (i < 2) generate
              output_vector(i) <= input_vector(i);
              output_vector(i+6) <= input_vector(i+6);
    end generate copy;

  end generate example;

end Behavioral;
```

Generic

supplies particular constant values to entities and components. A default value may be included which will be used if another value is not specified in the generic map. Section 2.5 is dedicated to generics with many simple examples. Generic map constructions permit default generic values in lower-level modules to be changed.

Guarded Signal

allows a concurrent assignment to be done only when the guard condition in the block with this assignment is true (see an example of the entity TestBlockGuarded in Sect. 2.4).

If

is a conditional statement which can be used in a process, function, and procedure Many simple examples are given in Sect. 2.3. This statement is demonstrated on a general template below:

```
if <condition1> then        <statements are executed if the condition1 is true>
elsif <condition2> then     < statements are executed if the conditions1 is false and 2 is true>
else                        < statements are executed if both the conditions are false >
end if;                     -- then, elsif, else, end if are reserved words
```

Impure

is an option for a function that extends the scope of variables and signals declared outside of the function that become available in the function. Thus, an *impure* function (in contrast to a *pure* function) may return different values for the same arguments (see Sect. 2.4 for details).

In

is a port mode allowing the port to be read only. If no mode is declared it is assumed to be **in**.

Inout

is a mode for bidirectional ports (for read and write). **Inout** is mainly aimed at indication of a tri-state port that can be both an output and an input. It can be applied to such signal MyBus that might be assigned like the following: MyBus <= MyIntBus **when** (MyWr = '1') **else** **(others** => 'Z');. Ports with **inout** type have to be used when bidirectional communications are actually needed. **Inout** type can also be used for procedures (see ***procedure***) and permits to return values of the respective arguments to the calling module and to read/write these arguments in the procedure.

Is

links identities to definitions in different constructions, for example: **architecture** Behavioral **of** TestTextFile **is**.

Library

permits the resources of a library to be used. The following examples demonstrate declaration of libraries IEEE and UNISIM:

```
library IEEE;                              -- "library", "use" and "all" are reserved words
use IEEE.STD_LOGIC_1164.all;               -- see details in section 2.6
use IEEE.STD_LOGIC_ARITH.all;              -- see details in section 2.6
use IEEE.STD_LOGIC_UNSIGNED.all;           -- see details in section 2.6
library UNISIM;  -- these lines have to be included if Xilinx primitives and the vendor-specific libraries are used
use UNISIM.VComponents.all;
```

The library work does not need to be declared. A user-defined library must be defined explicitly (see an example in Sect. 2.6).

Literal

is a value specified in the design that appears in expressions in form of a number, a character, a string, or a bit string:

- Numbers are represented by integer and real literals (a synthesizable object of type real cannot be defined). Examples of decimal integer literals are 45, 0, 1872. A base can be different from 10. In this case the number is enclosed in sharp characters # preceded by the base that can be any integer from 2 to 16. For example, the value 25 can be written as: 2#11_001#, or 16#19# or 5#100#. Separators (underscores _) are ignored in the literal and permit a more readable form to be used. The considered in the book convertions can be applied to numbers, for example leds(10 **downto** 2) <= conv_std_logic_vector(3#01_10#, 9);.
- A character literal is a single character enclosed in single quotation marks, for example: '3', 'f', 'S', ' ' (where in the last case the character is the space). There are a number of special (non-printable) characters that can be indicated by their predefined names (such as del), for example: **signal** special: character: = del; (see ***ASCII table***).
- Strings are sequences of characters enclosed in double quotation marks, for example: "this is a string", "" (where in the last case the string is empty). Two strings can be joined by the concatenation operator (&). The indices are positive numbers and the range may be either ascending or descending, although the majority of applications use an ascending range beginning with 1 [1]. The latter is also frequently used as the default initial index in a string range.

Table A.3 The truth table for binary logical operations

A	B	A and B	A nand B	A nor B	not A	not B	A or B	A xor B	A xnor B
False	False	False	True	True	True	True	False	False	True
False	True	False	True	False	True	False	True	True	False
True	False	False	True	False	False	True	True	True	False
True	True	True	False	False	False	False	True	False	True

- A bit string literal represents a string of binary values. For example, a binary vector can be described in the following form: B"1_001", where B (or b) is the base (*binary*). Other possible bases can be (O or o)—*octal* and (X or x)—*hexadecimal*. As before, separators (underscores _) are ignored. A bit string literal can be assigned to std_logic_vector with the proper size: SLV(2 **downto** 0) <= b"1_0_0".

Logical Operators

are summarized in Table A.3.

Map

associates names in a block (port or generic) with external names. Positional and named associations can be used. Let us consider some examples:

```
divider: entity work.clock_divider        -- positional association
-- clk and divided_clk are signals declared in the module which uses the clock_divider
port map (clk, divided_clk);
```

The entity clock_divider can be, for example, declared as:

```
entity clock_divider is              -- c and d_c are internal signals inside the entity clock_divider
port ( c        : in std_logic;      -- c corresponds to the signal clk in the upper level module
       d_c      : out std_logic);    -- d_c corresponds to the signal divided_clock
end clock_divider;
```

Named association may look like the following:

```
divider: entity work.clock_divider  -- names in upper and lower level modules may be the same
        port map (c=>clk, d_c=>divided_clk);   -- here external (clk, divided_clk)
                                               -- and internal (c, d_c) names are different
```

Modulo—mod

This is a binary operator that is predefined for integer types. The result has the sign of the second operand and absolute value less than the absolute value of the second operand. It is defined as: a **mod** b $= a - b \times n$, where n is some integer. See examples in Sect. 2.2.

Names

are used for identifiers. They may be composed of letters, digits, and underscores. Names are non-sensitive to case (i.e. AAa and aAA names are the same). Names must start with a letter, may not end with an underscore character, and may not include two successive underscore characters. VHDL reserved words cannot be used as names.

Next

terminates the current iteration (replication of logic) in loops and initiates a new iteration (replication). Much like the statement **exit**, **next** may have an optional label (e.g. **next** a;) which has the same interpretation as for the **exit**.

Null

indicates that no action is to be performed (normally used in **case** and **if** statements, for example, **when others** => **null**).

Of

links identities to names or types of elements, for example, **type** my_array **is array**(0 **to** 7) **of** std_logic_vector(15 **downto** 0).

Open

is used to leave the specified port unassociated. The IEEE VHDL specification does not allow unconnected input ports, but unconnected (open) output ports are permitted.

Operands

that are used in the book are: array aggregates, bit string literals, enumeration literals, function calls, integers, physical literals (only for behavioral simulation), record aggregates, string literals, static expressions, type conversions (see also ***literals*** above).

Operators

are divided into arithmetic (+, −, *, /, **abs**, **mod**, **rem**, sign + and −, **), concatenation (&), logical (**and**, **nand**, **nor**, **not**, **or**, **xnor**, **xor**), relational (=, /=, <, <=, >, >=), assignment (:=, <=) and shift (**rol**, **ror**, **sla**, **sll**, **sra**, **srl**). Some operators are represented by special symbols composed of individual or pairs of characters. A pair of characters, such as <= (if they correspond to an operator) must be typed next to each other without a space in between them. Logical operators can be combined with relational operators (ex. **if** ((a > b) **and** (c /= d)) or be bitwise (e.g. a **xor** b where a and b are std_logic_vector signals with equal sizes). All operators shown in **bold** font are reserved words. Operators are organized in the following groups according to their precedence (power − **), (**abs**), (**not**), (*), (/), (**mod**, **rem**), (+ identity, − negation), (+, −), (&), (**rol**, **ror**, **sll**, **srl**), (**sla**, **sra**), (=, /=), (<, <=, >, >=), (**and**, **nand**, **nor**, **or**, **xnor**, **xor**) with the highest priority in the first and with the lowest priority in the last group. Parentheses can be used to change the order of operators and are recommended for clarity.

On

is used in **wait** statements to introduce the sensitivity list as follows: **wait on** <SENSITIVITY LIST> **until** <BOOLEAN EXPRESSION>. An example is given below:

```
process          -- sequential process;
begin            -- for a combinational process the line below is changed to: wait on count;
  wait on count until rising_edge(divided_clk);
        led <= count;
end process;     -- support for wait statement is often limited (see, for example, restrictions in [2])
```

Others

is used as the last branch of **case** statement and the right part of a signal/variable assignments to cover not specified values in the **case** statement and to assign values to not-assigned array elements. Some examples are given below (see also ***aggregate***):

```
type memory is array (15 downto 0) of std_logic_vector (7 downto 0);
signal s_mem : memory := (others => (others => '0')); -- all elements of 2-dimensional array are zeros
-- beginning of a case statement
when others => null;
```

Out

is a port mode allowing the port to be only written.

Package

permits to describe functions, procedures, constants, and types in a separate file, which can be shared by different projects. See examples in Sect. 2.6. A general template for a package is shown below:

```
package <NAME OF THE PACKAGE> is
  type                          -- optional declaration of types
  constant                      -- optional declaration of constants
  function                      -- optional declaration of functions
  procedure                     -- optional declaration of procedures
end <NAME OF PACKAGE>;

package body <NAME OF THE PACKAGE> is
   -- definition of functions and procedures
end <NAME OF THE PACKAGE>;
-- some predefined packages are described in section 2.6
```

Port

is a signal enabling an entity to communicate with other upper-level modules. **Port map** construction defines mapping of signals from upper-level modules to signals from lower-level modules.

Procedure

differs from a function because it permits more than one signal to be produced. A general template for a procedure is:

```
procedure <NAME OF PROCEDURE> (<LIST OF INPUT, OUTPUT AND INOUT PARAMETERS>) is
      -- declarative part
begin
      -- sequential statements (the procedure body)
end <NAME OF PROCEDURE>;
```

Arguments of mode out and inout in procedures return their values to the calling module. Parameters can be unconstrained, i.e. they may have no bounds. The body of a procedure is similar to the body of a combinational process. Section 2.4 is dedicated to procedures and demonstrates simple examples.

Process

describes a level of hierarchy in a design. Different processes are executed concurrently (in parallel with other processes and concurrent signal assignments). A general template for a process is:

```
<OPTIONAL LABEL>: process (<SENSITIVITY LIST>)
        -- declarative part
begin
        -- sequential statements (the process body)
end process < OPTIONAL LABEL>;
```

Statements within a process are executed sequentially. Update of signals is done when the process suspends. Signal assignments (<=) inside processes do not take effect immediately as distinct from variables for which assignments (:=) of values are done immediately. The SENSITIVITY LIST is a set of signals written in parentheses after the word **process**. Any change in (any event on) these signals causes the process to be activated. The sensitivity list of a combinational process (for a combinational circuit) must contain all the input signals and the process must update all the output signals. Sequential processes include edge-triggered clocked timing. One process might change signals in a sensitivity list of another process. Processes without a sensitivity list should include a **wait** statement. Section 2.3 is dedicated to processes with many simple examples.

Pure

is an option for a function that does not allow the usage of signals or variables declared outside of the function. All functions are pure by default (see Sect. 2.4 and ***impure*** functions for details).

Qualified Expression

(**type**'(expression)) permits the type of expression to be explicitly indicated, for example:

```
architecture .....
signal user_signal : integer range 0 to 15 := 11;
begin
user_out <= unsigned'("0000") + user_signal;
```

Range

permits an interval of allowed values to be explicitly defined (see ***subtype***). The following example declares a range of integers:

```
signal user_signal : integer range -5 to 10; -- the range of integers is from -5 to 10
```

Record

permits a collection of data (with the same or different types) to be represented. A **record** can be declared as a type (see additional information in ***type***). Record type signals can be assigned using aggregates (see additional information in ***aggregates***). The following example demonstrates how a type **record** can be declared for a serial package that might be used in communications through RS232 interface:

```
type serial_package is record          -- type definition
        start_bit   : std_logic;
        data_bits   : std_logic_vector (7 downto 0);
        parity_bit  : std_logic;
        stop_bit    : std_logic;
        number      : integer range 0 to 127;
end record;
signal my_sp : serial_package;         -- declaration of my_sp signal of type serial_package
```

The following example shows how to access and assign individual fields:

```
my_sp.number <= 10;                    my_sp.start_bit <= '1';
my_sp.data_bits <= (others => '1');
```

Relational Operators

= (equal to), /= (not equal to), < (less than), <= (less than or equal to), > (greater than), >= (greater than or equal to).

Remainder—rem

is a binary operator for remainder that is predefined for any integer type. The result has the sign of the first operand and is defined as: a **rem** b = a−(a / b) × b. See examples in Sect. 2.2.

Report

is a statement for generating report messages (see Sect. 2.5 for details).

Return

terminates a function with this statement and passes control to the calling module. Any function must have a return statement.

Select

can be used in signal assignments in the body of an architecture. For example, the following architecture describes the full adder from Sect. 2.1:

```
architecture STRUCT of FULLADD is              -- another example is given in section 3.1.2
  signal three_bits : std_logic_vector(2 downto 0);
begin  three_bits <= A & B & CIN;
  with three_bits select SUM <= '1' when "100"|"010"|"001"|"111", -- SUM=1 for the listed vectors
                                '0' when others;                   -- SUM=0 for non-listed vectors
  with three_bits select COUT <= '1' when "011"|"101"|"110"|"111", -- COUT=1 for the listed vectors
                                 '0' when others;                   -- COUT=0 for non-listed vectors
end STRUCT;                                     -- another example is given in subtype
```

Severity

is a predefined type with the values note, warning, error and failure (see Sect. 2.5 for details).

Shared Variable

Shared keyword allows different processes to access the same variable. Shared variables can be declared only in entities, architectures, and generates (which are places where normal variables cannot be declared) according to the following syntax:

```
shared variable <VARIABLE_NAME> : <NAME OF TYPE> := <EXPRESSION>;
```

For example (N is the number of RAM words, M is the size of the words):

```
type type_of_the_RAM_block is array (0 to N-1) of std_logic_vector (M-1 downto 0);
shared variable RAM_block : type_of_the_RAM_block;
```

Xilinx recommends shared variables to be used to model a RAM with two write ports (examples are given in [1]).

Shift Operators rol, ror, sla, sll, sra, srl

The document [1] indicates that these operators are defined for a one-dimensional array with bit or Boolean elements. There are two arguments: A and B, where A is the array and B is the number of the array positions which are either shifted or rotated. Assuming that the operand A has N bits (N-1 **downto** 0) and the operand B is an integer, the following logical equivalence can be given:

- **rol** (rotate left): A(N-B-1 **downto** 0) & A(N-1 **downto** N-B);
- **ror** (rotate right): A(B-1 **downto** 0) & A(N-1 **downto** B);
- **sla** (shift left arithmetic): A(N-B-1 **downto** 0) & (B-1 **downto** 0 => A(0));
- **sll** (shift left logic): A(N-B-1 **downto** 0) & (B-1 **downto** 0 => '0');
- **sra** (shift right arithmetic): (B-1 **downto** 0 => A(N-1)) & A(N-1 **downto** B);
- **srl** (shift right logic): (B-1 **downto** 0 => '0') & A(N-1 **downto** B);

The following example provides an additional demonstration:

```
entity L_shift is                -- the library numeric_std has to be included (use ieee.numeric_std.all;)
  port(clk      : in std_logic;                -- system clock 100 MHz
       sw       : in unsigned(15 downto 0);    -- switches of the Nexys-4 or any other board
       led      : out unsigned(13 downto 0) ); -- LEDs of the Nexys-4 or any other board
end L_shift;
architecture Behavioral of L_shift is
  signal data_in          : unsigned(13 downto 0);  -- an input vector from the switches
  signal data_tmp         : unsigned(13 downto 0);  -- a temporary vector that is rotated
  signal sel              : unsigned(1 downto 0);   -- selects the number of positions to rotate
  signal divided_clk      : std_logic;              -- low frequency (1Hz) to make the rotations visible
begin
  data_in       <= sw(13 downto 0);         -- taking an input vector from the switches
  sel           <= sw(15 downto 14);        -- taking the sel value from the switches

process (divided_clk)         -- sequential process that demonstrates the use of the rol operator
begin
  if rising_edge(divided_clk) then
    case sel is               -- selection of the number of positions to rotate
      when "00" => data_tmp <= data_in ;        -- taking an initial vector from the switches
      when "01" => data_tmp <= data_tmp rol 1;-- rotate one position (B=1)
      when "10" => data_tmp <= data_tmp rol 2;-- rotate two positions (B=2)
      when "11" => data_tmp <= data_tmp rol 3;-- rotate three positions (B=3)
      when others => data_tmp <= data_in ;
    end case;    -- the operators ror, sll, srl can be used above instead of the operation rol
  end if;
end process;

led <= data_tmp;                        -- showing rotated data on leds
div: entity work.clock_divider          -- clock divider to reduce clock frequency from 100 MHz to  1Hz
      port map  (clk, '0', divided_clk); -- the reset signal is always deasserted ('0')

end Behavioral;
```

We would prefer to use logically equivalent operators shown above instead of the described here shift operators.

Signal

Signals model physical wires in hardware circuits. They are assigned with a pair of symbols <= and any assignment involves a delay (one delta delay by default). The latter applies when an assignment is done within a block or as a part of sequential statements within a process (see TestProc entity in Sect. 2.3.2 for details). Signals differ from variables which are assigned immediately. Signals are declared in architectures (and cannot be declared in processes, procedures or functions) in the following form:

```
signal <NAME OF SIGNAL> : <TYPE OF SIGNAL>;
signal <NAME OF SIGNAL> : <TYPE OF SIGNAL> := <INITIAL VALUE>;
```

Signals can be used in bodies of architectures, processes, procedures or functions and can be formal parameters of a function or a procedure. A sensitivity list of a process cannot include variables and includes only signals.

Concurrent signal assignments (<=), conditional signal assignments (**when** … **else**) and selected signal assignments (**with** … **select** … **when**) can be used in architecture body. In processes (procedures) normally only sequential signal assignments (<=) are allowed. The following rules are the most important:

1. Sequential signal assignments will be done in a process only when the process suspends.
2. If there are several assignments in the process to the same signal only the last one takes effect.

Subtype

Introduces constraints or subsets of values for the chosen base type. It is declared in the following general form:

```
subtype <NAME OF SUBTYPE> is <BASE_TYPE>
      range <VALUES IN RANGE>;
```

The use of subtypes is considered on an example below.

```
entity types_and_subtypes is
  port (          switches   : in   std_logic_vector(1 downto 0);         -- two switches
                  leds       : out  std_logic_vector (3 downto 0));       -- four LEDs
end types_and_subtypes;

architecture Behavioral of types_and_subtypes is
  subtype four_bits_std_logic_vector is std_logic_vector (3 downto 0);
  type my_pack is array (0 to 3) of four_bits_std_logic_vector;       -- a subtype of std_logic_vector
  constant set_of_lines : my_pack := (x"F", b"00_11", o"6"&'0', "0101"); -- defining a constant value
begin
  with switches select leds <= set_of_lines(0) when "00",    -- displayed value is "1111" = x"F"
                          set_of_lines(1) when "01",         -- displayed value is "0011" = b"00_11"
                          set_of_lines(2) when "10",         -- displayed value is "1100" = o"6"&'0'
                          set_of_lines(3) when "11",         -- displayed value is "0101" = "0101"
                          (others => '0') when others;
end Behavioral;
```

To

declares a direction in a range, for example, A(0 **to** 7).

Table A.4 Often used predefined VHDL data types

Type	Where declared	Possible values
bit	Standard in VHDL	'0', '1'
bit_vector	Standard in VHDL	Array of bits
boolean	Standard in VHDL	False, true
character	Standard in VHDL	7-bit ASCII codes in ISE
integer	Standard in VHDL	At least 32 bits (-2^{31} to $2^{31} - 1$)
natural	Standard in VHDL	Subtype of integer: at least from 0 to $2^{31} - 1$
positive	Standard in VHDL	Subtype of integer: at least from 1 to $2^{31} - 1$
real	There are many restrictions for synthesis	Floating-point values
signed	Packages: ieee.std_logic_arith, ieee.numeric_std	Array of std_logic
std_logic	Package: ieee.std_logic_1164	Resolved std_ulogic
std_logic_vector	Package: ieee.std_logic_1164	Array of std_logic
std_ulogic	Package: ieee.std_logic_1164	'U', 'X', '0', '1', 'Z', 'W', 'L', 'H', '-'
std_ulogic_vector	Package: ieee.std_logic_1164	Array of std_ulogic
string	Standard in VHDL	Array of characters
time	Standard in VHDL	Time units: hr, min, sec, ms, us, ns, ps, fs
unsigned	Packages: ieee.numeric_std, ieee.std_logic_arith	Array of std_logic

Type

is declared in the following general form:

```
type <NAME OF TYPE> is <SPECIFICATION OF TYPE>; -- see also enumerated type
```

Table A.4 summarizes information about types most commonly used in synthesizable VHDL (resolved type permits signals to be driven by more than one source). Note that there are many restrictions for using the type real.

Each type allows a set of values and a set of associated operations. There are several groups of predefined types such as scalar (bit, boolean, character, enumerated, integer, physical, real, severity) and composite (array, bit_vector, record, string).

Unsigned vector "1111" corresponds to integer 15 and signed vector "1111" corresponds to integer -1. The latter is represented in two's complement notation, i.e. the most significant bit indicates the sign (1 is minus '$-$' and 0 is plus '$+$') and has a negative weight -2^3, while all the remaining bits have positive weights (2^0, 2^1 and 2^2, accordingly) which are equal to 2^x where x is the index of the respective bit (the least significant bit has an index 0).

Type Conversions

are frequently required. They are provided either automatically, through type casts, or with the aid of conversion functions (see also Sect. 2.2). Type cast is used to convert equal sized signed or unsigned to std_logic_vector and vice versa:

```
signed_vector            <= signed(std_logic_vector_signal);
unsigned_vector          <= unsigned(std_logic_vector_signal);
std_logic_vector_signal  <= std_logic_vector(signed_vector);
std_logic_vector_signal  <= std_logic_vector(unsigned_vector);
```

The following assignments need conversion functions:

```
integer_signal           <= conv_integer (unsigned_vector);
integer_signal           <= conv_integer (signed_vector);
integer_signal           <= conv_integer (std_logic_vector_signal);
unsigned_vector          <= conv_unsigned (integer_signal, size_of_unsigned_vector);
signed_vector            <= conv_signed (integer_signal, size_of_signed_vector);
std_logic_vector_signal  <= conv_std_logic_vector (integer_signal, size);
```

Using conversion functions requires the relevant libraries to be included. For example the conv_integer function is defined in the library std_logic_unsigned (or std_logic_signed) and conv_std_logic_vector function is defined in the library std_logic_arith.

Until

is used in the condition of a **wait** statement (see ***on***). An example is given below. The support is limited [1].

```
process
begin
  wait until rising_edge(divided_clk) and BTNC = '1';
        count <= count + 1;
end process;
```

Use

enables functions, procedures, constants, and types of a package to become accessible (visible) in an associated entity/architecture.

Variable

Variables in VHDL are very similar to variables in general-purpose programming languages. They can be declared and used in processes, procedures, and functions. Assignments are allowed from signals to variables (<variable>:= <signal>;) and vice versa (<signal> <= <variable>;), however type match has to be satisfied and the proper operator (:= or <=) must be chosen. Variable assignments take effect without a delay (as distinct from signal assignments).

Wait

suspends a process. The document [1] recommends describing processes with a sensitivity list and indicate the following limitations: (1) only one **wait** statement is allowed and it must be the first in the process; (2) the condition in the **wait** statement has to describe a clock signal. See also ***on*** and ***until***.

When

can be used in **case** statements and in signal assignments (see examples in ***case*** and ***select***).

While

statement permits repeated operations to be implemented in replicated logic. It has the following general form:

```
<OPTIONAL LABEL>: while <CONDITION> loop
    -- sequential statements;
end loop <OPTIONAL LABEL>;
```

Let us consider an example:

```
process (vector) -- this process finds the position (from 1 to 8) of the first '1' in the vector
  variable first_right          : integer range 0 to N;
  variable i                    : integer range 0 to N;
begin
  first_right := 0;  -- variables have to be used here
  i := 0;
  while i < N loop              -- vector is declared as std_logic_vector (7 downto 0);
       if vector(i) = '1' then  first_right := i+1; exit;
       else                     i := i+1;
       end if;                  -- positions of the vector bits are: 8 for bit 7, 7 for bit 6, 6 for bit 5, etc.
  end loop;                     -- an optional label can be used for the loop
  led <=conv_std_logic_vector(first_right, 8); -- if vector = "00010100" then the result is 0011
end process;     -- the result 0011 indicates position 3 (for bit 2) which is the first '1' from the right
```

With

is used in a selected signal assignment (see ***select***).

References

1. Xilinx Inc. (2013) XST user guide for Virtex-6, Spartan-6, and 7 series devices. http://www.xilinx.com/support/documentation/sw_manuals/xilinx14_7/xst_v6s6.pdf. Accessed 17 Nov 2013
2. Ashenden PJ (2008) The designer's guide to VHDL, 3rd edn. Morgan Kaufmann, Boston

Appendix B
Coding Examples

Abstract Appendix B includes coding examples for frequently needed modules, any of which can easily be located by name. Entities have exactly the same names that are used for the relevant components described in Chaps. 3 and 4. VHDL codes, user constraints files, and bitstreams for all the projects are available online at http://sweet.ua.pt/skl/Springer2014.html.

Binary to BCD Converters (BinToBCD8)

The following VHDL code is a complete description of a module, which converts 8-bit binary numbers (binary) to binary coded decimal (BCD) numbers (BCD2, BCD1, BCD0):

```
library IEEE;                      -- a conversion can also be done on request and this will be
use IEEE.STD_LOGIC_1164.all;       -- shown after the next example BinToBCD16
use IEEE.STD_LOGIC_ARITH.all;
use IEEE.STD_LOGIC_UNSIGNED.all;

entity BinToBCD8 is        -- Binary to BCD converter for 8-bit numbers of std_logic_vector type
generic( size_of_data_to_convert :     integer := 8 );
port (  clk        : in  std_logic;
        reset      : in  std_logic;
        ready      : out std_logic;       -- ready is 0 when the number is being converted
        binary     : in  std_logic_vector (size_of_data_to_convert-1 downto 0);
        BCD2       : out std_logic_vector (3 downto 0);  -- BCD code for the most significant digit
        BCD1       : out std_logic_vector (3 downto 0);  -- BCD code for the digit in the middle
        BCD0       : out std_logic_vector (3 downto 0)); -- BCD code for the least significant digit
end BinToBCD8;

architecture Behavioral of BinToBCD8 is
  type state is (idle, op, done);
  signal c_s, n_s              : state;
  signal BCD2_c, BCD1_c, BCD0_c, BCD2_n, BCD1_n, BCD0_n : unsigned(3 downto 0);
  signal BCD1_tmp, BCD0_tmp                             : unsigned(3 downto 0);
  signal BCD2_tmp                                       : unsigned(2 downto 0);
  signal int_rg_c, int_rg_n   : std_logic_vector (size_of_data_to_convert-1 downto 0);
```

V. Sklyarov et al., *Synthesis and Optimization of FPGA-Based Systems*,
Lecture Notes in Electrical Engineering 294, DOI: 10.1007/978-3-319-04708-9,

```
    signal index_c, index_n    : unsigned(3 downto 0);
    signal get_outputs         : std_logic;
begin

process(clk, reset)
begin
  if rising_edge(clk) then
    if reset = '1' then
       c_s <= idle;              -- idle state at the beginning
       BCD2_c <= (others => '0'); BCD1_c <= (others => '0'); BCD0_c <= (others => '0');
       BCD0 <= (others=>'0'); BCD1 <= (others=>'0'); BCD2 <= (others=>'0');
    else c_s <= n_s;             -- next values are copied to current values
       BCD2_c <= BCD2_n; BCD1_c <= BCD1_n; BCD0_c <= BCD0_n;
       index_c <= index_n; int_rg_c <= int_rg_n;
       if (get_outputs = '1') then
                   BCD0 <= std_logic_vector(BCD0_n);
                   BCD1 <= std_logic_vector(BCD1_n);
                   BCD2 <= std_logic_vector(BCD2_n);
       end if;
    end if;
  end if;
end process;

process (c_s, BCD2_c, BCD1_c, BCD0_c, BCD2_tmp,
          BCD1_tmp, BCD0_tmp, binary, int_rg_c, index_c, index_n)
begin

  get_outputs <= '0'; n_s <= c_s;
  BCD2_n <= BCD2_c;       BCD1_n <= BCD1_c;
  BCD0_n <= BCD0_c;       index_n <= index_c;
  int_rg_n <= int_rg_c;   ready <= '0';

  case c_s is    -- at the beginning ready is 0
    when idle => n_s <= op; ready <= '0'; int_rg_n <= binary; index_n <= "1000";
    when op =>  ready <= '0';
       int_rg_n <= int_rg_c(size_of_data_to_convert-2 downto 0) & '0';
       BCD0_n <= BCD0_tmp(2 downto 0) & int_rg_c(size_of_data_to_convert-1);
       BCD1_n <= BCD1_tmp(2 downto 0) & BCD0_tmp(3);
       BCD2_n <= BCD2_tmp(2 downto 0) & BCD1_tmp(3);
       index_n <= index_c - 1;
       if (index_n = 0) then n_s <= done; get_outputs <= '1';
       end if;
    when done => n_s <= idle;
       BCD2_n <= (others => '0');
       BCD1_n <= (others => '0');
       BCD0_n <= (others => '0');
       ready <= '1'; -- now ready is 1, i.e. a new conversion can be done
  end case;
end process;

BCD0_tmp <= BCD0_c + 3 when BCD0_c > 4 else BCD0_c;
BCD1_tmp <= BCD1_c + 3 when BCD1_c > 4 else BCD1_c;
BCD2_tmp <= BCD2_c(2 downto 0) + 3 when BCD2_c > 4 else BCD2_c(2 downto 0);

end Behavioral;
```

Figure B.1a explains an interface with the module BinToBCD8. Signal ready is valid during one clock cycle and it indicates that the result of conversion is ready to be used. A new data item for conversion can be prepared when ready=0. The code

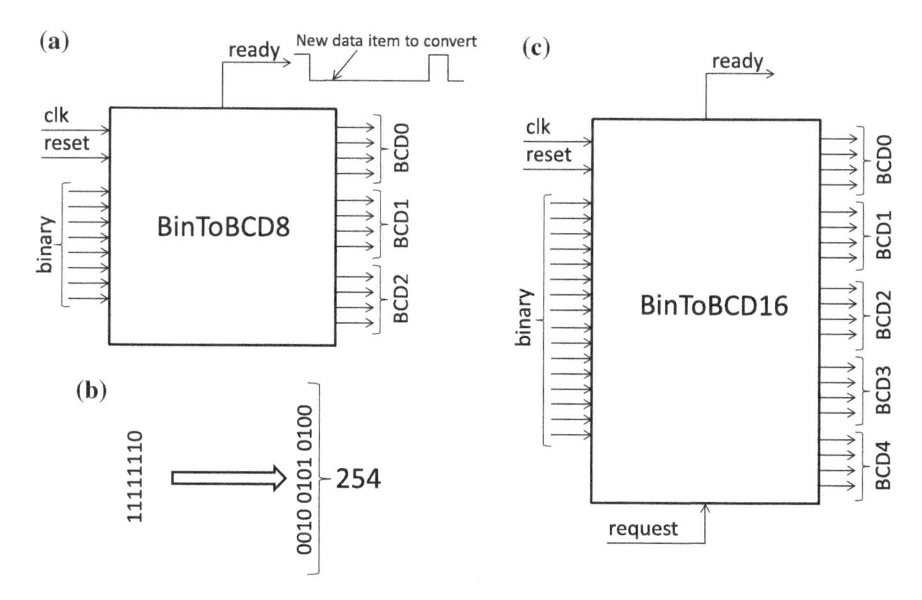

Fig. B.1 Interface with the BinToBCD8 module (**a**), an example of conversion (**b**), and interface with the BinToBCD16 module (**c**)

may be slightly changed in such a way that as soon as ready is active the FSM in the module above is continued to be in the idle state until a request for a new conversion is received. Additional details will be given after the next example.

Binary to BCD Converters (BinToBCD16)

The following VHDL code is a complete description of a module, which converts 16-bit binary numbers (binary) to binary coded decimal (BCD) numbers (BCD4, BCD3, BCD2, BCD1, BCD0):

```
library IEEE;
use IEEE.STD_LOGIC_1164.all;
use IEEE.STD_LOGIC_ARITH.all;
use IEEE.STD_LOGIC_UNSIGNED.all;

entity BinToBCD16 is        -- binary to BCD converter for 16-bit numbers of std_logic_vector type
generic( size_of_data_to_convert :    integer := 16 );
port ( clk          : in std_logic;
       reset        : in std_logic;
       ready        : out std_logic;        -- ready is 0 when the number is being converted
       binary       : in std_logic_vector (size_of_data_to_convert-1 downto 0);
       request      : in std_logic;         -- a request is assumed to be sent when ready is active (1)
       BCD4         : out std_logic_vector (3 downto 0); -- BCD code for the most significant digit
       BCD3         : out std_logic_vector (3 downto 0);
```

```
        BCD2     : out  std_logic_vector (3 downto 0);
        BCD1     : out  std_logic_vector (3 downto 0);
        BCD0     : out  std_logic_vector (3 downto 0)); -- BCD code for the least significant digit
end BinToBCD16;

architecture Behavioral of BinToBCD16 is
  type state is (idle, op, done);
  signal c_s, n_s               : state;
  signal BCD4_c, BCD3_c, BCD2_c, BCD1_c, BCD0_c, BCD4_n, BCD3_n, BCD2_n, BCD1_n,
        BCD0_n : unsigned(3 downto 0);
  signal BCD3_tmp, BCD2_tmp, BCD1_tmp, BCD0_tmp      : unsigned(3 downto 0);
  signal BCD4_tmp                                    : unsigned(2 downto 0);
  signal int_rg_c, int_rg_n   : std_logic_vector (size_of_data_to_convert-1 downto 0);
  signal index_c, index_n     : unsigned(4 downto 0);
  signal get_outputs          : std_logic;
begin
process(clk, reset)
begin
  if rising_edge(clk) then
    if reset = '1' then c_s <= idle;
       BCD4_c <= (others => '0'); BCD3_c <= (others => '0'); BCD2_c <= (others => '0');
       BCD1_c <= (others => '0'); BCD0_c <= (others => '0'); BCD0 <= (others=>'0');
       BCD1 <= (others=>'0'); BCD2 <= (others=>'0'); BCD3 <= (others=>'0');
       BCD4 <= (others=>'0');
    else c_s <= n_s;
       BCD4_c <= BCD4_n; BCD3_c <= BCD3_n; BCD2_c <= BCD2_n;
       BCD1_c <= BCD1_n; BCD0_c <= BCD0_n;
       index_c <= index_n; int_rg_c <= int_rg_n;
       if (get_outputs = '1') then
         BCD0 <= std_logic_vector(BCD0_n); BCD1 <= std_logic_vector(BCD1_n);
         BCD2 <= std_logic_vector(BCD2_n); BCD3 <= std_logic_vector(BCD3_n);
         BCD4 <= std_logic_vector(BCD4_n);
       end if;
    end if;
  end if;
end process;


    process (c_s, BCD4_c, BCD3_c, BCD2_c, BCD1_c, BCD0_c, BCD4_tmp, BCD3_tmp,
          BCD2_tmp, BCD1_tmp, BCD0_tmp, binary, int_rg_c, index_c, index_n, request)
    begin

      get_outputs <= '0';
      n_s <= c_s; BCD4_n <= BCD4_c; BCD3_n <= BCD3_c; BCD2_n <= BCD2_c;
      BCD1_n <= BCD1_c; BCD0_n <= BCD0_c; index_n <= index_c; int_rg_n <= int_rg_c;
      ready <= '0';

      case c_s is
        when idle =>
           n_s <= op; ready <= '1'; int_rg_n <= binary; index_n <= "10000";
           if request /= '1' then n_s <= idle; -- transition to the op state is
           end if;                      -- done as soon as the request is active (i.e. equal to 1)
        when op =>   ready <= '0';
           int_rg_n <= int_rg_c(size_of_data_to_convert-2 downto 0) & '0';
           BCD0_n <= BCD0_tmp(2 downto 0) & int_rg_c(size_of_data_to_convert-1);
           BCD1_n <= BCD1_tmp(2 downto 0) & BCD0_tmp(3);
           BCD2_n <= BCD2_tmp(2 downto 0) & BCD1_tmp(3);
```

```
            BCD3_n <= BCD3_tmp(2 downto 0) & BCD2_tmp(3);
            BCD4_n <= BCD4_tmp(2 downto 0) & BCD3_tmp(3);
            index_n <= index_c - 1;
            if (index_n = 0) then n_s <= done; get_outputs <= '1';
            end if;
        when done => n_s <= idle;
            BCD4_n <= (others => '0'); BCD3_n <= (others => '0'); BCD2_n <= (others => '0');
            BCD1_n <= (others => '0'); BCD0_n <= (others => '0');
            ready <= '1';          -- now ready is 1, i.e. a new conversion can be done
      end case;
   end process;

   BCD0_tmp <= BCD0_c + 3 when BCD0_c > 4 else BCD0_c;
   BCD1_tmp <= BCD1_c + 3 when BCD1_c > 4 else BCD1_c;
   BCD2_tmp <= BCD2_c + 3 when BCD2_c > 4 else BCD2_c;
   BCD3_tmp <= BCD3_c + 3 when BCD3_c > 4 else BCD3_c;
   BCD4_tmp <= BCD4_c(2 downto 0) + 3 when BCD4_c > 4 else BCD4_c(2 downto 0);

   end Behavioral;
```

The module BinToBCD16 operates slightly different comparing with the module BinToBCD8. Now the ready signal is active in the idle state (ready <= '1') and the module waits for a request for a new conversion. As soon as the signal request becomes active, new data item is taken and a new conversion will be done (see Fig. B.1c). To use the module BinToBCD16 in the entity TopForInteractingWitIPCores (see Sect. 4.1) the following small change can be done:

```
binTO_BCD3: entity work.BinToBCD16 – the request below is assumed to be always active (i.e. 1)
      port map (clk, reset, open, To_BCD, '1', BCD4, BCD3, BCD2, BCD1, BCD0);
```

The following changes can be done in the entity TopForInteractingWitIPCores that enable the request signal to be involved:

1. New signals have to be declared: **signal** request, ready: std_logic;
2. The request has to be generated, for example: request <= ready **and** BTNR;
3. Mapping in the **port map** is done as follows:

```
port map (clk, reset, ready, To_BCD, request, BCD4, BCD3, BCD2, BCD1, BCD0);
```

Now the conversion will be done on the request from the onboard button BTNR.

Clock Divider (*clock_divider*)

A *clock divider* permits system clock to be divided by 2^{how_fast+1}. The module with the reset signal is the following (from comments it is clear that reset can be removed if required):

```
library IEEE;
use IEEE.STD_LOGIC_1164.all;
use IEEE.STD_LOGIC_UNSIGNED.all;

entity clock_divider is
```

```
generic (how_fast : integer := 25 );
port ( clk, reset     : in std_logic;        -- similar circuit without the reset signal can also be used
       divided_clk    : out std_logic);
end clock_divider;

architecture Behavioral of clock_divider is
  signal internal_clock : std_logic_vector (how_fast downto 0);
begin

process(clk, reset)           -- remove reset if there is no reset in the circuit
begin
  if rising_edge(clk) then
    if reset = '1' then          -- remove reset if there is no reset in the circuit
       internal_clock <= (others=>'0');      -- remove this line if there is no reset in the circuit
    else internal_clock <= internal_clock+1;
    end if;                      -- remove else and end if keywords if there is no reset in the circuit
  end if;
end process;

divided_clk <= internal_clock(internal_clock'left) when falling_edge(clk);

end Behavioral;
```

DSP-Based Hamming Weight Counter/Comparator for N = 32 (*Test_HW32*)

The following VHDL code is a complete description of the Hamming weight counter/comparator (with a fixed threshold) for N = 32:

```
library IEEE;    -- The top-level module to test the 32-bit Hamming weight counter/comparator
use IEEE.STD_LOGIC_1164.all;             -- this circuit occupies 0 logical slices and 2 DSP48 slices
use IEEE.STD_LOGIC_UNSIGNED.all;         -- the maximum combinational path delay is 3.9 ns

entity Test_HW32 is        -- the project was tested in the Nexys-4 board
   port ( Sw         : in  std_logic_vector (15 downto 0);    -- Nexys-4 onboard switches
          led        : out  std_logic_vector (5 downto 0);    -- Nexys-4 onboard LEDs
          in16bit    : in std_logic_vector(15 downto 0);      -- signals from Nexys-4 PMod connectors
          led_comp : out std_logic);                          -- the result of comparison
end Test_HW32;

architecture Behavioral of Test_HW32 is
   signal threshold             : std_logic_vector(5 downto 0);
   signal HW1, HW2              : std_logic_vector(4 downto 0);
   signal remaining_inputs1     : std_logic_vector(11 downto 0);
   signal remaining_inputs2     : std_logic_vector(11 downto 0);
   signal remaining_outputs1    : std_logic_vector(5 downto 0);
   signal remaining_outputs2    : std_logic_vector(5 downto 0);
begin

threshold            <= not "011010" + 1;  -- this value of threshold was taken just for test
remaining_inputs1    <= '0' & HW1 & '0' & HW2;
remaining_inputs2    <= remaining_outputs1 & threshold;
led                  <= remaining_outputs1;

HWCC16_1: entity work.HW_counter_comparator_16bit      -- see the code below
```

```
        port map(Sw, HW1, remaining_inputs1, remaining_outputs1, open);

HWCC16_2: entity work.HW_counter_comparator_16bit     -- see the code below
        port map(in16bit, HW2, remaining_inputs2, open, led_comp);

end Behavioral;

library IEEE;      -- this is the component for the top-level module above
use IEEE.STD_LOGIC_1164.all;         -- this is 16-bit Hamming weight counter/comparator

entity HW_counter_comparator_16bit is -- this component is used as HWCC16_1 and HWCC16_2 above
   port ( Sw                   : in  std_logic_vector (15 downto 0);
          Hamming_weight       : out  std_logic_vector (4 downto 0);
          remaining_inputs     : in std_logic_vector(11 downto 0);
          remaining_outputs    : out std_logic_vector(5 downto 0);
          comp                 : out std_logic);
end HW_counter_comparator_16bit;

architecture Behavioral of HW_counter_comparator_16bit is
  signal A, B, Y   : std_logic_vector(47 downto 0); -- A and B are operands for DSP48E1, Y is the result
begin

process(Sw, Y, remaining_inputs)
begin
  A <= (others => '0'); B <= (others => '0'); -- at the beginning the operands are assigned zero values

  for i in 7 downto 0 loop   -- see also Fig. 4.10 and Fig. 4.11
    A(2*i) <= Sw(i);
    B(2*i) <= Sw(i+8);
  end loop;

  for i in 3 downto 0 loop
    A(16+3*i+1 downto 16+3*i) <= Y(2*i+1 downto 2*i);
    B(16+3*i+1 downto 16+3*i) <= Y(2*i+1+8 downto 2*i+8);
  end loop;

  for i in 1 downto 0 loop
    A(28+4*i+2 downto 28+4*i) <= Y(16+3*i+2 downto 16+3*i);
    B(28+4*i+2 downto 28+4*i) <= Y(16+3*i+2+6 downto 16+3*i+6);
  end loop;

  A(39 downto 36) <= Y(31 downto 28);
  B(39 downto 36) <= Y(35 downto 32);
  A(46 downto 41) <= remaining_inputs(5 downto 0);
  B(46 downto 41) <= remaining_inputs(11 downto 6);
end process;

Hamming_weight <= Y(40 downto 36);           -- the resulting Hamming weight
comp   <= Y(47);                             -- the result of comparison
remaining_outputs <= Y(46 downto 41);        -- the threshold is supplied here

DSP   : entity work.TesDSP48E1_HW16
        port map (A, B, "0000", Y);

end Behavioral;
```

Figure B.2a demonstrates a possible interface with the circuit.

Clearly, the value of the threshold can also be taken from outside and any mode from Fig. 3.30 can easily be added with just one additional look-up table (see Fig. B.2b). Xilinx primitive CFGLUT5 [1] is a runtime, dynamically reconfigurable 5-input LUT that enables the implemented logical function (configuration of the LUT) to be changed during the circuit operation. Hence, the bounds/thresholds can be modified during run-time if required.

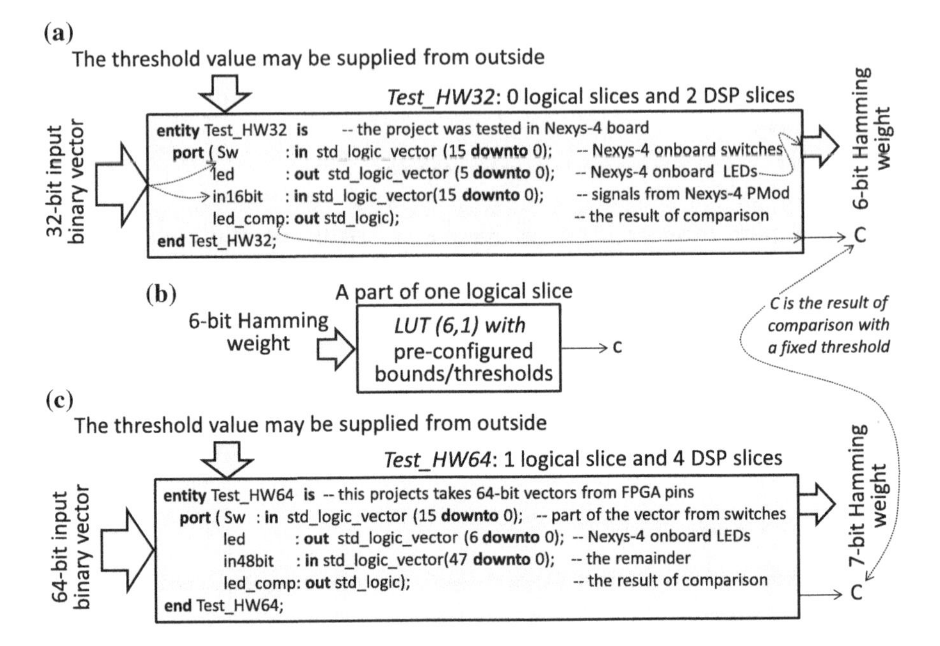

Fig. B.2 DSP-based 32-bit Hamming weight counter/comparator (**a**); using many bounds for comparison (**b**); DSP-based 64-bit Hamming weight counter/comparator (**c**)

DSP-Based Hamming Weight Counter/Comparator for N = 64 (Test_HW64)

The following VHDL code is a complete description of the Hamming weight counter/comparator (with a fixed threshold) for N = 64:

```
library IEEE;     -- the top-level module to test the 64-bit Hamming weight counter/comparator
use IEEE.STD_LOGIC_ARITH.all;             -- the project was tested in the Nexys-4 board
use IEEE.STD_LOGIC_UNSIGNED.all;          -- the maximum combinational path delay is 6.1 ns

entity Test_HW64 is -- this projects takes 64-bit vectors from FPGA pins
  port ( Sw      : in std_logic_vector (15 downto 0);    -- part of the vector from Nexys-4 switches
       led       : out std_logic_vector (6 downto 0);    -- Nexys-4 onboard LEDs
       in48bit   : in std_logic_vector(47 downto 0);     -- the rest of the vector from other pins
       led_comp  : out std_logic);                       -- the result of comparison
end Test_HW64;

architecture Behavioral of Test_HW64 is       -- this circuit occupies 1 logical slice and 4 DSP48 slices
  signal threshold              : std_logic_vector(6 downto 0);
  signal HW1, HW3               : std_logic_vector(4 downto 0);
  signal HW2, HW4               : std_logic_vector(5 downto 0);
  signal remaining_inputs1      : std_logic_vector(11 downto 0);
  signal remaining_inputs2      : std_logic_vector(11 downto 0);
  signal remaining_inputs3      : std_logic_vector(11 downto 0);
  signal remaining_inputs4      : std_logic_vector(11 downto 0);
```

```
  signal remaining_outputs1          : std_logic_vector(5 downto 0);
  signal remaining_outputs2          : std_logic_vector(6 downto 0);
  signal remaining_outputs3          : std_logic_vector(5 downto 0);
  signal comp                        : std_logic;
begin
  threshold              <= not "0110010" + 1; -- this value of threshold was taken just for test
  remaining_inputs1      <= '0' & HW1 & HW2;
  remaining_inputs2      <= remaining_outputs1 & remaining_outputs3;
  remaining_inputs3      <= '0' & HW3 & HW4;
  remaining_inputs4      <= remaining_outputs2(5 downto 0) & threshold(5 downto 0);
  led_comp               <= comp or remaining_outputs2(6);
  led                    <= remaining_outputs2;

HWCC16_1: entity work.HW_counter_comparator_16bit     -- see the code above
     port map(Sw, HW1, remaining_inputs1, remaining_outputs1, open);

HWCC16_2: entity work.HW_counter_comparator_16bit_m
     port map( in48bit(15 downto 0), HW2, remaining_inputs2, remaining_outputs2, open);

HWCC16_3: entity work.HW_counter_comparator_16bit     -- see the code above
     port map(in48bit(31 downto 16), HW3, remaining_inputs3, remaining_outputs3, open);

HWCC16_4: entity work.HW_counter_comparator_16bit_m -- the code above is slightly changed
     port map(in48bit(47 downto 32), HW4, remaining_inputs4, open, comp);

end Behavioral;
```

The next module may be helpful when only the onboard switches for the Nexys-4 board are used to supply 64-bit binary vectors as a sequence of four 16-bit fragments from the 16 available switches. Each fragment is saved when the associated onboard button is pressed (BTNL for in_16bit1, BTNC for in_16bit2, BTNR for in_16bit3, and BTND for in_16bit4).

```
library IEEE;    -- the top-level module to test the 64-bit Hamming weight counter/comparator
use IEEE.STD_LOGIC_1164.all;        -- the project was tested in the Nexys-4 board
use IEEE.STD_LOGIC_UNSIGNED.all;

entity Test_HW64 is
   port ( clk        : in std_logic;          -- for reading and saving 16-bit segments of 64-bit vector
          Sw         : in  std_logic_vector (15 downto 0);  -- segments of the vector from Nexys-4 switches
          led        : out  std_logic_vector (6 downto 0); -- Nexys-4 onboard LEDs
          BTNL, BTNC, BTNR, BTND   : in std_logic;    -- Nexys-4 onboard buttons
         led_comp : out std_logic);                  -- the result of comparison
end Test_HW64;

architecture Behavioral of Test_HW64 is
       -- the same signal declarations as in the previous example
signal in_16bit1, in_16bit2, in_16bit3, in_16bit4 : std_logic_vector(15 downto 0);
begin

process(clk)
begin -- reading and saving 16-bit fragments of 64-bit vector
  if rising_edge(clk) then
    if BTNL = '1' then         -- saving the first 16 bits if BTNL is pressed
       in_16bit1 <= Sw;
    elsif BTNC = '1' then      -- saving the second 16 bits if BTNC is pressed
```

```
        in_16bit2 <= Sw;
    elsif BTNR = '1' then      -- saving the third 16 bits if BTNR is pressed
        in_16bit3 <= Sw;
    elsif BTND = '1' then      -- saving the forth 16 bits if BTND is pressed
        in_16bit4 <= Sw;
    else null;
    end if;
  end if;
end process;

-- the code here is almost the same as in the example above. The only difference is in supplying the fragments
-- in_16bit1, in_16bit2, in_16bit3, and in_16bit4 to the 16-bit Hamming weight counters/comparators

end Behavioral;
```

Even–Odd Merge Sorting Network for N = 16 (EvenOddMergeSort16)

The network uses two components EvenOddMerge8Sort described in Sect. 3.4.1.

```
library IEEE;                  -- the project was tested in the Atlys board involving interactions with a host PC
use IEEE.STD_LOGIC_1164.all;   -- interactions with a host PC are not shown here
use work.set_of_data_items.all;  -- see the given below user-defined package

entity EvenOddMerge16Sort is   -- this circuit occupies 187 logical slices (including interactions)
generic (M      : integer := 4;    -- generic size of data items
         N      : integer := 16 ); -- generic number of data items (cannot be changed for this project)
  port ( input_1items      : in set_of_8items;
         input_2items      : in set_of_8items;
          sorted            : out set_of_16items );
  end EvenOddMerge16Sort;

architecture Structural of EvenOddMerge16Sort is
  signal sorted1,sorted2      : set_of_8items;
  signal out1_in2, out2_in3   : set_of_16items;
  signal out3_in4             : set_of_16items;
begin

sort8items1: entity work.EvenOddMerge8Sort    -- even-odd merge sorter for 8 items
        generic map (M => M, N => 8)
        port map(input_1items, sorted1);       -- the code of the sorter is given in section 3.4.1

sort8items2: entity work.EvenOddMerge8Sort    -- even-odd merge sorter for 8 items
        generic map (M => M, N => 8)
        port map(input_2items, sorted2);       -- the code of the sorter is given in section 3.4.1

stage4: for i in N/2-1 downto 0 generate
  group1stage4:  entity work.Comparator
        generic map (M => M)
        port map(sorted1(i), sorted2(i), out1_in2(i), out1_in2(i+8));
  step1stage4: if (i >= 4) generate
        group2stage4:  entity work.Comparator
        generic map (M => M)
        port map(out1_in2(i), out1_in2(i+4), out2_in3(i), out2_in3(i+4));
  end generate;

  step2stage4: if (i < 4) generate
        out2_in3(i) <= out1_in2(i);
        out2_in3(i+12) <= out1_in2(i+12);
  end generate;
  step3stage4: if (i < 3) generate
        incide_stage4: for j in 0 to N/8-1 generate
```

```
            group3stage4: entity work.Comparator
                    generic map (M => M)
                    port map(out2_in3(2+i*4+j), out2_in3(2+i*4+j+2), out3_in4(2+i*4+j),
                              out3_in4(2+i*4+j+2));
        end generate incide_stage4;
    end generate;
    step4stage4: if (i < 2) generate
        out3_in4(i) <= out2_in3(i);
        out3_in4(i+14) <= out2_in3(i+14);
    end generate;
    step5stage4: if (i < N/2-1) generate
    step5stage4: entity work.Comparator
        generic map (M => M)
        port map(out3_in4(1+i*2), out3_in4(1+i*2+1), sorted(1+i*2), sorted(1+i*2+1));
    end generate;
  end generate stage4;

  sorted(0) <= out3_in4(0);
  sorted(15) <= out3_in4(15);

  end Structural;
```

The following package set_of_data_items has been used:

```
library IEEE;
use IEEE.STD_LOGIC_1164.all;

package set_of_data_items is
  constant N    : integer := 8;
  constant M    : integer := 4; -- for different values of M this constant needs to be changed
  type set_of_8items is array (N-1 downto 0) of std_logic_vector (M-1 downto 0);
  type set_of_16items is array (2*N-1 downto 0) of std_logic_vector (M-1 downto 0);
end set_of_data_items;

package body set_of_data_items is
end set_of_data_items;
```

Hamming Weight Comparator for N = 15 (HammingWeightComparator)

The following VHDL code is a complete synthesizable specification of the Hamming weight comparator in Fig. 3.31a (for any module below the final comparison circuits from Fig. 3.25 can also be used):

```
library IEEE;                    -- the project was tested for the Nexys-4 board and occupies 3 slices
use IEEE.STD_LOGIC_1164.all;     -- maximum combinational path delay is 2.5 ns
-- the final comparator LUT6_I is configured for: if (3 < weight < 10) then LED if OFF otherwise - ON

entity HammingWeightComparator is
port ( Sw       : in std_logic_vector (14 downto 0); -- input 15-bit vector
       LedC     : out std_logic);                    -- the result of comparison
end HammingWeightComparator;

architecture Behavioral of HammingWeightComparator is
  signal Upper_bits, Middle_bits, Bottom_bits    : std_logic_vector(2 downto 0);
  signal ToLast                                  : std_logic_vector(5 downto 0);
  signal comp                                    : std_logic;
```

```
begin

LUT_5_3_upper  : entity work.LUT_5to3
        port map(Sw(14 downto 10), Upper_bits);

LUT_5_3_middle : entity work.LUT_5to3
        port map(Sw(9 downto 5), Middle_bits);

LUT_5_3_bottom : entity work.LUT_5to3
        port map(Sw(4 downto 0), Bottom_bits);

LUT6_1_comp: entity work.LUT6_1
        port map (ToLast, LedC);

FA_generate: for i in 0 to 2 generate
        FA: entity work.FullAdder       -- see entity FullAdder in section 3.7
        port map(Bottom_bits(i), Middle_bits(i), Upper_bits(i), ToLast(2*i), ToLast(2*i+1));
        end generate FA_generate;

        end Behavioral;
```

The following code is for the component LUT_5to3:

```
library IEEE;
use IEEE.STD_LOGIC_1164.all;
library UNISIM;                          -- for FPGA LUTs
use UNISIM.vcomponents.all;

entity LUT_5to3 is
   port ( fiveBitIn            : in  std_logic_vector (4 downto 0);
          ThreeBitOut          : out  std_logic_vector (2 downto 0));
end LUT_5to3;

architecture Structural of LUT_5to3 is
begin

LUT5_inst1 : LUT5
  generic map (INIT => X"E8808000") -- LUT Contents
  port map (ThreeBitOut(2), fiveBitIn(0), fiveBitIn(1), fiveBitIn(2), fiveBitIn(3), fiveBitIn(4));

LUT5_inst2 : LUT5
  generic map (INIT => X"177E7EE8") -- LUT Contents
  port map (ThreeBitOut(1), fiveBitIn(0), fiveBitIn(1), fiveBitIn(2), fiveBitIn(3), fiveBitIn(4));

LUT5_inst3 : LUT5
  generic map (INIT => X"96696996") -- LUT Contents
  port map (ThreeBitOut(0), fiveBitIn(0), fiveBitIn(1), fiveBitIn(2), fiveBitIn(3), fiveBitIn(4));

end Structural;
```

The following code is for the final comparator in Fig. 3.29a (LUT6_1):

```
library IEEE;
use IEEE.STD_LOGIC_1164.all;
library UNISIM;                       -- for FPGA LUTs
use UNISIM.vcomponents.all;

entity LUT6_1 is
  port ( SixIn    : in std_logic_vector (5 downto 0);
         Comp     : out std_logic);
end LUT6_1;

architecture Structural of LUT6_1 is
begin -- this LUT is used just for comparator and it is configured for two bounds

  LUT6_inst0 : LUT6                           -- if required the LUT contents can be configured
      generic map (INIT => X"ffffffffc00003f")  -- for different bounds (see table B.1 for details)
      port map (Comp, SixIn(0), SixIn(1), SixIn(2), SixIn(3), SixIn(4), SixIn(5));

end Structural;
```

Table B.1 below explains how to configure the LUT LUT6_1 for the final comparator in Fig. 3.29a.

The *SixIn* column shows input vectors that are represented by three 2-bit sub-vectors. The most significant sub-vector has the weight 4, the middle sub-vector—the weight 2 and the least significant sub-vector—the weight 1. Thus, the code 000101 has the value $0 \times 4 + 1 \times 2 + 1 \times 1 = 3$ and this is the settled bound 3. The next code 000110 has the value $0 \times 4 + 1 \times 2 + 2 \times 1 = 4$ and this is the value above the bound 3. All the subsequent values until the number 011001 are within the settled bounds (more than 3 and less than 10). The number 011010 has the first value $1 \times 4 + 2 \times 2 + 2 \times 1 = 10$ outside the bounds. Hexadecimal numbers from Table B.1 are used to configure the LUT. They have to be taken from the bottom right part to the upper left part giving the following constant: FFFFFFFFFC00003F that is used for the INIT statement: INIT => X"ffffffffc00003f".

The circuit has been tested in the Nexys-4 board. Input vectors were taken from 15 onboard switches: 14, 13, ..., 0 (switch 15 was not used). The result of comparison is shown on LED 0.

Hamming Weight Counter for N = 31 and Comparator for N = 32 (*HW31_HWC32*)

VHDL code below can be used directly for the circuit in Fig. 3.32 which counts the Hamming weight of any input binary vector for N = 31 (i.e. for $B = \{B_0,\ldots,B_{30}\}$)

Table B.1 Configuring the LUT6_1 for the final comparator (see also Fig. 3.29a)

SixIn	Comp			SixIn	Comp			SixIn	Comp			SixIn	Comp		
000000	1		F	010000	0		0	100000	1		F	110000	1		F
000001	1			010001	0			100001	1			110001	1		
000010	1			010010	0			100010	1			110010	1		
000011	1			010011	0			100011	1			110011	1		
000100		1	3	010100		0	0	100100		1	F	110100		1	F
000101		1		010101		0		100101		1		110101		1	
000110		0		010110		0		100110		1		110110		1	
000111		0		010111		0		100111		1		110111		1	
001000	0		0	011000	0		C	101000	1		F	111000	1		F
001001	0			011001	0			101001	1			111001	1		
001010	0			011010	1			101010	1			111010	1		
001011	0			011011	1			101011	1			111011	1		
001100		0	0	011100		1	F	101100		1	F	111100		1	F
001101		0		011101		1		101101		1		111101		1	
001110		0		011110		1		101110		1		111110		1	
001111		0		011111		1		101111		1		111111		1	

and provides comparison of any binary vector for N = 32 with fixed bounds (any set of bounds from Fig. 3.30 may be chosen).

```
library IEEE;                    -- the project was tested for the Nexys-4 board and occupies 14 logical slices
use IEEE.STD_LOGIC_1164.all;       -- the maximum combinational path delay is 4.4 ns
use IEEE.STD_LOGIC_ARITH.all;      -- constant compare configured for two bounds: 1) 10≥weight -
-- LedC is OFF; 2) 10<weight<20 - LedC is ON; 3) 20≤weight<30 - LedC is OFF;
use IEEE.STD_LOGIC_UNSIGNED.all;           -- and 30≤weight - LedC is ON;

entity HW31_HWC32 is    -- the names of used components are the same as in Fig. 3.32
  port ( Data_in  : in std_logic_vector (31 downto 0); -- 32-bit binary vector (Vector31_in in Fig. 4.6)
         led      : out std_logic_vector (4 downto 0);
        LedC      : out std_logic);
end HW31_HWC32;

architecture Mixed of HW31_HWC32 is
  signal HW15_1: std_logic_vector(3 downto 0);  -- the Hamming weight for Data_in(14 downto 0)
  signal HW15_2: std_logic_vector(3 downto 0);  -- the Hamming weight for Data_in(29 downto 15)
  signal LUT5_3 : std_logic_vector(3 downto 0);  -- LUT5_3 in Fig. 3.32 (see block D)
  signal LUT4_3 : std_logic_vector(3 downto 0);  -- LUT4_3 in Fig. 3.32 (see block C)
  signal Out5_3 : std_logic_vector(2 downto 0);  -- LUT5_3 in Fig. 3.32 (see block D)
  signal Out4_3 : std_logic_vector(2 downto 0);  -- LUT4_3 in Fig. 3.32 (see block C)
  constant compare : std_logic_vector(127 downto 0) :=    -- 128-bit constant for comparator
       X"FEE000077FFCC000FCC0000FFFF88000";
 -- there are five 64-bit constants below for the Hamming weight bits (4 downto 0) for a 31-bit binary vector
  constant bit0  : std_logic_vector(63 downto 0) := X"AAAAAAAAAAAAAAAA";
  constant bit1  : std_logic_vector(63 downto 0) := X"CCCCCCCCCCCCCCCC";
  constant bit2  : std_logic_vector(63 downto 0) := X"0FF00FF00FF00FF0";
  constant bit3  : std_logic_vector(63 downto 0) := X"0FFFF0000FFFF000";
  constant bit4  : std_logic_vector(63 downto 0) := X"0FFFFFFFF0000000";
begin
LUT_based1: entity work.HW15Counter                          -- see block B in Fig. 3.32
       port map (Data_in(14 downto 0), HW15_1);

LUT_based2: entity work.HW15Counter                          -- see block A in Fig. 3.32
       port map (Data_in(29 downto 15), HW15_2);

LUT4_3 <= HW15_1(3 downto 2) & HW15_2(3 downto 2);  -- see LUT4_3 lines in Fig. 3.32
LUT5_3 <= HW15_1(1 downto 0) & HW15_2(1 downto 0);   -- see LUT5_3 lines in Fig. 3.32

LUT_4_3: entity work.LUT4to3                     -- see block C in Fig. 3.32
       port map(LUT4_3, Out4_3);

LUT_5_3 : entity work.LUT5to3                    -- see block D in Fig. 3.32
       port map(LUT5_3, Data_in(30), Out5_3);

LedC <= compare(conv_integer(Data_in(31) & Out4_3 & Out5_3)); -- the result of comparison (block E)
-- 5-bit Hamming weight of the vector Data_in(30 downto 0) is copied to LED (this part is not shown in Fig. 3.32)
-- if necessary to get the Hamming weight for a vector Data_in(31 downto 0) an extra bit 31 can be added
led <= bit4(conv_integer(Out4_3 & Out5_3)) & bit3(conv_integer(Out4_3 & Out5_3)) &
       bit2(conv_integer(Out4_3 & Out5_3)) & bit1(conv_integer(Out4_3 & Out5_3)) &
       bit0(conv_integer(Out4_3 & Out5_3));  -- computation of Hamming weight is not shown in Fig. 3.32
end Mixed;
```

Table B.2 Preparing the constants for the module HW31_HWC32

SixIn	Hamming weight/comparator		SixIn	Hamming weight/ comparator	
000000 (0)	00000 (0)	0 000CA	100000 (16)	10000 (1)	F F00CA
000001 (1)	00001 (0)		100001 (17)	10001 (1)	
000010 (2)	00010 (0)		100010 (18)	10010 (1)	
000011 (3)	00011 (0)		100011 (19)	10011 (1)	
000100 (4)	00100 (0)	0 00FCA	100100 (20)	10100 (0)	0 F0FCA
000101 (5)	00101 (0)		100101 (21)	10101 (0)	
000110 (6)	00110 (0)		100110 (22)	10110 (0)	
000111 (7)	00111 (0)		100111 (23)	10111 (0)	
001000 (4)	00100 (0)	0 00FCA	101000 (20)	10100 (0)	0 F0FCA
001001 (5)	00101 (0)		101001 (21)	10101 (0)	
001010 (6)	00110 (0)		101010 (22)	10110 (0)	
001011 (7)	00111 (0)		101011 (23)	10111 (0)	
001100 (8)	01000 (0)	8 0F0CA	101100 (24)	11000 (0)	0 FF0CA
001101 (9)	01001 (0)		101101 (25)	11001 (0)	
001110 (10)	01010 (0)		101110 (26)	11010 (0)	
001111 (11)	01011 (1)		101111 (27)	11011 (0)	
010000 (8)	01000 (0)	8 0F0CA	110000 (24)	11000 (0)	0 FF0CA
010001 (9)	01001 (0)		110001 (25)	11001 (0)	
010010 (10)	01010 (0)		110010 (26)	11010 (0)	
010011 (11)	01011 (1)		110011 (27)	11011 (0)	
010100 (12)	01100 (1)	F 0FFCA	110100 (28)	11100 (0)	C FFFCA
010101 (13)	01101 (1)		110101 (29)	11101 (0)	
010110 (14)	01110 (1)		110110 (30)	11110 (1)	
010111 (15)	01111 (1)		110111 (31)	11111 (1)	
011000 (12)	01100 (1)	F 0FFCA	111000 (28)	11100 (0)	C FFFCA
011001 (13)	01101 (1)		111001 (29)	11101 (0)	
011010 (14)	01110 (1)		111010 (30)	11110 (1)	
011011 (15)	01111 (1)		111011 (31)	11111 (1)	
011100 (16)	10000 (1)	F F00CA	111100 (32)	100000 (1)	F 000CA
011101 (17)	10001 (1)		111101 (33)	100001 (1)	
011110 (18)	10010 (1)		111110 (34)	100010 (1)	
011111 (19)	10011 (1)		111111 (35)	100011 (1)	

Table B.2 explains how the constants compare, bit4, bit3, bit2, bit1, bit0 have been prepared.

The result of the comparison is changed twice on the left-hand side of Table B.2. Let us consider the first change: 001110 (10) and 001111 (11). The values in parenthesis (in the *SixIn* column) indicate the decimal numbers corresponding to the neighboring code. For the vector 001 110 the decimal number is formed as $1_{10} \times 4_{10} + 6_{10} = 10_{10}$ (see also Fig. 3.32). The value in parenthesis for the column *Hamming weight/comparator* indicates the result of comparison. It is equal to 0 for (10_{10}) and it is equal to 1 for (11_{10}). For the second vector 010 010 (10)

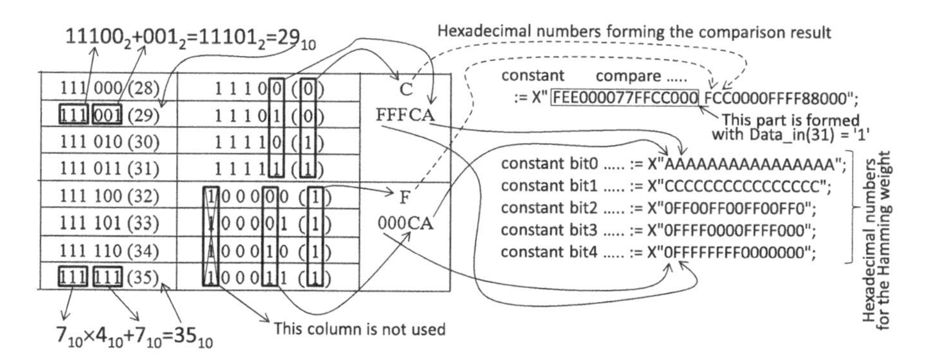

Fig. B.3 Preparing constant values

the value (10_{10}) is formed as $2_{10} \times 4_{10} + 2_{10} = 10_{10}$ (see also Fig. 3.32). Additional explanations are given in Fig. B.3. The upper one-digit hexadecimal numbers are used to configure the comparator and the lower five-digits hexadecimal numbers—to configure the counter. Figure B.3 explains how the constants have been prepared. This figure depicts the bottom right part of the Table B.2. Hexadecimal numbers for forming the comparison result are produced as shown in Fig. B.3. Thus, the 16-digit constant FCC0000FFFF88000 is defined. The most significant 16-digit part of the constant (FEE000077FFCC000) is formed using the same technique but considering also the most significant bit Data_in(31) (this part is not shown in Table B.2 and in Fig. B.3). Hexadecimal constants for the Hamming weight are built similarly, but now five columns with five-digit hexadecimal numbers are used (the rightmost column for bit0 and the leftmost column for bit4; the remaining constants are built from digits in the middle).

Thus, the following hexadecimal values are used: 000CA (the bottom right part of Table B.2), FFFCA, FFFCA, FF0CA, FF0CA, F0FCA, F0FCA, F00CA, F00CA, 0FFCA, 0FFCA, 0F0CA, 0F0CA, 00FCA, 00FCA, 000CA (the upper left part of Table B.2). Such constants have been defined only for five least significant digits in the column *Hamming weight/comparator* (because the Hamming weight is computed only for 31-bit vectors and five binary digits are sufficient). Now the constant is prepared for each hexadecimal digit. For the most significant digit the constant bit4 is 0FFFFFFFF0000000 (composed of the most significant digits in each hexadecimal number). The next constant for bit3 is: 0FFFF0000FFFF000, etc.

The component HW15Counter is very similar to the described above component HammingWeightComparator. The only difference is in computing the Hamming weight of a 15-bit input vector instead of the result of comparison. VHDL code below is used for the component HW15Counter that computes the Hamming weight:

```
library IEEE;
use IEEE.STD_LOGIC_1164.all;

entity HW15Counter is
  port ( Data_in   : in std_logic_vector (14 downto 0);       -- input binary vector
         HW15      : out std_logic_vector(3 downto 0));       -- Hamming weight of the input vector
end HW15Counter;

architecture Structural of HW15Counter is -- it is very similar to the entity HammingWeightComparator
  signal Upper, Middle, Bottom        : std_logic_vector(2 downto 0);
  signal ToLast                       : std_logic_vector(5 downto 0);
begin

LUT_5_3_upper  : entity work.LUT_5to3
       port map(Data_in(14 downto 10), Upper);

LUT_5_3_middle : entity work.LUT_5to3
       port map(Data_in(9 downto 5), Middle);

LUT_5_3_bottom : entity work.LUT_5to3
       port map(Data_in(4 downto 0), Bottom);

LUT6_4_comp_HW: entity work.LUT6_4
       port map (ToLast, HW15);

FA_generate: for i in 0 to 2 generate
  FA:  entity work.FullAdder                  -- see entity FullAdder in section 3.7
       port map(Bottom(i), Middle(i), Upper(i), ToLast(2*i), ToLast(2*i+1));
end generate FA_generate;

end Structural;
```

The component LUT_5to3 is the same as in the described above entity HammingWeightComparator. The component LUT6_4 has the following VHDL code:

```
library IEEE;
use IEEE.STD_LOGIC_1164.all;

library UNISIM;                               -- for FPGA LUTs
use UNISIM.vcomponents.all;

entity LUT6_4 is
   port ( Data_in   : in  std_logic_vector (5 downto 0);      -- input binary vector
          Data_out : out  std_logic_vector (3 downto 0));     -- Hamming weight for the input vector
end LUT6_4;

architecture Structural of LUT6_4 is
begin

LUT6_inst2: LUT6
  generic map (INIT => X"003f3fffffc0c000") -- LUT Contents
  port map (Data_out(3), Data_in(0), Data_in(1), Data_in(2), Data_in(3), Data_in(4), Data_in(5));

LUT6_inst3 : LUT6

generic map (INIT => X"c03f3fc0c03f3fc0") -- LUT Contents
port map (Data_out(2), Data_in(0), Data_in(1), Data_in(2), Data_in(3), Data_in(4), Data_in(5));

LUT6_inst4 : LUT6
  generic map (INIT => X"3c3c3c3c3c3c3c3c") -- LUT Contents
  port map (Data_out(1), Data_in(0), Data_in(1), Data_in(2), Data_in(3), Data_in(4), Data_in(5));

LUT6_inst5 : LUT6
  generic map (INIT => X"aaaaaaaaaaaaaaaa") -- LUT Contents
  port map (Data_out(0), Data_in(0), Data_in(1), Data_in(2), Data_in(3), Data_in(4), Data_in(5));

end Structural;
```

The component LUT4to3 has the following VHDL code:

```
library IEEE;
use IEEE.STD_LOGIC_1164.all;
library UNISIM;                          -- for FPGA LUTs
use UNISIM.vcomponents.all;

entity LUT4to3 is
   port ( Data_in   : in  std_logic_vector (3 downto 0);
          Data_out : out  std_logic_vector (2 downto 0));
end LUT4to3;

architecture Structural of LUT4to3 is
begin

LUT4_inst1 : LUT4
  generic map (INIT => X"EE80")
  port map (Data_out(2), Data_in(0), Data_in(1), Data_in(2), Data_in(3));

LUT4_inst2 : LUT4
  generic map (INIT => X"936C")
  port map (Data_out(1), Data_in(0), Data_in(1), Data_in(2), Data_in(3));

LUT4_inst3 : LUT4
  generic map (INIT => X"5A5A")
  port map (Data_out(0), Data_in(0), Data_in(1), Data_in(2), Data_in(3));

end Structural;
```

The component LUT5to3 (note that this component is not the same as the considered above component LUT_5to3 in the entity HammingWeightComparator) has the following VHDL code:

```
library IEEE;
use IEEE.STD_LOGIC_1164.all;
library UNISIM;                          -- for FPGA LUTs
use UNISIM.vcomponents.all;

entity LUT5to3 is
   port ( Data_in  : in  std_logic_vector (3 downto 0);
          Extra_bit : in std_logic;
          Data_out : out  std_logic_vector (2 downto 0));
end LUT5to3;

architecture Structural of LUT5to3 is
begin

LUT5_inst1 : LUT5
  generic map (INIT => X"FEC8EE80")
  port map (Data_out(2), Data_in(0), Data_in(1), Data_in(2), Data_in(3), Extra_bit);

LUT5_inst2 : LUT5
  generic map (INIT => X"C936936C")
  port map (Data_out(1), Data_in(0), Data_in(1), Data_in(2), Data_in(3), Extra_bit);

LUT5_inst3 : LUT5
  generic map (INIT => X"A5A55A5A")
  port map (Data_out(0), Data_in(0), Data_in(1), Data_in(2), Data_in(3), Extra_bit);

end Structural;
```

The respective project is ready to be tested and the circuit (with the maximum combinational path delay equal to 4.4 ns) occupies just 14 Artix-7 FPGA slices. It can be used as a Hamming weight counter and comparator. If just one from two such functions (i.e. either counting or comparison) is required then unnecessary fragment can be removed. The project above will also be used as a component of the last example in Appendix B.

Hamming Weight Counter for N = 36 (HammingWeightCounter36bits)

The following VHDL code is a complete synthesizable specification of the Hamming weight counter in Figs. 3.27 and 3.28 (any final comparison circuit from Fig. 3.25 can be used for the Hamming weight comparator):

```
library IEEE;                  -- the project was tested for the Nexys-4 board and occupies 15 slices
use IEEE.STD_LOGIC_1164.all;       -- the maximum combinational path delay is 3.5 ns

entity HammingWeightCounter36bits is
  generic ( N     : integer := 36 );
  port (Data_in   : in std_logic_vector (N-1 downto 0);      -- inputs a0,a1,...,a35 in Fig. 3.27
        Data_out : out std_logic_vector (5 downto 0));       -- the Hamming weight in Fig. 3.28
end HammingWeightCounter36bits;

architecture Behavioral of HammingWeightCounter36bits is
  type array_of_inputs is array (N/12-1 downto 0) of std_logic_vector(5 downto 0);
  signal Out18_bits : std_logic_vector(N/2-1 downto 0);       -- outputs of the layer 1 in Fig. 3.27
  signal In6_bits  : array_of_inputs;                         -- inputs of the layer 2 in Fig. 3.27
  signal Res9_bits : std_logic_vector(N/4-1 downto 0);        -- outputs of the layer 2 in Fig. 3.27
begin

generate_LUTs_at_level_0: for i in N/6-1 downto 0 generate
one_slice: entity work.LUT_6to3
        -- VHDL code of this component (LUT_6to3) is given in section 3.9
        port map(Data_in(6*i+5 downto 6*i), Out18_bits(3*i+2 downto 3*i));
 end generate generate_LUTs_at_level_0;

 generate_LUTs_at_level_1: for i in N/12-1 downto 0 generate
        In6_bits(i) <= Out18_bits(i) & Out18_bits(i+3) & Out18_bits(i+6) &
                       Out18_bits(i+9) & Out18_bits(i+12) & Out18_bits(i+15);

 one_slice:  entity work.LUT_6to3
        -- VHDL code of this component (LUT_6to3) is given in section 3.9
        port map(In6_bits(i), Res9_bits(3*i+2 downto 3*i));
 end generate generate_LUTs_at_level_1;

 FinalCircuit: entity work.Final_LUT_based_adders
        port map (Res9_bits(7 downto 0), Res9_bits(8), Data_out );

 end Behavioral;
```

The component Final_LUT_based_adders describes the functionality of the circuit in Fig. 3.28a and it is coded in VHDL as follows (similar but simpler circuit can be built for Fig. 3.28b):

```
library IEEE;
use IEEE.STD_LOGIC_1164.all;
use IEEE.STD_LOGIC_UNSIGNED.all;

entity Final_LUT_based_adders is      -- the mapping is described below by constants
port ( A_3bits_B_3bits_C_2bits: in std_logic_vector(7 downto 0);
       C_last_bit             : in std_logic;   -- C_last_bit is the symbol χ3 in Fig. 3.28 a
       Data_out               : out std_logic_vector(5 downto 0));  -- Hamming weight (N=36)
end Final_LUT_based_adders;

architecture Behavioral of Final_LUT_based_adders is
  type for_LUT is array (0 to 31) of std_logic_vector(3 downto 0); -- the first constant corresponds to
  -- INIT statements for ρ1,ρ0,γ2,γ1 in Fig. 3.28a and the second constant – to the INIT statement for γ5,γ4,γ3
  constant LUTs1 : for_LUT :=
      (x"0", x"1", x"2", x"3", x"1", x"2", x"3", x"4", x"2", x"3", x"4", x"5", x"3", x"4", x"5", x"6",
       x"2", x"3", x"4", x"5", x"3", x"4", x"5", x"6", x"4", x"5", x"6", x"7", x"5", x"6", x"7", x"8");
  -- only 3 least significant bits in 4-bit vectors are used below for the INIT statement for γ5,γ4,γ3 (see Fig. 3.28a)
  constant LUTs2 : for_LUT :=
      (x"0", x"1", x"1", x"2", x"2", x"3", x"3", x"4", x"1", x"2", x"2", x"3", x"3", x"4", x"4", x"5",
       x"2", x"3", x"3", x"4", x"4", x"5", x"5", x"6", x"3", x"4", x"4", x"5", x"5", x"6", x"6", x"7");
  -- A_3bits/B_3bits/C_2 bits are associated with the symbols α1α2α3/β1β2β3/χ1χ2 in Fig. 3.28 a
  signal A1A2A3, B1B2B3, C1C2C3   : std_logic_vector(2 downto 0);
  signal CmClO2O1                 : std_logic_vector(3 downto 0);
  signal O5_3                     : std_logic_vector(2 downto 0);
begin -- the lines below describe the circuit in Fig. 3.28a

-- (LUTs1 is the bottom block in Fig. 3.28a and LUTs2 is the upper block in Fig. 3.28a)
A1A2A3 <= C_last_bit & A_3bits_B_3bits_C_2bits(7 downto 6); -- signals α1α2 for the upper block
B1B2B3 <= A_3bits_B_3bits_C_2bits(5 downto 3);  -- signals β1 (upper block) and β2β3 (bottom block)
C1C2C3 <= A_3bits_B_3bits_C_2bits(2 downto 0);  -- signal χ3 (direct output) and χ1χ2 (bottom block)
O5_3   <= LUTs2(conv_integer(CmClO2O1(3 downto 2) &
        A1A2A3(2 downto 1) & B1B2B3(2)))(2 downto 0);
CmClO2O1 <= LUTs1(conv_integer(A1A2A3(0) & B1B2B3(1 downto 0) &
        C1C2C3(2 downto 1)));
Data_out <= O5_3 & CmClO2O1(1 downto 0)& C1C2C3(0); -- concatenation of (γ0) (γ1γ2) and (γ3γ4γ5)

end Behavioral;
```

The respective project is ready to be tested and the circuit (with the maximum combinational path delay equal to 3.5 ns) occupies just 15 Artix-7 FPGA slices. It only counts the Hamming weight of 36-bit vectors. A simple addition enables the same project to be used as a Hamming weight comparator.

Note that we have described many different projects and they may be chosen dependently on available embedded to FPGA components. Indeed, if embedded DSP slices are available then DSP-based projects are perhaps the best. If only logical slices can be used then one of the described here projects may be helpful.

Random Number Generator (RanGen)

The module generates random numbers with generic size width and it has the following VHDL code (for the default value width = 32):

```
library IEEE;
use IEEE.STD_LOGIC_1164.all;

entity RanGen is
  generic (width           : integer := 32 );      -- generic size of random numbers
  port (  clk               : in std_logic;         -- system clock
          random_num        : out std_logic_vector (width-1 downto 0) ); -- generated number
end RanGen;

architecture Behavioral of RanGen is
begin

process(clk)
  variable rand_temp : std_logic_vector(width-1 downto 0):=(width-1 => '1', others => '0');
  variable temp    : std_logic := '0';
begin
  if(rising_edge(clk)) then
    temp                            := rand_temp(width-1) xor rand_temp(width-2);
    rand_temp(width-1 downto 1)     := rand_temp(width-2 downto 0);
    rand_temp(0)                    := temp;
  end if;

  random_num <= rand_temp;

end process;

end Behavioral;
```

In each clock cycle a new 32-bit pseudorandom number is generated. The size width = 32 is generic and can easily be changed.

Segment Decoder (segment_decoder)

The decoder converts 4-bit binary codes in such a way that the respective digits become visible on 7-segment displays such as that are available on the Nexys-4 board. VHDL code for the decoder is given below:

```
library IEEE;
use IEEE.STD_LOGIC_1164.all;
entity segment_decoder is  -- any one hexadecimal or BCD code can be used as an input
port ( BCD        : in  std_logic_vector (3 downto 0);        -- decoder input
       segments   : out  std_logic_vector (7 downto 1));      -- decoder output
end segment_decoder;

architecture Behavioral of segment_decoder is
begin -- segment is active when the signal is '0' and passive when the signal is '1'
  segments <=   "1000000" when BCD = "0000" else              -- 0
                "1111001" when BCD = "0001" else              -- 1
                "0100100" when BCD = "0010" else              -- 2
                "0110000" when BCD = "0011" else              -- 3
                "0011001" when BCD = "0100" else              -- 4
                "0010010" when BCD = "0101" else              -- 5
                "0000010" when BCD = "0110" else              -- 6
                "1111000" when BCD = "0111" else              -- 7
```

```
                "0000000" when BCD = "1000" else          -- 8
                "0010000" when BCD = "1001" else          -- 9
                "0001000" when BCD = "1010" else          -- a
                "0000011" when BCD = "1011" else          -- b
                "1000110" when BCD = "1100" else          -- c
                "0100001" when BCD = "1101" else          -- d
                "0000110" when BCD = "1110" else          -- e
                "0001110" when BCD = "1111" else          -- f
                "1111111";        -- all segments are passive
end Behavioral;
```

Segment Display Control (*EightDisplayControl*)

This component controls eight 7-segment displays available on the Nexys-4 board. Functionality of the module is explained in Fig. B.4 and its VHDL code is given below:

```
library IEEE;    -- this code is for 8 7-segment displays available on the Nexys-4 board
use IEEE.STD_LOGIC_1164.all;         -- small changes permit the same code to be used for many
use IEEE.STD_LOGIC_UNSIGNED.all; -- prototyping boards, for example, Nexys-2/Nexys-3

entity EightDisplayControl is -- FourDisplayControl for Nexys-2/Nexys-3 can be also based on the code below
  port ( clk                  : in  std_logic;
        leftL, near_leftL     : in std_logic_vector (3 downto 0);
        near_rightL, rightL   : in std_logic_vector (3 downto 0);
        leftR, near_leftR     : in std_logic_vector (3 downto 0);
        near_rightR, rightR   : in std_logic_vector (3 downto 0);
        select_display        : out std_logic_vector (7 downto 0);
        segments              : out std_logic_vector (6 downto 0));
end EightDisplayControl;

architecture Behavioral of EightDisplayControl is
  signal Display  : std_logic_vector(2 downto 0);
  signal div      : std_logic_vector(16 downto 0);
  signal convert_me         : std_logic_vector(3 downto 0);
begin

div<= div + 1 when rising_edge(clk);
Display <= div(16 downto 14);

process(Display,  leftL, near_leftL, near_rightL, rightL, leftR, near_leftR, near_rightR, rightR)
begin -- sequential activation of the displays with proper control of the segments of the selected display
  if    Display ="111" then  select_display <= "11111110"; convert_me <= leftL;
  elsif Display ="110" then  select_display <= "11111101"; convert_me <= near_leftL;
  elsif Display ="101" then  select_display <= "11111011"; convert_me <= near_rightL;
  elsif Display ="100" then  select_display <= "11110111"; convert_me <= rightL;
  elsif Display ="011" then  select_display <= "11101111"; convert_me <= leftR;
  elsif Display ="010" then  select_display <= "11011111"; convert_me <= near_leftR;
  elsif Display ="001" then  select_display <= "10111111"; convert_me <= near_rightR;
  else                       select_display <= "01111111"; convert_me <= rightR;
  end if;          -- the display is active when the corresponding bit in 8-bit vector above is zero
end process;

decoder : entity work.segment_decoder                  -- segment decoder (see above)
      port map (convert_me, segments);

end Behavioral;
```

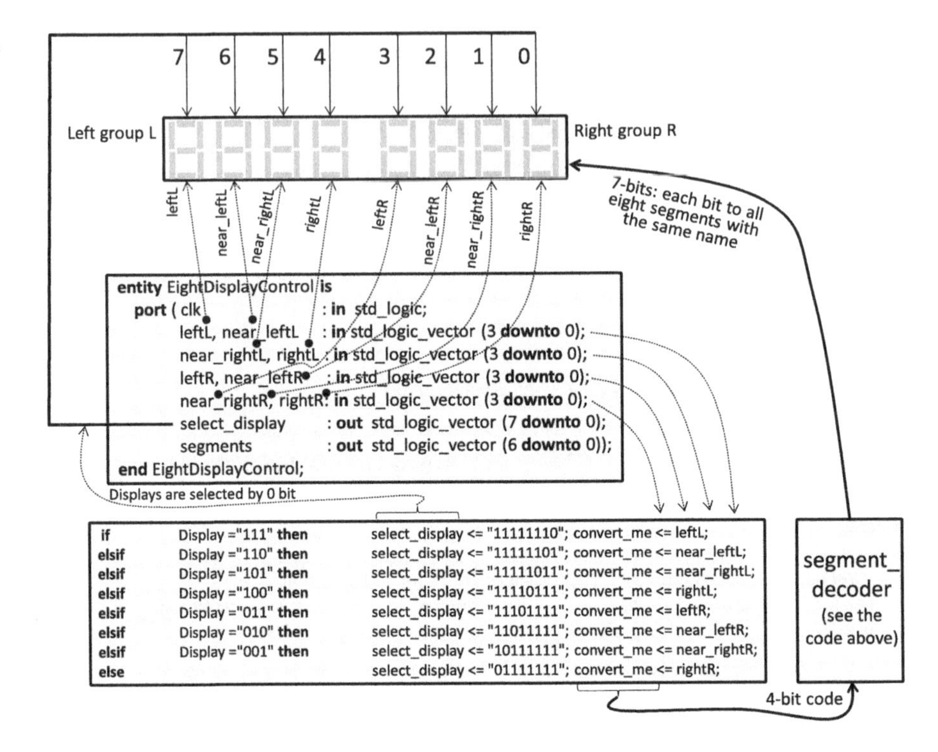

Fig. B.4 Functionality of the module EightDisplayControl

Four-bit codes (either BCD or binary) leftL, near_leftL, near_rightL, rightL, leftR, near_leftR, near_rightR, rightR are associated with different displays shown in Fig. B.4. These codes are sent to the inputs of the segment decoder. Since only one display is active at a time, scanning all the displays enables different numbers to be shown on each of them. Sequential activation of the displays is achieved with the aid of the lines: div<= div + 1 **when** rising_edge(clk) and Display <= div(16 **downto** 14). If a converter from binary to BCD codes is also used then binary numbers (see Fig. B.1b) will be displayed in decimal format.

Let us consider an example in which the considered above four components EightDisplayControl, segment_decoder, BinToBCD8, and HW31_HWC32 are used:

```
library IEEE;                    -- the project was tested for the Nexys-4 board and occupies 34 slices
use IEEE.STD_LOGIC_1164.all;            -- the project shows on segment displays the Hamming weight of
use IEEE.STD_LOGIC_UNSIGNED.all; -- 32-bit input binary vector and the result of comparison

entity HW32_HWC32 is -- in the experiments 32-bit input binary vector is received from onboard
  port ( clk      : in std_logic; -- switches of two Nexys-4 boards connected through PMod
       seg      : out std_logic_vector(6 downto 0); -- segments of onboard displays
       sel_disp : out std_logic_vector(7 downto 0); -- control of onboard displays
       Data_in  : in std_logic_vector (31 downto 0); -- 32-bit input binary vector
       LedC     : out std_logic); -- the result of comparison (see the entity HW31_HWC32 above)
end HW32_HWC32;
```

```
architecture Mixed of HW32_HWC32 is
  signal HW15_1,HW15_2  : std_logic_vector(3 downto 0);
  signal binary         : std_logic_vector(7 downto 0);
  signal BCD2,BCD1,BCD0 : std_logic_vector(3 downto 0);
  signal bits4_0        : std_logic_vector(4 downto 0);
begin
-- This line is used to compute the Hamming weight for 32-bit binary vector
binary <= "00" & (("00000"&Data_in(31)) + ('0'&bits4_0));

DispCont : entity work.EightDisplayControl
        port map(clk, "0000", "0000", "0000", "0000", "0000", BCD2, BCD1, BCD0,
                 sel_disp, seg);

BinToBCD : entity work.BinToBCD8
         port map (clk, '0', open, binary, BCD2, BCD1, BCD0);

HW_HWC_32 : entity work.HW_HWC32
            port map (Data_in, bits4_0, LedC);

end Mixed;
```

We mentioned at the end of Sect. 1.5 that almost all projects of the book are available at http://sweet.ua.pt/skl/Springer2014.html. They were implemented and tested in Xilinx ISE 14.7 and many of them were converted and tested in Xilinx Vivado 2013.4 design suite. The following project permits to examine *Test_HW16* entity from Sect. 4.2 that is a DSP-based Hamming weight (HW) counter. The resulting HW is shown on the leftmost display of the Nexys-4 in hexadecimal format. If HW=16 then the leftmost LED is ON and 0 appears on the display.

```
library IEEE;
use IEEE.STD_LOGIC_1164.all;
entity HW16 _DISPLAY is  -- Nexys-4 circuit occupies 4 logical slices and 1 DSP slice
port (ledL        : out std_logic;                    -- ledL is the leftmost LED
      seg          : out std_logic_vector(6 downto 0); -- from segment decoder
      sel_disp     : out std_logic_vector(7 downto 0); -- pins:N6,M6,M3,N5,N2,N4,L1,M1
      Sw           : in std_logic_vector(15 downto 0));-- input vector to count the HW
end HW16 _DISPLAY;

architecture Mixed of HW16 _DISPLAY is
signal HW16       : std_logic_vector(4 downto 0); -- represents the HW
begin
-- DSP-based computing of the Hamming weight (HW16) for 16-bit binary vector Sw from Sect. 4.2
HWCC        : entity work.Test_HW16   -- combining positional and named associations
              port map (Sw,led=>HW16,led_comp=> open);
-- segment display decoder for hexadecimal input numbers
seg_dec          : entity work.segment_decoder-- only named association is used
                   port map(BCD=>HW16(3 downto 0),segments=>seg);

ledL              <= HW16(4);           -- if HW16 = 16 then LedL is ON otherwise - OFF
sel_disp          <= "11111110";        -- only the leftmost display is chosen

end Mixed;
```

At the beginning let us test the project above in ISE and then make some conversions that enable the project to be synthesized, implemented and tested in Vivado. Firstly, Nexys-4 UCF file has to be converted to XDC file as follows:

1) Run Xilinx PlanAhead software and open the project created in ISE;
2) Run synthesis in the PlanAhead and open synthesized design;
3) Run the following command: *write_xdc c:/tmp/Nexys4.xdc* from Tcl console of the PlanAhead (note that sub-directory *tmp* has to be manually created).

Then the following steps have to be done:

4) Create a new RTL Vivado project for FPGA available on the Nexys-4;
5) Copy all VHDL files from ISE project (4 files have to be copied for our project above) and the newly created XDC file to the new Vivado project;
6) Run synthesis, implementation and generate bitstream in Vivado;
7) Open hardware manager in Vivado and program the FPGA of Nexys-4;
8) Test the project in the Nexys-4 board.

To simplify testing the projects in ISE and Vivado all necessary components can be found at http://sweet.ua.pt/skl/Springer2014.html (either in ISE or in Vivado subdirectories). They contain all necessary files that have to be included in ISE/Vivado projects and, thus, only the steps 4)-8) need to be done. Note that if a converted project has XCO files they may need to be upgraded (see Sect. 1.5). If a converted project has COE/TXT files, they have to be either copied to Vivado project or their locations have to be explicitly indicated, for instance:

```
signal array_name :  my_array := read_array("c:/tmp/data.txt");
```

Additional VHDL examples of different reusable blocks are available in [2,3]. Document [4] describes details about migration of ISE projects to Vivado projects.

References

1. Xilinx Inc. (2011) Xilinx 7 series FPGA libraries guide for HDL designs. http://www.xilinx.com/support/documentation/sw_manuals/xilinx13_3/7series_hdl.pdf. Accessed 21 Nov 2013
2. Sklyarov V, Skliarova I (2013) Parallel processing in FPGA-based digital circuits and systems. TUT Press, Tallinn
3. Skliarova I, Sklyarov V, Sudnitson A (2012) Design of FPGA-based circuits using hierarchical finite state machines. TUT Press, Tallinn
4. Xilinx Inc. (2013) Vivado Design Suite. ISE to Vivado Design Suite Migration Guide. http://www.xilinx.com/support/documentation/sw_manuals/xilinx2013_3/ug911-vivado-migration.pdf. Accessed 24 Jan 2014

Index

V. Sklyarov et al., *Synthesis and Optimization of FPGA-Based Systems*,
Lecture Notes in Electrical Engineering 294, DOI: 10.1007/978-3-319-04708-9,

CPSIA information can be obtained at www.ICGtesting.com
Printed in the USA
LVOW01*1154230314

378546LV00014B/221/P

9 783319 04707